CP OBGYN Clinical Perspectives in Obstetrics and Gynecology

Series Editor:

The late Herbert J. Buchsbaum, M.D.

CP OBGYN

Clinical Perspectives in Obstetrics and Gynecology

perspective *noun:* . . . the capacity to view subjects in their true relations or relative importance.

Each volume in Clinical Perspectives in Obstetrics and Gynecology will cover in depth a major clinical area in the health care of women. The objective is to present to the reader the pathophysiologic and biochemical basis of the condition under discussion and to provide a scientific basis for clinical management. These volumes are not intended as "how to" books, but as a ready reference by authorities in the field.

Though the obstetrician and gynecologist may be the primary provider of health care for women, this role is shared with family practitioners, pediatricians, medical and surgical specialists, and geriatricians. It is to all these physicians that the series is addressed.

Series Editor: The late Herbert J. Buchsbaum, M.D.

Published Volumes:

Buchsbaum (ed.): *The Menopause*
Aiman (ed.): *Infertility*
Futterweit: *Polycystic Ovarian Disease*
Lavery and Sanfilippo (eds.): *Pediatric and Adolescent Obstetrics and Gynecology*
Galask and Larsen (eds.): *Infectious Diseases in the Female Patient*
Buchsbaum and Walton (eds.): *Strategies in Gynecologic Surgery*
Szulman and Buchsbaum (eds.): *Gestational Trophoblastic Disease*
Cibils (ed.): *Surgical Diseases in Pregnancy*
Collins (ed.): *Ovulation Induction*
Altchek and Deligdisch (eds.): *The Uterus*

The Uterus

Pathology, Diagnosis, and Management

Albert Altchek
Liane Deligdisch
Editors

With a Foreword by Richard H. Schwarz

With 249 Illustrations

Springer-Verlag
New York Berlin Heidelberg London
Paris Tokyo Hong Kong Barcelona

Editors:
Albert Altchek, M.D.
Clinical Professor of Obstetrics, Gynecology, and Reproductive Science; Chief of Pediatric and Adolescent Gynecology, The Mount Sinai School of Medicine, New York, NY 10029, USA

Liane Deligdisch, M.D.
Professor of Pathology, Professor of Obstetrics, Gynecology, and Reproductive Science, The Mount Sinai School of Medicine, New York, NY 10029, USA

Series Editor:
The late Herbert J. Buchsbaum, M.D., Professor and Chairman, Department of Obstetrics and Gynecology, Medical College of Wisconsin, Milwaukee, WI 53202, USA

Library of Congress Cataloging-in-Publication Data
The Uterus / Albert Altchek, Liane Deligdisch, editors.
p. cm. — (Clinical perspectives in obstetrics and gynecology)
Includes bibliographical references and index.

ISBN-13: 978-1-4613-9088-6 e-ISBN-13: 978-1-4613-9086-2

DOI: 10.1007/978-1-4613-9086-2

(alk. paper)
1. Uterus—Diseases. 2. Uterus—Surgery. I. Altchek, Albert. 1925- . II. Deligdisch, Liane. III. Series.
[DNLM: 1. Uterine Diseases—diagnosis. 2. Uterine Diseases--therapy. 3. Uterus—pathology. WP 400 U894]
RG302.U85 1991
618.1'4—dc20
DNLM/DLC
for Library of Congress 90-10461
CIP

Printed on acid-free paper.

Typeset by Maryland Composition Company, Inc., Glen Burnie, MD.

9 8 7 6 5 4 3 2 1

In Loving Memory of my father, Solomon David Altchek,
Salonika, July 14, 1904, New York, September 19, 1981

In the memory of my parents Else and
Dr. med. Gustav Deligdisch

Foreword

Drs. Altchek and Deligdisch have taken a unique approach in formulating this scholarly work. They have eschewed the traditions of concentrating attention on a single facet of their field, such as pathology, diagnosis, or surgical treatment. At the same time, they have isolated the uterus from the remainder of the pelvic viscera for discussion purposes, albeit in many instances the chapter subjects of necessity involved the authors with the ovaries, tubes, and pelvic supports. Because of the architecture of this book, the editors have brought together not only a distinguished group of experts, but a group who are rather strange bedfellows. It is rare, indeed, to find reproductive endocrinologists, oncologists, pathologists, gynecologic surgeons, and, indeed, lawyers writing in the same text. The editors have also not avoided some of the more controversial subjects in modern-day gynecology. They have chosen to include discussions of the use of molecular technology in evaluation of cervical dysplasia, hysteroscopic endometrial ablation, and the uterus without ovaries as it relates to the new wave of assisted reproductive technology. For this, they are to be commended. It should also be noted that both editors made significant contributions themselves, each authoring three chapters.

This work, I believe, has something for everyone in gynecology and obstetrics, from the resident to the practitioner and the academic, and will be of interest to pathologists as well. The unique array of subjects and the thorough and scholarly way in which they are treated make most interesting and rewarding reading.

Richard H. Schwarz, M.D.
Provost and Vice President for Clinical Affairs
Professor, Obstetrics and Gynecology
SUNY Health Science Center at Brooklyn
Brooklyn, New York
President, American College of
Obstetricians and Gynecologists

Preface

Recent years have witnessed new concepts in physiology and pathology of the uterus, together with new technology and changing approaches in diagnosis and clinical management. Traditionally, books on this subject have focused exclusively either on basic science and pathology, or on clinical gynecology. The interested clinician therefore had to cull from basic science texts material that might be relevant for clinical gynecology and to interpret its possible application.

This new book on the uterus attempts to bring together in one volume some of the new pertinent basic-science knowledge of concern to the clinician as well as to review several contemporary aspects of clinical practice. Hopefully, *The Uterus* will serve as a broad yet concise update and review to help the gynecologist care for his or her patients.

With the clinical gynecologist in mind, the book has been co-edited by a clinician and a pathologist and is divided into three sections: basic science/pathology, diagnostic procedures, and contemporary management. It should be helpful for all who provide medical care for women. It is of course understood that medicine is a constantly evolving discipline and that changes are to be expected over time. We are always learning.

We are fortunate to be able to include chapters contributed by outstanding and dedicated experts in their respective fields in order to further medical education despite the time constraints imposed by many professional obligations. We deeply appreciate their work.

We wish to thank the editors of Springer-Verlag New York, Publishers, for their kind encouragement, and Fortune Uy for secretarial assistance.

Albert Altchek, M.D.
Liane Deligdisch, M.D.

Contents

II Diagnostic Procedures

III Contemporary Management

Contributors

ALBERT ALTCHEK, M.D.
Clinical Professor of Obstetrics, Gynecology, and Reproductive Science; Chief of Pediatric and Adolescent Gynecology, The Mount Sinai School of Medicine, New York, New York 10029; Obstetrics and Gynecology Staff, Lenox Hill Hospital, New York, New York 10021, USA

PATRICIA A. ATHEY, M.D.
Associate Professor, Department of Radiology, Baylor College of Medicine, Houston, Texas 77030, USA

HUGH R. K. BARBER, M.D.
Director, Department of Obstetrics and Gynecology, Lenox Hill Hospital, New York, New York 10021, USA

CLAUDE BLOCH, M.D., F.A.C.R.
Associate Clinical Professor, Department of Radiology, The Mount Sinai School of Medicine, New York, New York 10029, USA

MICHAEL L. BRODMAN, M.D.
Assistant Professor, Department of Obstetrics, Gynecology and Reproductive Sciences, The Mount Sinai School of Medicine, New York, New York 10029, USA

MAX BORTEN, M.D.
Associate Professor of Obstetrics and Gynecology, Harvard Medical School; Director, Division of Medical Gynecology, Beth Israel Hospital, Brookline, Massachusetts 02215, USA

LOUIS BURKE, M.D.
Associate Professor, Harvard Medical School; Senior Obstetrician-Gynecologist, Acting Obstetrician-Gynecologist-in-Chief, Emeritus, Director, Colposcopy/Laser Unit, Beth Israel Hospital, Boston, Massachusetts 02215, USA

CHRISTOPHER P. CRUM, M.D.
Director, Women's and Perinatal Pathology, Brigham and Women's Hospital; Associate Professor of Pathology, Harvard Medical School, Boston, Massachusetts 02115, USA

SAUL J. DAN, M.D.
Associate Professor, Department of Radiology, Mount Sinai School of Medicine, New York, New York 10029, USA

ALAN H. DECHERNEY, M.D.
John Slade Ely Professor of Obstetrics and Gynecology, Director of the Division of Reproductive Endocrinology, Department of Obstetrics and Gynecology, Yale University School of Medicine, New Haven, Connecticut 06510, USA

LIANE DELIGDISCH, M.D.
Professor of Pathology, Professor of Obstetrics, Gynecology, and Reproductive Science, The Mount Sinai School of Medicine, New York, New York 10029, USA

RITA IOVINE DEMOPOULOS, M.D.
Associate Professor, Department of Pathology, New York University College of Medicine, New York, New York 10016, USA

GUNTER DEPPE, M.D.
Professor of Obstetrics and Gynecology, Director of Gynecologic Oncology, Wayne State University School of Medicine, Hutzel Hospital, Department of Obstetrics and Gynecology, Detroit, Michigan 48201, USA

FREDERICK FRIEDMAN, JR., M.D.
Assistant Professor, Department of Obstetrics, Gynecology and Reproductive Sciences, New York, New York 10029, USA

NORBERT GLEICHER, M.D.
Chairman, Department of Obstetrics and Gynecology, Mount Sinai Hospital Medical Center of Chicago; Professor of Obstetrics and Gynecology, Associate Professor of Immunology/ Microbiology, Rush Medical College, Chicago, Illinois 60608, USA

MILTON H. GOLDRATH, M.D.
Associate Professor, Department of Obstetrics and Gynecology, Wayne State University School of Medicine; Chief, Section of Gynecology, Sinai Hospital, Detroit, Michigan 48235, USA

HARRY H. HATASAKA, M.D.
Department of Obstetrics and Gynecology, Section of Reproductive Endocrinology and Infertility, Northwestern University Medical School, Chicago, Illinois 60611, USA

Daniel M. Hays, M.D.
Professor of Surgery and Pediatrics, University of Southern California, School of Medicine, Los Angeles, California 90033, USA

Avner Hershlag, M.D.
Postdoctoral Associate, Division of Reproductive Endocrinology, Department of Obstetrics and Gynecology, Yale University School of Medicine, New Haven, Connecticut 06510, USA

Cynthia L. Janus, M.D.
Associate Professor, Department of Radiology, University of Virginia Medical Center, Charlottesville, Virginia 22908, USA

Jacqueline C. Johnson, M.D.
Chief Administrative Resident, Department of Obstetrics and Gynecology, Lenox Hill Hospital, New York, New York 10021, USA

Neela Lamki, M.D., F.R.C.P.
Coordinator, Medical Student Education, Department of Radiology, Radiologic Education Center, Baylor College of Medicine, Houston, Texas 77030, USA

Colonel Martin Lefkowitz, M.D., M.C., U.S.A.R.
Staff Pathologist, Department of Gynecologic and Breast Pathology, Armed Forces Institute of Pathology, Washington, D.C. 20306, USA

Vinay K. Malviya, M.D.
Associate Professor, Division of Gynecologic Oncology, Wayne State University School of Medicine; Vice Chief, Division of Gynecologic Oncology, Hutzel Hospital, Detroit, Michigan 48201, USA

Harold M. Maurer, M.D.
Jessie Ball duPont Professor, Chairman, Department of Pediatrics, Medical College of Virginia, Virginia Commonwealth University, Richmond, Virginia 23298, USA

Khushbakhat Rai Mittal, M.D.
Assistant Professor, Department of Pathology, New York University Medical Center, New York, New York 10016, USA

Daniel Navot, M.D.
Associate Professor, Department of Obstetrics, Gynecology and Reproductive Science, Mount Sinai School of Medicine, New York, New York 10029, USA

David H. Nichols, M.D., F.A.C.S., F.A.C.O.G.
Professor and Chairman, Department of Obstetrics and Gynecology, Brown University Program in Medicine, Providence, Rhode Island 02905, USA

Henry J. Norris, M.D.
Chairman, Department of Gynecologic and Breast Pathology, Armed Forces Institute of Pathology, Washington, D.C., 20306; Clinical Professor of Pathology, Uniformed Services University of the Health Sciences, Bethesda, Maryland 20892, USA

Gerard J. Nuovo, M.D.
Assistant Professor of Pathology, State University of New York at Stony Brook, Stony Brook, New York 11794, USA

Giglia A. Parker, M.D.
Clinical Assistant Professor, Department of Obstetrics and Gynecology, Brown University Program in Medicine, Providence, Rhode Island 02905, USA

Stanley J. Robboy, M.D.
Professor of Pathology and Obstetrics and Gynecology, UMDNJ–New Jersey Medical School, Newark, New Jersey 07103, USA

Alexander Sedlis, M.D.
Professor, Department of Obstetrics and Gynecology, State University of New York, Health Science Center at Brooklyn, Brooklyn, New York 11203, USA

Leon Speroff, M.D.
Section of Reproductive Endocrinology and Infertility, The Oregon Health Sciences University, Portland, Oregon 97201, USA

Peyton T. Taylor, M.D.
Richard and Louise Crockett Professor of Obstetrics and Gynecology, Director of Gynecologic Oncology, University of Virginia Health Sciences Center, Charlottesville, Virginia 22908, USA

Ilan E.Timor-Tritsch, M.D.
Director, Obstetrical and Gynecological Ultrasound, Department of Obstetrics and Gynecology, Columbia Presbyterian Medical Center, New York, New York 10032, USA

Ilan Tur-Kaspa, M.D.
Department of Obstetrics and Gynecology, Mount Sinai Hospital Medical Center of Chicago, Affiliate of The University of Health Sciences/Chicago Medical School, Chicago, Illinois 60608, USA

MARYANNE C. WILLIAMS, R.N., B.S.N.
Administrative Nurse Coordinator, Teaching Assistant, Department of Obstetrics, Gynecology and Reproductive Science, Mount Sinai School of Medicine, New York, New York 10029, USA

V. CECIL WRIGHT, M.D., F.R.C.S.(C)
Clinical Professor, Department of Obstetrics and Gynecology, The University of Western Ontario, London, Ontario, Canada N6A 3JI

ETAN Z. ZIMMER, M.D.
Visiting Associate Professor, Department of Obstetrics and Gynecology, Columbia University College of Physicians and Surgeons, New York, New York 10032, USA

1

Anatomy, Histology, and Physiology

RITA IOVINE DEMOPOULOS and KHUSHBAKHAT RAI MITTAL

The uterus serves the function of providing the developing fetus with a relatively safe and nutritive environment in which to mature. It shows considerable variation in gross and microscopic appearance from neonate to adulthood to senescence.

Embryology

The uterus is derived from caudally fused portions of paired müllerian or paramesonephric ducts. These ducts originate in the fifth and sixth week of intrauterine life as invaginations of celomic epithelium. The caudal portion of the two ducts fuse by the seventh week. This fused portion gives rise to the uterine corpus, cervix, and upper third of the vagina. Cervical glands appear by 15 weeks. Endometrial glands and smooth muscle differentiation are present by 19 weeks. Development of müllerian ducts is inhibited in the male by androgens and by a nonsteroidal circulating hormone known as müllerian-inhibiting substance (MIS), produced by the Sertoli cells of the testis.[1]

Congenital Anomalies

Congenital anomalies of the uterus are rare. Failure of the fusion of müllerian ducts can lead to a variety of congenital defects. If the ducts fuse but intervening walls persist, a septum can result. If the septum is present only in the fundus, a uterus arcuatus results. If the septum persists for the entire length, a double uterus (uterus didelphys) and double upper vagina result. Nonfusion of the uterine portion of the müllerian ducts results in uterus bicornis. These abnormalities may require surgery to treat infertility or recurrent abortions.[2,3]

Atresia of the müllerian ducts may be inherited as an autosomal recessive[4] or dominant[5] condition. It results in rudimentary development of all or part of the uterus or upper vagina. Rudimentary development of one müllerian duct may lead to uterus bicornis unicollis, with a rudimentary horn. This rudimentary horn may later present as a pelvic mass with pain during menses.

Anatomic and Physiologic Changes of the Uterus over the Life Span

Neonate

From birth to 9 years of age the body of the uterus remains relatively small, measuring only one fourth to one half the size of the uterine cervix; the two together measure 3 to 4 cm in length.

The endometrium at birth is usually in a resting phase and consists of a single layer of cuboidal epithelium with sparse glands and ill-defined stroma. However, a variety of appearances may result in the neonate from the influence of maternal hormones. Ober et al.[6] found proliferative endometrium in 16%, secretory endometrium in 27%, and predecidual or menstrual changes in 5% of cases. The endometrium reverts to a resting phase after delivery.

Menarche

Menarche usually occurs between 9 and 16 years of age, after secondary sex characteristics have begun to develop. From 9 years onward the body of the uterus grows more rapidly than the cervix, such that the uterus and cervix each measure approximately 3 cm in length at 13 years of age. This growth occurs as a consequence of estrogen stimulation. At puberty, the uterus together with the cervix measures 5.5 to 8 cm in length, 4.5 to 6 cm in width and 2.5 to 3.5 cm anteroposteriorly. It weighs from 40 to 50 g. The first menstrual cycles are often anovulatory and irregular, with intervals as long as 2 to 3 months. Once regularly established, cycles usually last 3 to 6 weeks, with a mean of 4 weeks. Because the time of menstruation is usually fixed at 2 weeks after ovulation, most variations in cycle duration result from variations in the preovulatory, proliferative phase of the cycle. The usual duration of menstrual periods is from 2 to 7 days. The menstrual blood flow does not usually clot because of the presence of fibrinolysins such as plasmin. Plasmin is generated from plasminogen by plasminogen activators, which in turn are released by degenerating endothelial cells.

Adulthood

The uterus is larger in parous women compared to nulliparous women and weighs 50 to 80 g. The ratio of uterus to cervix in adult life is about 2:1, with total length of 7 to 9 cm, width of 4.5 to 6 cm, and anteroposterior diameter of 2.5 to 3.5 cm. The corpus and the endometrial cavity are both symmetrical about a vertical axis. The endometrial cavity is a roughly triangular, collapsed space with the apex at the bottom. The thickness of the endometrium varies from 1 to 8 mm and the myometrium from 1.5 to 2.5 cm. The endometrial appearance in various phases of the menstrual cycle is described under Normal Histologic Response of the Endometrium to Sex Steroids.

Menopause

Failure of the ovary to respond to pituitary gonadotropins results in a lack of estrogenic stimulation of the endometrium, which becomes atrophic and thin and may measure less than 1 mm in thickness. This usually occurs in the fifth decade. There is some evidence that suggests that over the past several decades the average age at menopause has increased.[7] Premature ovarian failure is defined as 6 months of amenorrhea, not due to pregnancy, prior to age 35.

The menopausal endometrium is thin, measuring approximately 1 mm or less. Usually no mitotic or secretory activity is evident. A variety of metaplastic changes may be seen in senile endometrium. Usually the endometrial glands are scant in number but often show mild cystic dilatation. This is referred to as cystic atrophy and should not be confused with hyperplasia. The squamocolumnar junction is usually present higher in the endocervical canal compared with women in their child-bearing years. Since estrogen promotes maturation of the squamous epithelium, the postmenopausal cervical mucosa becomes thin and atrophic as a result of lack of estrogen in menopausal women.

Gross Anatomic Relationships and Uterine Ligaments

Serosa covers the entire posterior wall of the body of the uterus and then is reflected onto the anterior rectal wall to form the rectouterine cul-de-sac. The anterior lower portion of the uterus lies next to the posterior wall of the urinary bladder, with a thin layer of connective tissue in between. The upper anterior wall of the uterus is covered with serosa.

The broad ligaments extend from the lateral sides of the uterus to the pelvic side walls. The fallopian tubes, arteries, and veins are contained in these ligaments. The ureter is present at the base of these ligaments. The round ligaments arise below and anterior to the fallopian tubes and extend anteriorly to traverse the inguinal canal to end in the upper portion of the labia majora. These ligaments are composed of smooth muscle and connective tissue and increase considerably in size during pregnancy. The anatomic location of the round ligament, which is anterior to the fallopian tube origin, is useful in orienting the surgically removed uterus in the pathology laboratory.

The uterosacral ligaments arise posterolaterally from the upper cervix, form the lateral boundaries of the pouch of Douglas (rectouterine cul-de-sac), and pass lateral to the rectum to end on fascia over the second and third sacral vertebrae. These ligaments are also composed of smooth muscle and connective tissue and serve to retain the uterus in its anterior position.

Normal Histologic Response of the Endometrium to Sex Steroids

During adult life, the endometrium undergoes cyclic changes with periodic menstruation. Although the usual cycle range varies from 21 to 42 days, an average 28-day cycle has been used to describe cyclic changes in the endometrium. Day 1 is the first day of menses and ovulation is presumed to occur on day 14 in this model 28-day cycle. Because most variations in the normal cycle duration result from variations in the duration of the preovulatory proliferative phase, it has been recommended that endometrial dating be done according to days after ovulation. In this scheme, day 16 in the original system of dating would be considered postovulatory day 2 (PO day 2). In the postovulatory phase the dating can usually be accomplished to within 24 hours.

Menstrual Endometrium

Menstruation appears to result from constriction of spiral arteries with resultant ischemia, vascular stasis, necrosis of vessel walls, hemorrhage, and necrosis of tissue.[8] Degenerative changes can be seen in glands. The glands collapse and stroma appears condensed. An infiltrate of neutrophils occurs in response to tissue necrosis. Prostaglandins are believed to play a role in bringing about menstruation,[9] since they bring about menstruation and symptoms of dysmenorrhea if administered to nonpregnant women. Prostaglandin degrading enzyme 15-hydroxyprostaglandin dehydrogenase is detected at the highest level in endometrial glandular epithelium in the secretory phase. The activity of this enzyme is barely detectable or absent 2 days before the menses and during the menses. Thus levels of this enzyme parallel the progesterone levels. It has been suggested that a fall in the level of progesterone before the menses results in lowered levels of 15-hydroxyprostaglandin dehydrogenase, leading to accumulation of prostaglandins, which in turn induce vasoconstriction and menses.[9] Prostaglandin-induced myometrial contractions contribute to expulsion of menstrual blood.

Preovulatory Phase

The repair and regrowth of the menstrual endometrium starts between day 2 and 3 of a 4-day bleeding period and appears to be a local response to injury, independent of hormonal stimulation.[10] Regrowth of epithelium occurs from the basal glands as well as surface epithelium adjacent to denuded areas. This new epithelium acquires estrogen receptors on about day 5, and subsequently grows rapidly under the influence of circulating estrogens. Endometrial thickness increases from less than 1 to approximately 3 mm during this phase. Endometrial glandular and stromal cells express estrogen receptors in the proliferative phase and in the early secretory phase.[11]

During the proliferative stage, the gland outlines are tubular and mitoses are present at the luminal border of the glandular epithelium as well as in the stroma (Fig. 1-1). The glands are elongated and coiled in the late proliferative stage. A transient stage of stromal edema is seen in the midproliferative phase (days 8 to 10).

Postovulatory Phase

The first histologic evidence of ovulation appears 2 days after ovulation in the form of subnuclear secretory vacuoles in the columnar glandular epithelium. This is seen focally on day 16 (PO day 2) of a 28-day cycle in which ovulation has occurred on day 14. Postovulatory changes can be summarized as shown in Table 1-1.

The glands in the basal endometrium and in the lower uterine segment show these changes to a lesser degree than in the rest of the endometrium. The secretory changes in the postovulatory phase are caused by progesterone from the corpus luteum, and menstrual changes are mediated by loss of estrogen and progesterone activity.

TABLE 1-1. Postovulatory changes in endometrium

Cycle day	PO day	Postovulatory changes
16	2	Subnuclear vacuoles appear focally in glands
17	3	Subnuclear vacuoles in virtually all glands (Fig. 1-2); this is definite evidence that ovulation has occurred
18	4	Secretory vacuoles in supranuclear or luminal position
19	5	Luminal secretions with frayed edges to epithelial lining cells, which are now cuboidal; mitoses may still be present
20	6	Mitoses no longer present
21	7	Early stromal edema
22	8	Maximal stromal edema
23	9	Coiled blood vessels appear in clusters of two to five; predecidual changes (stromal cells have round nuclei and have more abundant, deeper staining cytoplasm) around the arterioles
24	10	Predecidual change beneath surface epithelium
25	11	Predecidua around glands (Fig. 1-3)
26	12	Sawtooth appearance of glands
27	13	Endometrial granulocytes and leukocytes are prominent in stroma
28	14	Nuclear debris at base of some glands (probably represents necrosis of glandular lining cells)

Pregnancy-Related Changes in the Uterus

The uterus enlarges markedly during pregnancy, reaching a weight of about 1,100 g at term. The uterine wall increases in thickness during the first few months but later thins out somewhat, measuring about 1.5 cm at term. The increase in uterus size results predominantly from hypertrophy of existing muscle fibers. The gravid uterus has a globular shape and soft consistency.

During pregnancy many endometrial glands appear exhausted, with flattened lining cells and wide lumina. Other endometrial glands become crowded and lined by enlarged cells with irregular hyperchromatic nuclei and vacuolated cytoplasm. This appearance is known as the Arias-Stella reaction (Fig. 1-4), and it may be confused with malignancy, particularly if the clinical setting is not known. The predecidual cells enlarge further and the plasma membrane becomes well defined. The sheets of decidual cells form a mosaic pattern of epithelial-like cells characteristic of pregnancy and are now referred to as decidua. In spite of this epithelial appearance the decidua shows immunoreactivity for vimentin (a marker for mesenchymal cells). This can be useful in distinguishing the decidua from intermediate trophoblasts, which can have a similar appearance. The intermediate trophoblastic cells show immunoreactivity for human placental lactogen and cytokeratin, but are negative for vimentin (Fig. 1-5). The syncytiotrophoblasts show immunoreactivity for human chorionic gonadotropin. These immunoreactivities can be useful in confirming the presence of a uterine gestation when clearly recognizable chorionic villi or tro-

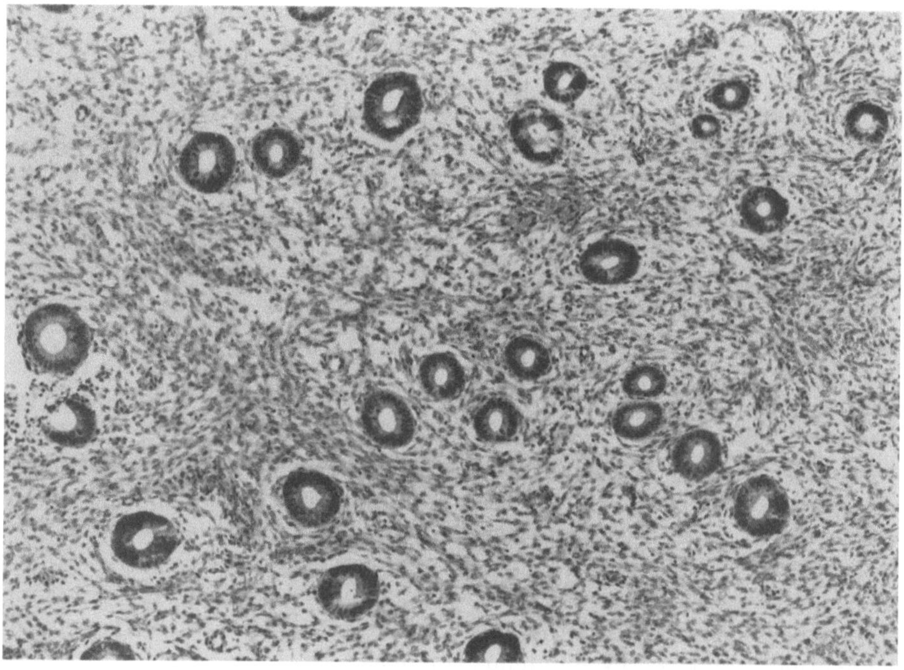

A

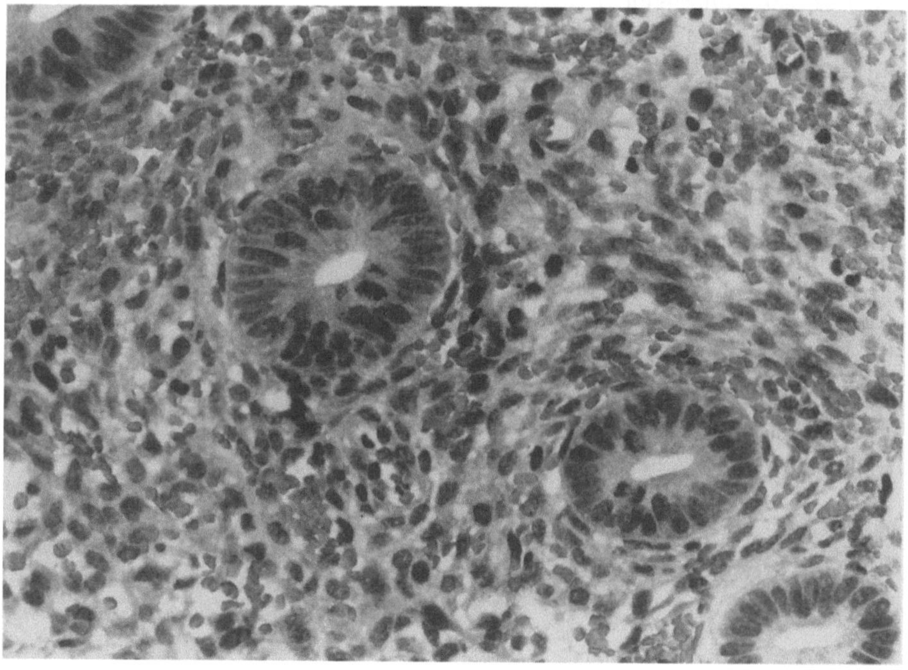

B

FIGURE 1-1. Proliferative endometrium with small tubular glands (A) and mitotic activity (best seen in glands in B).

phoblastic tissues are not seen on routinely stained slides prepared from endometrial curettings of a patient who is being evaluated for possible ectopic pregnancy.

During pregnancy there is a marked degradation of cervical collagen fibers, with accumulation of glycoprotein ground substance, edema, and some acute inflammation.[12] The cervix becomes soft and enlarged, allowing it to stretch during the delivery. Endocervical epithelium proliferates during pregnancy, and the term "microglandular endocervical hyperplasia" is used for this change. The majority of pregnant women have eversion of endocervical epithelium during pregnancy. This everted endocervical epithelium appears red and velvety and should be referred to as an ectropion. True erosion (characterized by denuded surface epithelium) is present in only about 10% of cases. The everted and/or eroded epithelium is usually eventually replaced by squamous epithelium. This squamous metaplasia of endocervical epithelium may extend a variable distance up into the endocervical canal, so that the squamocolumnar junction in parous women often lies within the cervical canal. The endocervical canal is filled with a thick mucous plug during pregnancy. Decidualization of cervical stroma is present in about one third of cases and this regresses by 2 months postpartum.[13]

At parturition oxytocin of both pituitary and ovarian origin reaches the uterus via the blood and initiates uterine contractions.[14] Oxytocin also binds to decidual cells, which in turn generate prostaglandins, further contributing to the uterine contraction.

Blood Supply, Innervation, and Lymphatic Drainage of the Uterus

Blood Supply

Arteries

The arterial supply is derived from the uterine and ovarian arteries. The uterine artery is a branch of the hypogastric artery. At about 2 cm lateral to the cervix, it crosses anterolaterally and close to the ureter to run up along the side of the uterus. Damage to the ureters may result during radical hysterectomy because of proximity of the ureters to the uterine arteries. The uterine artery gives a small cervicovaginal branch to supply the lower cervix and adjacent vaginal wall, a branch to supply the upper cervix, and numerous branches that penetrate the sides of uterus. The terminal part of the uterine artery anastomoses with the ovarian artery to supply parts of the fallopian tube and the fundus of the uterus.

The basal portion of the endometrium is supplied by straight basal arteries, which are not responsive to hormones. Most of the midportion of the endometrium and all of the superficial endometrium are supplied by coiled end-arteries, which respond to hormones.

Veins

Blood from the upper part of the uterus and ovary is collected by several veins that form the pampiniform plexus in the broad ligament. Vessels from the plexus drain into the ovarian vein. The right ovarian vein goes to the vena cava, while the left ovarian vein empties into the left renal vein. The lower uterus drains into the uterine veins, which empty into the hypogastric veins.

Innervation

Sympathetic innervation arises from the hypogastric plexus located anterior to the promontory of the sacrum and enters the uterovaginal plexus of Frankenhauser, which comprises ganglia of various sizes, many located on either side of the cervix and above the posterior fornix. Fibers from these ganglia terminate in the myometrium as well as in the endometrium. Parasympathetic supply is from pelvic nerves, which derive from the second, third, and fourth sacral nerves and end in the cervical ganglion of Frankenhauser.

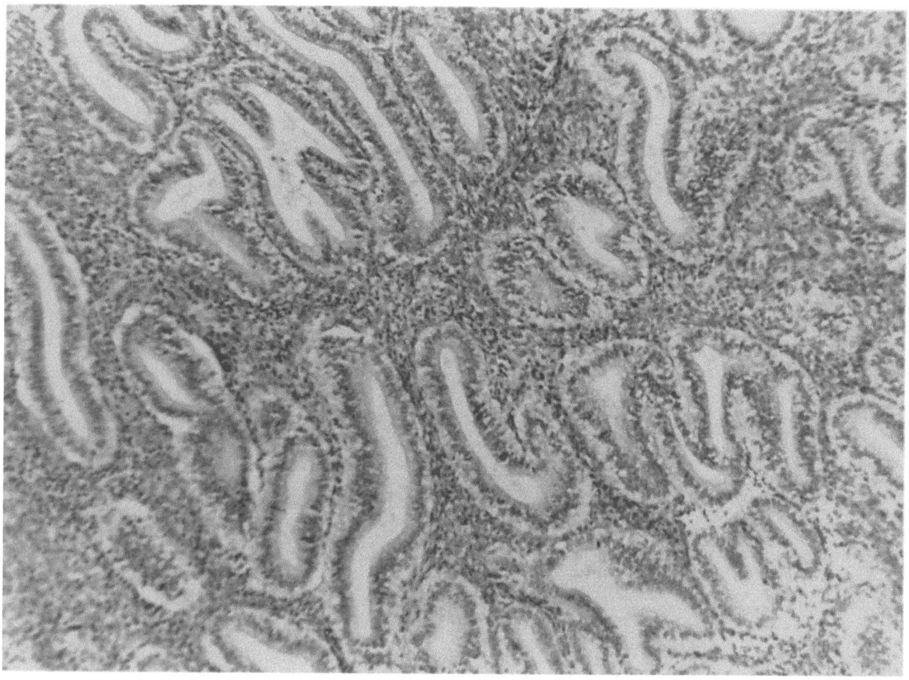

FIGURE 1-2. Secretory endometrium, day 17 (PO day 3), showing uniform presence of subnuclear vacuoles. This is definite evidence of ovulation.

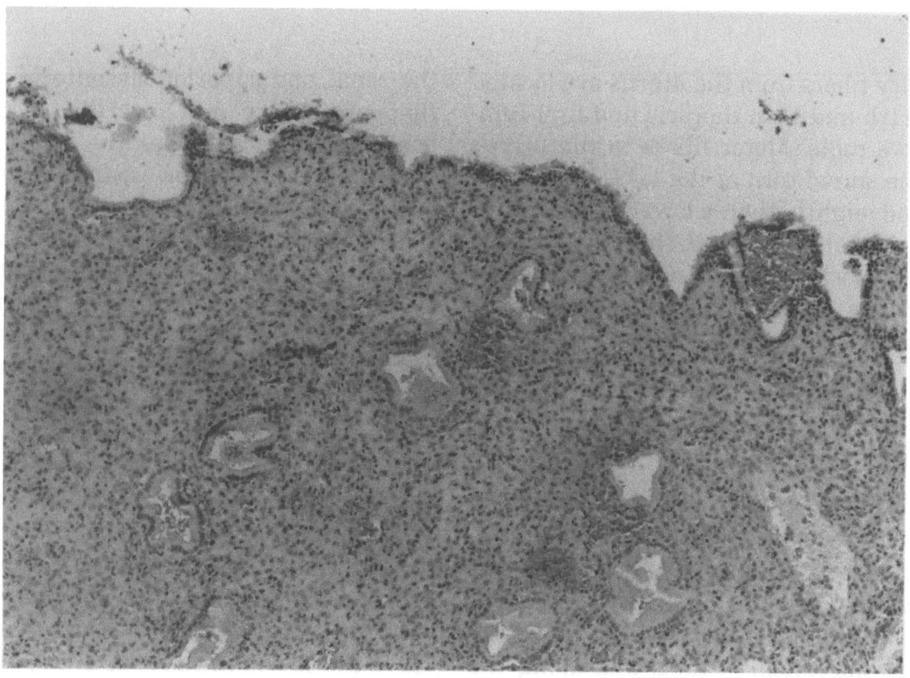

FIGURE 1-3. Secretory endometrium, day 25 (PO day 11). Predecidua is present under the surface, around vessels, and around the glands.

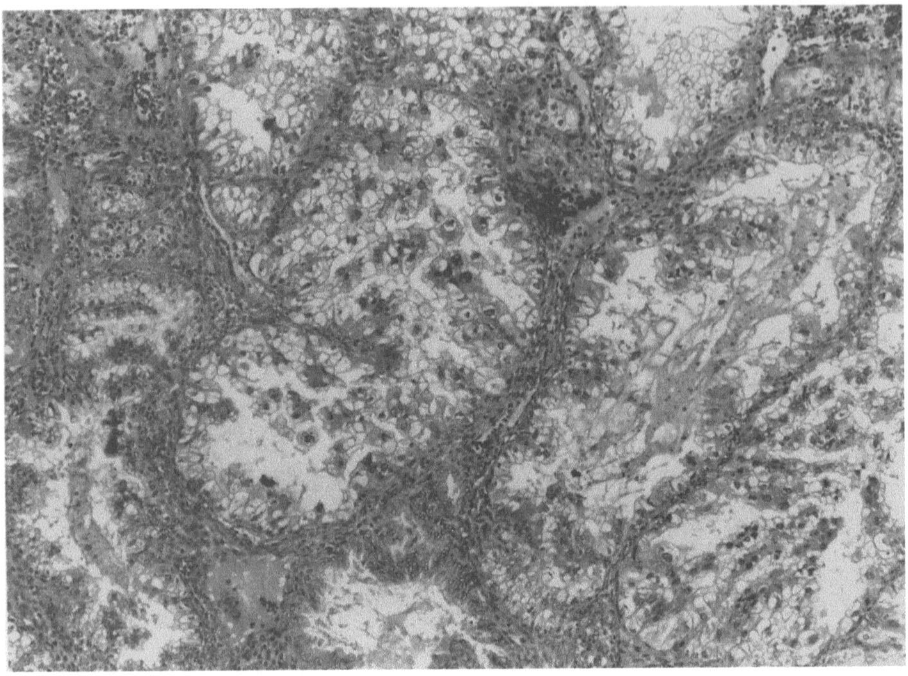

FIGURE 1-4. Endometrial glands during pregnancy showing Arias-Stella reaction. Note the crowding of glands, clearing of cytoplasm, and nuclear enlargement in glandular epithelium.

Sensory fibers from the uterus are located in the 11th and 12th thoracic and first lumbar nerve roots. Motor fibers to the uterus leave the spinal cord at the level of the seventh and eighth thoracic vertebrae. Thus anesthesia below the tenth thoracic vertebra achieves blockade of pain during uterine contractions without blocking the motor pathways.[15]

Lymphatic Drainage

The body of the uterus is drained by four sets of well-defined efferent lymphatic channels. These channels are present along the ovarian branch of the uterine artery, in folds of the broad ligament, between the mesosalpinx and the fallopian tube, and along the round ligament going toward the femoral area. The draining lymph nodes for these efferent lymphatics are the aortic nodes near the renal and superior mesenteric arteries, the parauterine and interiliac nodes, the aortic lymph nodes near the renal and superior mesenteric arteries, and the deep femoral lymph nodes. Analysis of the distribution of nodal metastases in patients with endometrial carcinoma with node involvement shows the aortic lymph nodes to be positive in about 40% of cases.[16] Hypogastric, external iliac, common iliac, obturator, and parametrial lymph nodes are involved in 9 to 17% of cases. Inguinal lymph nodes are involved in 6% of cases. Metastasis to mediastinal, clavicular, mesenteric, and axillary lymph nodes are seen in 2 to 0.5% of cases.[16]

The cervix is drained by three pairs of lateral lymphatics and anterior and posterior collecting trunks. The lateral lymphatics leave the cervix in the base of the broad ligament. The upper branches terminate in the upper interiliac nodes. The middle branches terminate in the lower interiliac, external

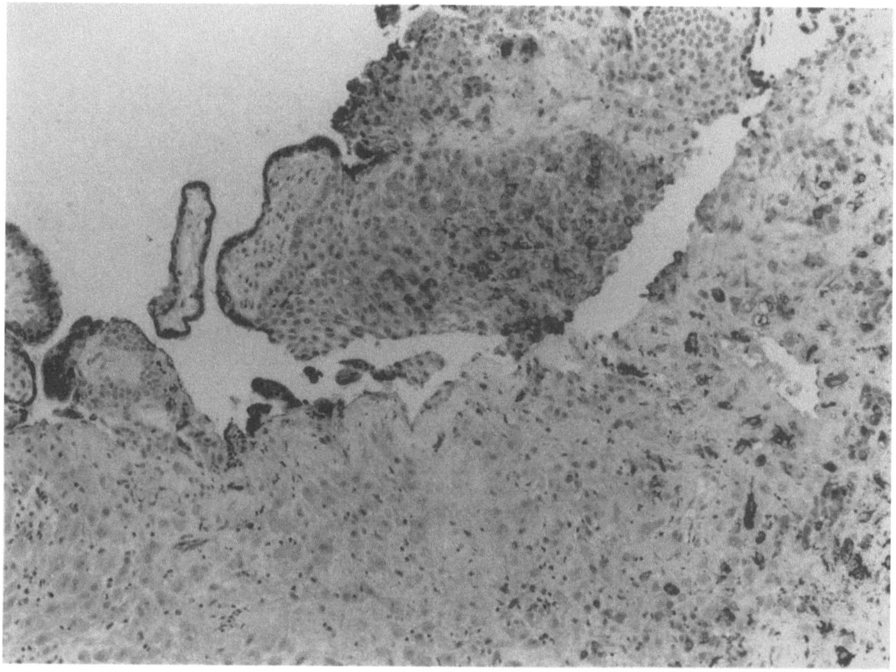

A

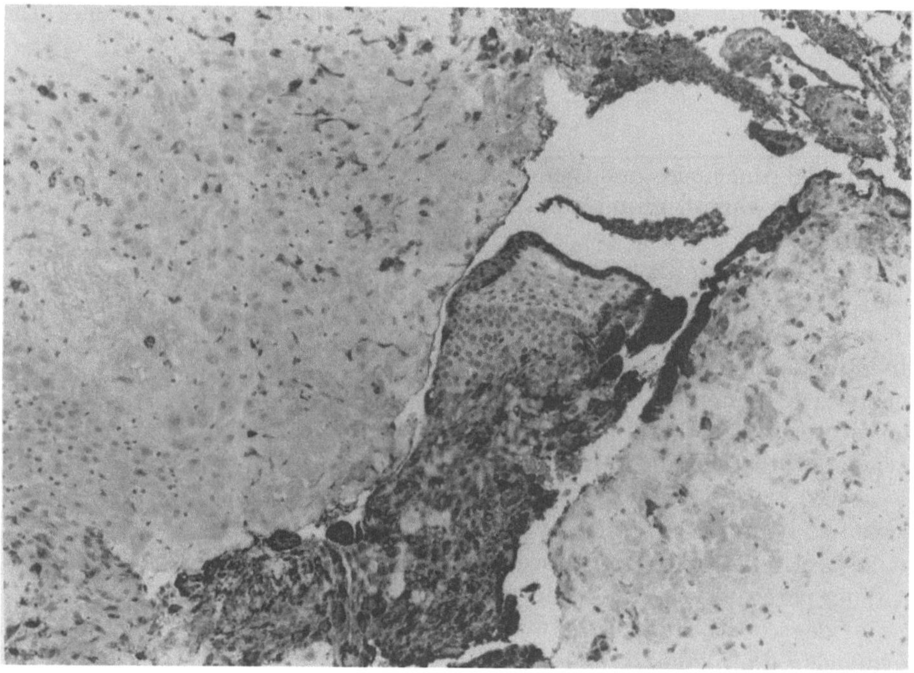

B

FIGURE 1-5. Use of immunohistochemistry for differentiating trophoblastic tissue from decidua. (A) Stain for human placental lactogen. The trophoblasts around the chorionic villus and focally in the decidua are positively stained. The decidua is negative. (B) Stain for keratin. The trophoblasts are positively stained. The decidua is negative. (*continued*)

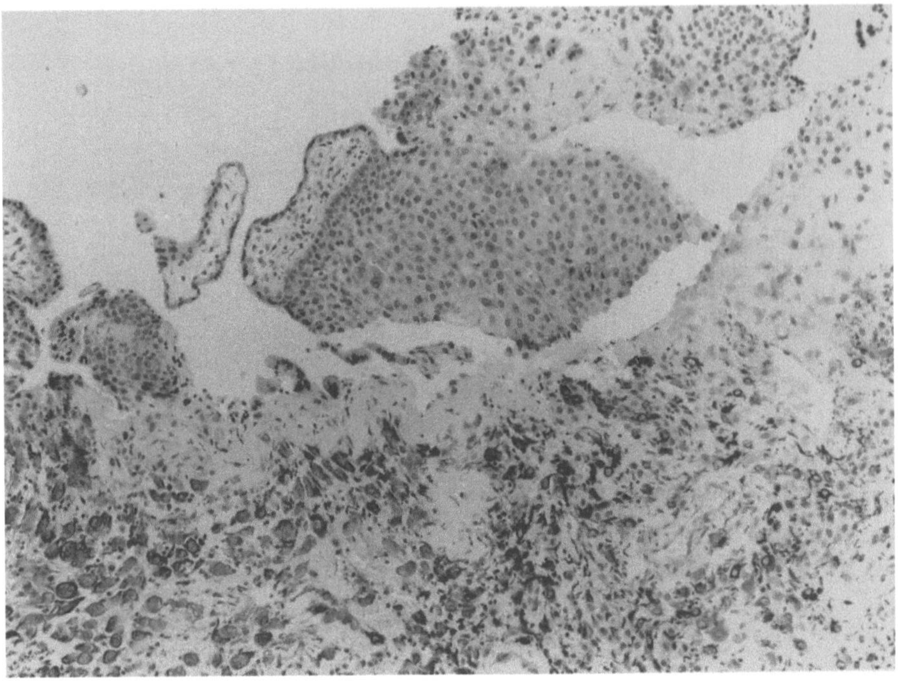

C

FIGURE 1-5. (*continued*) (C) Stain for vimentin. The decidual cells are positively stained. The trophoblasts are negative.

iliac, and common iliac nodes and the lower branches reach the sacral, gluteal and subaortic lymph nodes.

Posterior collecting trunks terminate in the superior rectal, common iliac, subaortic, and aortic nodes. Anterior collecting trunks terminate in the interiliac nodes. In contrast to endometrial carcinoma, aortic lymph nodes constitute only about 5% of lymph nodes involved with metastatic carcinoma from the cervix. External iliac, obturator, and hypogastric lymph nodes each account for about 20% of nodes with metastases.[16] Common iliac, parametrial, and paracervical lymph nodes constitute about 10% of nodes involved.[16] Paravaginal and sacral lymph nodes constitute less than 2% of lymph nodes with metastasis.[16] The lymphatics of the body of the uterus and the cervix anastomose, most markedly in the outer third of the cervical thickness.

Metaplastic Changes of Endometrial Epithelium

Epithelial metaplasia is defined as replacement of normal epithelium by a different type of epithelium. These changes are often focal. Commonly seen metaplasias include tubal, eosinophilic, squamous, papillary, hobnail, mucinous, and clear cell. These metaplasias should not be confused with neoplasia.[17] The vast majority (90%) of metaplasias occur in peri- and postmenopausal women and in inactive or proliferative endometrium.[17]

Tubal or Ciliated Metaplasia

In tubal metaplasia increased numbers of cells with cilia are seen. The proportion of ciliated cells is believed to increase under the influence of estrogen.[18] Ciliated cells may constitute 20% or more of endometrial epithelial cells in the late proliferative phase.[19] However, cilia of proliferative endometrium are not usually visible at light microscopic level. Cilial metaplasia is frequently seen in postmenopausal, atrophic endometrium and is occasionally seen in endometrial polyps.

Eosinophilic Metaplasia

The endometrial epithelium in this metaplasia shows a bright eosinophilic (pink stain on hematoxylin and eosin stain) appearance.

Squamous Metaplasia

Squamous metaplasia occurs in two forms—mature squamous metaplasia and immature squamous metaplasia. Mature squamous metaplasia may rarely occur as an isolated finding in peri- and postmenopausal women. More commonly, mature squamous metaplasia occurs in association with an intrauterine device,[20] endometritis, vitamin deficiencies,[21] endometrial polyps, and hyperplasia. Immature squamous metaplasia, or morule formation, occurs as oval, round or irregular collections of immature squamous cells in endometrial glands and/or stroma.[22] Although usually seen in association with endometrial hyperplasia, these may be seen less frequently in normal epithelium or in carcinoma. The cells of the morules have vesicular nuclei and finely granular cytoplasm with indistinct cell borders. Intercellular bridges and keratin are not usually seen. Central necrosis is occasionally seen in these morules.

Papillary Metaplasia

Uncommonly, the endometrial surface epithelium may have a papillary appearance in focal areas. The lining epithelial cells have bland oval nuclei with dispersed chromatin. Associated acute and chronic inflammation is often seen.

Hobnail Metaplasia

Rarely, endometrial glands may show isolated cells with large nuclei bulging toward the luminal side of the glands.

Mucinous Metaplasia

Occasionally endometrial epithelium may be replaced by epithelium resembling endocervical epithelium.[23] Focal mucinous metaplasia may be seen in some polyps.

Clear Cell Metaplasia

In this metaplasia the endometrial glandular epithelial cells show abundant clear cytoplasm as a result of the presence of large amounts of glycogen in the cytoplasm. This has been described in postmenopausal women on estrogens.[24]

Heterotropic Tissue

Skin adnexae (sebacous glands, hair, sweat glands) and mature cartilage have been described in the cervix on rare occasions.[13,25] Rare examples of occurrence of heterotopic bone[26–29] and cartilage[25] have been reported in the endometrium.

Developmental Nests

Remnants of the mesonephric duct may be seen in normal cervices as an incidental finding. These are located deep in the lateral wall of the cervix and show small tubular glands lined with cuboidal, nonciliated cells. These cells are distinguished from endocervical epithelium by the lack of mucin. Rare cases of mesonephric carcinoma of the cervix, presumably arising from these nests, have been reported.[30] These mesonephric adenocarcinomas show an immunoreactivity profile similar to normal mesonephric structures during ontogenesis.[31]

Uterine Immune System

Immunoglobulin A (IgA)-producing lymphoid cells in the endocervix[32] constitute the major source of immunoglobulins in the uterus. These IgA-producing lymphoid cells are increased in women with local infections and in those suffering from infertility without any obvious cause.

Langerhans cells have been described in the cervix and can be detected with immunoperixodase stains for S-100 and T6.[33] These cells have been shown to be decreased in women who smoke. This reduction of Langerhans cells with smoking has been postulated as one possible mechanism for an increased susceptibility to cervical cancer in smokers compared to nonsmokers.[34,35] A decrease in the number of Langerhans cells has also been reported in cervical-intraepithelial neoplasia with human papillomavirus types 16 and 18.[36]

In the late secretory phase many lymphoid cells infiltrate the endometrium. These cells bear early (OKT 11 and 3A1) but lack mature (UCHT 1, OKT 4 and OKT 8) T cell markers.[37] These cells are believed to correspond to stromal granulocytes. A much smaller number of IgA-, IgM-, or IgG-producing lymphoid cells are seen in the endometrium in all phases of the menstrual cycle.[38] These cells localize in lymphoid aggregates, which are seen in 50% of normal endometria from women of reproductive age. Premenstrual increase in leukocytes appears to consists of macrophages and neutrophils.[39]

References

1. Blanchard M, Jasso N: Sources of the anti-Mullerian hormone synthesized by the fetal testes: Mullerian inhibiting activity of fetal bovine Sertoli cells in tissue culture. Pediatr Res 1974;8:968–971.
2. Gray SW, Skandalakis JE: Embryology for Surgeons: The Embryological Basis for the Treatment of Congenital Defects. Philadelphia, WB Saunders Company, 1972, pp 663–664.
3. Wallach EE: The uterine factor in infertility. Fertil Steril 1972;23:138–158.
4. Winter JD, Kohn G, Mellinin WJ, et al: A familial syndrome of renal, genital, and middle ear anomalies. J Pediatr 1968;72:88–93.
5. Edwards JA, Gale RF: Camptobrachydactyly: a new autosomal dominant trait with two probable homozygotes. Am J Hum Genet 1972; 23:464–469.
6. Ober WB, Bernstein J: Observations on the endometrium and ovary in the newborn. Pediatrics 1955;16:445–460.
7. Frommer DJ: Changing age at menopause. Br Med J 1964;2:349–351.
8. Markee JE: Menstruation in intraoccular endometrial transplant in the rhesus monkey. Carnegie Institute of Washington Publ no 518. Contrib Embryol 1940;28:219–228.
9. Casey ML, Hamsell DL, MacDonald PC, Johnston JM: NAD$^+$ dependent 15-hydroxyprostaglandin dehydrogenase activity in human endometrium. Prostaglandins 1980; 19:115–122.
10. Ferenczy A, Bertrand G, Gefland M. Studies on the cytodynamics of human endometrial regeneration III—in vitro short-term incubation historadioautography. Am J Obstet Gynecol 1979;134:297–304.
11. Bur MA, Greene GL, Press MF: Estrogen receptor localization in formalin-fixed, paraffin embedded endometrium and endometriotic tissues. Int J Gynecol Pathol 1987;6:140–151.
12. Naftolin F, Stubblefield PG (eds): Dilatation of the Uterine Cervix. Connective Tissue Biology and Clinical Management. New York, Raven Press, 1980.
13. Dougherty CM, Moore WR, Cotton N: Histologic diagnosis and clinical significance of benign lesions of the nonpregnant cervix. Ann NY Acad Sci 1962;97:683–702.
14. Fuchs AR: Prostaglandin F2 alpha and oxytocin interactions in ovarian and uterine function. J Steroid Biochem 1987;27:1073–1080.
15. Pritchard JA, MacDonald PC, Gant NF: Wiliams Obstetrics, ed 7. Norwalk, CT, Appleton-Century-Crofts, 1985.
16. Plentl AA, Friedman EA: Lymphatic System of the Female Genitalia: The Morphologic Basis of Oncologic Diagnosis and Therapy. Philadelphia, WB Saunders Company, 1971.
17. Hendrickson MR, Kempson RL: Endometrial epithelial metaplasias: proliferations misdiagnosed as adenocarcinoma: report of 89 cases

and proposed classification. Am J Surg Pathol 1980;4:525–542.
18. Schueller EF: Ciliated epithelia of the human uterine mucosa. Obstet Gynecol 1968;31:215–223.
19. Fleming S, Tweeddale DN, Roddick JW: Ciliated endometrial cells. Am J Obstet Gynecol 1968;102:186–191.
20. Lane ME, Dacalos E, Sorrero AJ, Ober WB: Squamous metaplasia of the endometrium in women with an intrauterine contraceptive device. Follow-up study. Am J Obstet Gynecol 1974;119:693–697.
21. Baggish MS, Woodruff JD: The occurrence of squamous epithelium in the endometrium. Obstet Gynecol Surv 1967;22:69–115.
22. Dutra F: Intraglandular morules of the endometrium. Am J Clin Pathol 1959;31:60–65.
23. Solomon C, Polishuk W: Myxometra resulting from mucous metaplasia of the endometrium. Am J Obstet Gynecol 1954;68:1600–1603.
24. Hendrickson MR, Kempson RL: Surgical Pathology of the Uterine Corpus. Philadelphia, WB Saunders Company, 1980, p 209.
25. Roth E, Raylor HB: Heterotopic cartilage in the uterus. Obstet Gynecol 1966;27:838–844.
26. Adamson NE Jr, Sommers SC: Endometrial ossification—report of two cases. Am J Obstet Gynecol 1954;67:187–190.
27. Courpes AS, Morris JD, Woodruff JD: Osteoid tissue in utero. Report of 3 cases. Obstet Gynecol 1964;24:636–640.
28. Ganem K, Parsons L, Friedell G: Endometrial ossification. Am J Obstet Gynecol 1962; 83:1592–1598.
29. Hillenius L, Knutson F, Anberg A: Osseous tissue in the uterus. Acta Pathol Microbiol Scand 1953;33:387–392.
30. Hart WR: Cervix adenocarcinoma of mesonephric type. Cancer 1972;29:106–113.
31. Lang G, Dallenbach-Hellweg G: The histogenetic origin of cervical mesonephric hyperplasia and mesonephric adenocarcinoma of the uterine cervix studied with immunohistochemical methods. Int J Gynecol Pathol 1990;9:145–157.
32. Green FHY, Fox H: A study of secretory immune system of the female genital tract. Br Jr Obstet Gynaecol 1975;82:812–816.
33. Puts JG, Moesker O, Waal RMW, et al: Immunohistochemical identification of Langerhans cells in normal epithelium and epithelial lesions of the uterine cervix. Int J Gynecol Pathol 1986;5:151–162.
34. Winkelstein W Jr: Smoking and cancer of the uterine cervix: hypothesis. Am J Epidemiol 1977;106:257–259.
35. Grail A, Norval M: Significance of smoking and detection of serum antibodies to cytomegalovirus in cervical dysplasia. Br J Obstet Gynecol 1988;95:1103–1110.
36. Hawthorn RJS, Murdoch JB, MacLean AB, Machie RM: Langerhans cells and subtypes of human papillomavirus in cervical intraepithelial neoplasia. Br Med J 1988;297:643–646.
37. Marshall RJ, Jones DB: An immunohistochemical study of lymphoid tissue in human endometrium. Int J Gynecol Pathol 1988; 7:225.
38. Bulmer JN, Hagin SV, Browne CM, Billington WD: Localization of immunoglobulin-containing cells in human endometrium in the first trimester of pregnancy and throughout the menstrual cycle. Eur J Obstet Gynecol Reprod Biol 1986;23:31–44.
39. Kamat BR, Isaacson PA: The immunohistochemical distribution of leukocytic subpopulations in human endometrium. Am J Pathol 1987;127:66–78.

2

Pathophysiology of Diethylstilbestrol Changes

ALEXANDER SEDLIS and STANLEY J. ROBBOY

Diethylstilbestrol (DES) and other related nonsteroidal estrogenic substances, e.g., dienestrol and hexestrol, were commonly prescribed for treatment of threatened abortion, toxemia, diabetes, and other pregnancy complications between the 1940s and 1960s. Although the effectiveness of these compounds has never been documented, their use continued until the discovery in 1971 of a link between a clear cell adenocarcinoma of the cervix and vagina and the use of these drugs.[1] The impact of this discovery was much wider than just recognition of a new clinical entity. It reaffirmed the dangers of iatrogenic disease and reemphasized the need for proper screening before approval of any drug for treatment. It brought into a sharp focus the hazard to the fetus from transplacental transmission of medication and reaffirmed the importance of strict observance of limits to prescribing drugs in pregnancy. Furthermore, the discovery has stimulated interest in mechanisms of teratogenicity by pinpointing the exact site of interference of the drug with the normal development of an embryo. The DES–clear cell carcinoma connection also raises the question of whether DES is carcinogenic. Particularly intriguing is the latency of DES action, manifested by a long interval of many years between exposure to the compound and the development of cancer.

In addition to clear cell adenocarcinoma, in utero DES exposure is also responsible for benign structural and epithelial changes in the developing cervix and vagina.

Embryology of the Vagina

The cervix and vagina develop from the müllerian ducts, growing caudad, and the urogenital sinus, which advances from below (Fig. 2-1). At approximately day 54 of embryonal development, the ducts fuse medially and form a uterovaginal tube. The uterovaginal tube is composed of a mesenchymal wall and a mucosal lining of primitive ciliated cuboidal müllerian epithelial cells. The urogenital sinus is lined by squamous stratified epithelium of entodermal origin, closely resembling the lining of an adult vagina. As in the vagina, the sinus squamous cells show stratification, surface maturation, and glycogen-filled cytoplasm in the superficial cells. Maturation of the squamous epithelium and the cytoplasmic glycogen production in the embryo is promoted by stimulation with estrogen of placental origin. Before the tenth week of embryonic development, the fused müllerian ducts have advanced downward to the level of the future hymen.

After the tenth week of embryonic development, squamous epithelium from the urogenital sinus begins to advance craniad and to replace the original müllerian epithelium lining of the vaginal wall. First, squamous epithelium grows as a solid plate that

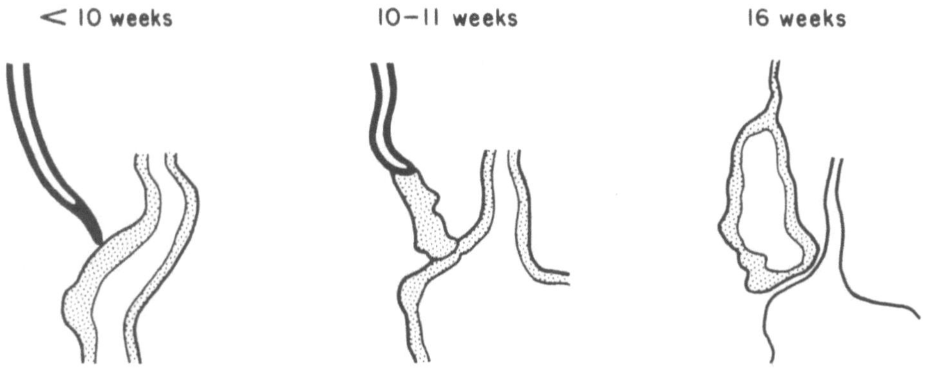

FIGURE 2-1. The development of the vagina in a normal embryo. By the end of 7 weeks, the müllerian ducts have reached the urogenital sinus and have fused. Beginning in the tenth week the squamous epithelium originating from the urogenital sinus ascends as a solid plate and replaces the müllerian epithelium. By 16 weeks, squamous epithelium from the urogenital sinus has reached the upper vagina and become canalized in the lower vagina. (Reprinted from ref. 2, with permission of Cambridge University Press.)

reaches half the length of the vagina by the 14th week. By the 16th week, the cranial tip of the advancing epithelium reaches the level of the external cervical os and the lower end becomes canalized to form the future vaginal cavity. By the 20th week, the developing vagina is fully canalized and completely lined by squamous stratified epithelium of sinus origin.[2] Squamous epithelium remains mature throughout embryonal and fetal life, but it changes after delivery to the atrophic type, similiar to that of postmenopausal state, when it is no longer stimulated by maternal estrogen.

In the meantime, in the müllerian duct the mesenchymal wall becomes stratified into at least two layers[3] (Fig. 2-2). The outer half will become the muscular wall. The inner portion throughout the entire length of the müllerian tube supports the growth of a tuboendometrial epithelium. In addition, in the region of the endocervix it also induces the transformation of the primitive müllerian epithelial lining into a mucinous type characterized by a single layer of tall columnar cells with basal nuclei and mucin-filled cytoplasm that in the adult lines the endocervical canal (Fig. 2-3A). The tuboendometrial type of lining is present in the fallopian tube, in the endometrium, deep in the

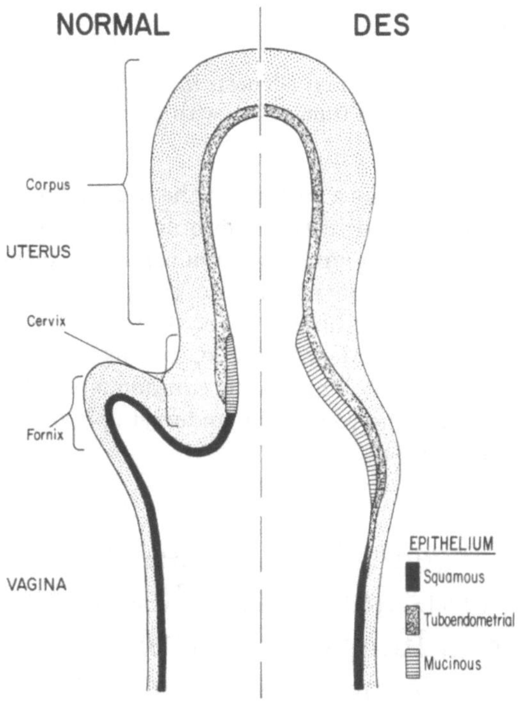

FIGURE 2-2. Hypothetical consequences of DES exposure on the vagina and cervix. (Left) Normal genital tract. (Right) Genital tract in patient exposed in utero to DES. (Reprinted from ref. 3, with permission.)

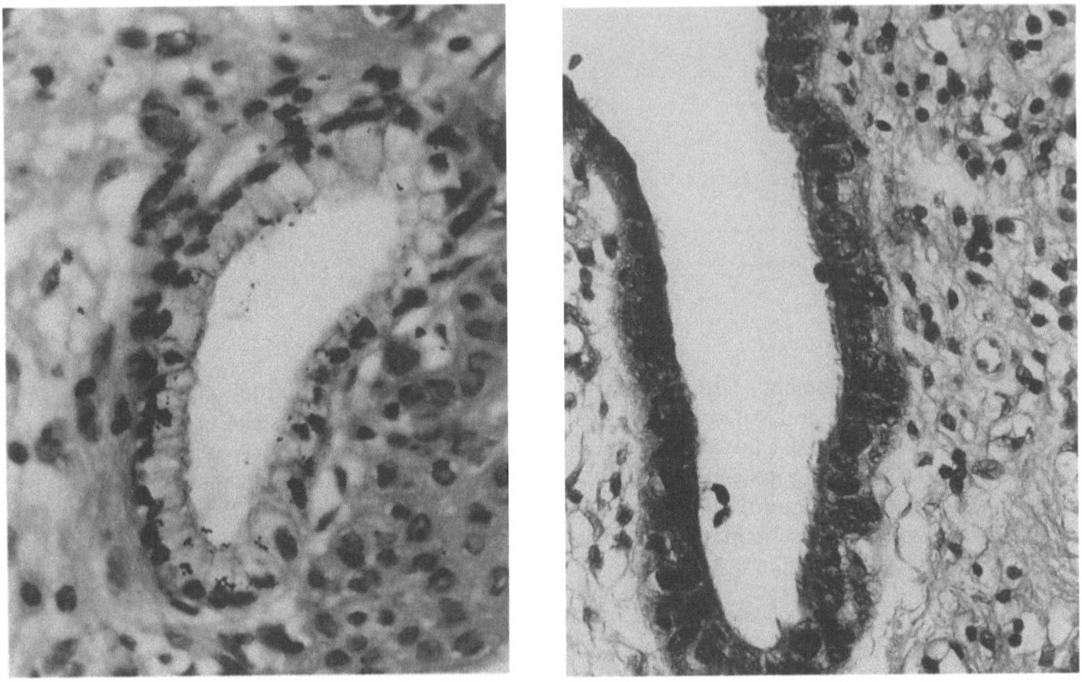

A B

FIGURE 2-3. Vaginal adenosis in which the glandular epithelium is mucinous (A) or tuboendometrial (B). (Reprinted from ref. 3, with permission.)

endocervix, as a cuff about the mucinous epithelium, and in the embronic vagina. It is arranged as a single layer of cells with large nuclei, copious eosinophic cytoplasm, and frequently a ciliated border (Fig. 2-3B). In a normally developed female, the vaginal wall is completely lined with squamous epithelium and the role of the underlying layers is not apparent.

Congenital Anomalies of Cervix and Vagina Resulting from Interference with Development of Vaginal Wall and Advance of Sinus Epithelium

Diethylstilbestrol taken in pregnancy reaches the embryo by a transplacental route and blocks certain phases of development of the fetal vagina by an as-yet unknown mechanism. For example, it blocks the upward migration of the stratified squamous epithelium of urogenital sinus origin that normally would replace the primitive müllerian lining of the vaginal tube. Consequently, the vaginal wall in DES-exposed females exhibits remnants of the müllerian epithelial lining of mucinous and endometrial tubal type (i.e., adenosis). In addition, DES affects the normal development of the mesenchymal tissue within the walls of the vagina, cervix, and uterus. The disturbed development of the mesenchyma is responsible for various structural anomalies of the cervix, uterus, and vagina found in DES-exposed females. Presumably, local proliferation of the mesenchyma may produce vaginal ridges and cervical hoods and collars. Inadequate development of the connective tissue, on the other hand, may account for obliterated lateral vaginal fornices and the hypoplastic cervix with asymmetrical os.

Similarly, the uterine structural abnormalities (e.g., the T-shaped uterine cavity) and the uterotubal and isthmic constrictions may develop by a similar mechanism.[2]

The extent of the lower genital tract anomalies depends on the timing of the DES administration during pregnancy. The most severe changes are produced when DES is administered before the tenth week of pregnancy, the time when the sinus epithelium begins to replace the müllerian lining. The later in pregnancy that the drug is first administered the fewer anomalies are found. After the 20th week, the effect of DES is negligible.

Adenosis similar to that caused by DES may be also produced by a mechanical block that impedes the ascent of the sinus squamous epithelium. A complete transverse vaginal septum, a rare congenital anomaly, is an example of such a mechanical obstacle.[4] In females with a transverse septum, the upper vagina and the proximal surface of the septum contain foci of müllerian epithelium of mucinous or tuboendometrial variety (Fig. 2-4). The squamous epithelium, found in the upper vagina in association with müllerian remnants, is presumably derived from squamous metaplasia of the mucinous cells.

Benign Changes in the DES-Exposed Females

Benign changes produced by DES are usually classified as (a) structural anomalies or (b) epithelial changes.

Structural Anomalies

Structural anomalies may involve the vagina, cervix, and uterus, and are probably caused by deranged development of the mesenchymal component of the lower müllerian tract.

The vagina in DES-exposed women is frequently shallow as a result of obliteration of the vaginal fornices. For the same reason, the cervix on palpation may appear flush with the vaginal vault, similar to the postmenopausal state. Another common anom-

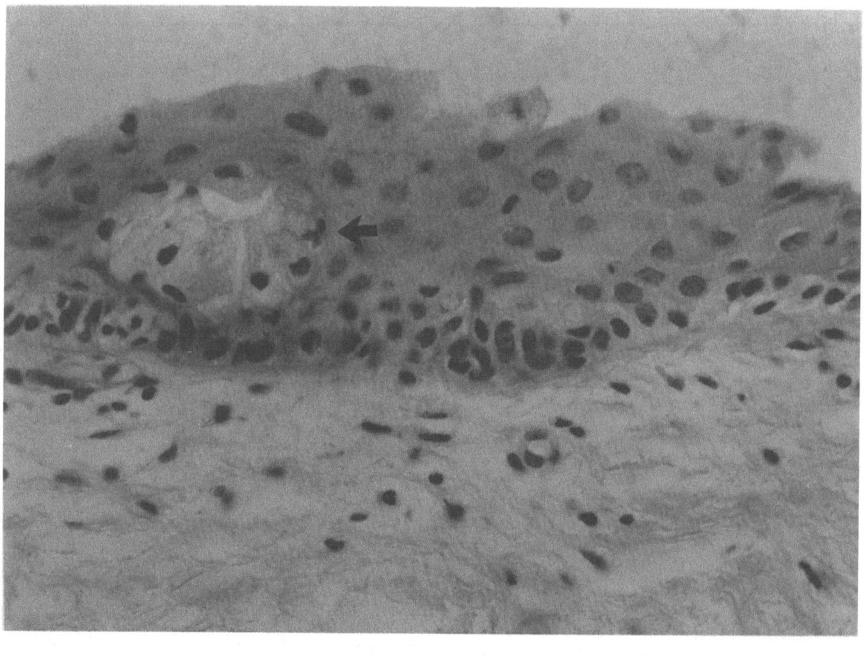

FIGURE 2-4. Complete transverse septum of the vagina. On microscopic examination, adenosis, represented by a crypt lined by columnar epithelium, is seen on the proximal side of the septum (arrow).

aly is a vaginal ridge or partial transverse septum. The semilunar border of the ridge may be appreciated on palpation, and the ridge also may make insertion of a vaginal speculum difficult. On speculum inspection, the ridge appears as a semilunar, thin membranous structure partially obstructing the view of the cervix (Fig. 2-5).

The most common cervical anomaly is a hood, also known as a "cock's comb." A hood is a triangular upward expansion of the anterior cervical lip (Fig. 2-6). A collar or circular fold surrounding the cervix in its peripheral portion is also very common. The collar may constrict the cervix and produce a bulging of the central portion of the cervix in the shape of a polyp, known as a "pseudopolyp." A pseudopolyp differs from a true polyp in that it contains the endocervical canal.

In the uterus, the best known anomaly is a T-shaped uterine cavity. The "T" distortion is characterized by bilateral expansion of the upper portion of the uterine cavity, including the interstitial and isthmic portions of the fallopian tubes. Annular constrictions may also be found around the interstitial areas. The isthmic portion of the uterine cavity may be either constricted or widened and square, and the capacity of the uterine cavity may be reduced by one to two thirds (Fig. 2-7).

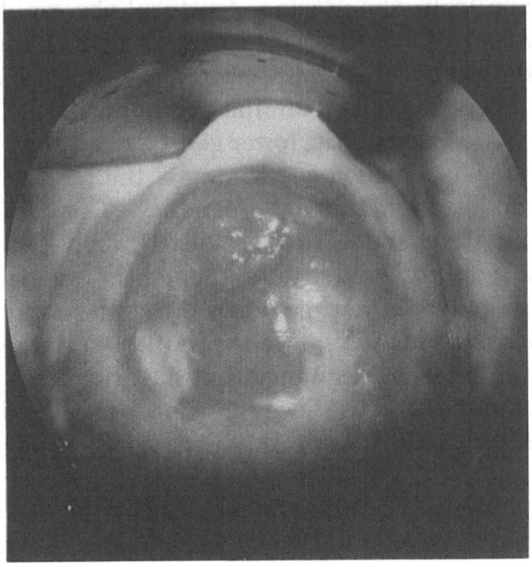

FIGURE 2-6. Cervical hood—a structural anomaly of the cervix. A triangular projection arises upward from the anterior cervical lip.

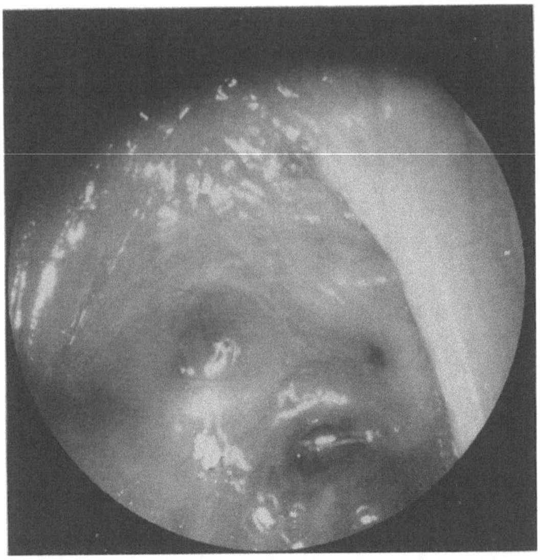

FIGURE 2-5. Vaginal ridge—a DES structural change. A membranous septum with semilunar border partially obstructs the view of a small hypoplastic cervix.

Epithelial Changes

Epithelial changes of the vagina are discussed separately from similar but not completely identical cervical epithelial changes.

Vaginal Epithelial Changes

Vaginal epithelial changes may be observed with an unaided eye, but are better appreciated with the use of colposcopy. Adenosis, or presence of müllerian epithelium in the vaginal wall, may often be seen with an unaided eye as red patches with irregular surfaces. With a colposcope, adenosis has the same appearance as columnar epithelium of the endocervical mucosa (Fig. 2-8). (Adenosis of the tuboendometrial type almost always appears as glands within the mesenchyma, and is therefore not seen with a colposcope.)

Adenosis features a characteristic grape-

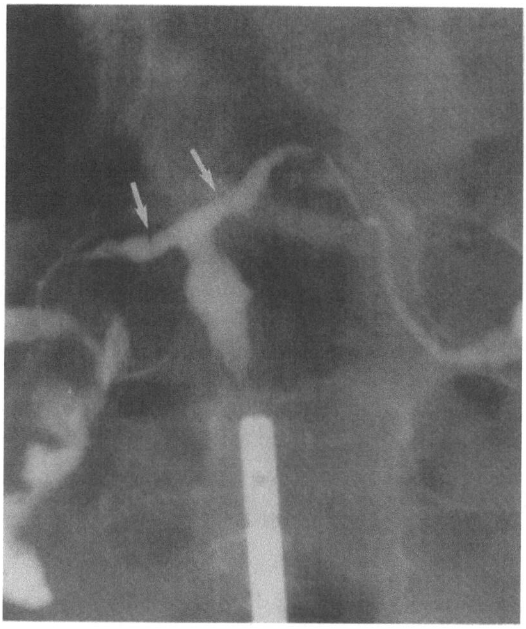

FIGURE 2-7. T-shaped uterus (hysterosalpingogram), showing annular constrictions of the interstitial portions of the tubes (arrows) and widening of the uterine isthmus.

like villous structure and a red color contrasting with the pink-gray tone of the surrounding squamous epithelium. In most instances, adenosis is associated with squamous metaplasia since mucinous epithelium in the vaginal location behaves in the same manner as that in the transformation zone of the cervix. Consequently, upon colposcopic examination various stages of metaplasia may be seen (e.g., flattening of the villi, aceto-whiteness, gland openings, and nabothian cysts). Frequently, active metaplasia in adenosis assumes a mosaic appearance characterized by interlacing horizontal vessels, or punctation with "end-on" vertical vessels against an aceto-white background. Characteristically, most mosaic and punctation patterns in adenosis are associated with benign metaplasia, although similar changes in the transformation of the cervix in non–DES-affected individuals usually denote neoplasia. As squamous metaplasia in adenosis becomes mature, it causes flattening of the epithelial surface and disappearance of the villous structure, and in advanced stages of maturation becomes indistinguishable from mature squamous epithelium. However, the difference between the squamous metaplastic epithelium during most of its stages and the native squamous epithelium becomes apparent upon application of iodine solution. The native squamous epithelium stains brown with iodine because it contains intracytoplasmic glycogen. Metaplastic epithelium, in contrast, remains unstained because glycogen is absent or present in insufficient quantity within its cytoplasm (Fig. 2-9A). In addition, on colposcopic examination and acetic acid application, metaplastic epithelium may be differentiated from the native epithelium because it becomes aceto-white.

On microscopic examination, the adenosis may be of the mucinous or tuboendometrial type. In the upper third of the vagina, in ab-

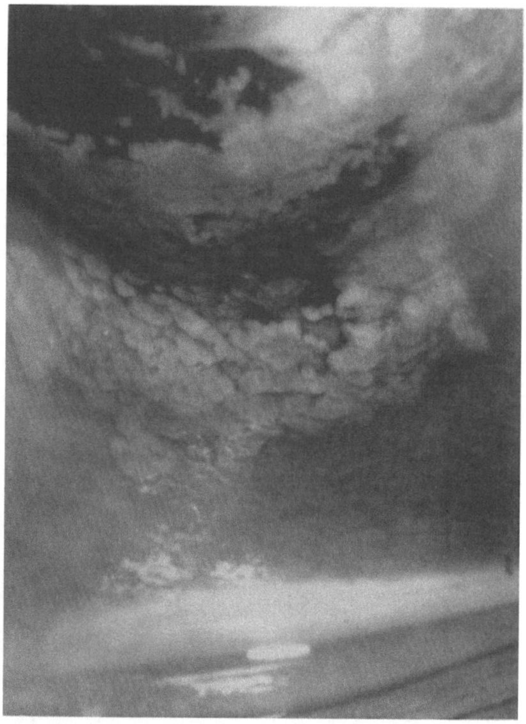

FIGURE 2-8. Adenosis (colposcopic view). Columnar epithelium characterized by villous, grapelike pattern is seen in the posterior vaginal fornix.

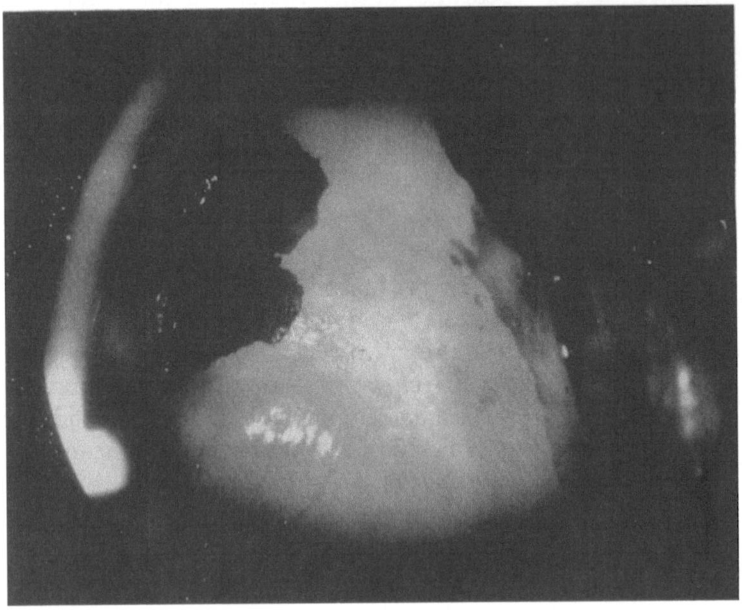

A

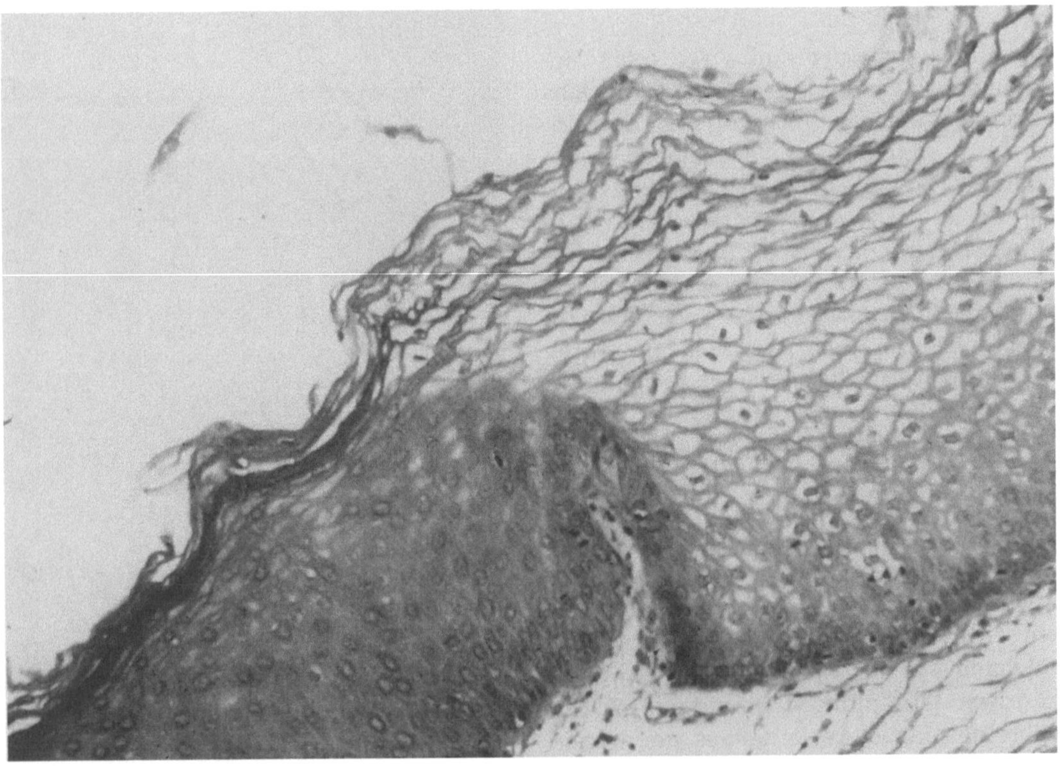

B

FIGURE 2-9. Squamous metaplasia replacing adenosis in the anterior vagina. (A) Colposcopic view. Metaplastic epithelium does not stain with iodine, in contrast to brown staining of the native squamous epithelium at the periphery. (B) Microscopic view. Metaplastic cells with dense cytoplasm (left) contrasting with clear, glycogen-filled cells in the adjacent native squamous epithelium.

solute numbers two thirds of biopsies with adenosis are of the mucinous type. Adenosis in the lower vagina, while quite rare, is composed almost exclusively of tuboendometrial epithelium. In the upper vagina, mucinous-type adenosis without squamous metaplasia presents as columnar epithelium with small basal nuclei and foamy cytoplasm. The cells are arranged in a single layer and may form tufts with fibrovascular cores corresponding to the villi observed on colposcopy. In addition to lining the surface, mucinous epithelial cells line the crypts or glandlike spaces in a manner analogous to that in the endocervix. Also as in the cervix, crypts become nabothian cysts when occluded by advancing metaplastic epithelium. Various stages of squamous metaplasia in adenosis may be observed analogous to those seen in the cervix. As metaplastic cells become mature, the glandular epithelium may completely disappear, leaving only residual droplets of mucus. In such cases, the peglike arrangement of squamous epithelium and occasional droplets of mucus are the only microscopic manifestations of adenosis (Fig. 2-10). During most, but the most advanced stages of maturation, metaplastic cells lack glycogen and are relatively easy to distinguish from the native squamous epithelium. Once metaplastic cells become totally mature and are filled with glycogen, they cannot be distinguished from native squamous cells (Fig. 2-9B).

On microscopic examination, the tuboendometrial type of lining differs from the mucinous epithelium by the presence of eosinophilic cuboidal cells with large nuclei and cilia (Fig. 2-3B). This type of lining also differs from the mucinous epithelium by its almost exclusive intracryptal location. Consequently, this type of adenosis is rarely detected on gross or colposcopic examination. Also, tuboendometrial epithelium is relatively more frequent than the mucinous type in the lower vagina versus the upper vagina

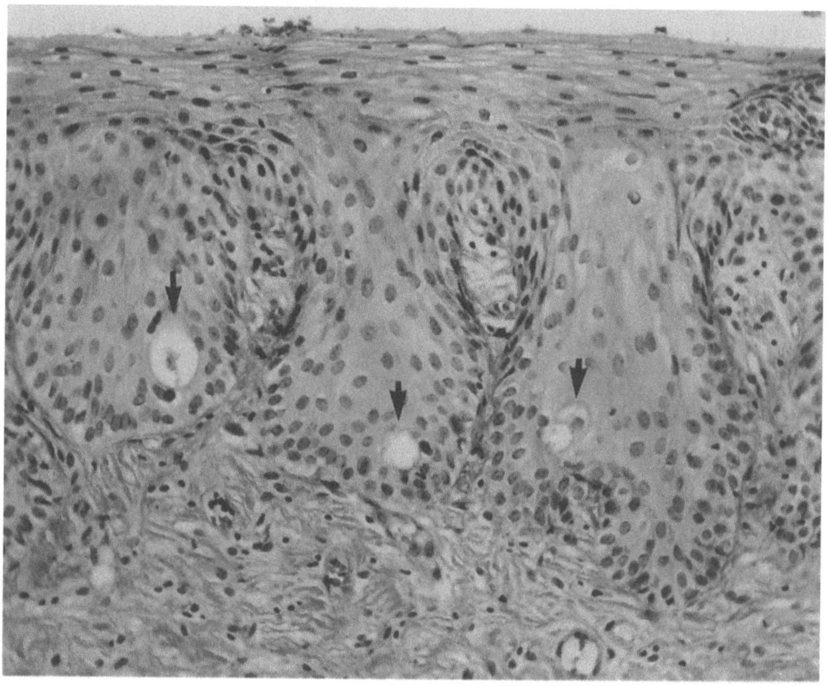

FIGURE 2-10. Adenosis with squamous metaplasia (microscopic view). Squamous epithelial "pegs" containing droplets of mucus (arrows) represent the crypt formerly occupied by columnar epithelium, which is now replaced by metaplastic squamous cells.

(21% occurrence in the upper; 31% in the middle, and 100% in the lower third of the vagina). Topographically, it is the tuboendometrial type of epithelium that is associated with clear cell adenocarcinoma.

Cervical Epithelial Changes

Epithelial changes in the cervices of DES-exposed women are similar to those in vaginal adenosis. Mucinous epithelium in these women persists outside of the external os, producing an ectropion that can be easily detected on gross examination and confirmed on colposcopy. (Fig. 2-11A). If the ectropion extends to the outer rim of the cervix it may blend with the adjacent adenosis in the vaginal fornix. Squamous metaplasia is almost always present in the ectropion and causes the usual colposcopic images, such as flattening of the cervix, aceto-whiteness, gland openings, and nabothian cysts. Similarly to adenosis, squamous metaplasia in the ectropion of DES-exposed women frequently assumes a mosaic or punctate pattern (Fig. 2-11B). This colposcopic pattern of benign squamous metaplasia may cause difficulty in the diagnosis of neoplasia (see Neoplasia). On microscopic examination cervical epithelial changes present mucinous epithelium in combination with various stages of metaplasia. Contrary to vaginal epithelial changes, the relative frequency of glandular epithelium is 97% mucinous to 3% tuboendometrial.[3]

Frequency and Natural History of Vaginal and Cervical Epithelial Changes

The frequency and the extent of the observed changes vary according to (a) week of pregnancy and dosage of DES, (b) the age at which patient was examined, and (c) the population group.

In women who were enrolled in the DESAD (DES-Adenosis) Project, 48% with history of exposure to DES before the eighth week of pregnancy had observable changes, whereas only 10% of women exposed after 20 weeks had vaginal and cervical epithelial changes. With increasing age, fewer DES changes are observed because of spontaneous regression of epithelial changes.[5] Spontaneous regression is not surprising be-

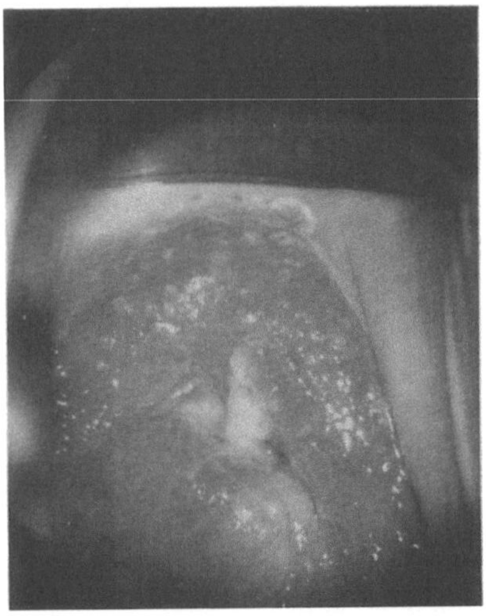

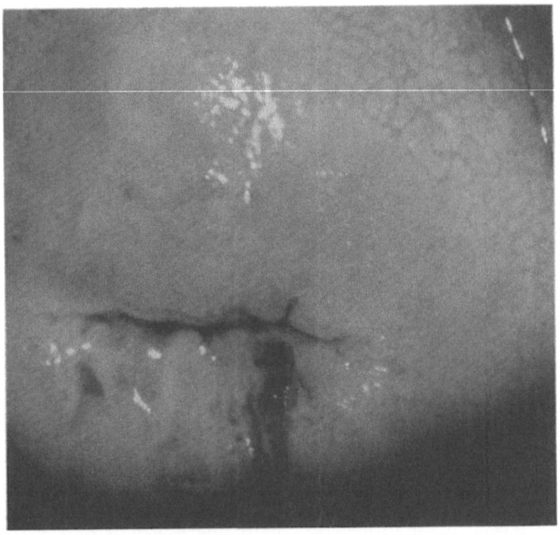

FIGURE 2-11. Ectropion—a cervical structural anomaly. (A) The entire surface of the exocervix is lined with columnar epithelial mucosa. (B) Metaplastic transformation with a mosaic pattern on colposcopy.

cause of almost universal occurrence of squamous metaplasia. According to DESAD experience, DES changes markedly decreased after 26 years of age. Spontaneous regression is observed not only in vaginal and cervical epithelial changes, but also in the structural changes, such as cervical hoods, and collars and vaginal septa. However, the mechanism of regression in the structural changes is less clear than that in regression of vaginal and cervical epithelial changes.[5]

Neoplasia

Clear cell adenocarcinoma is the neoplastic change most dramatically associated with prenatal DES exposure. Squamous cell neoplasia is also discussed.

Clear Cell Adenocarcinoma

These rare tumors of the vagina and cervix are classified as clear cell adenocarcinoma because of their predominant histologic pattern. Over 500 cases of these tumors have been accessioned into the Registry for Research on Hormonal Transplacental Carcinogenesis. More than 60% of these tumors were found in young women with a history of in utero exposure to nonsteroidal estrogenic substances (DES, hexestrol, and dianestrol). It is assumed that these compounds played an important role in the development of the adenocarcinoma for the following reasons. The prenatal exposure to DES was documented in 60% of the patients with tumors. Prior to the common use of DES, these tumors, especially in vaginal locations, have been exceedingly rare. The tumors arise from the area adjacent to vaginal or cervical epithelial changes. The vaginal epithelial changes in association with these tumors frequently include tuboendometrial lining. Both tubal and endometrial epithelia are capable of becoming transformed into tumors similar in appearance to clear cell carcinoma.[6]

The risk of cancer in the DES-exposed female is low, calculated as from 0.014 to 0.14%.[7] These calculations are based on the estimated number of women who have been exposed in utero to DES or similar compounds. Most patients were between 17 and 21 years old at the time of diagnosis; very few were less than 14 years. The upper age limit for these tumors is still unknown because sporadic occurrence of clear cell adenocarcinoma of the vagina is being reported in DES-exposed women over 35.

Clinical symptoms from these tumors, such as vaginal bleeding or discharge, may occur if the tumor surface is ulcerated. Large tumors may be palpated as an intravaginal mass. On inspection, these tumors may present a red, irregular, ulcerated surface, but in some instances they may be completely covered with intact vaginal mucosa if the tumor is located in the lamina propria of the vaginal wall. Colposcopic examination is not very helpful for the diagnosis because there is no specific colposcopic pattern associated with a clear cell carcinoma. Also of no use is colposcopic examination for those tumors that are enclosed in the lamina propria and completely covered with normal squamous epithelium.

Microscopically, several patterns may be recognized in these tumors. The clear cell pattern is characterized by solid sheets or glands lined by large cells with transparent cytoplasm and dark central nuclei (Fig. 2-12B). The tubal or cystic type is most common and presents tubules and cysts lined by flat or hobnail cells with the nuclei protruding into the gland lumen (Fig. 2-12A). Less frequent patterns are papillary and endometroid.

The tumor penetrates into the vaginal wall and spreads to neighboring structures such as the urinary bladder and rectum, as well as the regional lymph nodes. Distant metastases develop more frequently in clear cell adenocarcinoma than in a corresponding stage of squamous cell carcinoma. The 5-year survival of patients with this tumor is 80% overall and 90% in stage I.

The mechanism by which DES fosters the development of the cancer, if it is in fact more than a teratogen, is not clear. Although DES

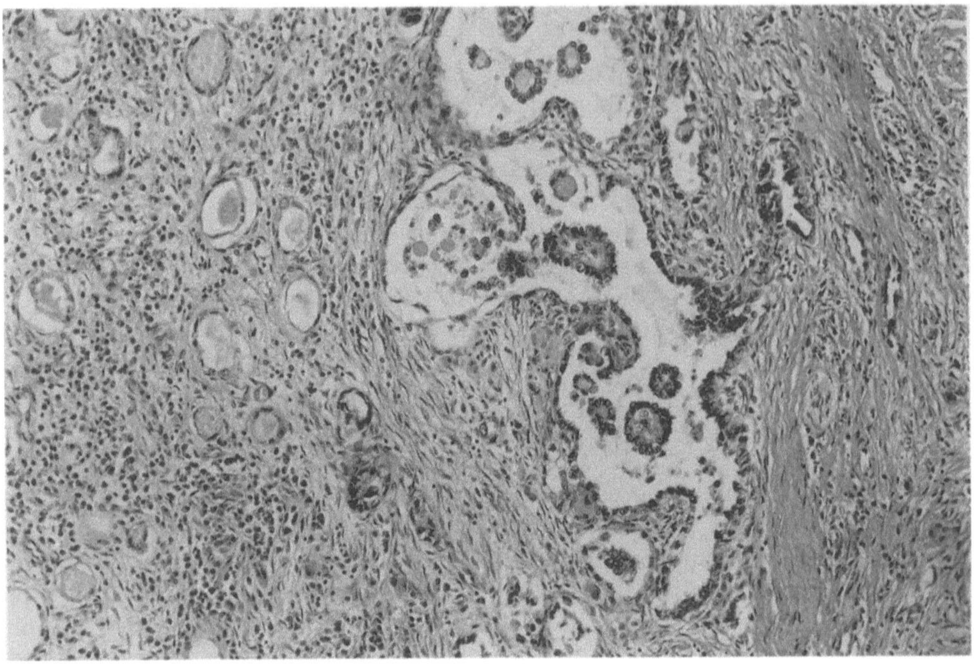

A

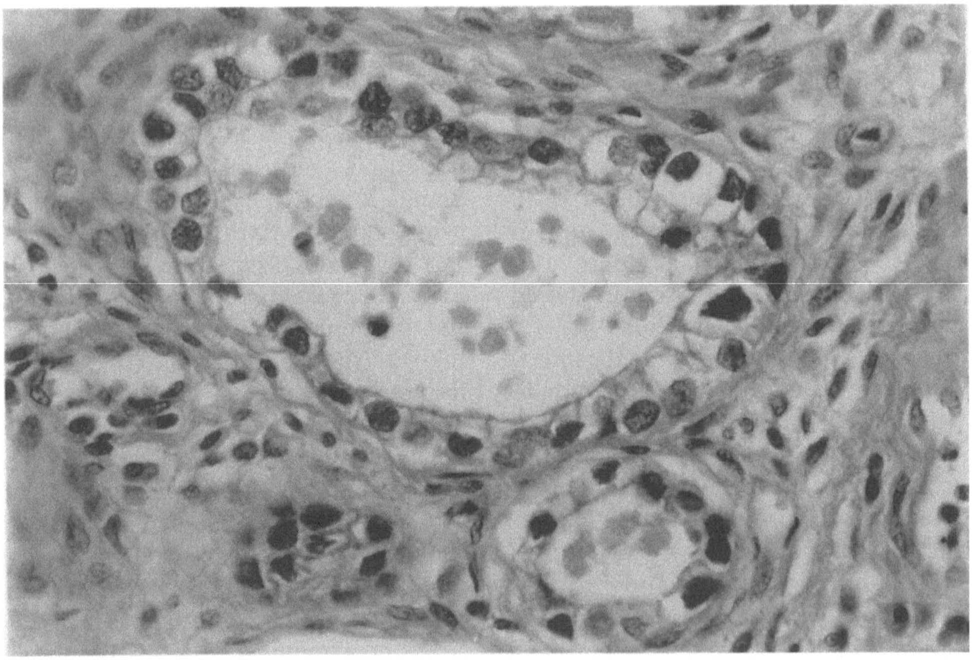

B

FIGURE 2-12. Clear cell carcinoma (microscopic view). (A) Tubulocystic (left) and papillary (right) patterns. (B) Clear cell pattern.

and estrogens produce cancer in experimental animals and promote the growth of endometrial carcinoma in humans, it is difficult to explain the latent interval of two decades between the DES exposure and the onset of vaginal cancer. Some have suggested that DES or one of its metabolites may produce the initial cellular insult in the embryo that may then undergo a malignant transformation under the influence of an as-yet unknown cofactor in postnatal life.[8] An alternate theory is that the role of DES is only teratogenic, displacing the immature müllerian epithelium to an abnormal location where it becomes more vulnerable to carcinogens.

Squamous Cell Neoplasia

The mucinous type of epithelium in vaginal and cervical epithelial changes undergoes squamous metaplasia analogous to that in the normal cervical transformation zone. Similarly, the metaplastic cells in adenosis and ectropion have the same potential to develop neoplastic changes as in the normal cervix. Therefore, squamous neoplasia occurring in ectropion and adenosis is being encountered in clinical practice. Specific problems of squamous neoplasia in DES-exposed women are (a) possibly increased risk of neoplasia, and (b) difficulty in diagnosis.

In theory, squamous neoplasia is expected to be more frequent in DES-exposed than in unexposed women because both the adenosis and the ectropion have more extensive squamous metaplasia, and hence increased numbers of metaplastic cells, the potential targets for human papillomavirus and other putative agents of the neoplastic process. Initial reports showing a high prevalence of cervical and vaginal dysplasia in DES-exposed patients seemed to support this theory and also prompted a prediction that the increased risk of malignancy in DES-exposed individuals is not only from the clear cell carcinoma but also from squamous cell cancer.[9] Although the data and conclusions of this early report subsequently have been challenged, the latest studies have documented a slight but increased incidence in dysplasia in DES-exposed patients.[10]

In part, the controversy about the frequency of neoplasia in DES-exposed women has been blamed on difficulties in diagnosis. Benign squamous metaplasia in both vaginal and cervical epithelial changes in DES-exposed women frequently resembles the colposcopic picture of mosaic and punctation characteristic of dysplasia. Moreover, microscopic findings of metaplasia in vaginal and cervical epithelial changes also have been frequently overinterpreted as neoplasia by the inexperienced pathologist because of the presence of cells with a high nucleus-to-cytoplasm ratio, and increased mitotic figures above the basal layers. True neoplasia, however, may be differentiated from immature metaplasia because it shows nuclear hyperchromasia, pleomorphism, and atypical mitotic figures. A correct diagnosis in these cases requires special expertise and meticulous attention to details. In some instances, differences of opinion even among experts may be resolved only with special techniques, such as determination of nuclear ploidy by means of flow cytometry.

Differentiation of dysplasia from benign metaplasia in DES-exposed women by means of colposcopy is perilous because there are no hard and fast rules for differential diagnosis. Colposcopic grading based on the intensity and tone of aceto-whiteness and the coarseness of the vascular markings may be of some help. In clinical practice, a biopsy from doubtful areas cannot be avoided.

Reproductive Function in DES Daughters

Structural anomalies of the uterus and cervix have been a cause for concern with regard to their role in reducing the reproductive performance of DES-exposed women. For example, the diminished capacity of the T-shaped uterus raises the probability of difficulty in accommodating the developing product of conception, which might result in abortion or premature birth. Annular constrictions of the isthmic and interstitial portions of the fallopian tubes might impede the migration of the fertilized egg into the uterus and thus promote an ectopic implantation. A short cervix may increase the risk of developing an incompetent os in the second trimester of pregnancy.

Available statistics on reproductive performance have indicated that DES daughters have the same conception rate as unexposed females. However, their ability to maintain a pregnancy may be reduced. The total rate of unfavorable outcomes of pregnancy in DES daughters is 38%, versus 22% in the unexposed females. These include a spontaneous abortion rate of 26% in exposed women versus 16% in unexposed women, ectopic pregnancy in 4% versus 1%, prematurity in 8% versus 4%, and still birth in 4% versus 1%.[11] Unfavorable outcome of pregnancy correlated with the hysteroscopically demonstrated structural changes according to one study in which normal pregnancy occurred in 58% of DES-exposed patients with normal hysterograms versus 35% with abnormal findings.[12]

References

1. Herbst AL, Ulfelder H, Poskanzer DC: Adenocarcinoma of the vagina: association of maternal stilbestrol therapy with tumor appearance in young women. N Engl J Med 1971;284:878–881.
2. Bulmer D: The development of human vagina. J Anat 1957;93:490–509.
3. Robboy SJ: A hypothetic mechanism of diethylstilbestrol (DES)-induced anomalies in prenatally exposed women. Hum Pathol 1983;14:831–833.
4. Brenner P, Sedlis A, Cooperman H: Complete imperforate transverse vaginal septum. Obstet Gynecol 1964;25:135–138.
5. Robboy SJ, Noller KL, Kaufman RH, et al: An Atlas of Findings in the Human Female after Intrauterine Exposure to Diethylstilbestrol. NIH Publ no 83-2344. Washington, DC, US Government Printing Office, 1984, 13.
6. Robboy SJ: Changes related to prenatal diethylstilbestrol exposure, in Kurman RJ (ed): Blaustein's Pathology of the Female Genital Tract, ed 3. New York, Springer-Verlag, 1987; page 121.
7. Herbst AL, Cole P, Colton T, et al: Age-incidence and risk of diethylstilbestrol-related clear cell adenocarcinoma of the vagina and cervix. Am J Obstet Gynecol 1977;128:43–50.
8. Vorherr H, Messer RH, Vorherr UF, et al: Teratogenesis and carcinogenesis in rat offspring after transplacental and transmammary exposure to diethylstilbestrol. Biochem Pharmacol 1977;28:1865–1877.
9. Stafl A, Mattingly RF: Vaginal adenosis—a precancerous lesion? Am J Obstet Gynecol 1974;120:666–677.
10. Robboy SJ, Noller KL, O'Brien PC, et al: Increased incidence of cervical and vaginal dysplasia in 3,980 diethylstilbestrol exposed young women. JAMA 1984;252:2979.
11. Barnes AB, Colton T, Gundersen JH, et al: Fertility and outcome of pregnancy in women exposed in utero to diethylstilbestrol: Preliminary findings from the DESAD Project. N Engl J Med 1980;302:509–613.
12. Kaufman RH, Adam E, Binder GL, et al: Upper genital tract changes and pregnancy outcome in offspring exposed in utero to diethylstilbestrol. Am J Obstet Gynecol 1980;137:299–308.

3

Immunology of the Uterus

ILAN TUR-KASPA and NORBERT GLEICHER

The relevance of immunology to the uterus and to reproduction has become apparent. The uterus is an immunologically dynamic environment with a unique local immune response. Embryos, which are biologically auto- as well as allografts, expressing transplantation antigens, are not rejected. Spermatozoa, bearing alloantigens, are not destroyed. However, infections of the uterus, which should occur as a result of chronic exposure to microorganisms, are apparently rare. Whenever those immunologic mechanisms are not well synchronized, a malfunction appears. Immunologic causes of infertility and immunologically caused fetal wastage are currently under expanding investigation by reproductive immunologists. Reproductive tumor immunology is a challenging area with potentially important clinical applications. This review is presented in an attempt to describe the unique immunologic environment of the uterus.

Basic Immunology

Normal immune function is specific.[1] *Antigen* (Ag) encounters cells that are capable of responding. This heterogeneous group of accessory cells, including *macrophages, monocytes*, and *dendritic cells*, capture and break microorganisms as well as soluble Ag molecules down into smaller molecular fragments that only in that form can react with lymphocyte receptors.[2] Such antigenic fragments (*epitopes*) can at times also be retained at the surface of accessory cells for prolonged periods. Accessory cell and epitopes then provoke an immune response at the level of regional lymph nodes.

The predominant cell in this afferent limb of the immune response is the macrophage, serving as an Ag-presenting as well as immunomodulating effector cell (Fig. 3-1). Monocytes have a circulating half-life of only 8 to 10 hours. They then enter tissues and become macrophages. As such, they may have a life span of months to years. Macrophages process Ag through phagocytic and pinocytic mechanisms and then present it to T lymphocytes.[3] Upon activation, macrophages release a variety of *monokines*, which regulate lymphocyte activities as well as having microbial and cytocidal (including tumoricidal) effects (Table 3-1).

The dendritic cell, a more recently described cell type, has Fc receptors and Ia proteins (Ia^+) on the surface and is also believed to be an Ag-presenting cell within the uterine endometrium.[4–6]

The efferent immune response involves two main types of lymphocytes: the thymus-derived *T cell*, responsible for cell-mediated immunity, and the bone marrow–derived *B cell*, which is responsible for the humoral immune response and antibody (Ab) formation (Figs. 3-1 and 3-2).

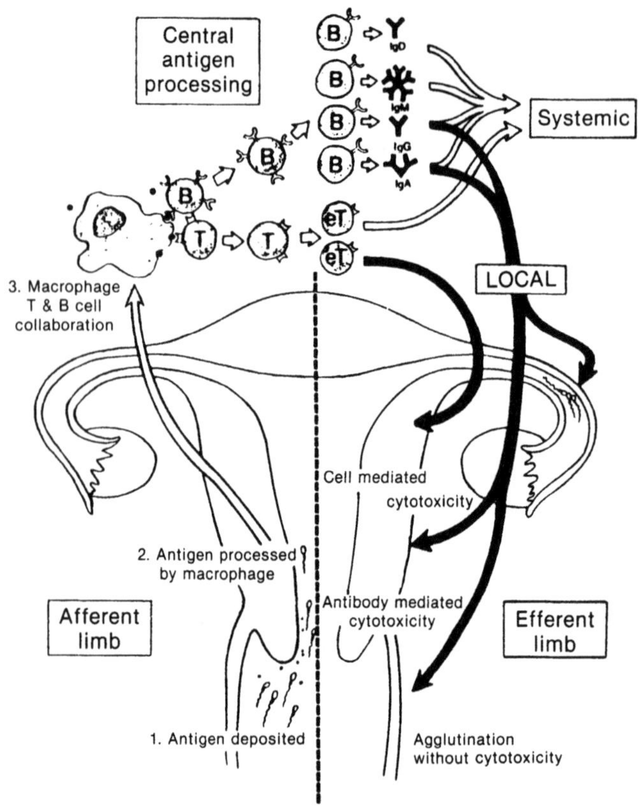

FIGURE 3-1. Elicitation and expression of systemic and local immune responses following local challenge of the reproductive tract of the female with a foreign antigen(s). (Modified from ref. 63, with permission of the American Fertility Society.)

TABLE 3-1. The main immunologic sources and functions of monokines and lymphokines

Mediator	Main sources	Principal functions
Interleukins (ILs)		
IL-1	Activated macrophages and natural killer (NK) cells	Stimulates helper and suppressor T and B cell proliferation and differentiation; induces IL-2 secretion by helper T cells; endogenous pyrogen
IL-2	Activated helper T cells	Autocrine/paracrine effect on T and B cell growth and other cytotoxic cells
IL-3	Activated helper T cells	Stimulates growth of bone marrow stem cells
IL-4	Activated helper T cells	B cell growth factor; induces Ia expression and IgG and IgE secretion; enhances cytotoxic T cells, macrophage activation.
IL-5	Activated helper T cells	Eosinophil differentiation; IgM and IgA synthesis; enhances B cell proliferation
IL-6	Monocytes	Induces production of acute-phase proteins and immunoglobulin secretions; growth induction of various cell lines
Tumor necrosis factor	Activated macrophages	Directs tumor cell necrosis; activation of T cells; angiogenic factor
γ-Interferon	Activated helper T and NK cells	Inhibits viral replication; enhances human leukocyte Ag expression and NK and macrophages activity

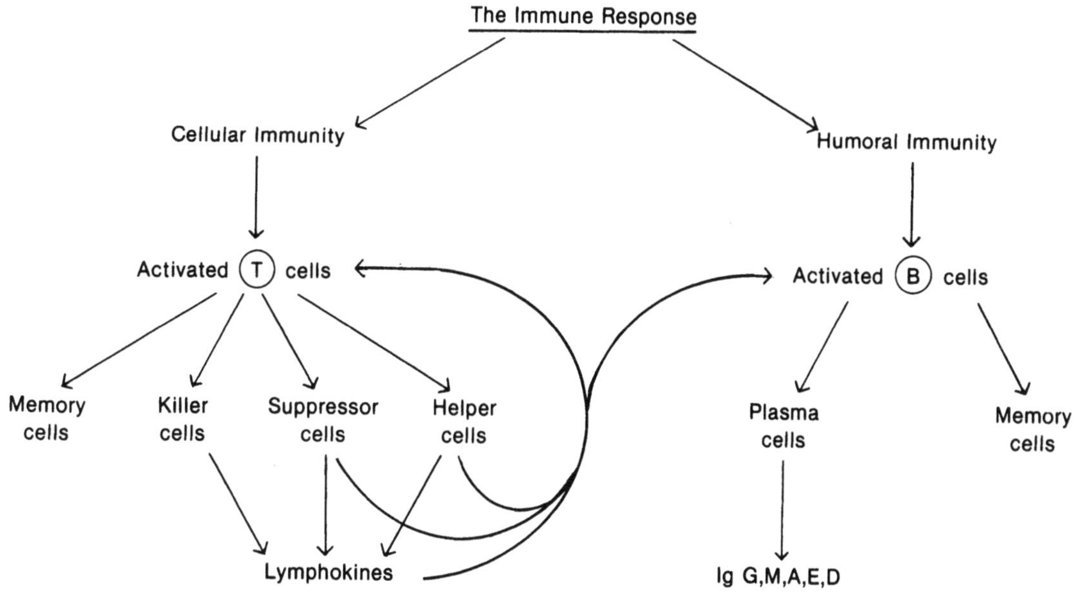

FIGURE 3-2. The cellular (T cells) and the humoral (B cells) immunoregulatory network of the immune response.

T Lymphocytes[2,7,8]

Thymus-derived lymphocytes (T cells) are responsible for cell-mediated immunity. Recent studies in cellular immunology, using monoclonal Ab, recombinant engineering, and modern cell culture technologies, revealed a complex cell-to-cell interaction with a unique effect of cellular mediators. T lymphocytes have the ability to secrete *lymphokines*, which regulate other T cell functions, regulate immunoglobulin production by B cells, and also have the ability to kill other cells (cytotoxicity) (Table 3-1 and Fig. 3-2).

Studies of the genes that encode the alpha and beta chains of the Ag receptor on the T cell membrane allow a better understanding of genetic mechanisms that can lead to the antigenic diversity among T cells. The genetic loci are present on the 14th and the 17th human chromosomes for the alpha and beta chains, respectively. The structure of the alpha and beta chains is derived separately by a combination of one of several variable genes with a joining gene—sometimes designated as a diversity gene—and a constant domain gene. During T cell differentiation specific gene translocations occur, which give each cell a unique specifity (Fig. 3-3). The whole process of T cell genesis and translocation takes place in the thymus. Thereafter, the cells migrate from the thymus to peripheral lymphoid tissues. When Ag activates particular lymphocytes, Ag-driven processes initiate a second cycle of multiplication and differentiation in which lymphocytes are turned into activated *effector T cells*.[2] Effector functions include delayed-type hypersensitivity, allograft rejection, tumor immunity, and graft-versus-host reactions.

The Ag receptor on the T cell surface is also associated with other molecules. T cells that carry the CD4 molecule are *helper/inducer T cells* (T4). CD8-positive cells are *suppressor/cytotoxic T cells* (T8). The normal ratio between peripheral blood helper and suppressor T cells is 2:1. This ratio is inverted with the acquired immunodeficiency syndrome (AIDS). Helper T cells enable B cells to respond to Ag that they would otherwise not recognize. T suppressor cells suppress antibody formation by B cells, thus modifying the humoral Ab response.

The basis for anamnestic responses to Ag is provided by so-called *memory T cells* (Fig. 3-2).

The cell Ag receptor, directed toward "foreign" epitopes, is held in a loose molecular association with "self" molecules that are coded by the major histocompatibility complex, termed the "human leukocyte antigen (HLA) system." HLA antigens have been divided into two classes that serve as restriction elements for the T4 and T8 T cell subsets. Class I molecules are the classic histocompatibility Ags and include HLA-A, -B, and -C. They serve as restriction elements for suppressor/cytotoxic T8 cells. Class II molecules include HLA-D, -DR, -DQ, and -DP, originally defined on the basis of typing by mixed lymphocyte culture (MLC). They assist helper/inducer T4 lymphocytes in fulfilling their function.

In the search for cytotoxic T cells, another cell that can kill a variety of tumor cells was discovered. This cell comprises a discrete population of large lymphocytes and has been called the *natural killer cell* (NK). Killer cells are also involved in allograft rejection.

B Lymphocytes and Immunoglobulins[2,8–11]

B cells develop from a pluripotential stem cell in adult human bone marrow equivalent to the bursa of Fabricius in birds, hence B (bursa-derived) cell. A B cell in a terminally differentiated state of high rate of Ab production and secretion is called a *plasma cell* (Fig. 3-2).

A major key for immune recognition is the Ab structure. Antibodies are multichain glycoproteins. Each immunoglobulin contains at least one basic unit or monomer, comprising one pair of heavy chains and one pair of light chains. The heavy chain is approximately twice the molecular weight of the light chain. Each chain has a "recognition" part. When two recognition parts come together, the part that actually combines with the epitope is formed, the *Ab combining site*.

The genes coding for immunoglobulins are also broken into *variable* (V) *diversity* (D), *joining* (J), and *constant* (C) sets. During the differentiation of B cells, a series of translocations takes place (Fig. 3-3). This process differs in every cell, giving each lymphocyte its unique expression of immunoglobulin. For the heavy chain, one of the D genes translocates with one of the J genes. Then, one of the V gene translocates to the DJ combination. Transcription of the rearranged genes follows and a unique heavy chain messenger RNA is created. Since a translocation occurs also for the light chain, the combinatorial possibilities are enormous. As this random process of gene translocation occurs, a whole repertoire of B cells is created, each with its own specificity. When this cell is activated by Ag, it secretes large amounts of the Ab for which its translocated immunoglobulin genes code. An ongoing or repeated immune response includes affinity maturation with somatic mutation, producing better and better Ab over time. When activated B cell clones revert to the resting state, *memory B lymphocytes* bear the traces of this mutational process.[2]

Unlike T cells, B cells recognize the Ag itself. Class I and II HLA are therefore not part of the genetic regulation of the humoral immune response. Immunoglobulins are bifunctional molecules that bind Ag (to the V regions of the heavy and light chains) and, in addition, initiate other biologic phenomena that are independent of Ab specifity and occur at the C region of the heavy chain.

Immunoglobulin G (*IgG*) constitutes approximately 75% of total serum immunoglobulins. It is the only isotype of immunoglobulin that can cross the human placenta in any significant amount. Most IgG is also capable of fixing serum complement and can bind to macrophage surface receptors. *IgM* is the primary humoral Ab response to an Ag. It represents about 10% of the total serum immunoglobulins. IgM and *IgD* are

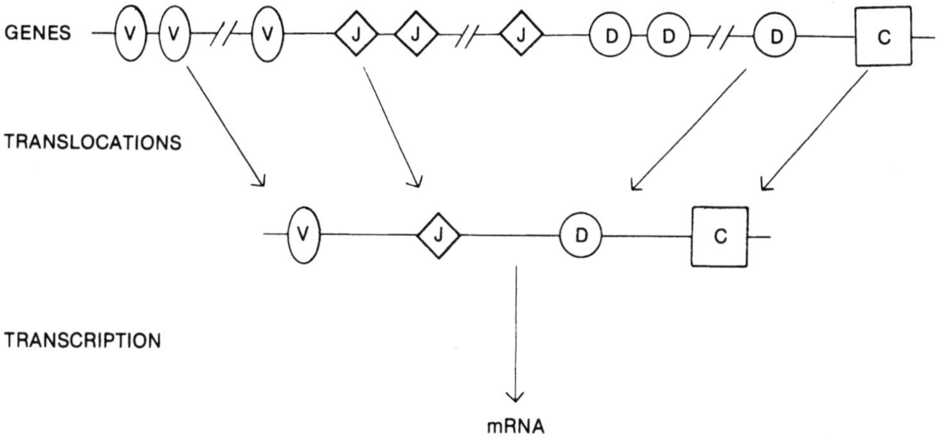

FIGURE 3-3. Graphic scheme of gene translocations of the variable (V), joining (J), diversity (D), and constant (C) genes, which give each lymphocyte a unique specificity.

the major immunoglobulins expressed on the surface of B cells. IgM is predominant in natural blood group Abs and is also the most efficient complement-fixing immunoglobulin. The predominant immunoglobulin class in body secretions is *IgA*. Secretory IgA provides the primary defense mechanism against mucosal infections. Secretory IgA consists of two basic units, a secretory component and the J chain. IgA constitutes approximately 15% of the total immunoglobulins in both monomeric and polymeric forms. *IgE* binds with very high affinity to mast cells and thus initiates an allergic response. It constitutes only a small amount of total serum immunoglobulins.

Monokines and Lymphokines

The secretory role of cells involved in the immune response is as important as their role in the inflammatory process. Mediators (*cytokines*) are secreted in the course of immunologic reactions by lymphocytes (lymphokines) and by monocytes or macrophages (monokines). These mediators are pleiotropic and serve as an autocrine/paracrine/endocrine control for the immune system (Fig. 3-2).

Monokines and lymphokines function as intercellular signals that regulate both local and systemic immune response. Table 3-1 depicts the main immunologic sources and functions of some better known monokines and lymphokines.[1,3,12–14] Anderson and Hill[3] classified the complex biologic activities of lymphokines and monokines into four broad categories:

1. Growth factors: interleukin (IL)-2, IL-3, and IL-4
2. Differentiation factor: granulocyte/macrophage–colony-stimulating factor (GM-CSF) and IL-5
3. Inducers of DNA transcription: IL-4, tumor necrosis factor (TNF), and α- and γ-interferons
4. Cytotoxic factors: TNF, lymphotoxin, γ-interferon, and IL-1

These polypeptide hormone–like cytokines affect not only the immune system but also the reproductive system. Monokines and lymphokines have been shown to affect fertilization,[15] embryonic development,[16] sperm motility,[17] and expression of class II HLA molecules on endometrial epithelium.[18] Human decidua can produce IL-1, which may be involved in the physiologic response to implantation and may also mediate the onset of labor in patients with intrauterine infections.[19] IL-1 may be also an important local regulator of human chorionic gonadotropin (hCG) secretion by the first-trimester trophoblast.[20] Blocking of the cytolytic effector response to IL-2[21] or lack of ability to stimulate IL-2 production[22] may also explain the role of lymphokines in the active suppression of the host-versus-graft reaction during pregnancy.

Immunologic Tolerance

The immune response, which an allograft elicits in a normal host, is the consequence of complex interactions between foreign antigens and various subsets of lymphocytes as well as accessory cells of the host. The acceptance of the fetal allograft by the apparently normal maternal immune system represents one of the most fascinating unsolved problems of modern biology.

In an attempt to explain this immunologic tolerance, the uterus has been considered as an immunologically privileged site.[23] Immunologically privileged sites in the body have been recognized as such even before the science of immunogenetics emerged. The anterior chamber of the eye and the wall of the hamster's cheek pouch represent the most widely known examples. What most immunologically privileged sites seem to have in common is the lack of lymphatic drainage.[24]

The uterine mucosa has no recognizable lymphoid tissue for the initial processing of Ag. This is in clear contrast to, for example, the intestinal (Peyer's patches) and bronchial mucosa.[4] Although an intact lymphatic drainage is necessary for the afferent pathway of the immune response, it is not absolutely necessary for graft rejection. Accessory cells, acting as Ag-presenting cells, in the endometrium and decidua have the capacity to process Ag and either carry it to draining lymph nodes or interact locally with Ag-reactive host lymphocytes.[3,24,25] Dendritic (Ia^+) cells may have a significant role in local Ag processing within the uterine endometrium and stroma.[4]

Recent evidence suggests that more than one mechanism may be involved in the immunologic acceptance of the fetal allograft. Proposed mechanisms include regulation of placental alloantigen expression and therefore antigenicity[26–28]; the placenta as an effective immunologic barrier, filter, or Ab sponge[29,30]; and blocking Ab and suppressor cells in the endometrium and decidua.[25,31] Blocking Abs have been suggested to coat the surface of fetal paternal antigens and prevent their recognition by other components of the maternal immune response. They also prevent the binding of complement-fixing Ab and/or cytotoxic cells by competitive inhibition.[23]

The uterus is an immunologically dynamic region. The endometrium and decidua of the human uterus contain numerous macrophages and lymphocytes.[3,31] It seems that the tolerance of the fetal allograft does not occur by "depression" of the intrauterine maternal immune response, but rather occurs as a result of an "activation" of suppressive cells within the uterus.

Suppressor Cells in the Endometrium/Decidua

Cells that apparently are capable of providing a protected environment for fetal development are components of the uterine endometrium and maternal decidua. Suppressor cells have been found in endometrium and decidua of murine models[25,32–34] and in the human.[31,35,36] These suppressor cells are, however, not T8 suppressor/cytotoxic ($CD8^+$) T cells.[5,31]

The main types of suppressor cells that have been identified in the uterus are macrophages[16] and small and large lymphocytes.[15,31] Macrophages are present in the strata basalis and functionalis throughout the menstrual cycle, and their number does not appear to fluctuate substantially between proliferative and secretory phases.[5] The endometrial macrophages may have a role in Ag presentation and/or may act in an immunosuppressive fashion.[34]

Clark and collegues[25,31–33] have provided evidence about two types of suppressor lymphocytes in the endometrium after ovulation. An early-phase large cell is hormone dependent and is associated with preimplantation suppression. The later-phase small cell, which appears following implantation, is probably trophoblast dependent. These phenotypically unusual lymphocytes may differentiate in situ within the endometrium, either from undifferentiated stem cells or from mature T cells. Alternatively, these lymphocytes may migrate in the secretory phase of the menstrual cycle directly from the circulation to the endometrium or may multiply in situ.[5] The recruitment process that results in the appearance of two types of nonspecific suppressor lymphocytes in the uterine endometrium and decidua may be useful for the understanding of early pregnancy wastage and of habitual abortions.[31]

The suppressor cells in the decidua must inhibit the effector phase of the graft rejection response, the infiltration of the grafts by sensitized maternal T cells. The finding by Nebel[37] that failed in vitro fertilization (IVF) embryos were infiltrated by maternal lymphocytes supports the concept that human fetal allograft rejection by the maternal immune system may occur. Toder et al.[38] also demonstrated, in the investigation of the contraceptive action of intrauterine devices, that the presence of various mononuclear cell populations within the uterus were related to failed implantation (Fig. 3-4).

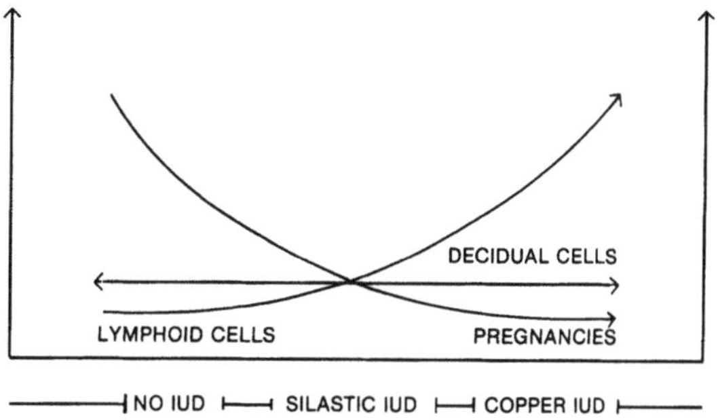

FIGURE 3-4. Graphic scheme of inverted correlation between cell counts and occurrence of pregnancy. (Reprinted from ref. 38, with permission of Butterworth Publishers.)

Immunologic Pregnancy Wastage

The fetus is an auto- as well as an allograft, with half of its genome representing "allo" antigens and half maternal "self" antigens. Based on immunologic transplantation principles, the fetal alloantigen, exposed to maternal immune cells, should be the target of maternal immunologic attack. Conversely, the fetal autograft may also elicit immune responses via autoimmune mechanisms.[39]

Considerable evidence has accumulated that supports the concept that allogenic stimulation of the maternal immune system is required in order to achieve activation of a variety of immunologic mechanisms that, in turn, allow for implantation to occur.[40] Most of these mechanisms appear to be concentrated at the local uterine level, and especially within the decidua.[31,41] Recently, it has been suggested that normal pregnancy requires maternal recognition and modulation of foreign fetal antigens by Ab. These blocking Abs are necessary for successful implantation and pregnancy. Based on the assumption that some fault in maternal "immunization" is responsible for recurrent spontaneous abortions, a variety of stimulation protocols for maternal immunization are currently under investigation.

Taylor and Faulk[42] attempted to produce cross-reacting antibodies by immunization with buffy coat leukocytes from unrelated donors. Beer et al.[43] and Mowbray et al.[44] immunized with paternal lymphocytes. Whether presently ongoing immunization studies suggest a benefit from such experimental treatment has remained controversial. The best suited stimulant remains to be established. Some doubts about the validity of immunization protocols has arisen, since neither HLA sharing, conversion to antipaternal lymphocytotoxic Ab positivity, nor mixed lymphocyte blocking reactivity is a good predictor for successful pregnancy outcome after paternal leukocyte immunization.[45] A number of controlled studies are presently underway and should finally resolve the efficacy of such immunization treatment protocols.

In parallel to alloantigens, autoantigens enter the maternal circulation. Although clearly "auto" in maternal genetic origin, they represent "foreign" Ags since they are derived from the fetus.[39] Since human pregnancies with donor oocytes have been successful,[46] it appears that this autoantigenic component is not essential to normal pregnancy. However, abnormal autoimmune function in clinically asymptomatic patients can lead to reproductive failure. The occurrence of reproductive autoimmune failure syndrome[47] may be teleologically related to the woman's need for increased self-tolerance in the face of antigenic exposure to the maternal haplotype of the fetus during normal pregnancy. Pregnancy not only calls for tolerance of paternal antigenicity, but also of an enormous volume of "self" that is never required from men.[48] This need for increased self-tolerance is documented by higher "normal" Ab levels in women than in men[49] and may also be responsible for the highly increased incidence of autoimmune diseases in women in comparison with men. Autoimmune conditions leading to pregnancy loss have been associated with the presence of antiphospholipid antibodies, such as anticardiolipin and lupus anticoagulant.[50–52] A much broader B cell abnormality in these clinically asymptomatic women has also been described.[53,54] It has been suggested that autoantibodies may produce vascular abnormalities in the decidua and placenta. Successful pregnancies can be achieved in most of these women by treatment with corticosteroids and low-dose aspirin.[40]

Infertility and Sperm Antibodies

Antisperm Abs that react with surface membranes of motile spermatozoa may interfere with in vivo and in vitro fertilization.[55,56] Infertility has been associated with the presence of antisperm Abs in serum and/or cervical mucus.[55] Impaired fertility is probably a combination of antisperm Ab effects, such as sperm agglutination, reduced motility, enhanced phagocytosis of sperm, impaired cervical mucus penetration, inefficient egg-sperm fusion, pre- or postimplantation embryo loss.[57,58] Uterine fluids of patients with sperm-agglutination antibodies and cervical dysmucorrhea may provide a better milieu for spermatozoa than cervical mucus.[59] This may be a rationale for intrauterine inseminations to achieve pregnancy.

It has been suggested that antisperm antibodies in sera from females may be caused either by a deficiency in the ability of their T suppressor/cytotoxic cells to respond to inducer factors in semen or by the occurrence of microorganisms, antibodies, or other factors in the ejaculate that induce T helper lymphocytes.[60] Activated T cells produce gamma interferon, which induces Ia Ag expression on macrophages and allows the female's T helper cells to recognize processed sperm antigens.[60–62]

The production of sperm Ab in females is inhibited both by physical isolation of spermatozoa from the systemic immune system (the thick stratified vaginal squamous epithelia and the decrease in the antigenic load by expelling the majority of ejaculated spermatozoa from the vagina shortly after their deposition)[60] and by active immunosuppression mechanisms. Thus, spermatozoa, which are immunologically foreign to women, usually do not stimulate a humoral immune response.

The local uterine immune response behaves quite different from the systemic response. Local-cell mediated immunity also affects reproductive efficiency. Antibody activity in serum correlates poorly with Ab activity in secretions of the reproductive tract.[3,63] This reemphasizes that systemic immunity may not necessarily reflect local immunity, and local immunity may be more relevant in regard to reproduction.

Immunization against spermatozoa antigen as a method of contraception is under investigation. The only testis-specific autoantigen that has so far been isolated is LDH-C_4.[64] Immunization of female rabbits, mice, and baboons with this protein interfered with both pre- and postimplantation events and reduced fertility. Significant efforts are being expanded by the World Health Organization (WHO) in attempts to identify other sperm, oocyte, and zona pellucida antigens as contraceptive vaccines.[64]

Tumor Immunology

Tumor immunology is the study of the host immune response to transformed tumor cells. The ability of malignant cells to evade an immune response and the regulation of normal implantation may involve similar mechanisms. Both of these processes probably depend on cell surface antigenic modulation, blocking factors, or Ab and/or suppressor cells.

In order for the immune system to differ between a malignant and normal cell, tumor cells must express unique tumor antigens. An apparent tumor-specific T and B cell in tumor-draining lymph nodes of patients with breast cancer and melanoma has been identified. However, an unequivocal unique tumor Ag from these human tumors has not yet been isolated or characterized.[65]

By contrast, tumor-associated antigens are found on tumor cells and also on normal cells. These antigens may be expressed on the normal cell surface at embryonic stages of differentiation but are absent or very difficult to detect in normal adult cells. They are thus called *oncofetal antigens*. Such oncofetal antigens include α-fetoprotein (AFP), which can serve as a marker for germ cell and liver tumors, and carcinoembryonic Ag (CEA), formed in fetal gut, which serves as a marker for human colon and ovarian cancer. Because tumor-associated antigens have also been found in nonmalignant tissues, the clinical significance has remained controversial. For example, CA125, a widely used Ag in the clinical diagnosis of ovarian malignancies,[66] is also found in normal endometrium, endometriosis,[67] and decidua.[68]

H and Lewis-b blood group antigens have recently been found in endometrial cancer.[69] Since these antigens were expressed in fetal but not in normal adult endometrium, they can also be considered to represent oncofetal antigens. An association of ABO blood groups and choriocarcinoma was also demonstrated in studies on immunogenetic predisposition to malignancies.[70]

Variations in the regulation of normal cell surface components and antigenic modulation, rather than production of abnormal cell antigens, may help tumors to escape the immune response. Changes in carbohydrates of cell surface antigens may play a functional role during embryogenesis or carcinogenesis. Carbohydrates may regulate cellular antigenicity, modulate receptor functions and membrane proteins, and control cell-to-cell recognition.[69,71]

Sharing of HLA is another potential mechanism by which tumor cells may evade the host's immune response. Although there is no specific HLA associated with either a molar pregnancy or choriocarcinoma, couples who share antigens, encoded by HLA-A and HLA-B loci, demonstrate an increased incidence of choriocarcinoma.[70,71] Human Leukocyte Ag sharing has also been reported with recurrent spontaneous abortions. Although these data are not yet conclusive, they would be compatible with a current hypothesis that suggests an etiologic relationships between the expression of HLA on cell surface and the development of cancer. By presenting a much less immunogenic profile, the tumor may be able to escape the host's immune response. In less HLA-compatible matings, the immune response to foreign HLA could play a more effective role in the rejection of choriocarcinoma and/or result in a more favorable response to chemotherapy.[72]

Developing a specific immunologic attack against malignant tumors without damaging the adjacent normal cells is the theoretical basis of immunotherapy. At present, immunotherapy cannot yet produce a cure. Moreover, as in other fields, it is still experimental in gynecologic oncology. Using monoclonal Ab technology, immunochemotherapy may become the principle future cancer therapy. Specific antitumor Abs will then direct a cytotoxic drug to destroy only a malignant cell, thus avoiding the known side effects of chemotherapy.

Immunoendocrine System of the Uterus

The immunology of the uterus seems to be as important to reproduction as its endocrinology. Increasing evidence suggests, in fact, that the immune and endocrine systems may have to be considered as one. The immune response is apparently at least partially under endocrine control. Sex steroid hormones may regulate local cellular immunity to alloantigens,[73] immunoglobulins in uterine secretions,[74] and monocyte IL-1 activity.[75] Conversely, reproductive failure may be caused by polyclonal B cell activation.[47,76]

In subsequent editions of this book, immunology of the uterus may be found in the chapter on reproductive endocrinology or gynecologic oncology, or perhaps things will be exactly the other way around.

References

1. Fauci AS, Rosenberg SA, et al: NIH Conference: Immunomodulators in clinical medicine. Ann Intern Med 1987;106:421–433.
2. Nossal GJV: The basic components of the immune system. N Engl J Med 1987;316:1320–1325.
3. Anderson DJ, Hill JA: Cell-mediated immunity in infertility. Am J Reprod Immunol Microbiol 1988;17:22–30.
4. Head JR, Billingham RE: Concerning the immunology of the uterus. Am J Reprod Immunol Microbiol 1986;10:76–81.
5. Bulmer JN, Lunny DP, Hagin SV: Immunohistochemical characterization of stromal leucocytes in nonpregnant human endometrium. Am J Reprod Immunol Microbiol 1988;17:83–90.
6. Oksenberg JR, Mor-Yosef S, Ezra Y, Brautbar C: Antigen presenting cells in human decidual tissue: III. Role of accessory cells in the activation of suppressor cells. Am J Reprod Immunol Microbiol 1988;16:151–158.
7. Barrett JT: The T lymphocytes, in Barrett JT (ed): Textbook of Immunology. St. Louis, CV Mosby Company, 1988, pp 121–136.
8. Stobo JD: Lymphocytes, in Stites DP, Stobo JD, Wells JV (eds): Basic & Clinical Immunology. East Norwalk, CT, Appleton & Lange, 1987, pp 65–81.
9. Barrett JT: The B lymphocyte, in Barrett JT (ed): Textbook of Immunology. St. Louis, CV Mosby Company, 1988, pp 107–120.
10. Barrett JT: The immunoglobulins, in Barrett JT (ed): Textbook of Immunology. St Louis, CV Mosby Company, 1988, pp 47–75.
11. Goodman JW: Immunoglobulins 1: structure & function, in Stites DP, Stobo JD, Wells JV (eds): Basic & Clinical Immunology. East Norwalk, CT, Appleton & Lange, 1987, pp 27–36.
12. Miyajima A, Miyatake S, Schreurs J, et al: Coordinate regulation of immune and inflammatory responses by T cell-derived lymphokines. FASEB J 1988;2:2462–2473.
13. Tartakovsky B, Finnegan A, Muegge K, et al: IL-1 is an autocrine growth factor for T cell clones. J Immunol 1988;141:3863–3867.
14. Oppenheim JJ, Ruscetti FW, Faltynck CR: Interleukins & interferons, in Stites DP, Stobo JD, Wells JV (eds): Basic & Clinical Immunology. East Norwalk, CT, Appleton & Lange, 1987, pp 82–95.
15. Hill JA, Cohen J, Anderson DJ: The effects of lymphokines and monokines on human sperm fertilizing ability in the zona-free hamster egg penetration test. Am J Obstet Gynecol 1989;160:1154–1159.
16. Hill JA, Haimovici F, Anderson DJ: Products of activated lymphocytes and macrophages inhibit mouse embryo development in vitro. J Immunol 1987;139:2250–2254.
17. Hill JA, Haimovici F, Politch JA, Anderson DJ: Effects of soluble products of activated lymphocytes and macrophages (lymphokines and monokines) on human sperm motion parameters. Fertil Steril 1987;47:460–465.
18. Tabibzadeh SS, Satyaswaroop PG: Differential expression of HLA-DR, HLA-DP, and HLA-DQ antigenic determinants of the major histocompatibility complex in human endometrium. Am J Reprod Immunol Microbiol 1988;18:124–130.
19. Romero R, Wu YK, Brody DT, et al: Human decidua: A source of interleukin-1. Obstet Gynecol 1989;73:31–34.
20. Yagel S, Lala PK, Powell WA, Casper RF: Interleukin-1 stimulates human chorionic gonadotropin secretion by first trimester human trophoblast. J Clin Endocrinol Metab 1989;68:992–995.
21. Clark DA, Falbo M, Rowley RB, et al: Active

suppression of host-vs graft reaction in pregnant mice IX. Soluble suppressor activity obtained from allopregnant mouse decidua that blocks the cytolytic effector response to IL-2 is related to transforming growth factor-β. J Immunol 1988;141:3833–3840.
22. Blank M, Nebel L, Toder V: Trophoblast does not cause the cytotoxic T lymphocyte generation due to the lack of ability to stimulate interleukin-2 production. Am J Reprod Immunol Microbiol 1987;14:49–53.
23. Gurka G, Rocklin RE: Reproductive immunology. JAMA 1987;258:2983–2987.
24. Head JR, Billingham RE: Immunologically privileged sites in transplantation immunology and oncology. Perspect Biol Med 1985;29:115–131.
25. Clark DA, Slapsys R, Croy A, Krcek J, Rossant J: Local active suppressor cells in the decidua: a review. Am J Reprod Immunol 1984; 5:78–83.
26. Head JR, Drake BL, Zuckerman FA: Major histocompatibility antigens on trophoblast and their regulation: implications in the maternal-fetal relationship. Am J Reprod Immunol Microbiol 1987;15:12–18.
27. Chaouat G: Immunoregulatory placental functions in normal and pathological pregnancies. Am J Reprod Immunol Microbiol 1988;17:18–21.
28. McIntyre JA: In search of trophoblast-lymphocyte crossreactive (TLX) antigens. Am J Reprod Immunol Microbiol 1988;17:100–110.
29. Redman CWG: Immunology of the placenta. Clin Obstet Gynecol 1986;13:469–499.
30. Innes A, Stewart GM, Thomson MAR, et al: Human placenta—an antibody sponge? Am J Reprod Immunol Microbiol 1988;17:57–60.
31. Daya S, Clark DA, Devlin C, Jarrell J: Preliminary characterization of two types of suppressor cells in the human uterus. Fertil Steril 1985;44:778–785.
32. Daya S, Rosenthal KL, Clark DA: Immunosuppressor factor(s) produced by decidua-associated suppressor cells: a proposed mechanism for fetal allograft survival. Am J Obstet Gynecol 1987;156:344–350.
33. Brierley J, Clark DA: Characterization of hormone-dependent suppressor cells in the uterus of mated and pseudopregnant mice. J Reprod Immunol 1987;10:201–217.
34. Hunt JS, Manning LS, Wood GW: Macrophages in murine uterus are immunosuppressive. Cell Immunol 1984;85:499–510.
35. Khong TY: Immunohistologic study of the leukocytic infiltrate in maternal uterine tissues in normal and preeclamptic pregnancies at term. Am J Reprod Immunol Microbiol 1987;15:1–8.
36. Kamet BR, Isaacson PG: The immunocytochemical distribution of leukocytic subpopulations in human endometrium. Am J Pathol 1987;127:66–73.
37. Nebel L: Malimplantation a cause of failure after IVF and ET. Am J Reprod Immunol 1984;6:56–57.
38. Toder V, Madanes A, Gleicher N: Immuologic aspects of IUD action. Contraception 1988;37:391–403.
39. Gleicher N: IVF and immunological infertility, in Proceeding of the VIth World Congress In Vitro Fertilization and Alternate Assisted Reproduction. New York, Plenum Press, 1991, in press.
40. Scott JR, Rote NS, Branch DW: Immunologic aspects of recurrent abortion and fetal death. A review. Obstet Gynecol 1987;70:645–656.
41. Parhar RS, Yagel S; LaLa PK: PGE_2-mediated immunosuppression by first trimester human decidual cells blocks activation of maternal leukocytes in the decidua with potential anti-trophoblast activity. Cell Immunol 1989;120:61–74.
42. Taylor C, Faulk WP: Prevention of recurrent abortion with leukocyte transfusions. Lancet 1981;2:68–69.
43. Beer AE, Quebbeman JF, Ayers JW, Haines RF: Major histocompatibility complex antigens, maternal and paternal immune responses, and chronic habitual abortions in humans. Am J Obstet Gynecol 1981;141:987–997.
44. Mowbray JF, Gibbings C, Liddell H, et al: Controlled trial of treatment of recurrent spontaneous abortion by immunization with paternal cells. Lancet 1985;1:941–943.
45. Smith JB, Cowchock FS: Immunological studies in recurrent spontaneous abortion: effects of immunization of women with paternal mononuclear cells on lymphocytotoxic and mixed lymphocyte reaction blocking antibodies and correlation with sharing of HLA and pregnancy outcome. J Reprod Immunol 1988;14:99–113.
46. Rosenwaks Z: Donor eggs: their application in modern reproductive technologies. Fertil Steril 1987;47:895–909.
47. Gleicher N, El-Roeiy A: The reproductive autoimmune failure syndrome. Am J Obstet Gynecol 1988;159:223–227.

48. Gleicher N: Pregnancy and autoimmunity. Acta Haematol 1986;76:68–77.
49. El-Roeiy A, Gleicher N: Definition of normal autoantibody levels in an apparently healthy population. Obstet Gynecol 1988;72:596–602.
50. Lubbe WF, Butler WS, Palmer SJ, et al: Lupus anticoagulant in pregnancy. Br J Obstet Gynaecol 1984;91:357–363.
51. Branch DW, Scott JR, Kochenour NK, Hershgold E: Obstetric complications associated with the lupus anticoagulant. N Engl J Med 1985;313:1322–1326.
52. Barbui T, Cortelazzo S, Galli M, et al: Antiphospholipid antibodies in early repeated abortions: a case-controlled study. Fertil Steril 1988;50:589–592.
53. Gleicher N, Friberg J: IgM gammopathy and the lupus anticoagulant syndrome in habitual aborters. JAMA 1985;253:3278–3281.
54. Confino E, Gleicher N: Pregnancy wastage with abnormal autoimmunity. Immunol Allerg Clin North Am 1990;10:103–108.
55. Bronson R, Cooper G, Rosenfeld D: Sperm antibodies: their role in infertility. Fertil Steril 1984;42:171–183.
56. Clarke GN: Sperm antibodies and human fertilization. Am J Reprod Immunol Microbiol 1988;17:65–71.
57. Alexander NJ, Anderson DJ: Immunology of semen. Fertil Steril 1987;47:192–205.
58. Moghissi KS, Sacco AG, Borin K: Immunologic infertility I. Cervical mucus antibodies and postcoital test. Am J Obstet Gynecol 1980;136:941–947.
59. Confino E, Friberg J, Gleicher N: Improved penetration of spermatozoa into uterine fluids of women with abnormal postcoital tests. Obstet Gynecol 1988;72:655–658.
60. Witkin SS: Mechanisms of active suppression of the immune response to spermatozoa. Am J Reprod Immunol Microbiol 1988;17:61–64.
61. Witkin SS: Failure of sperm-induced immunosuppression: association with antisperm antibodies in women. Am J Obstet Gynecol 1989;160:1166–1168.
62. Witkin SS: Production of interferon gamma by lymphocytes exposed to antibody-coated spermatozoa: a mechanism for sperm antibody production in females. Fertil Steril 1988;50:498–502.
63. Beer AE, Neaves WB: Antigenic status of semen from the viewpoints of the female and male. Fertil Steril 1978;29:3–22.
64. Spieler J: Development of immunological methods of fertility regulation. Bull WHO 1987;65(6):779–783.
65. Greenberg PD: Tumor immunology, in Stites DP, Stobo JD, Wells JV (eds): Basic & Clinical Immunology. East Norwalk, CT, Appleton & Lange, 1987, pp 186–196.
66. Malkasian GD Jr, Podratz KC, Stanhope CR, et al: CA 125 in gynecologic practice. Am J Obstet Gynecol 1986;155:515–518.
67. Kabawat SE, Bast RC Jr, Bhan AK, et al: Tissue distribution of a coelomic epithelium related antigen recognized by the monoclonal antibody OC125. Int J Gynecol Pathol 1984;2:275–285.
68. Kobayashi F, Sagawa N, Nakamura K, et al: Mechanism and clinical significance of elevated CA125 levels in the sera of pregnant women. Am J Obstet Gynecol 1989;160:563–566.
69. Inoue M, Sasagawa T, Saito J, et al: Expression of blood group antigens A, B, H, Lewis-a, and Lewis-b in fetal, normal, and malignant tissues of the uterine endometrium. Cancer 1987;60:2985–2993.
70. Buckley JD: The epidemiology of molar pregnancy and choriocarcinoma. Clin Obstet Gynecol 1984;27(1):153–159.
71. Gill TJ: Immunological and genetic factors influencing pregnancy, in Knobil E, Neill J, et al (eds): The Physiology of Reproduction. New York, Raven Press, 1988, pp 2023–2042.
72. Berkowtiz RS, Goldstein DP, Hock EJ, Anderson DJ: Immunobiology of molar pregnancy and gestational trophoblastic tumors. J Reprod Med 1984;29:796–801.
73. Taylor E, Snow C, Hilgard HR: Sex-dependent local cellular immunity to alloantigens. Transplantation 1986;41:384–387.
74. Tauber PF, Wettich W, Nohlen M, Zaneveld LJD: Diffusable proteins of the mucosa of the human cervix, uterus, and fallopian tubes: Distribution and variations during the menstrual cycle. Am J Obstet Gynecol 1985; 151:1115–1125.
75. Polan ML, Daniele A, Kuo A: Gonadal steroids modulate human monocyte interleukin-1 (IL-1) activity. Fertil Steril 1988;49:964–968.
76. Maier DB, Parke A: Subclinical autoimmunity in recurrent aborters. Fertil Steril 1989;51:280–285.

4

Papillomaviruses and Cervical Neoplasia

GERARD J. NUOVO AND CHRISTOPHER P. CRUM

The evolution of molecular biology and related technologies and their application to medical sciences has rapidly altered our perception of the pathogenesis and therapeutic approach to intraepithelial squamous lesions of the lower female genital tract. As with any new technology, the knowledge can be applied on several different levels. For example, it may broaden our understanding of a disease process without changing the therapeutic approach, or it may change our perceptions of disease and indirectly benefit the patient by providing a more knowledgeable physician. Ultimately these techniques, such as viral DNA detection, may have clinical utility, provided they are tested and proven useful within the context of scientific inquiry. The purpose of this chapter is to review the body of newly acquired information on the molecular biology of human papillomavirus (HPV)-associated disease and relate it to the problems of clinical management of HPV-related diseases of the lower female genital tract.

Epidemiologic Perspectives

Approximately 2 to 3% of young women yearly develop a cervical abnormality related to HPV infection, and approximately 1 million women per year are treated for genital warts in this country.[1,2] The link between HPV infection and cervical intraepithelial and invasive neoplasia has been established primarily by observations from molecular biology.[3] For example, condylomata and intraepithelial neoplasia have been produced by infecting normal squamous cells with HPVs 11 and 16, respectively, in a nude mouse and tissue culture model.[4,5] However, it is important to emphasize that the epidemiologic data linking HPV infection to cancer are far from clear-cut, and one of the most interesting aspects in this regard is the somewhat ambiguous nature of the epidemiologic information.[6] To better understand this ambiguity, it is important to understand that the molecular diagnosis of HPV infection is based upon the mere presence of HPV DNA, which may or may not signal the presence of replicating virus.

The techniques used to detect HPV include Southern blot hybridization, dot or slot blot hybridization, in situ hybridization, and the polymerase chain reaction.[6–11] The first two techniques analyze cellular DNA that has been bound to nitrocellulose membranes using radioactive HPV DNA probes. In situ hybridization involves the actual detection of the HPV nucleic acids in a tissue section by visualizing a reagent that marks the site of the HPV probe-target complex. The polymerase chain reaction makes it possible to amplify the target HPV DNA several hundred thousand-fold in vitro prior to binding the DNA to a membrane and performing hybridization.[10,12] Hence this technique is the most sensitive, being able to detect as few as 10 viral particles per tissue section, which is about three orders of magnitude more sensitive than Southern or slot blot hybridization,[13,14] albeit most prone to false positivity resulting from contamination of samples. Using these techniques, investigators have been able to define populations harboring HPV DNA. Detection rates of HPV DNA in the general population range as high as 29% (in pregnant women) using Southern blot hybridization,[15] and up to 80% in repeatedly screened women using the polymerase chain reaction[16–18] (L. Gissman, personal communication). It has been presumed that such populations may be at higher risk for developing HPV-related diseases, but this has not been established with certainty in follow-up studies. In summary, the high sensitivity of Southern/slot blot hybridization and polymerase chain reaction add a confounding variable when doing epidemiologic studies with HPV and genital neoplasia, because many patients will have the virus without any clinical or pathologic evidence of the infection. The relationship between "occult" infection and neoplasia will be addressed below.

The most prominent risk factors for cervical cancer include number of sexual partners (including number of partners for the male consort), age at first intercourse, parity, smoking, barrier contraception, immune status, socioeconomic status, and possibly oral contraceptive use.[19–28] In this country, nonwhite women have two- to threefold higher incidence of and mortality from cervical cancer.[29] Predictably, the above factors influence not only the risk of cancer but also the acquisition of cervical precancers and genital warts, although the link between parity and cervical intraepithelial neoplasia (CIN) is less firm.[20] In contrast, factors influencing the acquisition of "occult" HPV infection (in the form of HPV DNA only) are less clear. For example, detection of HPV DNA is not related directly to the number of sexual partners, smoking, or oral contraceptive use.[27,30] There is some evidence that barrier contraceptives correlate negatively and that the number of sexual partners in the previous year may influence the detection of HPV DNA.[27] In addition, studies of high-risk populations have shown that the higher risk is not related to a higher prevalence of HPV infection in that population.[31] Hence, although there is a close relationship between the presence of HPV nucleic acids and cervical neoplasia, there are obviously a host of factors that will determine why only a small proportion of women with HPV DNA in their genital tract develop invasive cancer (Fig. 4-1).

It is important to stress that, although most genital squamous cell neoplasms are HPV related, not all are. Although intraepithelial and invasive cancers of the cervix are strongly associated with sexual activity and with HPV nucleic acids, invasive cancers of the vulva, in particular those in older women, may not be, an observation that is consistent with the epidemiology of these neoplasms.[32] Rather, other factors, including inflammatory diseases of the vulva, may play a role.[32,33]

In summary, the pathogenesis of many cervical neoplasms appears to include HPV infection. However, at this time the interplay of factors producing clinical HPV infection from occult infection, or invasive cancer from an intraepithelial neoplasm, are not clear.

HPV Type-Specific Neoplasia

The accumulated body of evidence implicates specific HPV types in the genesis of genital neoplasia. For example, HPV types 6 and 11, the first genital HPVs characterized, are strongly associated with exophytic warts on the external genitalia and vagina.[9,34,35] In contrast, HPV-16, which was cloned from an invasive cancer, has been associated with precancerous conditions (intraepithelial neoplasms) of the vulva, vagina, and cervix.[36–38] HPV-16 is the most frequently detected HPV type in cervical cancer precursors, in addition to the closely related HPV-31.[9,39–41] Accordingly, HPV-16 is the most frequent HPV nucleic acid re-

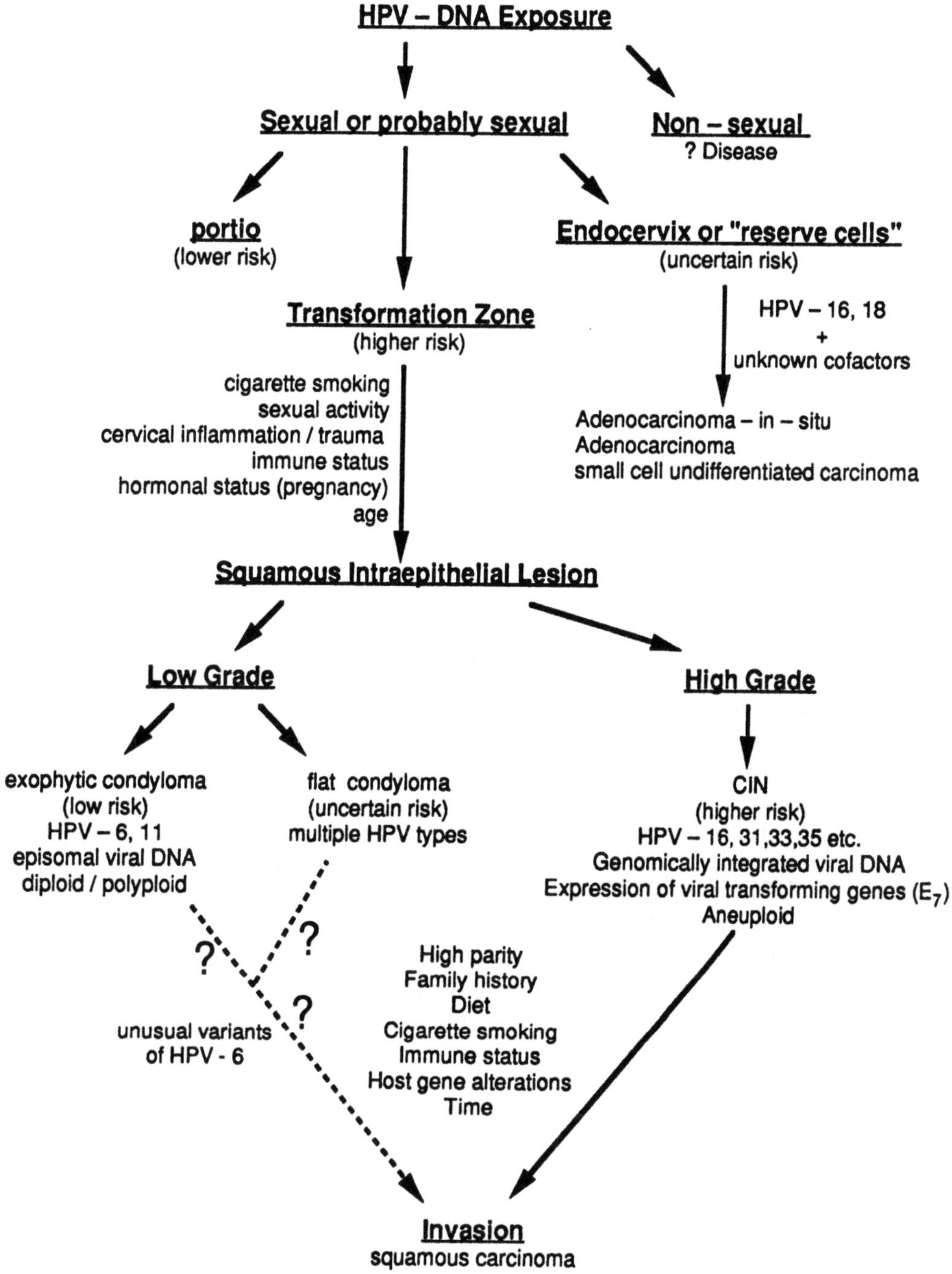

FIGURE 4-1. Schematic of potential factors influencing the genesis of cervical neoplasia. (Reprinted from Crum CP, Taylor PT: Intraepithelial squamous lesions of the cervix, in Knapp RC, Berkowitz RS (eds): Gynecologic Oncology. New York, Macmillan, 1990.)

covered from invasive cancers[6,38] from the genital tract as well as the periungual area of the finger and the conjunctiva.[42,43] This has formed the basis for assuming that infection by HPV-16 (at least that which produces a morphologic lesion) initiates a spectrum of intraepithelial changes that may progress to invasive cancer. HPV-16 is found rarely in typical exophytic condylomata of the genital tract.[9,34] Conversely, HPV-6 or HPV-11 are found rarely in intraepithelial neoplasms or invasive carcinomas.[9,34]

The relationship of these viruses to so-called flat condylomata of the cervix is less clear, as is the biologic potential of these lesions. In contrast to findings in the vulva/vagina, HPV types 6 and 11 are uncommonly found in the cervix, except in exophytic condylomata that occasionally occur there. Flat condylomata contain HPV-16 nucleic acids in from 15 to 30% of cases, and the remainder tend to hybridize with multiple types of DNA probes, suggesting their association with other viral types.[44,45] Recently, Nuovo et al.[9] showed that about 30% of low-grade "flat" condylomata of the cervix contain HPV types 31, 33, or 35, and another 15% that are negative for these types and for HPV types 6, 11, or 16 contain HPV types 42, 43, 44, 45, 51, 52, or 56. Clearly, there is a much more heterogeneous distribution of HPV types in CIN 1 lesions compared to vulvar condylomata and vulvovaginal intraepithelial neoplasia (VIN) as well as compared to "high" grade CIN, for reasons that are as yet unclear.

HPV-18 is an enigmatic viral DNA that is recovered uncommonly from intraepithelial neoplasms.[34,38,46,47] In contrast, it is found in approximately 10% of invasive cancers and at least 20% (and probably closer to 50%) of adenocarcinomas of the cervix.[48–50] Because of this association with cancer, it is suggested that HPV-18–related neoplasms may evolve more rapidly, bypassing a period of stable intraepithelial growth,[51] and that cervical cancers that contain HPV-18 may have a worse prognosis.[52] Additional support for this is provided by the recent link between HPV-18 and small cell carcinomas of the cervix. Stoler et al.[53] found recently that eight of nine small cell cervix lesions contained HPV-18 nucleic acids. This has led to speculation that HPV-18 may preferentially involve (infect?) endocervical reserve cells, although the cofactors involved in the production of a neoplasm are not known.

Molecular Basis for Type-Specific Neoplasia

Studies of the molecular biology of HPV infection have unearthed a variety of mechanisms by which cancer-associated viruses may exert their influence over the host cells. These include: (a) the process of integration, (b) derepression of the viral genome, (c) expression of specific viral oncogenes, (d) interaction of viral gene products with host genes, and (e) expression of host oncogenes.

Integration

Integration is defined as the covalent attachment of viral DNA to chromosomal DNA, in contrast to episomal DNA, which is free DNA that is capable of entering the viral life cycle and producing virions. Integration of the viral DNA is found most frequently in high-grade precancers and invasive cancers and with few exceptions is peculiar to those HPV types that are associated with neoplasia.[54–57] In contrast, HPV types 6 or 11 are present as episomal and not as integrated DNA in genital warts.[35] Unlike some viruses, HPV does not integrate into a specific region of the chromosomal DNA. However, the HPV DNA is usually interrupted in a specific region of the viral genome, specifically that which encodes the E1 or E2 proteins.[3]

Derepression of the Viral Genome

Integration in the region of the E1 and E2 coding areas effectively eliminates the normal viral life cycle and disconnects the viral

DNA such that expression of the virus "downstream" to the integration site is silenced. This is thought to be particularly important because the E2 open reading frame encodes proteins that bind to the upstream regulatory region and influence expression of other proteins, including those that may be involved in neoplastic transformation, such as E6 or E7. Whereas expression of these early genes may be repressed during the normal viral life cycle, integration may free them from control and lead to the production of potentially oncogenic proteins in susceptible cells.[3,58,59]

Expression of Specific Viral Oncogenes

The genes that will be preferentially expressed after integration include the E6 and E7 reading frames (genes), which have been demonstrated to be responsible for HPV-16–mediated transformation in cell culture.[59] In addition, it is possible that unique variants of these proteins may be produced by internal "splicing" in the E6 gene, producing variations of the E6 protein.[60] This potential for internal splicing is unique to cancer-associated viruses, but it is not known whether the unique protein products actually play a role in the genesis of neoplasia.

The potential importance of integration is illustrated by several studies of clinical material. Le and Defendi[59] found that integrated DNA, including both the cellular and viral components, was more efficient for transforming cells in vitro than either the viral or the cellular component alone. This raises questions about the actual role of integration and the importance of the flanking cellular DNA in the process of transformation. Studies of cell culture by Schlegel et al.,[61] studies of keratinocytes by McCance et al.,[5] primary culture of precancerous lesions by Schneider-Manoury et al.,[62] and studies by Lehn et al.[54] suggest, either directly or indirectly, that the integration of HPV-16 into the cellular genome may occur relatively early in the development of genital neoplasia.

Interaction with Host Genes

Recently, the E7 gene product has been demonstrated to bind the retinoblastoma gene product in vitro, suggesting that this may be one mechanism by which the E7 product influences cell transformation.[58] The retinoblastoma gene product is characterized as an "antioncogene" that is inactivated in familial retinoblastoma and other neoplasms.

Involvement of Host Oncogenes

Both C-*myc* and H-*ras* oncogenes have been demonstrated to be expressed in greater quantities in cancers of the cervix. DiPaulo et al.[63] recently were able to immortalize human keratinocytes with HPV-16 DNA, and when both HPV-16 and H-ras were cotransfected into the cells, the cells were not only immortalized but were capable of producing squamous cell carcinomas in nude mice. This is evidence in support of a multistep mechanism for inducing invasive carcinoma of the cervix and has been supported by other investigators.[64–66]

Latent or Occult Infection

The Concept

The discovery of histologically occult or "latent" papillomavirus infection has broadened the perspective of HPV-related disease, raising questions about both the virus-host relationship and the practical issue of detecting HPV infection. Occult HPV infection is defined as the presence of HPV nucleic acids in the absence of morphologic or clinical abnormalities, and is diagnosed primarily by Southern blot or slot blot hybridization methods because it is unusual to detect HPV DNA by in situ hybridization in tissues that lack the clear-cut features of an HPV infection.[67,68] Using these and other technologies with cervical lavage material, investigators have documented the presence of papillomavirus DNA in the absence of histologic or cytologic abnormalities. Estimates on the frequency of infection in the population go as high as 29%, the latter figure reported in pregnant patients.[15,18,69–72]

The Threshold for Diagnosis

A by-product of the latency concept has been the increased focus on the use of molecular biologic techniques to define the lower limits of morphologically detectable HPV infection. Currently, management protocols for women with abnormal Papanicolaou smears dictate that if a cervical lesion is found, it should be removed.[73] Defining exactly what a lesion is has been difficult. As expected, attention to koilocytotic atypia as a diagnostic feature of both warts and precancerous lesions may have resulted in the overuse of this term, with both cytology and histology. Broadening the criteria for the diagnosis of this cellular change has immediate impact on clinicians and their patients, who are burdened by the information that HPV infection is widespread and is associated with neoplasia, and thus are "held hostage" to the uncertainties of cytologic and histologic interpretation. Although it might be tempting to assume that mild cytologic atypia in the biopsy or Papanicolaou smear indicates a low-grade or subtle condyloma, studies designed to correlate such findings with HPV infection have failed to consistently support this concept. For example, when classical criteria for koilocytotic *atypia* are present, the molecular data are supportive, revealing HPV DNA in about 70 to 90% of cases.[34,38,74] However, when histologic changes are subtle, the association with HPV as determined by Southern blot analysis drops to approximately 15%, similar to the detection rate of HPV DNA in normal tissues.[74–76]

This observation has been relevant on both conceptual and therapeutic grounds. For example, it suggests that because normal squamous epithelium is morphologically "transformed" by the virus into a wart, this process cannot be reliably identified until there are clear-cut morphologic changes present. Hence, it is arguable that a spectrum of subtle changes that develops over an extended period of time characterizes the initial development of cervical precancers. On the morphologic level, it is obvious that it is not acceptable to make a diagnosis of HPV-related disease in the presence of very subtle morphologic changes, because in these cases the detection of HPV DNA is so low relative to tissues with clear-cut histologic changes. In clinical management, overdiagnosis carries the threat of both pain and stigma of having a sexually transmitted disease. Because the distinction between clear-cut koilocytotic atypia and "subtle," equivocal changes may be somewhat subjective and because almost 100% of "true" condylomata are HPV positive, HPV testing by in situ hybridization analysis can serve as a useful adjunct for the pathologist to the histologic examination.[9]

Molecular Biology of Recurrent Disease

The goal of treating HPV-related lesions of the genital tract is to remove them and avoid recurrence. This is the single most emotionally charged issue involved in treating women with genital warts, because time-consuming and painful therapeutic approaches will demoralize the patient if they do not meet with success. Two issues of importance when managing women with genital lesions, including those on the cervix, are the mechanisms of recurrence and the significance of recurrence from the standpoint of natural history.

A component of immediate recurrences following therapy seems to be the recrudescence of occult or latent infection.[51] The potential clinical importance of latent infection was demonstrated by Ferenczy et al.,[77] who found that after laser therapy of lesions, recurrences were more likely in cases where HPV nucleic acids were detected in the adjacent normal tissue at the time of laser therapy. The value of this information for therapeutic decision-making is clouded by the difficulty in determining in what normal-appearing cells the HPV DNA is present. Attempts to perform extensive ablation of both the lesion and surrounding tissue, in an effort to eradicate occult infection, have met with failure.[78] Thus, although the association of latent infection with clinical disease is important conceptually, it is of limited value therapeutically.

Recently, a study by Nuovo and Pedemonte[79] determined that a large proportion of recurrent cervical lesions are low grade, irrespective of the grade of the original cervical lesion, although most of the original lesions were also CIN 1. One might anticipate that if recurrent CIN is commonly due to reinfection from the untreated partner, then the HPV type in the initial lesion might often be the same as in the recurrent CIN. However, this study noted that the HPV type in CIN that appeared 3 to 12 months after cryotherapy ablation of the initial lesion was a different HPV type than that detected in the initial CIN.[79] Clarification of this issue may change the perception of the mechanisms of "recurrence."

Conceivably, recurrences in the cervix may be a function not only of inadequately treated disease, or latent infection, but of occult infections with *other* HPV types that produce diverse morphologies depending upon the nature of the virus involved. While this is not surprising given the plurality of HPV types in the cervix[9] and the tendency for more than one type to be present in the genital tract at one time,[34] it points to the importance of considering this when managing recurrences after treating CIN. Detection of another HPV type during cervix pretreatment may identify a population at high risk for recurrent disease. This observation also implies that a type-specific resistance might develop after ablation of a CIN. Consistent with this possibility is the observation that when HPV is identified in an unremarkable cervix after laser ablation of CIN, the HPV type is usually different from the type present in the CIN.[80,81] If true, this has widespread implications with respect to a better understanding of the host response to HPV infection and the development of a possible vaccine.

Immunology of Papillomaviruses

Because genital warts are a viral-induced disease, attention has focused on the host response to HPV-related disease. As a general statement, many external condylomata of the vulva/vagina will regress spontaneously, and the time course of regression appears linked to the development of an immune response on the part of the host. This holds true particularly for those lesions associated with HPV-6 or HPV-11 (conventional warts), although precancerous lesions containing HPV-16 may regress as well. It is generally assumed that the longer the patient has had the warts the easier it will be to achieve success after the initial ablation. This may be true as well for the cervix, but a significant proportion of higher grade lesions will persist and coincidentally, many are associated with HPV-16 nucleic acids.[36,82] Because of the unique problems associated with HPV—in particular the lack of a system for propagating virus in culture—the technology has been developed for producing HPV-related proteins in bacterial systems (in vitro).[83,84] This has made it possible both to explore expression of HPV proteins in tissues and to acquire reagents to evaluate the human immune response.

The most comprehensive analysis of HPV types 6 and 16 proteins to date has been performed by Firzlaff et al. and Jenison et al.[84–86] In analyses of the human immune response to HPV proteins, they have identified serologic immune responses principally to "capsid" proteins, specifically the L1 protein of HPV 6.[83] Interestingly, they have not succeeded in detecting antibodies to HPV-16 L1 proteins, although they found antibodies to HPV-16 L2 proteins in nearly half of adults, including those attending sexually transmitted disease clinics (S. A. Jenison, personal communication).

Further studies of the host immune response to HPV types 16 and 18 proteins are in progress and may prove critical, because of the association between these types and precancers and a subset of aggressive neoplasms in the cervix.[49,51,52] Differences in host response to viral subtypes (associated with neoplasia or benign disease) may reveal important clues to why certain HPV-related abnormalities, specifically HPV-16–related precancers, are less likely to regress spontaneously. Understanding these differences ultimately may be critical to developing strategies for immunizing patients. Here the limitations imposed by the absence of convenient nonhuman models for HPV infection will be most obvious, but not insurmountable. For example, in vivo systems for propagating HPV-11 infection have been developed in mice[4] and are currently being exploited to evaluate the neutralization potential of anti–HPV-11 antibodies (J. Kreider, personal communication). Systems such as this for exploring HPV-16 or HPV-18 infection have not yet been reported, but in vitro systems using human foreskin keratinocytes have evolved that closely mimic the in vivo state.[5] With these advances, it is not inconceivable that within the next few years the components of viral infection, propagation, and neutralization observed principally in vivo will be produced and studied in vitro, and the testing of potential vaccines will become feasible. The ultimate success of such strategies will be dependent, however, upon the strength of the association between certain HPV infections and neoplasia, and particularly the association between HPV gene products and biologic events necessary for establishing the neoplastic phenotype.

The Male Sexual Partner

The female genital tract has received the majority of attention regarding the association of HPV with neoplasia. The main reasons for this are the widespread use of the Papanicolaou smear to detect cervical precancers and the relatively lower frequency of large HPV-related lesions of the penis. However, given the marked increase in HPV-associated lesions over the last several decades[87,88] and the evidence favoring a venereal mode of spread of the virus,[1,89] investigators have begun to focus on the male sexual partner. Brinton et al.[90] found that the relative risk of cervical cancer was 2 when the corresponding male had a large number of sexual partners.

Several features of disease in males should be stressed: (a) it is prevalent; (b) it is subtle in clinical appearance; (c) it is subtle histologically; (d) lesions from pairs may share similar histology and viral type, but also frequently do not; and (e) the value of managing the male partner is completely unknown.[89,91]

Approximately 80% of male sexual partners of women with genital lesions have penile lesions.[89,92] A fact that becomes quickly apparent when reviewing the topic of the clinical detection of HPV-related lesions of the penis is that they are usually subtle.[89,92] The majority of men whose sexual partners have CIN will not have a visible penile lesion on routine examination.[89] The standard approach to detect these lesions is to combine application of acetic acid and peniscopy (examination with a colposcope). The detection rate of penile HPV-related lesions with this technique in the partners of women with CIN ranges from 53 to 100%. Barrasso et al.[89] noted a threefold increase in the detection rate after the application of acetic acid. The majority of the lesions were present on the shaft just proximal to the glans. However, another study noted that most of the abnormal tissues seen after magnification and acetic acid that were 1 mm or smaller were not associated with HPV DNA.[91]

The histology of penile lesions, like that of cervical lesions, may be confusing. The diagnostic feature of condylomata, as discussed previously, is koilocytotic atypia, which are perinuclear halos in conjunction with variation in nuclear size, shape, and chromaticity. HPV DNA will be detected in 90 to 100% of penile tissues that demonstrate perinuclear halos in conjunction with nuclear atypia.[91] If these histologic features are not seen, then 11% of the penile tissues had HPV DNA as detected by in situ hybridization in these "high"-risk men. Because these tissues contain at least 20 (and usually many more) viral particles in many squamous cells, the patient is presumably infectious, whereas it is assumed, although not proven, that the inability to detect HPV by this methodology in these "nondiagnostic" tissues implies a noninfectious lesion.[91]

Given the venereal mode of spread of the virus and the rarity of double infections,[34,38] it is reasonable to predict that there would be some correlation between the histology and virology of the penile lesion and the genital tract lesion of the partner. Most of the work analyzing noninvasive HPV-related penile lesions and their correlation with cervical lesions in the man's partner is to be found in the extensive and excellent study of Barrasso et al.[89] They noted that there was a strong correlation between the histologic features of the penile lesions and the corresponding lesion in the man's partner. Specifically, histologically defined condylomata of the penis was found in 41% of men whose partners had cervical condylomata (CIN 1) but in only 5% of men whose partners had cervical intraepithelial neoplasms (CIN 2 or 3). Alternatively, in men with histologically defined penile intraepithelial neoplasms, 33% had a partner who had CIN 2 or 3 whereas only 1% had partners with cervical condylomata. Thus, their study provided further evidence not only for the venereal transmission of these lesions but also for the segregation of different HPV types with condylomata and the precancerous intraepithelial neoplasms. A salient question follows: will the men with penile intraepithelial neoplasms be at an increased risk for developing penile squamous cell carcinoma? Based on the observation that approximately 12% of high-grade cervical intraepithelial neoplasms will progress to cancer over 10 years,[73] it is reasonable to assume that these men will be at an increased risk for developing a penile malignancy. However, given the rarity of this lesion compared to its counterpart in the cervix, the absolute risk appears to be small.

Pediatric HPV Infection

Anogenital Condylomata

Because genital warts are transmitted primarily via sexual contact, the prospect of anogenital condylomata occurring in children is a particularly emotional issue. Prior to 1980 genital warts in children were rarely reported. Stumpf[93] reported 3 cases and reviewed 22 from the English language literature between 1940 and 1980 (see also G. J. Nuovo[107]). However, an equal or greater number have been reported since that time.[94–96]

The three issues in relation to genital warts occurring in childhood are: (a) the association with child abuse, (b) the potential for transmission by other modes of contact, and (c) the potential for malignant transformation. The potential for child abuse is real, judging from the number of reported instances of genital lesions in which a clear sexual history could be obtained. The American Academy of Dermatology Task Force on Pediatric Dermatology reviewed the literature and concluded that approximately 50% of cases reported were associated with sexual contact.[95] In a review of 30 cases prior to 1982, DeJong et al.[97] reported history of sexual abuse in none of 4 children less than 2 months of age and 1 of 16 under 7 years.

In cases in which there is no clear sexual history, other modes of transmission have been suggested. These include acquiring the virus during birth,[98] direct transmission by innocent contact between parent and infant, and autoinoculation by the infants themselves. The first two explanations appear to be the best for infants under 2 years old who present with anogenital lesions, although given how common genital condylomata are in pregnant women, one might anticipate that, if transmission could occur this way, more children would have genital warts. The potential for autoinoculation has been suggested in the past, but was supported primarily by early studies indicating that the viral DNAs in genital warts were heterogeneous, and might include cutaneous wart viruses as well.[99] However, molecular epidemiologic studies in recent years indicate that a high percentage of genital lesions are associated with viruses that exclusively (or at least predominantly) occur at that site (G. J. Nuovo[107]). Vallejos et al.[94] confirmed this when they detected HPV types 6, 11, or 16 DNA sequences in seven of eight genital lesions from prepubertal girls by in situ hybridization; Rock et al.[95] reported similar findings. Thus, genital warts in young girls must be assumed to originate from genital papillomaviruses (though, see below) and, despite the relative absence of a history of sexual abuse in the very young, this must be considered. It is important to keep in mind that a significant proportion of cases occurring in infants have been associated with minimal historical information. In some there may be a family situation that could exclude neither poor hygiene nor abuse, leaving the management of young children with genital warts a thorny problem.

Nuovo et al. have analyzed a series of genital and nongenital warts for HPV DNA in children and adults.[107] They noted that HPV-2 was the most common type in nongenital warts in children and adults. As expected, HPV-6/11 were the most common types in genital warts in each group. No HPV-2 was detected in any genital wart in an adult, although it was detected in genital warts in children, as has been noted by others.[96] It follows that detection of HPV-2 in a genital wart in a child implies nongenital transmission, whereas detection of HPV-6/11 DNA implies genital transmission.

The potential for malignancy in children with genital warts is a concern because of their youth and the relatively longer period in which they will be at risk for subsequent neoplasia. Cases of intraepithelial neoplasia have been reported in children, although progression to cancer has not been reported in these young patients who have been prospectively followed. Cases of vulvar carcinoma have been reported in young women,[100] but a connection with preexisting condylomata has not been appreciated.

Laryngeal Papillomatosis

Laryngeal papillomatosis may be manifest in infancy or young adulthood, and the former has been linked to genital papillomavirus infection in the mother.[101] It is characterized by multiple papillary lesions in the laryngeal area, but may include the tracheobronchial tree as well.[102] Because of airway obstruction the disease may be life threatening, and the infants are often doomed to multiple procedures aimed at clearing the airway. Treatment may remove the lesions but the virus may persist as it does in other mucosal sites and be associated with recurrences.[103] Presumably the infants are exposed to the virus from the maternal genital tract during delivery. Shah et al.[104] noted a relatively small proportion of cases associated with cesarean section; however, the actual risk of contracting the disease after vaginal delivery is very low. Shah et al. estimated that approximately 1:1,500 children "at risk" would develop laryngeal papillomatosis.[104] Because immunity is acquired with time and may be passed from mother to fetus, conceivably risk is related to the duration of infection in the mother. However, this remains unproven.

Role of Molecular Diagnosis

With the development of technologies for detecting and classifying HPV DNA in the genital tract, there is considerable interest in the feasibility of HPV testing as an adjunctive technique for diagnosing HPV-related diseases of the cervix. This can be approached on at least four levels: (a) the use of HPV-DNA testing to reduce the false-negative rate of screening Papanicolaou smears,

(b) evaluation of screening smears with squamous atypia of uncertain significance, (c) detection of occult or latent HPV infection, and (d) modification of therapy of patients with morphologically distinct HPV-related lesions.

Reducing False-Negative Papanicolaou Smears

The Papanicolaou smear clearly carries a significant false-negative rate, as described above, and the component of false-negatives that is most preventable is the misreading of the smears. The addition of HPV DNA testing to reduce this would employ rescreening of negative smears that were HPV-DNA positive. Studies indicate that approximately 20% of such smears will be found to contain abnormal cells that were missed on the first analysis[95,105] (B. E. Ward and C. P. Crum, unpublished). However, it is not clear (a) how efficiently routine viral testing would reduce false-negative rates for Papanicolaou smear interpretation, (b) the impact of such a program on morbidity and mortality related to cervical cancer, and (c) the cost of achieving an incremental improvement in morbidity and mortality. On the one hand, approximately one third of the nearly 7,500 women who die of cervical cancer–related causes each year had a negative Papanicolaou smear in the past 5 years prior to diagnosis, and conceivably may have benefitted from more intensive screening.[29] On the other hand, the cumulative cost of delivering this test indiscriminately would be extremely high and its value to women who already receive regular examinations is questionable.

HPV DNA Testing to Evaluate Borderline Papanicolaou Smears

In this scenario, HPV DNA testing would be used to "fine tune" the management of women with Papanicolaou smear abnormalities of uncertain significance. In a recent study of "nondiagnostic" squamous atypias, it was found that those associated with HPV DNA were more likely to remain abnormal on follow-up and correlate with the presence of CIN.[106] The principal objections to using HPV DNA testing to resolve these borderline abnormalities, however, are (a) the inherent false-negative rate of HPV DNA sampling and (b) the requirement that any persistent cytologic abnormality be evaluated by colposcopy under any circumstances.

Detecting Occult or Latent HPV Infection

By definition, programs aimed at detecting occult infection would require sampling large populations. Unlike the approach outlined for reduction of false-negative smears, HPV-positive patients would be notified that they carried the infection and counseled to return for routine cytologic follow-up. However, at present, the significance of HPV DNA in the absence of cytologic or histologic abnormalities is unknown. Furthermore, the rate of conversion of subclinical infection to clinical disease and the factors influencing this conversion are unknown. There is also the risk of considerable emotional morbidity in the knowledge that one carries occult HPV infection with no alternative but to adhere to currently accepted practice of a regular Papanicolaou smear.

HPV DNA Testing to Modify Therapy

Because the accepted approach to cervical intraepithelial lesions is to remove them, HPV DNA testing has little value in the management of women prior to ablative therapy. In cases in which lesions are extensive or the patient has concomitant immunosuppression, HPV DNA testing may be useful with "low-risk" HPV types for identifying extensive infections, possibly providing for an approach centering on a period of follow-up rather than proceeding directly to extensive ablation. However, there is no evidence that making such decisions requires more than careful histologic interpretation. Moreover, most commercially available techniques for HPV DNA detection categorize HPV DNA positives using groups of more than one probe (e.g., 6/11, 16/18, or 31/33/35). Depending upon the hybridization conditions specified by the detection technique, specifying precisely what HPV type is present may be difficult. Because there is no evidence that the probe types in some groups identify infections with the same natural history, the "prognostic" value of categorizing lesions as positive for "6/11", "16/18," and "31/33/35" is uncertain at this time.

Summary

HPV DNA testing remains an experimental technique that has provided insights into the distribution of papillomaviruses in the population and their association with neoplasia. As such it has been a useful tool for gathering information about epidemiology, the significance of certain morphologic changes, and follow-up studies. At present this technology may be producing as many questions as answers concerning the relationship between HPV and the risk of disease. It is reasonable to assume that widespread testing under optimal conditions would reduce the false-negative Papanicolaou smear rate and possibly increase the rate of early detection. Whether such an ambitious (and expensive) program would alter the death rate from cervical cancer is unknown, particularly because it would address only one component of risk (screening), leaving other risk factors, including smoking, sexual activity, and barrier contraceptive methods, unaltered. This is well illustrated in the study by Kjaer et al.[31] wherein the risk of cancer between two populations is related not to screening but to sexual practices. Clearly the nature of the population at risk will determine the effectiveness of individual intervention programs, whether aimed at simply increasing screening, increasing effectiveness of screening, or introducing technologies that will improve screening programs.

Acknowledgments. Supported by grants from the National Cancer Institute (CA47676) and American Cancer Society (MV-395), and an institutional support grant. Dr. Crum is a recipient of a Physician Scientist Award from the National Institute of Allergy and Infectious Disease (AI 00628). Dr. Nuovo is a recipient of a grant from the Lewis Foundation.

References

1. Meisels A, Morin C: Human papillomavirus and cancer of the uterine cervix. Gynecol Oncol 1981;12:S111–S123.
2. MMWR 1983;23:306.
3. Broker TR, Botchan MT: Papillomavirus—retrospectives and prospectives, in Botchan MT, Grodzicker T, Sharp PA (eds): Cancer Cells 4. New York, Cold Spring Harbor, 1986, pp 7–35.
4. Kreider JW, Howett MK, Leure-Dupree AE, et al: Laboratory production in vivo of infectious human papillomavirus type 11. J Virol 1987;61:590–593.
5. McCance DJ, Kopan R, Fuchs E, Laimins LA: Human papillomavirus type 16 alters human epithelial cell differentiation in vitro. Proc Natl Acad Sci USA 1988; 85:7169–7173.
6. Koutsky LA, Galloway DA, Holmes KK: Epidemiology of genital human papillomavirus infection. Epidemiol Rev 1988;10:122–163.
7. Beckmann AM, Myerson D, Daling JR, et al: Detection and localization of human papillomavirus DNA in human genital condylomas by in situ hybridization with biotinylated probes. J Med Virol 1983;16:265–273.
8. Crum CP, Nagai N, Levine RU, Silverstein SJ: In situ hybridization analysis of human papillomavirus 16 DNA sequences in early cervical neoplasia. Am J Pathol 1986; 123:174–182.
9. Nuovo GJ, Friedman D: In situ hybridization analysis of HPV DNA segregation patterns in lesions of the female genital tract. Gynecol Oncol 1990;36:256–262.
10. Shibata D, Fu YS, Gupta JW, et al: Detection of human papillomavirus in normal and dysplastic tissue by the polymerase chain reaction. Lab Invest 1988;59:555–559.
11. Melchers WJG, Schift R, Stolz E, et al: Human papillomavirus detection in urine

samples from male patients by the polymerase chain reaction. J Clin Microbiol 1989; 27:1711–1714.
12. Saiki NK, Grelfond DM, Stoffel S, et al: Primer directed amplification of DNA with thermostable DNA polymerase. Science 1988;239:487–491.
13. Lorincz AT: Human papillomavirus testing. Diagn Clin Test 1989;27:28–37.
14. Lorincz AT: Detection of human papillomavirus infection by nucleic acid hybridization. Obstet Gynecol Clin North Am 1987;14:451–469.
15. Burk RD, Kadish AS, Calderin S, Romney SL: Human papillomavirus infection of the cervix detected by cervicovaginal lavage and molecular hybridization: correlation with biopsy results and Papanicolaou smear. Am J Obstet Gynecol 1986;154:982–989.
16. Tidy JA, Parry GCN, Ward P, et al: High rate of human papillomavirus type 16 infection in cytologically normal cervices. Lancet 1989;1:434.
17. Tidy JA, Vousden KH, Farrell PJ: Relation between infection with a subtype of HPV 16 and cervical neoplasia. Lancet 1989;1:1225–1227.
18. Schneider A, Hotz M, Gissman L: Increased prevalence of human papillomaviruses in the lower genital tract of pregnant women. Int J Cancer 1987;40:198–201.
19. Kjaer SK, de Villiers EM, Haugaard BJ, et al: Human papillomavirus, herpes simplex virus and cervical cancer incidence in Greenland and Denmark. A population-based cross-sectional study. Int J Cancer 1988; 41:518–524.
20. Brinton LA, Reeves WC, Brenes MM, et al: Parity as a risk factor for cervical cancer. Am J Epidemiol 1989;130:486–496.
21. Slattery ML, Overall JC, Abbot TM, et al: Sexual activity, contraception, genital infections, and cervical cancer: support for a sexually transmitted disease hypothesis. Am J Epidemiol 1989;130:248–258.
22. Hellberg D, Valentin J, Nilsson S: Smoking and cervical intraepithelial neoplasia. An association independent of sexual and other risk factors? Acta Obstet Gynecol Scand 1986;65:625–631.
23. Kessler II: Perspectives on the epidemiology of cervical cancer with special reference to the herpes virus hypothesis. Cancer Res 1974;34:1091–1110.
24. Kjaer SK, Teisen C, Haugaard BJ, et al: Risk factors for cervical cancer in Greenland and Denmark: a population-based cross-sectional study. Int J Cancer 1989;44:40–47.
25. Layde PM: Smoking and cervical cancer: cause or coincidence. JAMA 1989;261:1631–1632.
26. Layde PM, Broste SK: Carcinoma of the cervix and smoking. Biomed Pharmacother 1989;43:161–165.
27. Burkett B, Peterson C, Ward BE, et al: The relationship between contraceptives, sexual practices, and cervical human papillomavirus infection among a college population. J Clin Epidemiol, in press.
28. Vessey M, Grice D: Carcinoma of the cervix and oral contraceptives: epidemiological studies. Biomed Pharmacother 1989; 43:157–160.
29. Dunn JE Jr, Crocker DW, Rube IF, et al: Cervical cancer occurrence in Memphis and Shelby County Tennessee during 25 years of its cervical cytology screening program. Am J Obstet Gynecol 1984;150:861–864.
30. Kiviat NB, Koutsky LA, Paavonen JA, et al: Prevalence of genital papillomavirus infection among women attending a college student health clinic or a sexually transmitted disease clinic. J Infect Dis 1989;159:293–302.
31. Kjaer SK, deVilliers EM, Haugaard BJ, et al: Human papillomavirus, herpes simplex virus and chemical cancer incidence in Greenland and Denmark. A population-based cross-sectional study. Int J Cancer 1988;41:518–524.
32. Mabuchi K, Bross DS, Kessler II: Epidemiology of cancer of the vulva: a case control study. Cancer 1985;55:1843–1848.
33. Buscema J, Stern J, Woodruff JD: The significance of the histologic alterations adjacent to invasive vulvar carcinoma. Am J Obstet Gynecol 1980;137:902–909.
34. Reid R, Greenberg M, Jenson AB, et al: Sexually transmitted papillomaviral infections I. The anatomic distribution and pathologic grade of neoplastic lesions associated with different viral types. Am J Obstet Gynecol 1987;156:212–222.
35. Gissman L, Wolnick L, Ikenberg H, et al: Human papillomavirus type 6 and 11 DNA sequences in genital and laryngeal papillomas and in some cervical cancer. Proc Natl Acad Sci USA 1983;80:560–563.

36. Crum CP, Ikenberg H. Richart RM, Gissmann L: Human papillomavirus type 16 and early cervical neoplasia. N Engl J Med 1984; 310:880–883.
37. Durst M, Croce CM, Gissmann L, et al: Papillomavirus sequences integrate near cellular oncogenes in some cervical carcinomas. Proc Natl Acad Sci USA 1987;84:1070–1074.
38. Lorincz AT, Temple GF, Kurman RJ, et al: Oncogenic association of specific human papillomavirus types with cervical neoplasia. JNCI 1987;79:671–677.
39. Crum CP, Mitao M, Levine RU, Silverstein SJ: Cervical papillomaviruses segregate within morphologically distinct precancerous lesions. J Virol 1985;54:675–681.
40. Willett GD, Kurman RJ, Reid R, et al: Correlation of the histologic appearance of intraepithelial neoplasia of the cervix with human papillomavirus types. Int J Gynecol Pathol 1989;8:18–25.
41. Beaudenon S, Kremsdorf D, Croissant O, et al: A novel type of human papillomavirus associated with genital neoplasias. Nature 1986;321:246–249.
42. Moy RL, Eliezri YD, Nuovo GJ, et al: Squamous cell carcinoma of the finger is associated with human papillomavirus type 16 DNA. JAMA 1989;261:2669–2673.
43. McDonnell JM, Mayr AJ, Martin WJ: DNA of human papillomavirus type 16 in dysplastic and malignant lesions of the conjunctiva and cornea. N Engl J Med 1989; 320:1442–1446.
44. Franquemont D, Ward B, Andersen W, Crum CP: Prediction of "high risk" cervical papillomavirus infection by biopsy morphology. Am J Clin Pathol 1989;92:577–582.
45. Willett GD, Kurman RJ, Reid R, et al: Correlation of the histological appearance of intraepithelial neoplasia of the cervix with human papillomavirus types. Int J Gynecol Pathol 1989;8:18–25.
46. Nuovo GJ: Correlation of histology with human papillomavirus DNA detection in the female genital tract. Gynecol Oncol 1988; 31:176–181.
47. Nuovo GJ: A comparison of slot blot, Southern blot and in situ hybridization analyses for human papillomavirus DNA in genital tract lesions. Obstet Gynecol 1989;74:673–677.
48. Wilczynski SP, Bergen S, Walker J, et al: Human papillomaviruses and cervical cancer: analysis of histopathologic features associated with different viral types. Hum Pathol 1988;19:697–704.
49. Walker J, Bloss JD, Liao S, et al: Human papillomavirus genotype as a prognostic indicator in carcinoma of the uterine cervix. Obstet Gynecol 1989;74:781–785.
50. Tase TT, Okagaki T, Clark BA, et al: Human papillomavirus types and localization in adenocarcinoma and adenosquamous carcinoma of the uterine cervix: a study by in situ DNA hybridization. Cancer Res 1988; 48:993–998.
51. Kurman RJ, Schiffman MH, Lancaster WD, et al: Analysis of individual human papillomavirus types in cervical neoplasia: a possible role for type 18 in rapid progression. Am J Obstet Gynecol 1988;159:293–296.
52. Barnes W, Delgado G, Kurman RJ, et al: Possible prognostic significance of human papillomavirus type in cervical cancer. Gynecol Oncol 1988;29:267–273.
53. Stoler MH, Walker AN, Mills SE: Small cell neuroendocrine carcinoma of the cervix: a human papillomavirus type 18 associated cervix cancer. Lab Invest 1989;60:92A.
54. Lehn H, Villa LL, Marziona F, et al: Physical state and biological activity of human papillomaviruse genomes in precancerous lesions of the female genital tract. J Gen Virol 1988;69:187–196.
55. Schwarz E, Durst M, Demankowski C, et al: DNA sequence and genome organization of genital human papillomavirus type 6b. EMBO J 1983;2:2341–2348.
56. ElAwady MK, Kaplan JB, O'Brien SJ, Burk RD: Molecular analysis of integrated human papillomavirus 16 sequences in the cervical cancer line SiHa. Virology 1987;159:389–398.
57. Schwartz E, Freese UK, Gissman L, et al: Structure and transcription of human papillomavirus sequences in cervical carcinoma cells. Nature 1985;314:111–114.
58. Dyson N, Howley PM, Munger K, Harlow E: The human papilloma virus-16 E7 oncoprotein is able to bind to the retinoblastoma gene product. Science 1989;243:934–936.
59. Le J, Defendi V: A viral-cellular junction fragment from a human papillomavirus type 16-positive tumor is competent in transformation of NIH 3T3 cells. J Virol 1988; 62:4420–4426.
60. Schneider-Gadicke A, Schwartz E: Different human cervical carcinoma cell lines show similar transcription patterns of human pap-

illomavirus type 18 early genes. EMBO J 1986;5:2285–2292.
61. Schlegel R, Phelps WC, Zhang Y-L, Barbosa M: Quantitative keratinocyte assay detects two biological activities of human papillomavirus DNA and identifies viral types associated with cervical carcinoma. EMBO J 1988;7:3181–3187.
62. Schneider-Manoury S, Croissant O, Orth G: Integration of human papillomavirus type 16 DNA sequences: a possible early event in the progression of genital tumors. J Virol 1987;61:3295–3298.
63. DiPaolo JA, Woodworth CD, Popescu NC, et al: Induction of human cervical squamous cell carcinoma by sequential transfection with human papillomavirus 16 DNA and viral Harvey ras. Oncogene 1989;4:395–399.
64. Matlashewski G, Osborn K, Banks L, et al: Transformation of primary human fibroblast cells with human papillomavirus type 16 DNA and EJ-ras. Int J Cancer 1988;42:232–238.
65. Cook T, Morgenstern JP, Crawford L, Banks L: Continued expression of HPV 16 E7 protein is required for maintenance of the transformed phenotype of cells cotransformed by HPV 16 plus EJ-ras. EMBO J 1989;8:513–519.
66. Cook T, Almond N, Murray A: Constitutive expression of c-myc oncogene confers hormone independence and enhanced growth-factor responsiveness on cells transformed by human papillomavirus type 16. Proc Natl Acad Sci USA 1989;86:5713–5717.
67. Crum CP, Friedman D, Nuovo GJ, Silverstein SJ: Morphological correlates of genital human papillomavirus infection: viral replication, transcription, and gene expression, in Gallo R, Hazeltine W, Klein G, zur Hausen H (eds): Viruses and Human Cancer. New York, Alan R. Liss, 1986, pp 355–369.
68. Nuovo GJ, Nuovo MA, Cottral S, et al: Histological correlates of clinically occult human papillomavirus infection of the uterine cervix. Am J Surg Pathol 1988;12:198–204.
69. Fuchs PG, Girardi F, Pfister H: Human papillomavirus DNA in normal, metaplastic, preneoplastic and neoplastic epithelia of the cervix uteri. Int J Cancer 1988;41:41–45.
70. deVilliers EM, Schneider A, Miklaw H, et al: Human papillomavirus infections in women with and without abnormal cervical cytology. Lancet 1987;1:703–706.
71. Nuovo GJ, Cottral S, Richart RM: Occult infection of the uterine cervix by human papillomavirus in postmenopausal women. Am J Obstet Gynecol 1989;160:340–344.
72. Schneider A, Kirchmayr R, deVilliers EM, Gissmann L: Subclinical human papillomavirus infections in male sexual partners of female carriers. J Urol 1988;140:1431–1434.
73. Richart RM: Causes and management of cervical intraepithelial neoplasia. Cancer 1987; 60:1951–1959.
74. Nuovo GJ, Blanco JB, Silverstein SJ, Crum CP: Histologic correlates of papillomavirus infection of the cervix. Obstet Gynecol 1988; 72:770–774.
75. Nuovo GJ, O'Connell M, Blanco JB, et al: Correlation of histology and human papillomavirus DNA detection in condyloma acuminatum and condyloma-like vulvar lesions. Am J Surg Pathol 1989;13:700–706.
76. Nuovo GJ: Human papillomavirus (HPV) DNA in genital tract lesions histologically negative for condylomata: analysis by in situ, Southern blot hybridization and the polymerase chain reaction. Am J Surg Pathol 1990;14:643–651.
77. Ferenczy A, Mitao M, Nagai N, et al: Latent papillomavirus and recurring genital warts. N Engl J Med 1985;313:784–788.
78. Riva J, Sedlacek T, Cunnane M, Mangan C: Extended carbon dioxide laser vaporization in the treatment of subclinical papillomavirus infection of the lower genital tract. Obstet Gynecol 1989;73:25–30.
79. Nuovo GJ, Pedemonte BA: Human papillomavirus types and recurrent genital warts. JAMA 1990;263:1223–1226.
80. Byrne MA, Wickenden C, Coleman DV: Prevalence of human papillomavirus types in the cervices of women before and after laser ablation. Br J Obstet Gynaecol 1988; 95:201–202.
81. Wickenden C, Malcolm ADB, Byrne MA, et al: Prevalence of HPV DNA and viral copy numbers in cervical scrapes from women with normal and abnormal cervices. J Pathol 1987;153:127–135.
82. Nasiell K, Nasiell M, Vaclavinkova V: Behavior of moderate cervical dysplasia during long term follow-up. Obstet Gynecol 1983; 61:609–614.
83. Firzlaff JM, Kiviat NB, Beckmann AM, et al: Detection of human papillomavirus capsid antigens in various squamous epithelial lesions using antibodies directed against the

L1 and L2 open reading frames. Virology 1988;164:467–477.
84. Firzlaff JM, Hsia C-NL, Halbert CPHL, et al: Polyclonal antibodies to human papillomavirus type 6b and type 16 bacterially derived fusion proteins, in Steinberg BM, Brandsma JL, Taichman LB (eds): Cancer Cells 5, "Papillomaviruses." New York, Cold Spring Harbor, 1987, pp 105–113.
85. Jenison SA, Firzlaff JM, Langenberg A, Galloway DA: Identification of immunoreactive antigens of human papillomavirus type 6b using *Escherichia coli*-expressed fusion proteins. J Virol 1988;62:2115–2123.
86. Jenison SA, Yu X-P, Valentine JM, Galloway DA: Human antibodies react with an epitope of the human papillomavirus type 6b L1 open reading frame which is distinct from the type-common epitope. J Virol 1989; 63:809–818.
87. Oriel JD: Natural history of genital warts. Br J Venereal Dis 1971;47:1–13.
88. Oriel JD: Condylomata acuminata as a sexually transmitted disease. Dermatol Clin 1983;1:93–102.
89. Barrasso R, DeBrux J, Croissant O, Orth G: High prevalance of papillomavirus associated penile intraepithelial neoplasia in sexual partners of women with cervical intraepithelial neoplasia. N Engl J Med 1987; 317:916–923.
90. Brinton LA, Reeves WC, Brenes MM, et al: The male sexual factor in the etiology of cervical cancer among sexually monogamous women. Int J Cancer 1989;44:199–203.
91. Nuovo GJ, Hochman H, Eliezri YD, et al: Human papillomavirus DNA in penile lesions histologically negative for condylomata: analysis by in situ hybridization and the polymerase chain reaction. Am J Surg Pathol 1990;14:829–836.
92. Levine RU, Crum CP, Herman E, et al: Cervical papillomavirus infection and intraepithelial neoplasia: a study of male sexual partners. Obstet Gynecol 1984;64:16–20.
93. Stumpf PG: Increasing occurrence of condylomata acuminata in premenarchal children. Obstet Gynecol 1980;56:262–264.
94. Vallejos H, DelMistro A, Kleinhaus S, et al: Characterization of human papilloma virus types in condylomata acuminata in children by in situ hybridization. Lab Invest 1987; 56:611–615.
95. Rock B, Naghashfar Z, Barnett N, et al: Genital tract papillomavirus infection in children. Arch Dermatol 1986;122:1129–1132.
96. Fleming KA, Venning V, Evans M: DNA typing of genital warts and diagnosis of sexual abuse in children. Lancet 1987;1:454.
97. DeJong AR, Weiss JC, Brent RL: Condyloma acuminata in children. Am J Dis Child 1982; 136:704–706.
98. Tang CK, Shermeta DW, Wood C: Congenital condylomata acuminata. Am J Obstet Gynecol 1978;131:912–913.
99. Krzyzek RA, Watts SL, Anderson DL, et al: Anogenital warts contain several distinct species of human papillomavirus. J Virol 1980;36:236–244.
100. Bergeron C, Naghashfar Z, Canaan C, et al: Human papillomavirus type 16 in intraepithelial neoplasia (bowenoid papulosis) and coexistant invasive carcinoma of the vulva. Int J Gynecol Pathol 1987;6:1–11.
101. Quick CA, Watts SL, Krzyzek RA, Faras AJ: Relationship between condylomata and laryngeal papillomata. Ann Otol 1980;89:467–471.
102. Abramson AL, Steinberg BM, Winkler B: Laryngeal papillomatosis: clinical, histopathologic and molecular studies. Laryngoscope 1987;97:678–685.
103. Steinberg BM, Topp WC, Schneider PS, Abramson AL: Laryngeal papillomavirus infection during clinical remission. N Engl J Med 1983;308:1261–1264.
104. Shah K, Kashima H, Polk BF, et al: Rarity of cesarian delivery in cases of juvenile-onset respiratory papillomatosis. Obstet Gynecol 1986;68:795–799.
105. Ritter DB, Kadish AS, Vermund SH, et al: Detection of human papillomavirus deoxyribonucleic acid in exfoliated cells as a predictor of cervical neoplasia in a high-risk population. Am J Obstet Gynecol 1988; 159:1517–1525.
106. Nuovo GJ, Blanco JS, Leipzig S, et al: Human papillomavirus detection in cervical lesions histologically negative for cervical intraepithelial neoplasia: correlation with Pap smear, colposcopy, and occurrence of cervical intraepithelial neoplasia. Obstet Gynecol 1990;75:1006–1011.
107. Nuovo GJ, Lastania D, Smith S, et al: Human papillomavirus segregation patterns in genital and non-genital warts in prepubertal children and adults. Am J Clin Pathol 1990; in press.

5

Intraepithelial and Invasive Squamous Cell Lesions of the Uterine Cervix

CHRISTOPHER P. CRUM AND PEYTON T. TAYLOR

The purpose of this chapter is to provide insight into a select group of pathologic entities that involve the uterine cervix and that are clinically relevant. Given the explosion in recombinant DNA technology, intraepithelial and invasive squamous cell neoplasms of the cervix are being viewed from new perspectives, in particular the relationship with papillomaviruses, as discussed in Chapter 4. The value of this new information to clinical management still remains to be determined. This chapter will explore several areas of interest and controversy, including the diagnostic approach to precursor lesions, the differential diagnosis of precursors with emphasis on borderline changes, the exclusion of invasive cancer, microinvasion, and variants of squamous cell carcinoma, including a separate entity, small-cell carcinoma. The goal of this chapter is to bring the clinician into the sphere of pathologic diagnosis. Given the complexities of the disease processes mentioned above, appropriate clinical management will increasingly become more dependent upon an educated clinician who has more than a casual exposure to pathology and is able to understand its strengths and limitations.

Anatomic Considerations

At the external os the ectocervix joins the endocervical canal, and at this point the "original" squamocolumnar (S-C) junction exists. During the reproductive years the original S-C junction tends to be found on the ectocervix because of the process of "eversion," which is a consequence of hormonal factors that influence the conformation of the cervix during fetal life, puberty, and particularly pregnancy. The process of eversion exposes the endocervix of the anterior (and to a lesser extent the posterior) cervix to the acidic pH of the vagina, which, in concert with other inflammatory stimuli, results in the replacement, or "transformation," of the endocervical columnar epithelium by squamous epithelium. As expected, this phenomenon occurs principally on the anterior and posterior cervical lips, in parallel with the site of the most prominent cervical eversion.[1,2] The gradual reduction of cervical eversion, combined with the advancement of the transformation zone, eventually places the S-C junction deep within the endocervical canal by the time of menopause.[3] The transformation zone is the principal site of origin for precancers and invasive squamous cell carcinomas of the cervix. Predictably, these neoplasms occur most commonly on the anterior and posterior cervical lips.[3]

Squamous Intraepithelial Lesions of the Cervix

In this chapter, we will introduce the recently described Bethesda Classification for cytologic diagnosis and adapt it to histology as well to classification of cervix precursor lesions. As discussed in Chapter 4, human papillomaviruses (HPV) have been associated with over 90% of intraepithelial and invasive squamous cell lesions of the cervix. Before discussing terminology for squamous precancers, it is useful to review the natural history of these lesions.

Natural History of Squamous Intraepithelial Lesions

The various intraepithelial lesions have traditionally been considered to comprise a continuum of change rather than actually representing morphologically or biologically distinct steps with inevitable progression between them[4] (Fig. 5-1). This view has been changed somewhat by the knowledge that specific HPV types are generally associated with higher grade lesions. Moreover, many lower grade lesions have been characterized morphometrically and by DNA microspectrophotometry; they are principally diploid or polyploid in DNA content, and have a tendency to regress.[5–8] In contrast, higher grade lesions frequently are biologically aneuploid, demonstrate greater degrees of cytologic atypia, and are more likely to progress.[5–9] Nevertheless, the consistent distinction of true precancerous lesions from cytologically and histologically similar lesions that are benign is not readily accomplished, notwithstanding the capacity of the above techniques to segregate lesions into general groups. The variability in natural history of histologically (and biologically) similar lesions may well be influenced by a multitude of factors, as outlined in Figure 4-1 (see Chap. 4).

Nasielle et al.[10,11] observed that slightly less than two thirds of mild dysplasia and one third of moderate dysplasia lesions regressed during follow-up. Problems encountered in most studies, however, include length of follow-up, arbitrary criteria for determining the grade of an individual lesion, and the potential alteration of the natural history of a lesion by biopsy.[10–12]

The time required for a lesion to "progress" from low grade to high grade or eventually to invasive cancer is not known and is impossible to determine by direct observation. It is not ethically permissible or pragmatically possible to take a large group of women who have what is believed to be a potentially dangerous and progressive epithelial lesion, *not* interrupt the natural history by biopsy or treatment, and follow them over many years to see what will happen. In order to try to understand what *might* happen, mathematical models based on data from several studies in women of different ages with different epidemiologic risk factors have been developed. Barron and Richart[13] have calculated the mean time to progression between grades of cervical intraepithelial neoplasia (CIN) to be approximately 5 years. The mean time to progression of high-grade (CIN III) intraepithelial lesions to invasive disease is also uncertain, and has been estimated to be from 1 to 30 years, with a reasonable estimate being 10 to 13 years.[14–16]

Role of Pathology in Clinical Management

The Papanicolaou smear is the principal tool by which women are selected for colposcopic examination, biopsy, and further study. By combining this with guided cervical biopsies, it is possible to segregate patients into those with unsuspected invasive cancer, those with lesions amenable to office/outpatient treatment, and those who require diagnostic and often therapeutic conization.

The process of diagnosing precursor lesions begins with analysis of the Papanicolaou smear. The value of the Papanicolaou smear lies in the sampling of the cervix at regular intervals.[17] It has been estimated that the mortality from cervical cancer can be reduced by at least 70% with screening every 3 years.[18–20] There is controversy over what the accepted interval is, although it is currently accepted that if a woman has two or three consecutive negative yearly smears, she may be followed at less regular intervals, depending upon her risk. However, it is es-

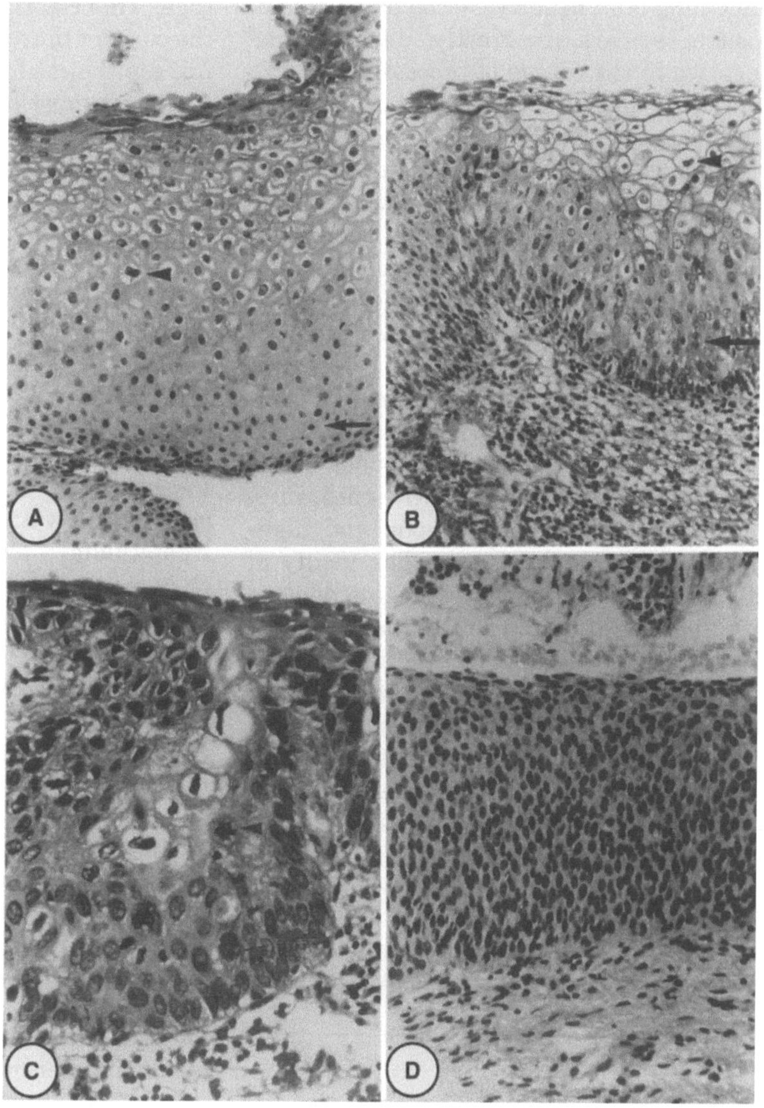

FIGURE 5-1. Spectrum of low- and high-grade squamous intraepithelial lesions (SIL). (A) A portion of exophytic cervical condyloma (low-grade SIL) with koilocytotic atypia and multinucleated cells in the upper cell layers (arrowhead). Note the evenly distributed parabasal cells with minimal nuclear atypia (arrow) (B). Flat condyloma (low-grade SIL) of the cervix, with koilocytotic atypia (arrowhead). In this lesion, the parabasal cells exhibit slightly greater nuclear atypia (arrow). (C) High-grade SIL (CIN II). Koilocytotic atypia is present in the center of the lesion. However, the parabasal cells contain prominent variation in nuclear size, shape, and staining, with a higher mitotic index (arrowheads). (D) High-grade SIL (CIN III).

timated that 37% of women with invasive cancer fall into this category.[18] Approximately one third of women with invasive cancer, however, have had a negative Papanicolaou smear in the past 5 years.[21] It is assumed that sampling error or reader error contributes to this lack of correlation in most cases, although it is possible that a subset of invasive cancers develop rapidly in the absence of a precursor lesion of long standing.[19,20]

The false-negative rates due to sampling have been reduced by the use of both the Ayers spatula and endocervical sampling, such as aspiration.[23] Recently the endocervical brush has become the preferred method for obtaining endocervical material, reducing the incidence of inadequate smears to less than 2%.[24] Determining the laboratory "false-negative" rate depends upon the parameters used in its calculation. For example, if errors are calculated based upon total smears processed, false-negative rates are usually less than 1%. However, if false negatives are calculated by dividing the false negatives by the total number of positives (including false negatives), the proportion of positive smears incorrectly diagnosed as negative in the course of processing a large volume of smears is placed at from 11 to 23%.[25]

Classification of Papanicolaou smear abnormalities has traditionally included normal (group I), uncertain (group II), abnormal consistent with precancer (group III), high-grade CIN (group IV), and invasive cancer (group V). This has recently been replaced by the Bethesda classification, in which groups II through V are divided into nonspecific atypia and abnormal,[26] as outlined in Table 5-1. The latter includes three groups: abnormal of uncertain significance, low-grade intraepithelial lesion and high-grade intraepithelial lesion. The utility of this approach is to clearly segregate nonspecific abnormalities and reduce the number of classifications for intraepithelial lesions. The latter terminology is designed to avoid the use of "intraepithelial neoplasia," which has been increasingly difficult to define in the face of HPV-related changes and condylomata. However, such terms as CIN or dysplasia may be included to clarify the message. This classification conforms loosely to the concept that there are low- and high-risk intraepithelial lesions as defined above, although it must be stressed that it is impossible to make these distinctions consistently by cytology alone.[26]

TABLE 5-1. Comparison of cytologic classifications for cervical intraepithelial lesions

Finding	By group	By grade	Bethesda classification
Negative	Group I	Negative	Negative
Uncertain	Group II	Narrative	Narrative
Abnormal	Group III	CIN	Abnormal
	IIIa	CIN I/condyloma[a]	Low-grade SIL[b]
	IIIb	CIN II	High-grade SIL
	IIIc	CIN III	
	Group IV	CIN III	
	Group V	Cancer	Cancer

From Crum CP, Taylor PT: Intraepithelial squamous lesions of the cervix, in Knapp RC, Berkowitz RS (eds): Gynecologic Oncology. New York, Macmillan, 1990. Copyright © 1990 Robert C. Knapp, M.D., and Ross S. Berkowitz, M.D. Reprinted by permission of Macmillan Publishing Company, a division of Macmillan, Inc.

[a] Flat or exophytic.

[b] SIL = squamous intraepithelial lesion.

Diagnosis

Because the distinction of precursors with low versus high risk is not always possible, the principal goal of the pathologist is to identify lesions that should be removed, rather than be concerned with nuances of grading. In fact, with the simplification in Papanicolaou smear interpretation to include just low and high grade squamous intraepithelial lesions (SILs), concurrent changes in the classification of biopsy material is timely. Hence, low-grade SILs will include flat and exophytic condylomata, and high-grade SILs will encompass those in the CIN II and CIN III categories (Table 5-2). This classification is useful for two reasons. First, it eliminates the use of two terms that are becoming increasingly confusing; condyloma and CIN. Some lesions in the former group fail to behave like benign warts, and some in the latter are not true precancers. The term "lesion" relieves us of this problem. In our laboratory, we do not use the term "koilocytotic atypia," because it may be found in either group and is not an important factor in determining therapy.

Low-Grade Squamous Intraepithelial Lesions

Low-grade SILs include flat and exophytic condylomata. The typical cervical condyloma may vary in appearance from flat to slightly raised to that resembling condylomata acuminata of the vulva. Accordingly, the degree of papillomatosis observed histologically will vary. The distinguishing features consist of thickening of the epithelium (acanthosis) and koilocytotic atypia in the middle and upper portions of the epithelium. Koilocytotic atypia is defined as the presence of nuclear atypia with variation in nuclear size and shape, wrinkling of nuclei, poly-

TABLE 5-2. Selection of criteria for cervical diagnosis

Findings	Diagnosis	Management
Negative, no transformation zone seen	Descriptive	Repeat if indicated
Acanthosis, parakeratosis, nonspecific halos, atrophy	Descriptive	Follow
Severe inflammation, reparative atypia	Descriptive	Culture, follow
Koilocytosis, maturation, minimal basal atypia	Low-grade squamous intraepithelial lesion (flat or exophytic condyloma, (CIN I)	Remove
Koilocytosis, maturation, diffuse atypia	High-grade squamous intraepithelial lesion (CIN II)	Remove
Minimal koilocytosis or maturation, diffuse atypia	High-grade squamous intraepithelial lesion (CIN III)	Remove
Neoplastic epithelium in the ECC[a]	Strips of neoplastic squamous epithelium	Cone biopsy
Condyloma in the ECC	Descriptive	Optional[b]
Very scant condyloma or neoplastic epithelium in the ECC	Descriptive	Optional[b]

From Crum CP, Taylor PT: Intraepitheleal squamous lesions of the cervix, in Knapp RC, Berkowitz RS (eds): Gynecologic Oncology. New York, Macmillan, 1990. Copyright © 1990 Robert C. Knapp, M.D. and Ross S. Berkowitz, M.D. Reprinted by permission of Macmillan Publishing Company, a division of Macmillan, Inc.

[a] ECC = endocervical curettage.

[b] Management will depend upon the clinical assessment of the endocervical canal and the lesion.

chromatism and binucleate forms, and perinuclear halos. The perinuclear halos tend to vary as well in appearance, shape, and conformation, with a distinct zone of clearing between the nucleus and cytoplasmic membrane. Important features of exophytic condyloma are the presence of modest nuclear atypia in the lower half of the epithelium, a low mitotic index, and absence of bizarre or abnormal mitoses (Fig. 5-1A). Flat condylomata have a similar appearance, although the parabasal epithelium may be slightly more atypical (Fig. 5-1B). Accordingly, this group of lesions may be associated with HPV type 16 nucleic acids, as described in Chapter 4.[8,27,28]

High-Grade Squamous Intraepithelial Lesions

High-grade SILs include two variants of the same spectrum. The first is CIN with koilocytotic atypia (CIN II, moderate dysplasia). Lesions in this category exhibit features of both condyloma and aneuploid epithelium (Fig. 5-1C). Cell maturation is present in the upper half of the epithelium, usually with koilocytotic atypia. The koilocytes may be identical to those in flat condyloma, but commonly differ by the presence of smaller, more concentric halos, and dense, hyperchromatic and pleomorphic nuclei. The epithelial surface of CIN II and CIN III lesions frequently contains horizontally arranged parakeratotic cells with abnormal nuclei. In addition, nuclear atypia is present in the lower half of the epithelium in at least a portion of the lesion. The picture may be of a mixed lesion, with a combination of flat condyloma and CIN.[29–32]

The second variant is CIN without koilocytotic atypia (CIN III, severe dysplasia–carcinoma in situ). These lesions contain minimal evidence of maturation or koilocytotic atypia, although they may be combined with areas resembling CIN I or II (Fig. 5-1D). Cells with a high nucleus-to-cytoplasm ratio are present throughout the epithelium. Paradoxically, high-grade CIN lesions often exhibit a more homogeneous population of neoplastic cells, with less variation in nuclear size and staining than lower grade CIN lesions. However, the nuclei are crowded, enlarged, and hyperchromatic and contain a higher mitotic index than conventional basal cells (Figs. 5-1C and 5-1D).

Differential Diagnosis

Nonspecific Cellular Changes Mimicking Koilocytotic Atypia

The increased emphasis on recognizing cervical HPV infections has brought with it the expected problems of differentiating HPV-specific changes from nonspecific halos in the superficial cells. Here the reader should picture the scenario that results when a patient is referred for suspected cervical or vaginal warts, based either on a Papanicolaou smear or direct visualization of wartlike lesions. The pathologist will carefully scrutinize the biopsy based upon the clinical information, and may make a diagnosis of suspected HPV infection (condyloma, low-grade SIL, etc.) based upon the presence of epithelial acanthosis or cytoplasmic halos resembling koilocytes. Because of the clinical and social ramifications of a diagnosis of HPV-related disease, it is important that minimal changes not be overcalled as HPV infection or a related lesion. There is no evidence that such lesions are dangerous, and they frequently do not contain HPV nucleic acids.[33,34] Examples are illustrated in Figures 5-2A and 5-2B. The important feature that justifies a diagnosis of a condyloma or low-grade SIL includes the presence of nuclear atypia. The terms "koilocytosis" and "koilocytotic atypia" should be held synonymous and applied only when nuclear atypia is present (Figs. 5-1A and 5-1B).

Inflammation and Infection

The inflammatory reaction produced in either acute or chronic cervicitis is usually nonspecific and the diagnosis is made following analysis of wet mounts, direct smear, culture, or the cytology report. Cervicitis is characterized by ulceration, reparative atypia, and a mixed inflammatory infiltrate in the stroma consisting of polymorphonuclear leukocytes, plasma cells, lymphocytes, and other lymphoreticular cells (Figs. 5-2C and 5-2D). Lymphoid follicles may also be present, prompting the diagnosis of follicular cervicitis. When severe inflammation is present, and in particular when it consists of lymphoreticular cells, the diagnosis of chlamydia should be considered and excluded.[35–37] Chlamydia infect endocervical columnar and immature metaplastic cells, executing their life cycle in inclusion vacuoles that eventually produce cell lysis. It is invariably associated with marked inflammation. Despite the fact that chlamydia produces cytoplasmic vacuoles in infected cells, the presence of cytoplasmic vacuoles in biopsies or Papanicolaou smears is not diagnostic, because of the nonspecific nature of such changes.[38] The diagnosis of chlamydia is confirmed either by culture or by direct immunofluorescence for chlamydia organisms.

Inflammation will induce epithelial changes with cytologic atypia, the most common of which is anaplasia or repair or reparative atypia (Fig. 5-2C and 5-2D). This is characterized by enlarged nuclei, inflammatory cells in the epithelium, and the presence of mitoses. In contrast to CIN, nucleoli are present and there is minimal nuclear crowding or overlap on the histologic section.[3]

Other forms of reactive atypia that may or may not be associated with inflammation include acanthosis, parakeratosis, and occasionally perinuclear halos and mild nuclear atypia. These lesions may be difficult to distinguish from flat condyloma (Fig. 5-2A and 5-2B).

Postmenopausal Epithelial Changes

Another group of lesions that may be confusing diagnostically occur in postmenopausal women. In these patients there are three abnormalities that may be associated with abnormal Papanicolaou smears. The first is *intraepithelial lesions of all grades*. Postmenopausal patients may be sexually active and it is not uncommon to see such lesions in these patients. The second group are *atrophic lesions*. In contrast to intraepithelial lesions, atrophy is characterized by a uniform population of immature cells with a very low mitotic index.[3] The third group are those associated with *low-grade abnormal-*

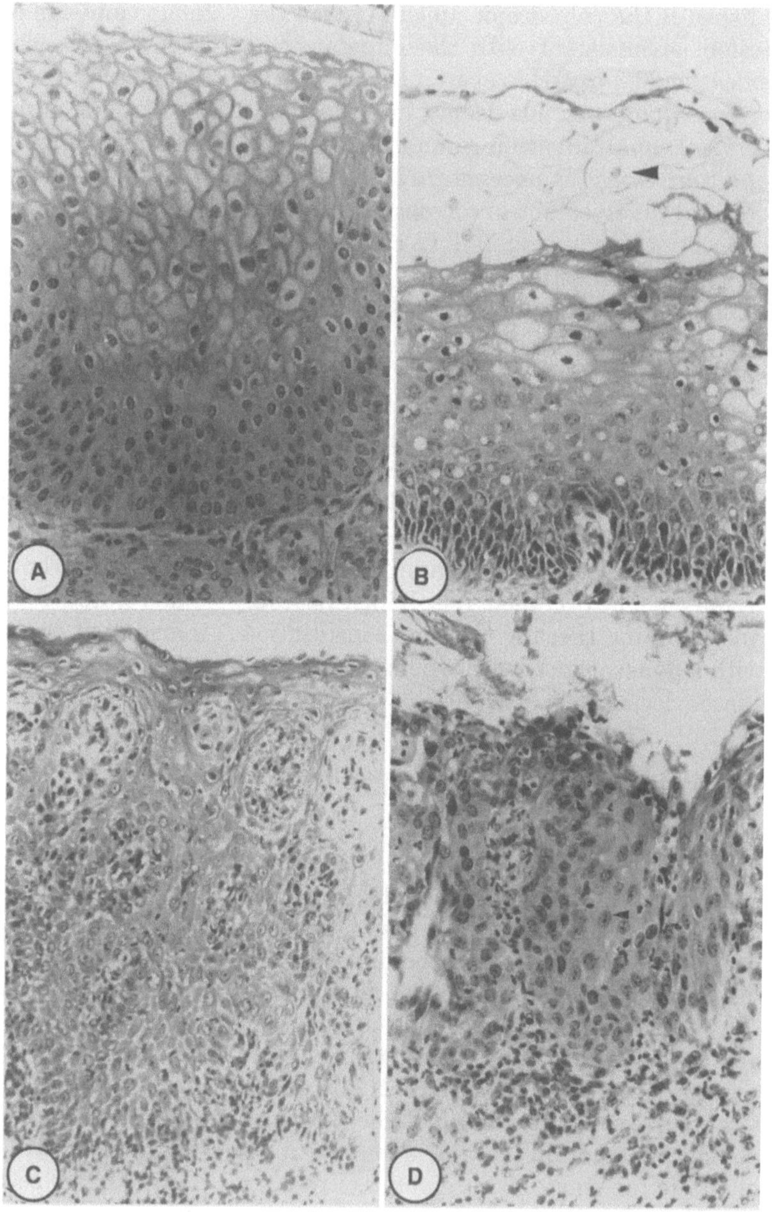

FIGURE 5-2. Differential diagnosis of squamous intraepithelial lesions. (A) Cytoplasmic halos with minimal nuclear atypia. A portion of these lesions are associated with HPV, but the association is not consistent. Hence these changes are diagnosed as nonspecific. (B) Altered maturation of portio epithelium with cytoplasmic halos but no nuclear atypia (i.e., nonspecific). (C) Inflamed transformation zone epithelium with minimal nuclear atypia in the surface cells. (D) Reparative changes in endocervical epithelium. The nuclei tend to be well spaced and, although atypical, do not exhibit coarse chromatin. Discrete chromocenters (small dots) are frequently present in individual nuclei (arrowhead).

ities on the Papanicolaou smear, unremarkable colposcopic exam, and a vaginal smear suggesting a high estrogen effect. The pathogenesis of these changes is not clear. In our experience many do not appear to be HPV related, but will be associated with repeated low-grade abnormal Papanicolaou smears and may be particularly frustrating to the clinician. Whether they represent epithelial alterations peculiar to menopause remains to be proven.

Endocervical Curettage

The clinical management of cervical precancers depends upon both the identification of disease that must be removed and the exclusion of invasive cancer. This can be accomplished if the colposcopic appearance of the lesion is consistent with the Papanicolaou smear and biopsy. However, if the precancer or condyloma extends deeply into the endocervical canal, invasion cannot be ruled out and cone biopsy is necessary. The suspicion of canal involvement is corroborated by the endocervical curettage. If it contains free strips of neoplastic squamous epithelium, cone biopsy is necessary. If the curettage contains strips of condyloma, conization should be considered seriously because associated invasive cancer cannot be excluded. One possible exception is a curettage that contains very small amounts of neoplastic or condylomatous epithelium in the context of a clinically negative endocervical canal. In such a case inadvertent sampling of a portio lesion must be excluded. Resolving this issue and planning therapy requires consultation with a gynecologist, with whom the therapy can be decided.

Microinvasive Squamous Cell Carcinoma

Approximately 4 to 7% of CIN lesions are associated with superficial invasion.[39,40] The basis for designating a subset of invasive cervix carcinoma as microinvasive carcinoma is the assumption that a portion of early invasive cancers can be identified reproducibly and can be approached differently from conventional invasive cancer. Radical hysterectomy or radiation therapy is the treatment of choice for stage 1B carcinoma of the cervix. Identification of a subset that carries no risk of lymph node metastases (stage IA) and that can be treated by simple hysterectomy carries a decided advantage. Currently, approximately 10% of invasive carcinomas are microinvasive when diagnosed.[41–43]

There are several issues of importance or interest concerning microinvasive carcinoma of the cervix. They include (a) identifying invasion, (b) distinguishing it from mimics, and (c) applying correctly the criteria for microinvasion. It is important to emphasize that the diagnosis of microinvasion is histologic and can only be made on a cone specimen. The specimen must have negative margins and the pathologist must examine a sufficient number of sections, usually one for every 2 mm of cone thickness, preferably with three levels from each block.

Diagnosis of Invasion

Identifying invasion in the routine biopsy specimen is the principal goal of the pathologist. The classical criteria for invasion include a desmoplastic response in the adjacent stroma, focal conspicuous maturation of the neoplastic epithelium with prominent nucleoli, blurring of the epithelial-stromal interface, and loss of palisading nuclei at the epithelial-stromal border. Two additional and related features are scalloping of the margins at the epithelial-stromal interface, and the apparent "folding or duplication" of the neoplastic epithelium. These features are helpful when faced with an intense inflammatory response, which may obscure desmoplasia on one hand and blur the epithelial-stromal interface on the other[44] (Fig. 5-3).

Differential Diagnosis of Invasion

In a review of 265 cases of presumed microinvasion sent to the Gynecologic Oncology Group it was found that nearly 40% were overdiagnosed examples of intraepithelial lesions.[45] The most important mimics of microinvasion are gland involvement that is tangentially sectioned, cautery or crush artifact, and previous biopsy sites. Intraepithelial lesions associated with underlying inflammation, either from secondary infection or previous biopsy, must be evaluated carefully to avoid the overdiagnosis of invasion when the epithelial-stromal interface is disrupted by the inflammatory process.

Confirming "Microinvasion"

Cone biopsy is necessary if (a) the lesion does not appear grossly invasive clinically or colposcopically and (b) it is not clearly deeper

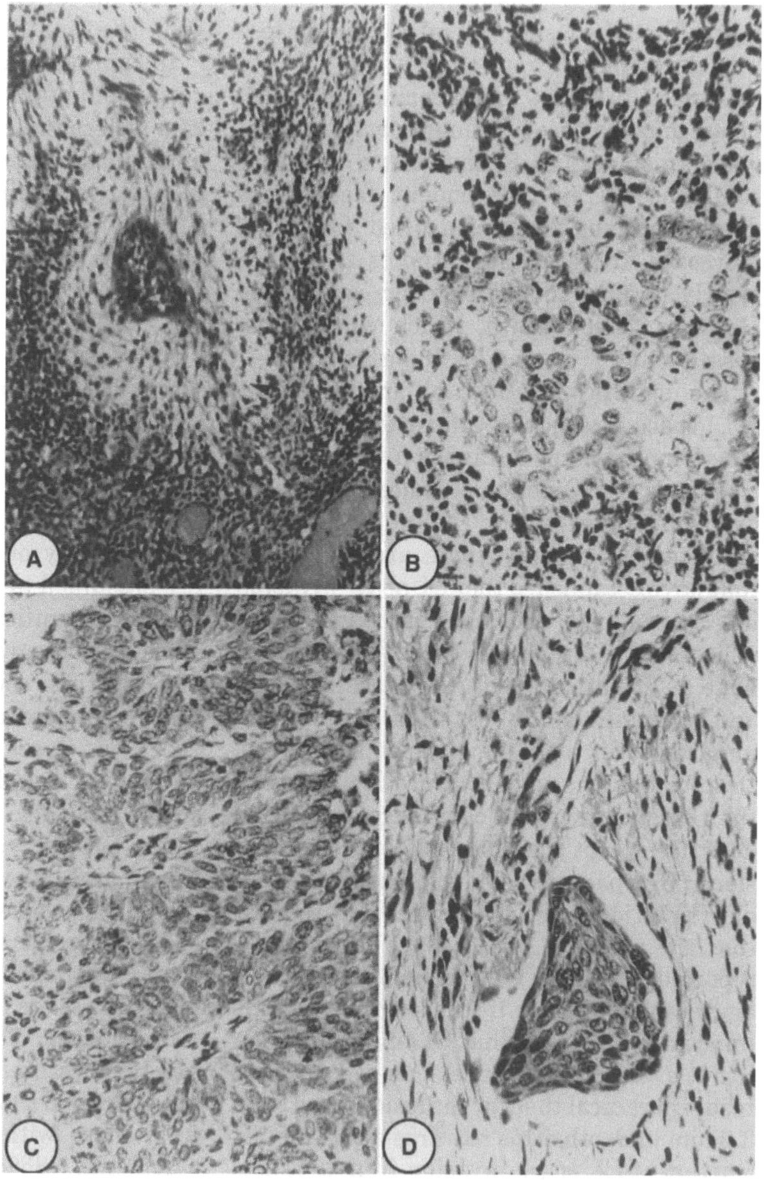

FIGURE 5-3. Diagnosis of invasive squamous cell carcinoma. (A) A prominent desmoplastic response surrounding a small focus of invasion (arrowheads) is characterized by young, widely spaced fibroblasts. (B) A focus of invasion that illustrates the loss of cell polarity at the margins of the neoplastic epithelium. Compare to the stromal epithelial interface in Figures 5-1C and 5-1D. (C) Folding and duplication of epithelium, signaling invasion. (D) Capillary-lymphatic space invasion. The role of this phenomenon in management remains unclear. (Figures 5-1A, 5-1B, and 5-1C from Crum CP, Nuovo G: The cervix, in Sternberg S (ed): Diagnostic Surgical Pathology. New York: Raven Press, 1989, pp. 1571, 1573. Reprinted by permission.)

than 3 mm in the original biopsy. Once the cone biopsy is performed, the measurement of depth of invasion should be made from the most superficial epithelial-stromal interface of the adjacent intraepithelial process. This is best accomplished using an ocular micrometer. It is not always possible, however, to accurately estimate depth because of specimen distortion.

Three issues of potential concern when considering microinvasion are tumor depth, confluence of growth pattern, and capillary-lymphatic space invasion. Although microinvasion was originally defined as any lesion less than 5 mm in depth, the risk of lymph node metastases increases with depths over 3.0 mm. The risk is from 0 to 0.9% for lesions invading equal to or less than 3.0 mm, in contrast to 0 to 13.9% for lesions invading between 3.1 and 5.0 mm.[46–

[49] Hence, a diagnosis of microinvasion requires that the lesion extend 3.0 mm or less into the stroma.

Despite a report emphasizing the prognostic importance of confluent patterns of invasion, some authors have not found that confluence is an independent factor once depth of invasion is controlled for.[46,48,49] Moreover, there are few studies that have clearly defined the criteria for confluence. However, lesion width, which will be discussed below,[45,50] may be important.

The significance of capillary-lymphatic (CL) space invasion is controversial. Capillary-lymphatic space invasion increases in frequency as a function of lesion depth and, by itself, increases the risk of lymph node metastases. However, there are no studies that have established that CL space invasion is a critical factor in lesions of 3.0 mm depth or less.[47–49] Van Nagell et al.[47] found that none of 17 patients with CL space invasion and lesions invading less than 3.0 mm had metastases in their lymph node specimens. Despite this, the nomenclature committee of the Society of Gynecologic Oncology (SGO) does not accept the diagnosis of microinvasion if CL space invasion is present. In standard practice, most oncologists will request that the presence of CL space invasion be reported, and they will probably opt for an aggressive approach if it is seen. Thus, it is important to ensure that CL space invasion is not overcalled, because such reports may result in radical therapy for lesions less than 3.0 mm in depth. This requires a careful distinction of CL space invasion from artifacts due to inflammation or retraction.[48]

The concept of tumor volume as a more accurate predictor of recurrence and metastases has been proposed by Bughardt and Holzer. They have found that lesions less than 420 mm^3 rarely recur.[50] Unfortunately, determining tumor volume is tedious and not universally accepted. In lieu of this approach the greatest width of the lesion can be determined and reported by examining the histologic sections. There is an association between width, recurrence, and metastases, although the precise limits are unclear. A width of 10 mm is proposed by some as a limit for conservative therapy with lesion less than 3.0 mm in depth.[51]

In summary, a diagnosis of microinvasive carcinoma requires an invasive lesion of 3.0 mm or less in depth with no evidence of CL space invasion and free margins on the cone biopsy. Determination of greatest width is also advisable, although the precise width that should serve as a cutoff is unclear.

Squamous Cell Carcinoma and Its Variants

Conventional Squamous Cell Carcinoma

The majority of squamous cell carcinomas are presumed to evolve from a precancerous lesion. A significant proportion of CIN III lesions will progress to cancer if untreated, and the time course for this evolution has been estimated to range from 3 to over 20 years.[15,22] Cases of squamous cell carcinoma developing rapidly without a defined precursor have also been reported rarely.[53] The mean age for patients with cancer is approximately 51 years, in contrast to approximately 28 years for those with CIN III. A subset of invasive cancers has been termed "occult carcinoma." These are clinically inapparent, stage IB lesions of greater than 3.1 mm in depth. The mean age for this group has been estimated at 43 years, and the 5-year survival (96%) distinguishes this group from those with clinical stage IB invasive carcinoma (86%).[39] Accordingly, patients with occult disease can be managed with radical hysterectomy and lymph node dissection.

Survival is most closely related to the stage of the disease when diagnosed, which correlates closely with the risk of regional lymph node metastases. Spread of the carcinoma occurs principally through the lymphatic drainage of the cervix, which consists of superficial and deep lymphatics draining to the iliac and obturator, hypogastric and common iliac, and sacral lymph nodes, as well as lymph nodes in the posterior bladder wall.[1] Approximately two thirds of invasive squamous carcinomas are stage I or II when diagnosed. The actuarial 5-year survival drops abruptly from over 70% for stage II to 30 to 35% for stage III neoplasms.[2]

Squamous cell carcinomas have been classified according the degree of squamous differentiation (grades I through III) or according to cell type. Reagan et al.[54] subdivided squamous cancer of the cervix into (a) large-cell keratinizing, (b) large-cell nonkeratinizing, and (c) small-cell carcinoma. The basis for this distinction has been the observation that large-cell keratinizing carcinomas are radioresistant relative to nonkeratinizing carcinomas, and that small-cell carcinomas have the worst overall prognosis. In a review of five large series by Reagan and Fu,[55] the average 5-year survival for stage I tumors treated by radiation therapy was 54, 84, and 42% for keratinizing, nonkeratinizing, and small-cell carcinomas. The distinction between large-cell keratinizing and nonkeratinizing squamous cell carcinoma is based primarily upon the presence of intercellular bridges and keratin pearls in the former, although focal individual cell keratinization may be present in the latter.[56] The small-cell group has been more precisely defined and now consists of neoplasms that are morphologically and functionally identical to small-cell undifferentiated carcinoma (oat cell carcinoma, argyrophillic carcinoma, neuroendocrine carcinoma).

Excepting small-cell undifferentiated carcinomas, classification according to cell type is not universally accepted. Not all authors have observed differences in survival between keratinizing and nonkeratinizing tumors. Randall et al.[56] found that keratinizing tumors had a greater tendency to recur locally after radiotherapy, but that the frequency of distant metastases was the same for both cell types. An alternative, and accepted, approach to grading is to classify squamous cell carcinomas into well, moderate, and poorly differentiated categories.

Verrucous Carcinoma

Verrucous carcinomas are very rare lesions.[57] The criteria for diagnosis are the same as in the vulva, and the diagnosis is one of exclusion. Verrucous carcinomas usually present as large sessile lesions resembling condylomata. Histologically, they consist of hyperplastic-appearing lesions lacking the more delicate architecture of condylomata, and demonstrating instead columns of well-differentiated epithelium expanding into the underlying stroma. The pattern of invasion is blunt with minimal nuclear atypia at the epithelial-stromal interface. An intense inflammatory infiltrate has been associated with verrucous carcinomas of the cervix, but is, in itself, nonspecific. The differential diagnosis includes large exophytic condylomata with crypt involvement and well-differentiated squamous cell carcinomas. The latter usually exhibit finger-like or angulated invasive tongues. The presence of filiform papillary projections or marked nuclear atypia at the epithelial-stromal interface rules out the diagnosis of verrucous carcinoma. Given the extreme rarity of this lesion, the diagnosis of verrucous carcinoma must be made with caution.

Because verrucous carcinomas present grossly as large, sessile, wartlike growths, the diagnosis may be difficult without multiple biopsies or hysterectomy. Local excision is not usually possible and extension into the adjacent pelvic tissues may occur. Very few cases are available to determine metastatic potential, although the few reported cases did not metastasize to lymph nodes. Human papillomavirus DNA sequences, including types 6 and 16, have been isolated from verrucous carcinomas, primarily in the vulva and larynx.[58,59]

Papillary Neoplasms

Papillary lesions have been described in the cervix that range from those resembling transitional cell papillomas to those diagnosed as papillary carcinoma in situ.[60–62] Although the former are not invariably associated with invasive cancer, a number of cases have been observed of filiform papillomas of the cervix in which deeper sampling disclosed invasive cancer.[62,63] For this reason, any diagnosis of transitional papilloma should be made carefully, and be based upon local excision of the lesion, to rule out invasion. Papillary lesions containing marked squamous atypia (carcinoma in situ) may be associated with invasion and probably should be considered invasive until proven otherwise. The differential diagnosis, as with verrucous carcinoma, includes condylomata, although the latter do not exhibit delicate finger-like proliferations. An occasional diagnostic problem is the presence of immature metaplasia overlying endocervical papillae, or in association with condylomata. In this case, the diagnosis of papillary neoplasia can be excluded by the presence of endocervical columnar cells in areas of the lesion, which identify an origin in metaplastic epithelium rather than a papillary neoplasm (Fig. 5-4A).

Deep Endocervical/Endometrial Condylomata with Invasive Cancer

Rare reports have described classical (or almost classical)–appearing condylomata extending deeply into the endocervical canal or endometrium, some of which have been associated with invasive cancer[64] (B. Stein-

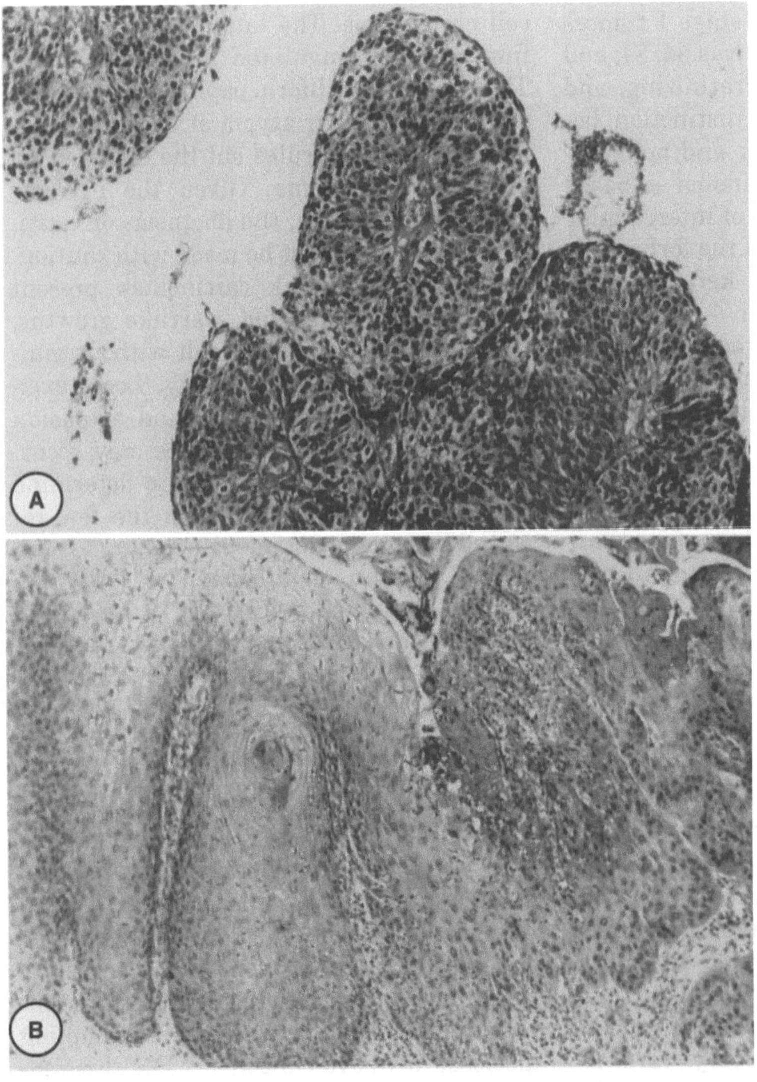

FIGURE 5-4. Two rare lesions that may present as a difficult diagnosis. (A) Papillary "carcinoma-in-situ," which when present frequently is associated with an underlying invasive cancer. If lesions such as this are found on superficial biopsy, rebiopsy may be necessary to rule out invasive cancer. (From Crum CP, Nuovo G: The cervix, in Sternberg S (ed): Diagnostic Surgical Pathology. New York: Raven Press, 1989, p. 1574. Reprinted by permission.) (B) A deep endocervical "condyloma" (left) that merges abruptly into an invasive cancer. Although rare, such lesions underscore the importance of critically evaluating any lesion that extends deeply into the endocervical canal.

berg and B. W. Winkler, personal communication). Another rare variant is CIN extending into the endometrial cavity.[65] Lesions of this type are usually picked up on the endocervical or endometrial curettage. Hence, if the endocervical curettage contains abundant fragments of condyloma, or any papillary neoplasm, a cone biopsy or hysterectomy should be strongly considered to confirm or exclude invasive cancer (Fig. 5-4B).

Undifferentiated Small-Cell Carcinoma

Interest in undifferentiated small-cell carcinoma has increased in recent years, primarily because it exhibits specific histologic features and may be associated with peptide hormone production, a phenomenon commonly found in oat cell carcinomas of the lung. Moreover this carcinoma, like its counterpart in the lungs, appears to have a propensity for rapid metastasis and radiosensitivity and carries a high mortality.[66]

Small-cell carcinoma also shares many of the histologic features of small-cell carcinoma of the lung. It is composed of a small, uniform cell population with hyperchromatic nuclei and a very high nucleus-to-cytoplasm ratio. The tumor cells form irregular, loose aggregates, often with little cohesion. The nuclei contain a coarse chromatin pattern and, because there is little cytoplasm, they often appear to "mold" as they are artifactually compressed in tissue sections, a feature common in small-cell carcinoma of the lung (Fig. 5-5A). The lesions tend to be small, but they extensively infiltrate the underlying cervical stroma. Hence, they may appear more as an indurated mass than as a fungating lesion. Additional histologic features that distinguish this entity are vascular invasion, observed in up to 93% of cases by Van Nagell and coworkers,[66] and conspicuous lack of coexisting inflammation in contrast to most cases of conventional squamous cell carcinoma. Furthermore, by virtue of their rapid growth, these neoplasms often are characterized by broad zones of necrosis, a feature less common in well-differentiated squamous cell carcinoma[66] (Fig. 5-5B).

One of the more intriguing aspects of small-cell carcinoma is its association with peptide hormone production, which can be exhibited by ultrastructural or immunohistochemical techniques in a portion of the neoplasms.[67] Such neoplasms have been called argyrophilic, oat cell, or neuroendocrine carcinomas, or poorly differentiated carcinoids.[68–71] These observations have led to the assumption that both small-cell undifferentiated carcinomas of the cervix and carcinoid tumors of the cervix are similar to neoplasms capable of amine precursor uptake and decarboxylation.[72] These so-called APUDomas are characterized by peptide hormone induction, the propensity to take up silver salts in special stains (argyrophilia) and, in a portion, dense core granules on electron microscopy.[70] As a result of peptide hormone production they have the ability to induce certain clinical syndromes. The best known of these is the carcinoid syndrome caused by the production of serotonin, but neoplasms of the cervix, like their counterparts elsewhere, have also been associated with the production of adrenocorticotropic hormone, insulin, and parathormone.[68,72]

The origin of these neoplasms has been a matter of debate. Originally, it was believed that all APUDomas may arise from cells with a common embryonic source—the neural crest.[72] Evidence that this process could occur in the cervix was supported by Fox et al.'s[73] observation that such cells did indeed normally inhabit the cervix. Two other observations suggest, however, that small-cell undifferentiated carcinomas of the cervix, including those functional neoplasms, not only do not arise from preexisting, normal argyrophilic cells of the cervix, but also probably do not arise from cells originating in the neural crest. First, certain neoplasms occurring de novo may contain argyrophilic cells, including those mucinous cystadenomas of the ovary that contain intestine-like epithelium.[72] Second, carcinoids

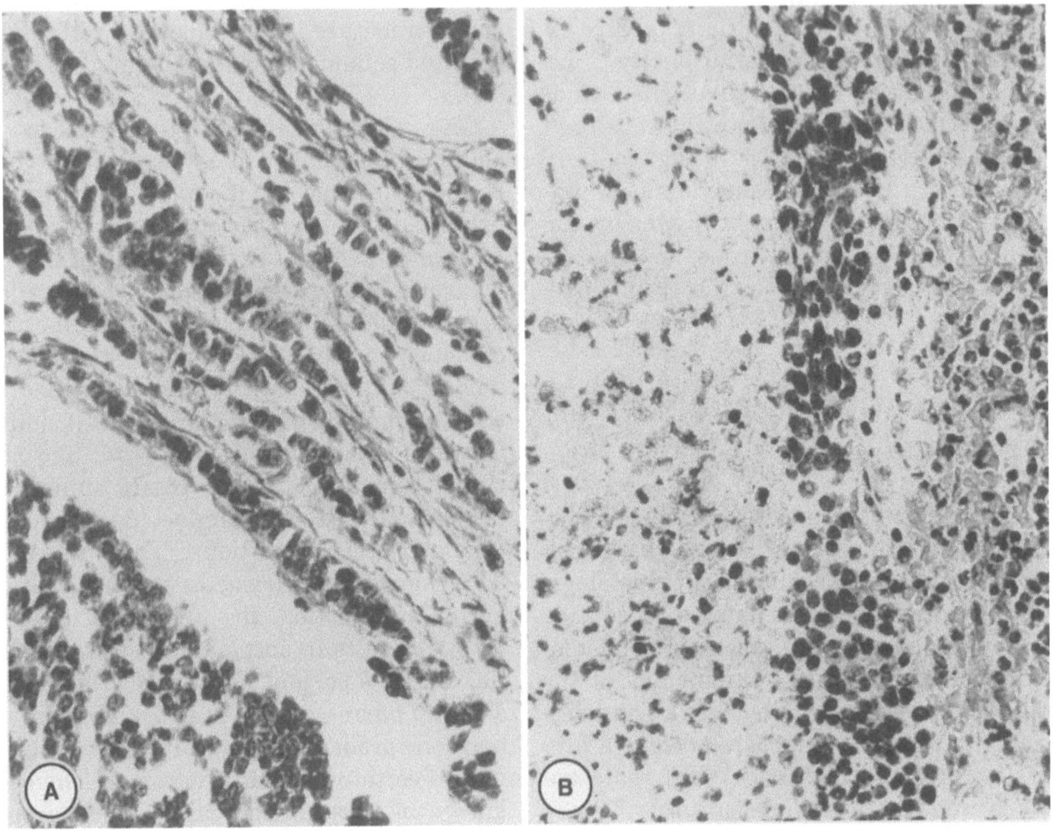

FIGURE 5-5. Small-cell undifferentiated carcinoma of the cervix. (A) The neoplasm consists of small cells arranged in cords, nests, or irregular aggregates with poor cohesion. Cells may abut closely on one another, producing nuclear molding. (From Crum CP, Nuovo G: The cervix, in Sternberg (ed) Diagnostic Surgical Pathology. New York: Raven Press, 1989, p. 1583. Reprinted by permission.) (B) Extensive necrosis surrounding a central focus of viable appearing tumor. Small-cell undifferentiated tumors frequently contain these features, frequently with the notable absence of an inflammatory cell infiltrate.

and small-cell carcinomas of the cervix have been observed in association with conventional neoplasms, including adenocarcinomas in situ, invasive adenocarcinomas, and CIN.[74] Thus, it appears that, rather than originating from a specific argyrophilic cell, small-cell carcinomas and carcinoids probably represent "selective differentiation," often within a preexisting, more conventional carcinoma. This observation has been supported by findings in other organs where conventional carcinomas blend with the argyrophilic neoplasms.[75]

Although certain small-cell neoplasms of the cervix may exhibit functional characteristics, the major criterion for distinguishing them from conventional squamous cell carcinomas is histologic. Small-cell morphology is not itself the most important consideration in making the diagnosis of small-cell carcinoma. Many squamous cell carcinomas are composed of relatively small cells, but these maintain the infiltrating pattern of a conventional squamous cell cancer. In contrast, small-cell undifferentiated carcinomas of the cervix not only are composed of small cells, but also have the cellular characteristics and pattern of infiltration described above. Intermediate and well-differentiated (carcinoids) variants have also been described, but with the exception of anecdotal reports the prognosis is poor.[69]

It is important to emphasize that Grimelius stains are not uniformly positive in small-cell carcinomas and ultrastructural studies may not demonstrate the dense core cytoplasmic granules associated with these neoplasms.[67,69] In most instances, application of histologic criteria alone will identify small-cell carcinomas. The recognition of this tumor is important, given its radiosensitivity, potential chemosensitivity, and metastatic potential.

Recently, Stoler et al.[76] have identified HPV-18 nucleic acids in a high proportion of small-cell carcinomas of the cervix. As mentioned in Chapter 4, the association of HPV-18 with these neoplasms suggests that this virus may be a marker for more aggressive neoplasms of the cervix.

Acknowledgments. Supported in part by grants from the National Institutes of Health AI00628 and American Cancer Society (to Dr. Crum). Dr. Crum is a recipient of a Physician Scientist Award from the National Institute of Allergy and Infectious Disease (AI06628).

References

1. Krantz KE: The anatomy of the human cervix, gross and microscopic, in Blandau RJ, Moghissi K (eds): The Biology of the Cervix. Chicago, University of Chicago Press, 1973, pp 57–59.
2. Ferenczy A: Anatomy and histology of the cervix, in Blaustein A (ed): Pathology of the Female Genital Tract. New York, Springer-Verlag, 1982, pp 119–135.
3. Richart RM: Cervical intraepithelial neoplasia, in Sommers SC (ed): Pathology Annual. New York, Appleton-Century-Crofts, 1973, pp 301–328.
4. Ferenczy A, Winkler B: Cervical intraepithelial neoplasia and condyloma, in Kurman, R (ed): Blaustein's Pathology of the Female Genital Tract. ed 3. New York, Springer Verlag, 1987, pp 177–217.
5. Bibbo M, Dytch HE, Alenghat E, et al: DNA ploidy profiles as prognostic indicators in CIN lesions. Am J Clin Pathol 1989;92:261–265.
6. Fu YS, Reagan JW, Richart RM: Definition of precursors. Gynecol Oncol 1981;12:s220–s231.
7. Fu YS, Braun L, Shah KV, et al: Histologic, nuclear DNA and human papillomavirus study of cervical condylomas. Cancer 1983; 52:1705–1711.
8. Fu YS, Huang I, Beaudenon S, et al: Correlative study of human papillomavirus DNA, histopathology, and morphometry in cervical condyloma and intraepithelial neoplasia. Int J Gynecol Pathol 1988;7:297–307.
9. Wilbanks GD, Richart RM, Terner JY: DNA content of cervical intraepithelial neoplasia studied by two-wavelength Feulgen cytophotometry. Am J Obstet Gynecol 1967;98:792–799.
10. Nasielle K, Nasielle M, Vaclavinkova V: Behavior of moderate cervical dysplasia during long term follow-up. Obstet Gynecol 1983; 61:609–614.
11. Nasielle K, Roger V, Nasielle M: Behavior of mild cervical dysplasia during long term followup. Obstet Gynecol 1986;67:665–669.
12. Richart RM, Barron BA: A followup study of patients with cervical dysplasia. Am J Obstet Gynecol 1969;105:386–393.
13. Barron BA, Richart RM: A statistical model of the natural history of cervical carcinoma based on a prospective study of 557 cases. JNCI 1968;41:1343–1353.
14. Barron BA, Cahill MC, Richart RM: A statistical model of the natural history of cervical neoplastic disease: the duration of cervical carcinoma-in-situ. Gynecol Oncol 1978; 6:196–205.
15. Coppelson LW, Brown B: Observations on a model of the biology of carcinoma of the cervix. Am J Obstet Gynecol 1975;122:127–136.
16. Gustafsson L, Adami H-O: Natural history of cervical neoplasia: consistent results obtained by an identification technique. Br J Cancer 1989;60:132–141.
17. Guzick DS: Efficacy of screening for cervical cancer: a review. Am J Public Health 1978; 68:125–134.
18. MMWR: Cervical cancer control–Rhode Island.
19. MMWR: Chronic disease reports: deaths from cervical cancer–United States, 1984–1986.
20. National Cancer Institute Cancer control ob-

jectives for the nation: 1985–2000 Bethesda, MD, US Department of Health and Human Services, publication No (NIH) 86-2880 (NCI Monographs, no 2). Public Health Service, 1986.
21. Dunn JE, Crocker DW, Rube IF, et al: Cervical cancer occurrence in Memphis and Shelby County, Tennessee, during 25 years of its cervical cytology screening program. Am J Obstet Gynecol 1984;150:861–864.
22. Kurman RJ, Shiffman RM, Lancaster WD, et al: Analysis of individual human papillomavirus types in cervical neoplasia: a possible role for type 18 in rapid progression. Am J Obstet Gynecol 1988;159:293–296.
23. Richart RM: An evaluation of the "true" false-negative rate in cytology. Am J Obstet Gynecol 1964;89:723–726.
24. Taylor PT, Andersen WA, Barber SR, et al: The screening Papanicolaou smear: contribution of the endocervical brush. Obstet Gynecol 1987;70:734–738.
25. Quality assurance in cervical cytology: the Papanicolaou smear. JAMA. 1989;262:1672–1679.
26. National Cancer Institute Workshop: The 1988 Bethesda System for reporting cervical/vaginal cytologic diagnoses. JAMA 1988; 262:931–934.
27. Willett GD, Rurman RJ, Reid R, et al: Correlation of the histological appearance of intraepithelial neoplasia of the cervix with human papillomavirus types. Int J Gynecol Pathol 1989;8:18.
28. Durst M, Gissman L, Ikenberg H, zur Hausen H: A papillomavirus DNA from a cervical carcinoma and its prevalence in cancer biopsy samples from different geographic regions. Proc Natl Acad Sci USA 1983;80:3812–3815.
29. Crum CP, Ikenberg H, Richart RM, Gissman L: Human papillomavirus type 16 in early cervical neoplasia. N Engl J Med 1984; 310:880–883.
30. Mitao M, Nagai N, Levine RU, et al: Human papillomavirus type 16 infection of the uterine cervix: a morphological spectrum with evidence of late gene expression. Int J Gynecol Pathol 1986;5:287–296.
31. Crum CP, Mitao M, Levine RU, Silverstein S: Cervical papillomaviruses segregate within morphologically distinct precancerous lesions. J Virol 1984;54:675–681.
32. Winkler B, Brum CP, Fujii T, et al: Koilocytotic lesions of the cervix: the relationship of mitotic abnormalities to the presence of papillomavirus antigens and nuclear DNA content. Cancer 1984;53:1081–1087.
33. Nuovo GJ, Nuovo MA, Cottral S, et al: Histological correlates of clinically occult papillomavirus infection of the uterine cervix. Am J Surg Pathol 1988;12:198–204.
34. Nuovo G, Blanco J, Richart RM, et al: Histological correlates of papillomavirus infection of the vagina. Obstet Gynecol 1988; 72:770.
35. Kiviat NB, Paavonen JA, Brockway J: Cytologic manifestations of cervical and vaginal infection I: epithelial and inflammatory cellular changes. JAMA 1985;253:989–996.
36. Crum CP, Mitao M, Winkler B, et al: Localizing chlamydial infection in cervical biopsies with the immunoperoxidase technique. Int J Gynecol Pathol 1984;3:191–197.
37. Winkler BW, Crum CP: *Chlamydia trachomatis* infection of the female genital tract: pathogenetic and clinico-pathologic correlations, in Rosen PP, Fechner R, (eds): Pathology Annual. New York, Appleton-Century Crofts, 1985, pp 193–223.
38. Giampaolo C, Murphy J, Benes S, McCormack WM: How sensitive is the Papanicolaou smear in the diagnosis of *Chlamydia trachomatis?* Am J Clin Pathol 1983;80:844–849.
39. Boyes DA, Worth AJ, Fidler HK: The results of treatment of 4389 cases of preclinical squamous cell carcinoma. J Obstet Gynecol Br Commonw 1973;77:769–780.
40. Savage EW: Microinvasive carcinoma of the cervix. Am J Obstet Gynecol 1972;113:708–717.
41. Ng ABP, Reagan JW: Microinvasive carcinoma of the uterine cervix. Am J Clin Pathol 1969;52:511–529.
42. Rubio CA, Soderberg G, Einhorn N: Histological and followup studies in cases of microinvasive carcinoma of the uterine cervix. Acta Pathol Microbiol Scand [A] 1974;82:397–410.
42. Leman MH, Benson WL, Kurman RJ, et al: Microinvasive carcinoma of the cervix. Obstet Gynecol 1976;48:571–578.
44. Wilkinson EJ, Komorowski RA: Borderline microinvasive carcinoma of the cervix. Obstet Gynecol 1977;51:472–476.
45. Sedlis A, Sall S, Tsukada Y, et al: Microinvasive carcinoma of the uterine cervix: a clinicopathologic study. Am J Obstet Gynecol 1979;133:64–74.
46. Hasumi K, Sakamoto A, Sugano H: Microinvasive carcinoma of the uterine cervix. Cancer 1980;45:928–931.

47. VanNagell JR, Greenwell N, Powell DF: Microinvasive carcinoma of the cervix. Am J Obstet Gynecol 1983;145:981–991.
48. Benson WL, Norris HJ: A critical review of the frequency of lymph node metastasis and death from microinvasive carcinoma of the cervix. Obstet Gynecol 1977;49:632–638.
49. Roche WD, Norris HJ: Microinvasive carcinoma of the cervix: the significance of lymphatic invasion and confluent patterns of growth. Cancer 1975;36:180–186.
50. Burghardt E, Holzer E: Diagnosis and treatment of microinvasive carcinoma of the uterine cervix. Obstet Gynecol 1977;49:641–653.
51. Fidler HK, Boyes DA, Worth AJ: Cervical cancer detection in British Columbia. J Obstet Gynecol Br Commonw 1968;75:392–404.
52. Barron BA, Cahill MC, Richart RM: A statistical model of the natural history of cervical neoplastic disease. The duration of carcinoma in situ. Gynecol Oncol 1978;6:196–205.
53. Schiller W, Daro AF, Gollin HA, et al: Small pre-ulcerative invasive carcinoma of the cervix—the spray carcinoma. Am J Obstet Gynecol 1953;65:1088–1098.
54. Reagan JW, Mamonic MS, Wentz WB: Analytical study of the cells in cervical squamous cell cancer. Lab Invest 1957;6:241–250.
55. Reagan JW, Fu YS: Histologic types and prognosis of cancers of the uterine cervix. Int J Radiat Oncol Biol Physics 1979;5:1015–1020.
56. Randall ME, Constable WC, Hahn SS, et al: Results of the radiotherapeutic management of carcinomas of the cervix with emphasis on the influence of histologic classification. Cancer 1988;62:48–53.
57. Spratt DW, Lee SC: Verrucous carcinoma of the cervix. Am J Obstet Gynecol 1977; 129:699–700.
58. Abramson A, Brandsma J, Steinberg B, et al: Verrucous carcinoma of the larynx: possible human papillomavirus etiology. Arch Otolaryngol 1985;111:709–715.
59. Gissman L, de Villiers E-M, zur Hausen H: Analysis of genital warts and other genital tumors for human papillomavirus type 6 DNA. Int J Cancer 1982;29:143–146.
60. Kister RW, Hertig AT: Papillomas of the uterine cervix—the malignant potentiality. Obstet Gynecol 1955;6:147–161.
61. Qizilbash A: Papillary squamous tumors of the uterine cervix: a clinical and pathological study of 21 cases. Am J Clin Pathol 1974; 61:508–520.
62. Randall M, Anderson W, Mills S, et al: Papillary squamous cell carcinoma of the uterine cervix: a clinicopathologic study of nine cases. Int J Gynecol Pathol 1986;5:1–10.
63. Walker AN, Mills SE: Unusual variants of uterine cervical carcinoma, in Rosen P, Fechner R (eds): Pathology Annual. Norwalk, CT, Appleton-Century-Crofts, 1987, pp 277–310.
64. Venkataseshan VS, Woo TH: Diffuse viral papillomatosis (condyloma) of the uterine cavity. Int J Gynecol Pathol 1985;4:370–377.
65. Ferenczy A, Richart RM, Okagaki T: Endometrial involvement by cervical carcinoma *in-situ*. Am J Obstet Gynecol 1971;110:590–592.
66. Van Nagell JR, Donaldson ES, Wood EG, et al: Small cell cancer of the uterine cervix. Cancer 1977;40:2243–2249.
67. Barrett RJ, Davos I, Leuchter RS, Lagasse LD: Neuroendocrine features in poorly differentiated and undifferentiated carcinomas of the cervix. Cancer 1987;60:2325–2330.
68. Albores-Saavedra J, Larraza P, Poucell S, et al: Carcinoids of the uterine cervix: additional observations in a new tumor entity. Cancer 1976;38:2328–2342.
69. Groben P, Reddick R, Askin F: The pathologic spectrum of small cell carcinoma of the cervix. Int J Gynecol Pathol 1985;4:42–57.
70. Mullins JD, Hilliard GD: Cervical carcinoid ("Argyrophil cell carcinoma") associated with an endocervical adenocarcinoma: a light and ultrastructural study. Cancer 1981;47:785–790.
71. MacKay B, Osborne BM, Wharton JT: Small cell tumor of the cervix with neuroepithelial features: ultrastructural observations in two cases. Cancer 1979;43:1138–1145.
72. Pearce AGE: The APUD cell concept and its implications in pathology. Pathol Annu 1974; 9:27–41.
73. Fox H, Kazzaz B, Langley FA: Argyrophil and argentaffin cells in the female genital tract and in ovarian mucinous cysts. J Pathol Bacteriol 1964;88:479–488.
74. Stassart J, Crum CP, Yordan EL, et al: Argyrophilic carcinoma of the cervix: a report of a case with coexisting cervical intraepithelial neoplasia. Gynecol Oncol 1982;13:247–251.
75. Taxi JB, Tischler AS, Insalaco SJ, et al: "Carcinoid" tumor of the breast A variant of breast cancer? Hum Pathol 1981;12:170–179.
76. Stoler MH, Walder AN, Mills SE: Small cell neuroendocrine carcinoma of the cervix: a human papillomavirus type 18 associated cervix cancer. Lab Invest 1989;60:92A.

6

Endocervical Carcinoma

LIANE DELIGDISCH

Incidence and Etiopathogenesis

The incidence of invasive squamous cell carcinoma of the uterine cervix has declined dramatically over the past decades because of the use of preventive methods that detect its precursors. Neoplasms of the endocervix arising in the glandular epithelium have not benefited to the same extent from the methods used for early diagnosis of cervical cancer, such as exfoliative cytology and colposcopy. The general incidence of endocervical cancer is rising, as is its relative proportion among total cases of cervical cancer, which has evolved from 5% in the 1950s[1] to 10, 15, and even 25% according to more recent reports.[2–4] The highest series reported is 34%.[5] This increase is probably a result of improved screening and diagnosis of precursors of exocervical carcinoma.

Endocervical cancer generally has a poor prognosis because it is rarely diagnosed in its early stages. It often remains elusive to physical and colposcopic examination because of the secluded location of the endocervical glands. Histologic criteria for diagnosis of cancer precursors or early neoplasia, such as dysplasia,[6,7] carcinoma in situ, and microinvasion, are still unclear, despite the important work that has been recently done using cytology, histopathology, morphometry, and immunocytochemistry.[8–15] Criteria for the diagnosis of glandular atypia, and of precancerous lesions of lesser severity than adenocarcinoma in situ, have not yet been well defined. The association of these conditions with squamous intraepithelial neoplasia has been described.[2,8,16] This coexistence is still poorly understood because invasive neoplasms of the cervix, squamous cell carcinoma, and adenocarcinoma involve different types of patients. As opposed to patients with cervical squamous cell carcinoma, patients with endocervical adenocarcinoma are less sexually promiscuous at a younger age, belong to higher socioeconomic groups,[2,3] and are often multiparous.[17] A controversial but undeniable risk factor is the association with oral contraceptives.[2,8,18–20] The fact that Rhesus monkeys treated with high doses of medroxyprogesterone acetate developed endocervical carcinoma supports this hypothesis.[8]

Dallenbach-Hellweg reported that 82% of patients with endocervical adenocarcinoma who were below 50 years of age had taken oral contraceptives.[8] She considered that the use of oral contraceptives contributes to the transformation of microglandular hyperplasia to adenomatous hyperplasia and to adenocarcinoma. These conditions originate in the glandular epithelial and reserve cells of the endocervix, which are normally stimulated to proliferate by gestagens. The most potent progestational agents, presumably involved in the histogenesis of endocervical adenocarcinoma, are norgestrel and norethisterone acetate, which are derivatives of 19-nortestosterone.[8]

The mean age of patients with cervical adenocarcinoma is estimated to be higher than that of patients with squamous cell carcinoma. It seems, however, to be decreasing according to reports relating a rising incidence of this neoplasm in premenopausal women.[8]

The anatomic origin of endocervical adenocarcinoma is considered to be the subcolumnar reserve cells located beneath the columnar epithelium lining the endocervical cavity and glands. Normally, the endocervical glands branch and their bottoms may be seen deep in the muscular wall of the cervix. This anatomic peculiarity makes the differential diagnosis between in situ adenocarcinoma and invasive well-differentiated adenocarcinoma often difficult.

Recent virologic studies using DNA in situ hybridization have yielded the surprising result that this neoplasm, generally thought to be hormone rather than infection related, often displays a human papillomavirus (HPV) type 18 DNA.[21]

Pathology

Adenocarcinoma In Situ

Adenocarcinoma in situ (ACIS) is considered a precursor of endocervical carcinoma and is diagnosed in women 10 to 15 years younger on average than those with invasive adenocarcinoma. Adenocarcinoma in situ is often seen at the margin of invasive cancer and sometime associated with squamous cervical intraepithelial noeplasia (CIN).[7] Although the criteria for this diagnosis are still not well defined, there is an increasing awareness by pathologists that this entity must be diagnosed by both cytology and histopathology.[9–12] Histologically, the architecture of the endocervical glands is generally maintained. The ACIS is often superimposed on a pattern of microglandular hyperplasia, a common condition often related to pregnancy. The individual cells show an increased nucleus-to-cytoplasm ratio, nuclear hyperchromasia with prominent nucleoli, and mitotic activity (Figs. 6-1 and 6-2). These changes were evaluated by morphometry and a correlation between the severity of the changes in the nuclei and the biologic behavior of the lesions was noted.[10] As opposed to invasive adenocarcinoma, the epithelial changes are confined to the glands and there is no surrounding stromal reaction. The breaking of the basement membrane, which is a strong criterion for evaluating carcinoma in situ versus microinvasion in squamous cell neoplasias, cannot be used in differentiating between ACIS and adenocarcinoma because even invasive adenocarcinomas have a basement membrane. Therefore, microinvasive adenocarcinoma is not yet defined by histologic criteria; it has been reported, however, to be an accepted cytologic diagnosis.[12]

The diagnosis of ACIS, based on architectural changes of the glands (irregularity, outpouching, budding, cribriform pattern) and on the cytologic characteristics described above, should only be made on cervical cones or hysterectomy specimens. Punch biopsy samples rarely offer enough tissue for this diagnosis, which can only be suspected.

Microglandular endocervical hyperplasia is an important differential diagnosis. This lesion involves the endocervical canal and/or the squamocolumnar junction area. It consists of hyperplastic glands varying in size and shape, crowded next to each other, often with no intervening stroma displaying basally located bland nuclei (Fig. 6-3). The epithelial cells lining these glands are generally uniform, tall to cuboidal, mucin secreting, or flattened.[20,22] Squamous metaplasia is often present along with subcolumnar reserve cell hyperplasia. It may be grossly noticeable as an endocervical polyp, occasionally forming an exophytic cervical lesion that may look suspicious for malignancy.

Clinically, endocervical microglandular hyperplasia is known to be associated with a history of oral contraceptive use or pregnancy. It seems that hormonal stimulation with a predominant progesterone component is causally related to endocervical gland proliferation.[8] During the 1960s, when oral con-

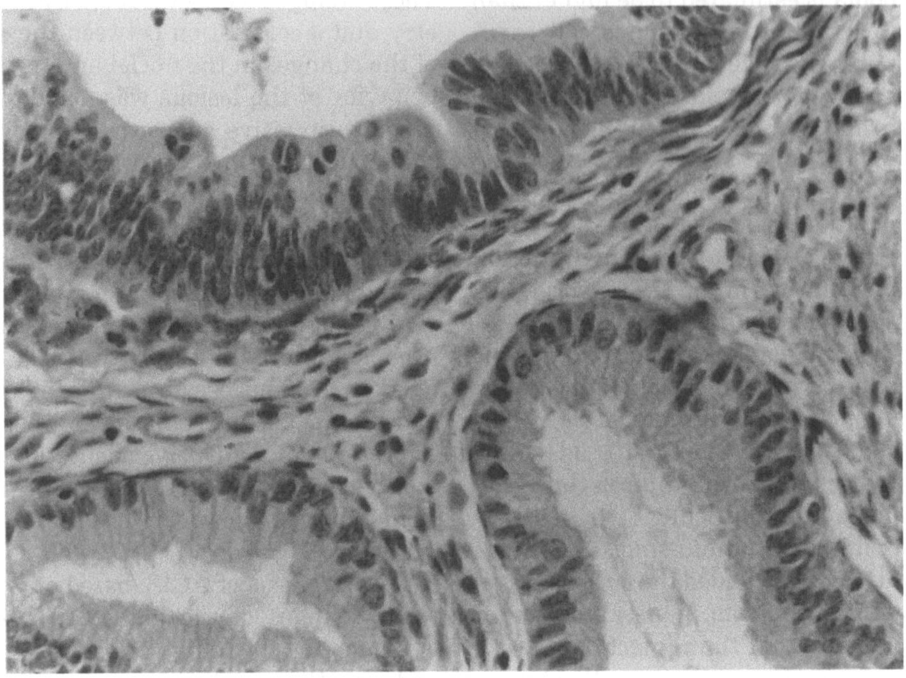

FIGURE 6-1. Adenocarcinoma in situ (ACIS) of cervix. Two normal glands are seen at the bottom and ACIS is present in the gland on top. (H&E, × 200.)

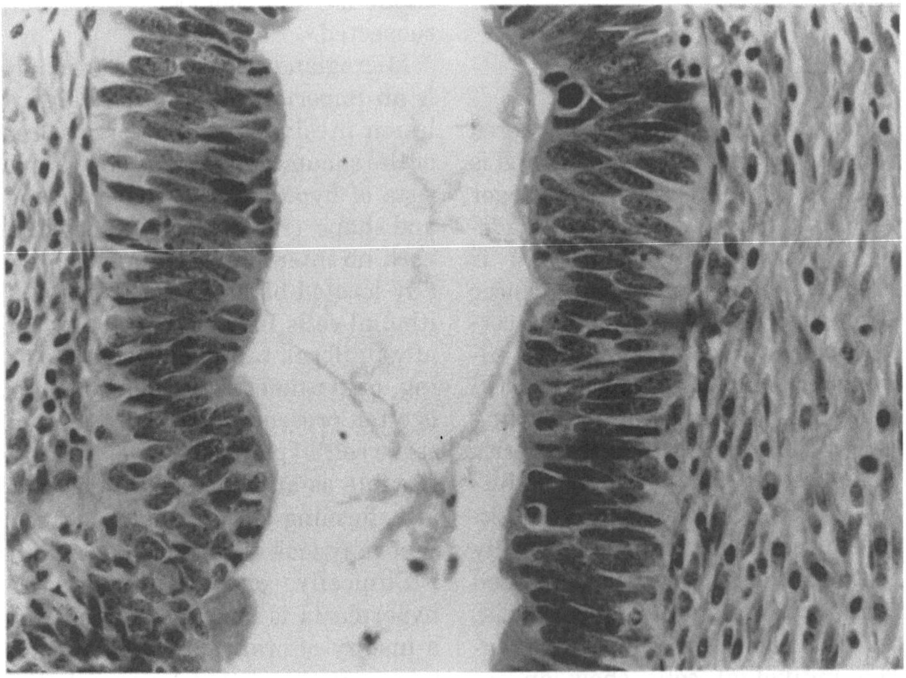

FIGURE 6-2. Adenocarcinoma in situ shows piling up of glandular epithelium, loss of mucus secretion, increased nucleus-to-cytoplasm ratio, and marked nuclear pleomorphism. Note well-preserved basement membrane. (H&E, × 400.)

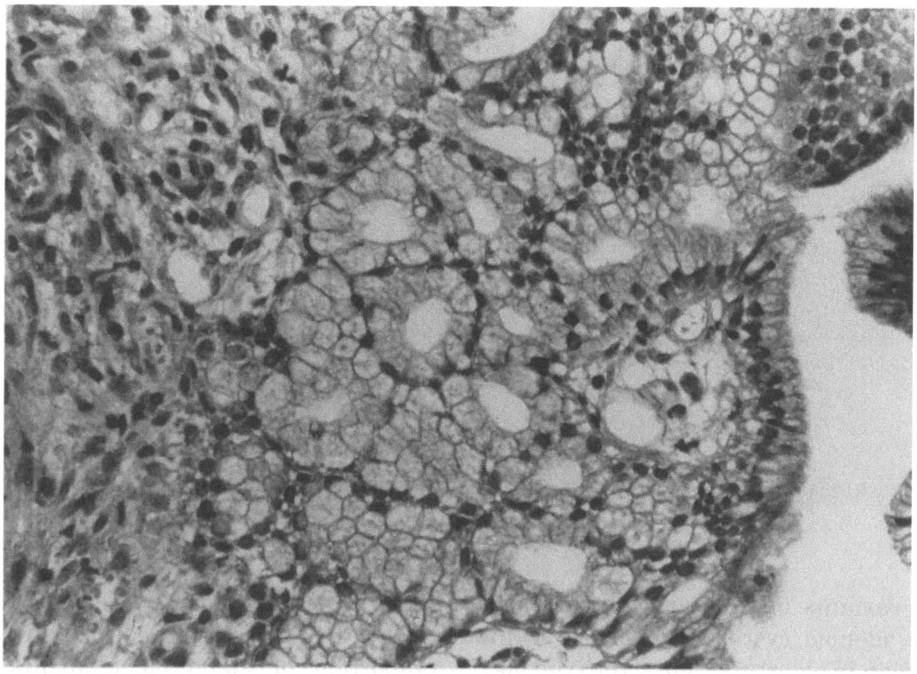

FIGURE 6-3. Microglandular endocervical hyperplasia with crowded endocervical mucin-secreting glands. Note small, uniform, basally located nuclei. (H&E, × 100.)

traceptives contained higher doses of hormones, an atypical endocervical hyperplasia had been described.[20,23] The differential diagnosis from endocervical adenocarcinoma is based on the scant or absent mitotic activity, the relatively bland nuclear features, and the absence of invasion.[22] As opposed to endometrial hyperplasia, in which a back-to-back glandular pattern is seen in more severe forms, in endocervical hyperplasia the lack of intervening stroma between the glands is not significant in the absence of cellular malignant characteristics.

Invasive Adenocarcinoma

Invasive adenocarcinoma of the uterine cervix usually presents with abnormal vaginal bleeding. The accuracy of the cytologic diagnosis is variable and generally lower than that for squamous cell carcinoma. The use of aspiration with an endocervical pipette enhances the diagnostic accuracy, but definitive diagnosis is made on histopathologic slides. Grossly, the lesion may be visible at the external os as a polypoid, fungating, or ulcerated mass; it may also be hidden in the endocervical canal, making an early diagnosis very difficult (Fig. 6-4). According to the extent of the tumor, the entire cervix may be distorted, indurated, or softened by superimposed inflammation or necrosis.

Histologically, endocervical cancers are mostly epithelial and arise from the columnar lining of the clefts and glands located in the endocervical canal and external os. Mucinous or typical adenocarcinoma is the most common histologic type. Other histologic types are endometrioid, clear cell, adenosquamous (or mixed) papillary carcinomas and adenoma malignum (or minimal deviation adenocarcinoma). The heterogeneous cellular elements reflect the multipotential differentiation of the endocervical epithelium. Similar histologic variants can be seen in other areas of the müllerian and "extended" müllerian tract, such as the endometrium and ovary.

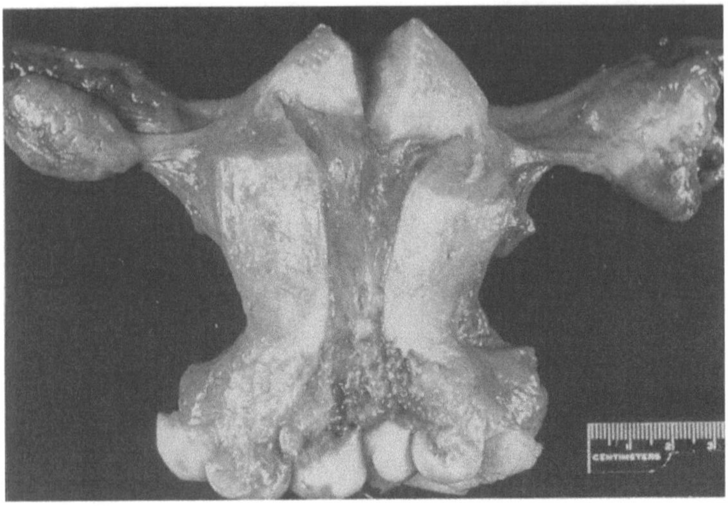

FIGURE 6-4. Endocervical carcinoma, hysterectomy specimen.

Rare variants of endocervical carcinoma are the adenoid cystic carcinoma or cylindroma seen in elderly patients, the true mesonephric adenocarcinoma originating in the wolffian remnants and located deep in the lateral cervical wall in continuity with wolffian (mesonephric) duct remnants, and the neuroendocrine carcinomas (carcinoid, small-cell, or oat cell carcinomas). Nonepithelial tumors such as malignant melanoma or malignant lymphoma are extremely uncommon in the endocervix.

The diagnosis may be a difficult one in well-differentiated tumors because of their invasive pattern. Unlike adenocarcinomas of the endometrium, which invade the myometrium in a "broad front," in the endocervix the early invasive tumor nests are scattered and separated by fibromuscular tissue, and therefore are reminiscent of the deep crypts of normal endocervical glands. The diagnosis is based on the finding of complex, crowded glandular structures lined by tall columnar epithelial cells with a mucin-secreting cytoplasm and generally basally located nuclei (Figs. 6-5A and 6-5B).

The prognosis of endocervical carcinoma is not dependent on the histologic subtype.[24] Tumor grade, size, depth of invasion, and staging, including lymph node status, are significant factors.[24,25] Staging is the same as for squamous cell carcinoma.[2] Grading is based on the histologic degree of differentiation, which is established according to the proportion of glandular elements in the tumor and the degree of cellular anaplasia. The well-differentiated (grade 1) cancers show a 75% or more glandular component, the moderately differentiated (grade 2) cancers are approximately 50% solid tumor, and the poorly differentiated (grade 3) cancers are mostly solid tumors with few glandular structures.[2] The degree of nuclear pleomorphism and mitotic activity is generally correlated with architectural differentiation; in poorly differentiated adenocarcinoma, the cells show bizarre and irregular nuclei and a high number of atypical asymmetric mitoses (Fig. 6-6). The degree of differentiation as determined by the nuclear features is useful in predicting the clinical outcome.[26] Nuclear DNA analysis has shown that high-ploidy stem lines are biologically more aggressive, regardless of the degree of differentiation.[27]

Very young patients (less than 30 years of age) seem to have a poorer survival rate.[25]

Mucin-Secreting Adenocarcinomas

Mucin-secreting adenocarcinomas are the most common malignant neoplasms of the endocervix. In well-differentiated tumors, the glands resemble the normal mucin-se-

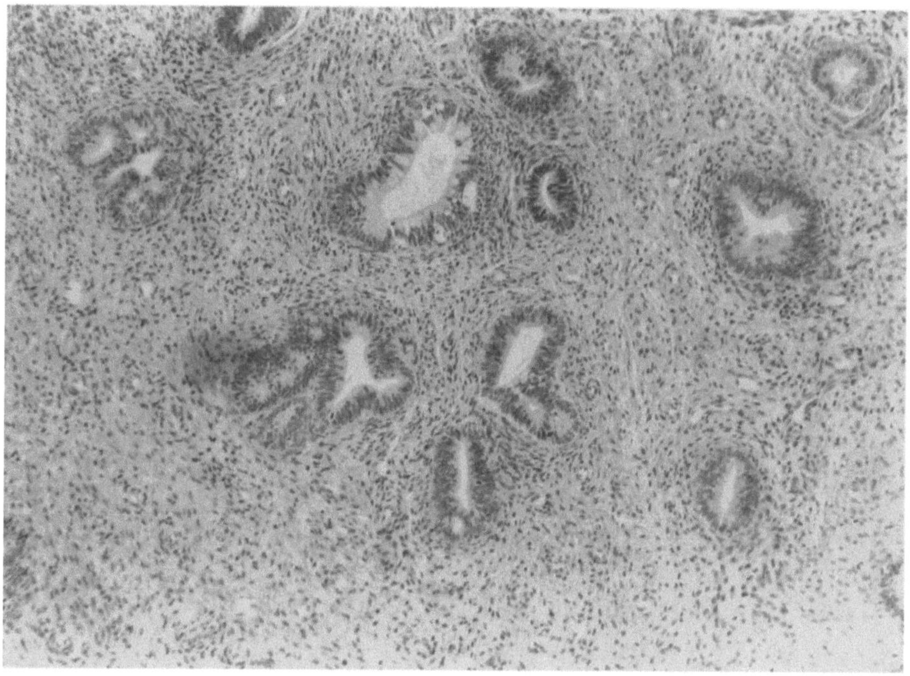

A

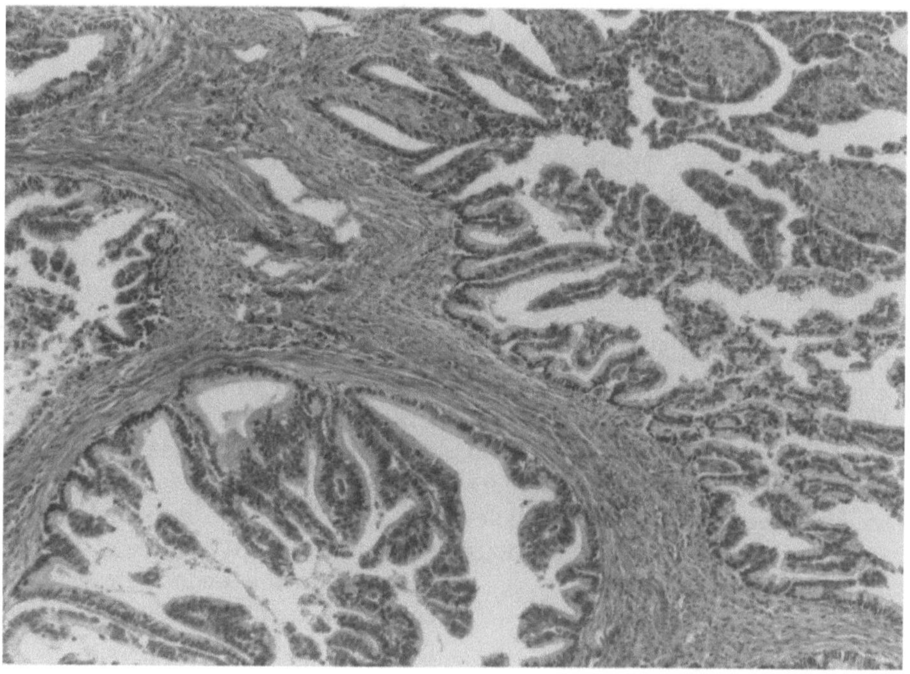

B

FIGURE 6-5. (A) Early invasive well-differentiated endocervical adenocarcinoma. The neoplastic glands are separated by normal fibromuscular tissue. (H&E, × 40.) (B) Invasive well-differentiated adenocarcinoma of endocervix. (H&E, × 100.)

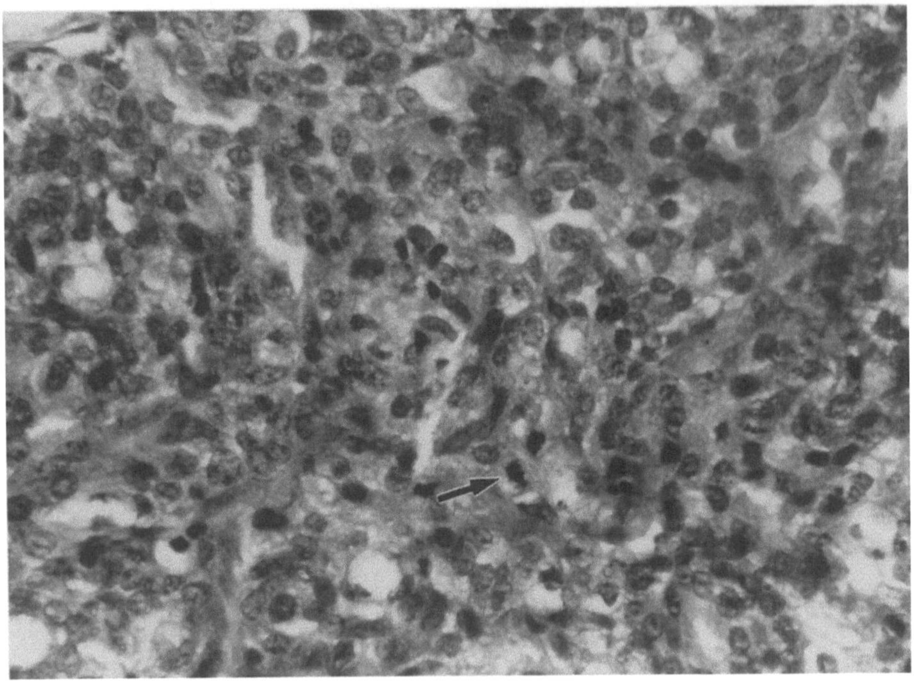

FIGURE 6-6. Poorly differentiated adenocarcinoma of endocervix showing mostly solid tumor and atypical mitotic activity (arrow). (H&E, × 200.)

creting endocervical glands. The tumors contain goblet cells and are similar to colonic adenocarcinomas. Some tumors show signet ring cells or have a solid appearance. Colloid carcinomas, similar to those of the intestine, show pools of mucin with scattered tumor cells (Fig. 6-7). In less well-differentiated tumors, the amount of mucin and its supranuclear location decreases, the polarity of the nuclei is lost, and nuclear pleomorphism along with mitotic activity becomes more prominent. A number of cases have been described as being associated with mucinous adenocarcinomas of the ovary.[28]

Endometrioid Carcinomas

Endometrioid carcinomas resemble the glandular cancers most commonly seen in the endometrium. The neoplastic glands are tubular in the well-differentiated carcinomas, but show various degrees of irregularity and solid configuration in the moderately or poorly differentiated adenocarcinomas (see Chap. 7). The epithelial cells usually do not secrete mucus. On special stains for mucopolysaccharides (periodic acid–Schiff reagent) the apical, luminal border stains positively, thus resembling the müllerian epithelium in other locations. Histologically, it is impossible to differentiate the endometrial carcinoma of the endocervix from the endometrioid carcinoma of the corpus uteri. The diagnosis of endocervical carcinoma in this case requires documentation that the endometrial cavity is free of tumor and evidence that the endocervical endometrioid carcinoma is in continuity with endocervical glands (Fig. 6-8).

Clear Cell Carcinoma

Clear cell carcinoma is composed of tubules, glands, papillary structures, or solid nests with cells displaying a clear cytoplasm that appears on special stains to contain glycogen

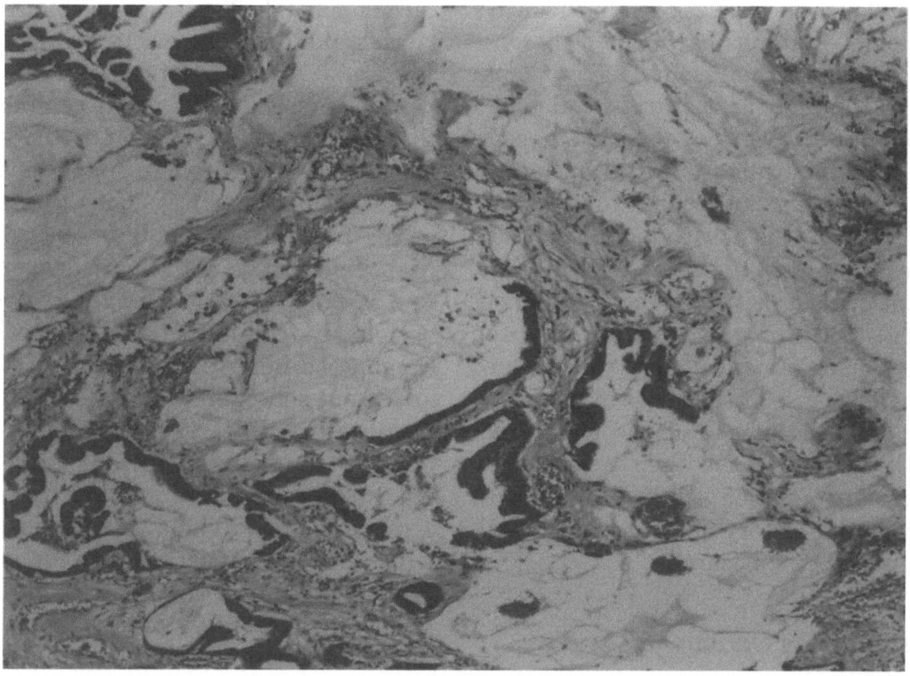

FIGURE 6-7. Mucinous adenocarcinoma of endocervix. Neoplastic glands are seen in pools of mucin. (H&E, × 100.)

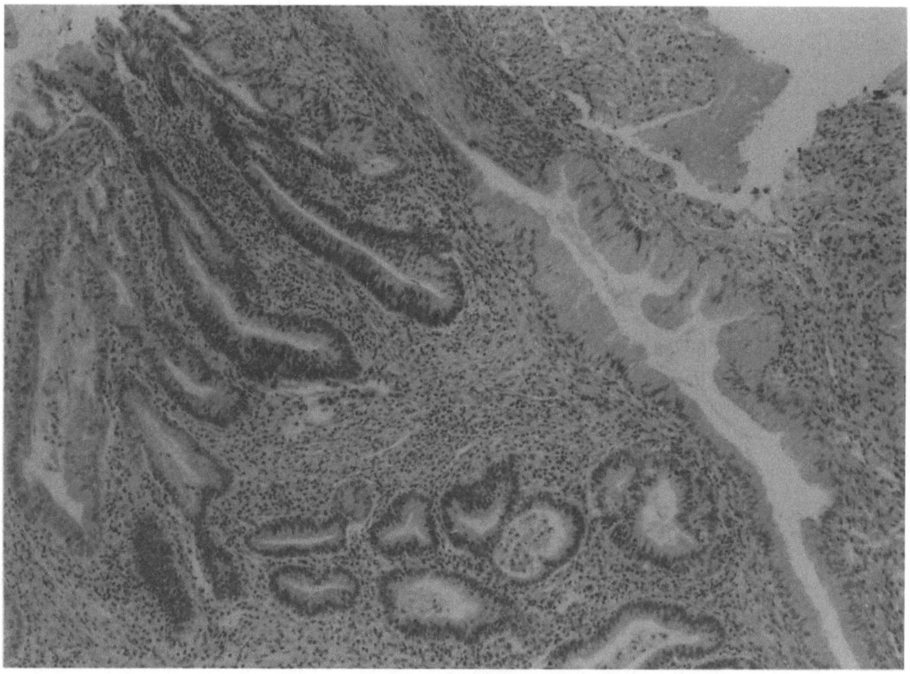

FIGURE 6-8. Endometrioid carcinoma of endocervix. The neoplastic glands are not mucin secreting and are seen in continuity with endocervical glands. (H&E, × 100.)

granules. This cancer, also seen in the ovary and endometrium, has been called mesonephroid because of the resemblance to renal cell carcinoma. Ultrastructural studies and histochemistry have demonstrated a completely different structure in the müllerian-derived carcinomas as opposed to the nephrogenic tumors. It is now thought that their origin is müllerian[29] in the majority of cases. A variant of clear cell carcinoma is the hobnail type of cells in which the nuclei of the clear-appearing tumor cells protrude toward the lumen because of their loss of polarity.

Clear cell carcinomas of the endocervix and of the vagina have been described to occur in younger patients in association with adenosis and a history of maternal diethylstilbestrol (DES) exposure.[30]

Adenosquamous Carcinoma

Adenosquamous carcinoma is one of the most common variants of invasive adenocarcinoma and often, on punch biopsies, may be mistaken for exocervical carcinoma because of the predominant squamous cell component. Also named "mixed carcinomas,"[26] these tumors have been reported to occur more frequently in younger patients[17] and in pregnant women.[31] They consist of a mixed pattern of glandular and squamous neoplasms (Fig. 6-9). Their histogenetic origin is probably from the reserve cells underlying the columnar epithelium, which are capable of differentiating into both columnar and squamous cells. An association with intrauterine DES exposure has been reported.[32] A histologic variant is the mucoepidermoid carcinoma, in which there is keratinization of individual squamous cells and which has been reported to occur mostly in patients who were treated with oral contraceptives.[8] According the some authors, mixed (adenosquamous) carcinomas have a worse prognosis than pure adenocarcinoma.[26,27]

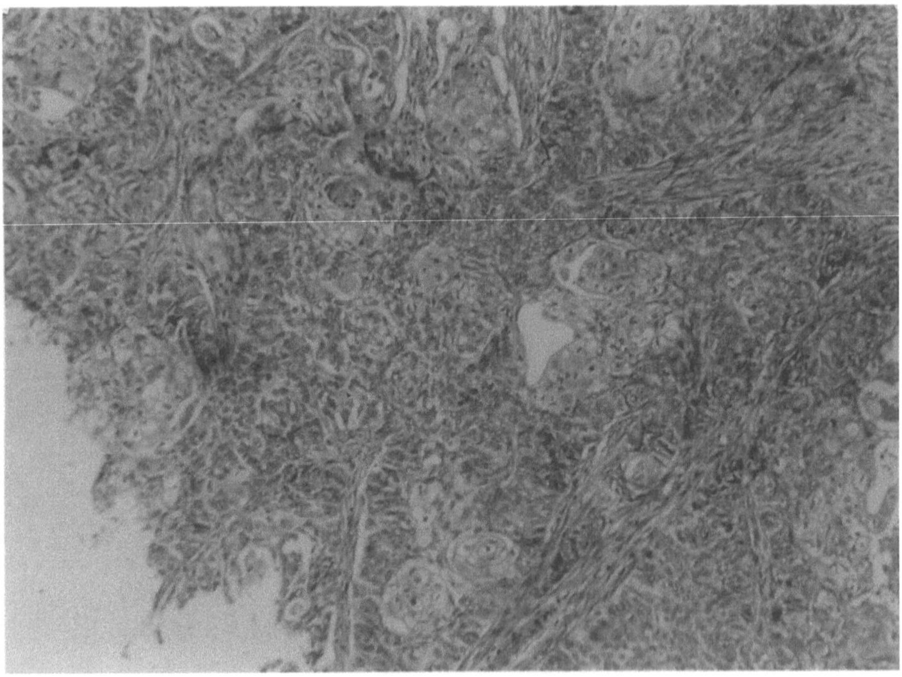

FIGURE 6-9. Infiltrating adenosquamous ("mixed") carcinoma of endocervix. The patient is a 36-year-old grand multipara. (H&E, × 40.)

Papillary Carcinomas

Papillary carcinomas are rather uncommon in the endocervix. They may resemble ovarian papillary carcinoma and can be associated with clear cell or endometrioid carcinomas. They consist of finger-like projections or tufts in which a fibrovascular stalk is covered by neoplastic epithelial cells.

"Adenoma Malignum"

Adenoma malignum is an uncommon but extremely important histopathologic entity. Also called minimal deviation adenocarcinoma (MDA),[33] this lesion resembles the normal histologic structure of the endocervix to the point of making the diagnosis on punch biopsies very difficult. There is no stratification or anaplasia of the epithelium and the glands are usually separated by a normal amount of stroma (Fig. 6-10). A back-to-back glandular pattern is uncommon. Findings of mitotic activity and extension of the lesion deeper than the crypts of normal endocervical glands, as well as angular outpouching of the glands, are characteristics that are helpful for the diagnosis of adenoma malignum; these changes can be documented only on cervical cones and hysterectomy specimens. The glandular epithelium can be mucinous or endometrioid or show clear cells. The biologic behavior of the MDA is unfavorable, in sharp contrast with its relatively innocuous histologic appearance. It has been found to be associated with the Peutz-Jeghers syndrome[34] and with mucinous carcinoma of the ovary.[35] It has been documented that immunohistologic staining for carcinoembryonic antigen is positive in MDA, as opposed to microglandular endocervical hyperplasia, which MDA may resemble histologically.[36,37] A correct and early identification and treatment of this tumor is associated with a better outcome.[38]

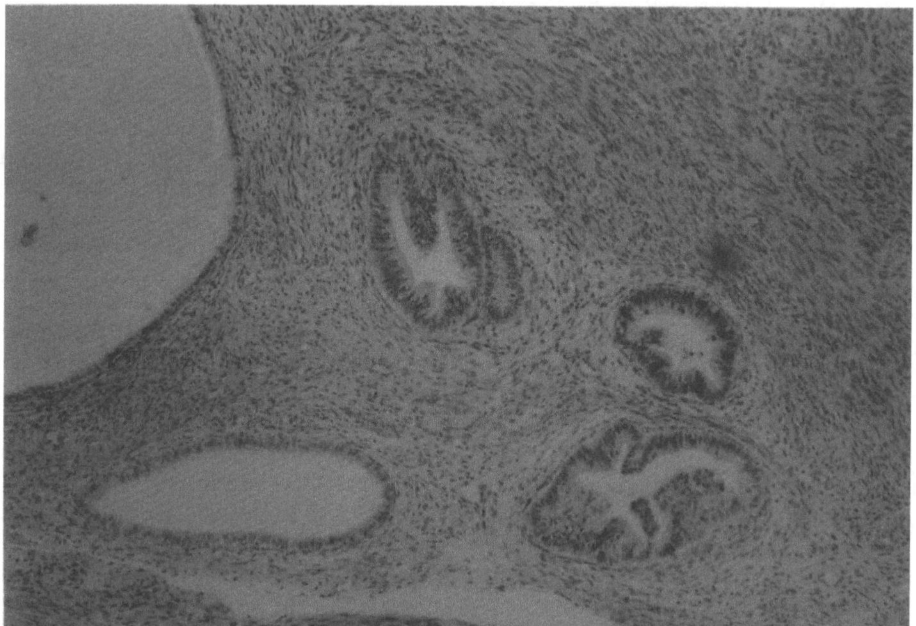

FIGURE 6-10. Minimal deviation adenocarcinoma ("adenoma malignum") of endocervix. Minimal histologic changes are seen in endocervical glands separated by a fair amount of fibromuscular stroma. (H&E, × 40.)

Conclusion

Because of the increasing frequency of endocervical adenocarcinoma (both absolute and relative to squamous cell carcinoma) and its generally poor prognosis, a careful clinical and histopathologic search is required in order to exclude this possibility when cervical neoplasia is considered.[25,39] The precursors of this neoplasm are still elusive, despite the numerous histologic reports attempting to define endocervical gland dysplasia and ACIS and the considerable progress made by immunohistochemistry, morphometry, and electron microscopy. Achieving an early diagnosis of endocervical carcinoma is a goal of major priority because this cancer, when diagnosed in an advanced stage, is highly lethal.

As far as the etiopathogenesis is concerned, a causal relationship to hormonal agents (mostly progesterone) is highly probable, although definite proof is still lacking. The effect of viral oncogenes associated with progesterone activity is suggested by recent research.[8,21] It seems that HPV type 18 has a certain propensity to infect the glandular epithelium in the cervical canal. The finding of this viral type in ACIS as well as in invasive cervical cancers suggests the possibility of a more rapidly progressing lesion[40]; therefore the finding of HPV-18 by in situ DNA hybridization on cervical biopsies may indicate high-risk lesions.

References

1. Hepler TK, Dockerty MB, Randal LM: Primary adenocarcinoma of the cervix. Am J Obstet Gynecol 1952;63:800–808.
2. Ferenczy A, Winkler B: Carcinoma and metastatic tumors of the cervix, in Kurman RJ (ed): Blaustein's Pathology of the Female Genital Tract, ed 3. New York, Springer-Verlag, 1987, pp 218–256.
3. Horowitz IR, Jacobson LP, Zucker PK, et al: Epidemiology of adenocarcinoma of the cervix. Gynecol Oncol 1988;31:25–31.
4. Ireland D, Hardiman P, Monaghan JM: Adenocarcinoma of the uterine cervix: a study of 73 cases. Obstet Gynecol 1985;65:82–85.
5. Davis JR, Moon LB: Increased incidence of adenocarcinoma of uterine cervix. Obstet Gynecol 1975;45:79–83.
6. Alva J, Lauchlan SC: The histogenesis of mixed cervical carcinoma. Am J Clin Path 1975;64:20–25.
7. Gloor E, Ruzicka J. Morphology of adenocarcinoma in-situ of the uterine cervix. A study of 14 cases. Cancer 1982;49:294–302.
8. Dallenbach-Hellweg G: On the origin and histological structure of adenocarcinoma of the endocervix in women under 50 years of age. Pathol Res Pract 1984;179:38–50.
9. Jaworksi RC, Pacey NF, Greenberg ML, et al: The histologic diagnosis of adenocarcinoma in-situ and related lesions of the cervix uteri. Adenocarcinoma in-situ. Cancer 1988;61: 1171–1181.
10. Boon MA, Baak JPA, Kurver PJH, et al: Adenocarcinoma in-situ of the cervix. An underdiagnosed lesion. Cancer 1981;48:768–772.
11. Clark AH, Betsill WL Jr: Early endocervical glandular neoplasia II: Morphometric analysis of the cells. Acta Cytol 1986;30:127–134.
12. Ayer B, Pacey F, Greenberg M: The cytologic diagnosis of adenocarcinoma in-situ of the cervix uteri and related lesions. II microinvasive adenocarcinoma. Acta Cytol 1988;3:318–324.
13. Ostor AG, Pagano R, Davoren RAM, et al: Adenocarcinoma in-situ of the cervix. Int J Gynecol Pathol 1984;3:179–190.
14. Tobon H, Dave H: Adenocarcinoma in-situ of the cervix. Clinicopathologic observations of 11 cases. Int J Gynecol Pathol 1988;7:139–151.
15. Hopkins MP, Roberts JA, Schmidt RW: Cervical adenocarcinoma in-situ. Obstet Gynecol 1988;6:842–844.
16. Brown J, Wells B: Cervical glandular atypia associated with squamous intra-epithelial neoplasia: a pre-malignant lesion? J Clin Pathol 1986;39:22–28.
17. Deligdisch L, Martinez-Escay E, Cohen CJ: Endocervical carcinoma: A study of 23 cases with clinical-pathological correlation. Gynecol Concol 1984;18:326–333.
18. Czernobilsky B, Kessler Lancet M: Cervical adenocarcinoma in a woman on long-term contraceptives. Obstet Gynecol 1980;43:517–521.
19. Valente PT, Hanjani P: Endocervical neoplasia in long-term uses of oral contraceptives: clinical and pathologic observation. Obstet Gynecol 1986;24:695–704.

20. Kyriakos M, Kempson RL, Konikov NF: A clinical and pathologic study of endocervical lesions associated with oral contraceptives. Cancer 1968;22:99–110.
21. Tase T, Okagaki T, Clark BA, et al: Human papilloma virus types and localization in adenocarcinoma and adenosquamous carcinoma of the uterine cervix: a study by in-situ DNA hybridization. Cancer Res 1988;48:993–998.
22. Young RH, Scully RE: Atypical forms of microglandular hyperplasia of the cervix simulating carcinoma. A report of five cases and review of the literature. Am J Surg Pathol 1989;1:50–56.
23. Taylor HB, Irey NS, Norris HJ: Atypical endocervical hyperplasia in women taking oral contraceptives. JAMA 1967;202:637.
24. Kilgone LC, Soong SJ, Gore H, et al: Analysis of prognostic features in adenocarcinoma of the cervix. Gynecol Oncol 1988;31:137–148.
25. Wells M, Brown LJ: Glandular lesions of the uterine cervix: the present state of our knowledge. Histopathology 1986;8:772–792.
26. Fu YS, Reagan JW, Hsiu JG, et al: Adenocarcinoma and mixed carcinoma of the uterine cervix I. A clinicopathologic study. Cancer 1982;49:2560–2570.
27. Fu YS, Reagan JW, Fu A, et al: Adenocarcinoma and mixed carcinoma of the uterine cervix. II: Prognostic value of nuclear DNA analysis. Cancer 1982;49:2571–2577.
28. Young RH, Scully RE: Mucinous ovarian tumors associated with mucinous adenocarcinomas of the cervix. A clinicopathological analysis of 16 cases. Int J Gynecol Pathol 1988;7:99–111.
29. Hasumi K, Ehrman RL: Clear cell carcinoma of the uterine endocervix with an in-situ component. Cancer 1978;42:2435–2438.
30. Herbst AL, Cole P, Norusis MJ, et al: Epidemiologic aspects and factors related to survival in 384 registry cases of clear cell adenocarcinoma of the vagina and cervix. Am J Obstet Gynecol 1979;135:876–883.
31. Glucksman A: Relationship between hormonal changes in pregnancy and the development of "mixed carcinoma" of the uterine cervix. Cancer 1957;10:831–837.
32. Vandrie DM, Puri S, Upton RT, et al: Adenosquamous carcinoma of the cervix in a woman exposed to diethylstilbesterol in utero. Obstet Gynecol 1983;61:845–875.
33. Silverberg SG, Hurt WG: Minimal deviation adenocarcinoma ("adenoma malignum") of the cervix: a re-appraisal. Am J Obstet Gynecol 1975;121:971–975.
34. Kaku T, Hachisuga T, Toyoshima S, et al: Extremely well-differentiated adenocarcinoma ("Adenoma malignum") of the cervix in a patient with Peutz-Jeghers syndrome. Int J Gynecol Pathol 1985;4:266–273.
35. Li Volsi V, Merino MJ, Schwartz PE: Coexistent endocervical adenocarcinoma and mucinous adenocarcinoma of the ovary: a clinicopathologic study of four cases. Int J Gynecol Pathol 1983;1:391–402.
36. Nanbu Y, Fujii, Konishi I, et al: Immunohistochemical localizations of CA125, carcinoembryogenic antigen and Ca 19-9 in normal and neoplastic glandular cells of the uterine cervix. Cancer 1988;62:2580–2588.
37. Steeper TA, Wick MR: Minimal deviation adenocarcinoma of the uterine cervix ("adenoma malignum"): an immunohistochemical comparison with microglandular endocervical hyperplasia and conventional endocervical adenocarcinoma. Cancer 1986;58:1131–1138.
38. Kaminski PF, Norris HJ: Minimal deviation carcinoma (adenoma malignum) of the cervix. Int J Gynecol Pathol 1983;2:141–152.
39. Bertrand M, Likrish GM, Colgan TJ: The anatomic distribution of cervical adenocarcinoma in situ: implications for treatment. Am J Obstet Gynecol 1987;157:21–25.
40. Farnsworth A, Laverty C, Stoler MH: Human papilloma virus. Messenger RNA expression in adenocarcinoma in-situ of uterine cervix. Int J Gynecol Pathol 1989;8:321–330.

7

Endometrial Hyperplasia and Endometrial Adenocarcinoma

LIANE DELIGDISCH

The endometrium is one of the most sensitive target tissues responding to hormonal influences. It is capable of undergoing profound structural changes with an astonishing promptness, under the stimulation of sex hormones. These changes are manifested by the changes taking place with the menstrual cycle during the reproductive years, when the morphology of the endometrial tissue has different characteristics every day. Estrogens stimulate proliferation within the endometrial glands and stroma. The response to ovulation and subsequent secretion of progesterone is a dramatic arrest of any proliferative activity in the endometrial glands and a complex process of secretion, with maturation of the stroma and development of blood vessels, aiming to prepare the endometrium to be the host tissue for a possible implanting conceptus.

Failure of ovulation, after the menopause, deprives the endometrium of progesterone stimulation. The estrogenic stimulation, however, may continue because of the conversion of androgens, which are secreted by the menopausal ovary and by the cortical adrenal gland, into estrogens by peripheral aromatization. Obesity, diabetes, and other metabolic disorders may enhance the extragonadal estrogen production, and the presence of high estrogen levels, especially of estradiol (E_2), may often be associated with endometrial hyperplasia as a result of the binding of the hormone to receptor sites in the nuclei of endometrial cells.[1–3]

The endometrial response to a prolonged stimulation by estrogens is continuous proliferation and hyperplasia. A continous and unopposed (by progesterone) effect of estrogen, either of endogenous, metabolic origin or from exogenous sources (i.e., estrogen therapy), may induce endometrial hyperplasia in postmenopausal women. Endometrial hyperplasia and neoplasia are less commonly encountered in younger, premenopausal women, and are associated with anovulatory cycles, such as in polycystic ovarian disease (PCOD) with or without Stein-Leventhal syndrome.[4–6] Prolonged treatment with estrogens in gonadal dysgenesis (Turner's syndrome, for example) may also result in endometrial hyperplasia.[7] Endometrial carcinoma is known to be associated with the same risk factors (anovulation, obesity, nulliparity, estrogen therapy), but despite the well-known association between hyperestrogenism and endometrial cancer a causal relationship between the two has not been demonstrated.[8,9] It was reported that the estradiol growth-promoting ability is associated with an impaired growth control mechanism, possibly related to genetic abnormalities consisting of the absence of a uterine "chalone" when progesterone is absent.[2]

Endometrial Hyperplasia

Endometrial hyperplasia has been produced in animal experiments by estrogenic treatment.[10] In humans it is definitely a hormone-related, estradiol-mediated condition, never seen in the absence of female gonads or without estrogen therapy. In prepubertal females, endometrial hyperplasia is extremely uncommon but is encountered with pseudoprecocious puberty due to feminizing ovarian tumors. Most frequently, endometrial hyperplasia is diagnosed in peri- and postmenopausal women who present with vaginal bleeding either as menometrorrhagia or postmenopausal bleeding. Endometrial biopsy sampling is mandatory since the diagnosis is a histologic one. Gross examination of the hyperplastic endometrium does not offer characteristic features. The dilation and curettage specimens and the tissue obtained by endometrial aspiration biopsy consists usually of tannish-gray fragments often mixed with blood clots, uniformly soft to rubbery. The histologic changes in endometrial hyperplasia range from an exaggerated proliferative endometrium to findings somewhat similar to those of neoplasia.

The most widely accepted classification of endometrial hyperplasia includes cystic-glandular hyperplasia, mild adenomatous or simple hyperplasia, moderate adenomatous or complex hyperplasia, and severe adenomatous or atypical hyperplasia (Fig. 7-1). Our classification combines the classic Gusberg classification[11] with the newer one proposed by Silverberg.[12] By no means is a progression to the more severe forms of endometrial hyperplasia inevitable in the individual patient. Most hyperplasias, including even severe or atypical, regress either spontaneously, by surgical curettage, or with progesterone therapy. Approximately 23% of severe atypical endometrial hyperplasia progresses to, or is seen concomitantly with, endometrial adenocarcinoma.[13]

The histologic diagnosis of endometrial biopsies should be based on a number of criteria that make a correct diagnostic classification possible in the majority of cases. These criteria concern the architecture of the tissue, including the relative proportions of glands and stroma, and the cytologic characteristics of individual cells.

Cystic-Glandular Hyperplasia

Cystic-glandular hyperplasia is known by its description as a Swiss cheese configuration of the glands. There is cystic dilatation of numerous glands alternating with normal-sized, tubular proliferative glands similar to those seen in the first part (estrogenically stimulated) of the menstrual cycle. The cystically dilated glands vary in size and shape

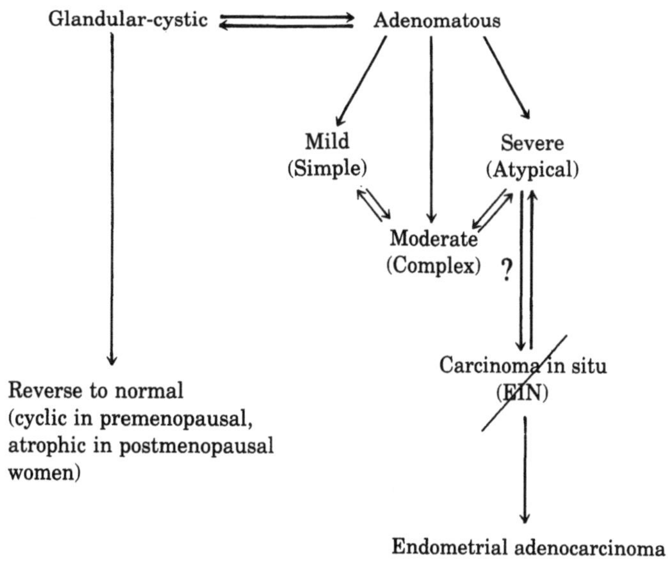

FIGURE 7-1. Classification of endometrial hyperplasia.

and are lined by a tall, columnar, and sometimes stratified epithelium, as opposed to the atrophic cuboidal or flat epithelium characteristic of endometrial cystic atrophy normally seen in the late postmenopausal endometrium. The stroma is present in normal amount, representing approximately half of the examined tissue. Occasional dilatation and thrombosis of blood vessels is seen, concomitant with bleeding episodes. The epithelial cells are regular and perpendicular on the basement membrane, and resemble those seen in the late proliferative endometrium (Fig. 7-2). Ciliated cells are not uncommon, resembling those seen in the fallopian tube lining, and are often reported as "tubal metaplasia." They are benign cells that develop generally under estrogenic influence.

Mitotic activity is present in both normally sized and cystically dilated glands, and in the stromal cells.

Adenomatous Hyperplasia

Mild Adenomatous Hyperplasia

Mild adenomatous, or simple glandular hyperplasia is an increase of both glands and stroma, similar to an exaggerated late proliferative endometrium, with tortuous tubular glands displaying a round or oval shape on cross-section. The glands are crowded with occasional "gland-in-gland" invaginations. The epithelial lining is composed of cuboidal to tall, columnar cells with large nuclei, containing a crisp, coarse chromatin network with prominent nucleoli. Mitoses are frequent and typical, resulting in symmetrical daughter cells. There is pseudostratification of the epithelium, but the main axis of the cells is perpendicular on the basement membrane and the cells are well oriented and mostly parallel to each other (Fig. 7-3). One an ultrastructural level, there is abundance of cilia, free ribosomes, rough

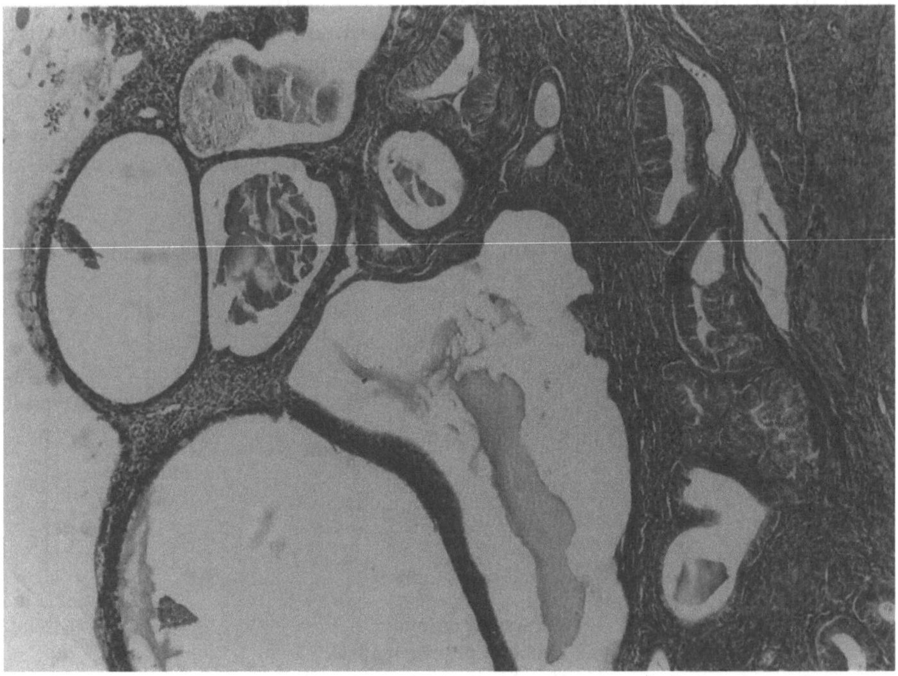

FIGURE 7-2. Cystic glandular hyperplasia of the endometrium (right) with adjacent adenomatous hyperplasia (left). (H&E, × 40.)

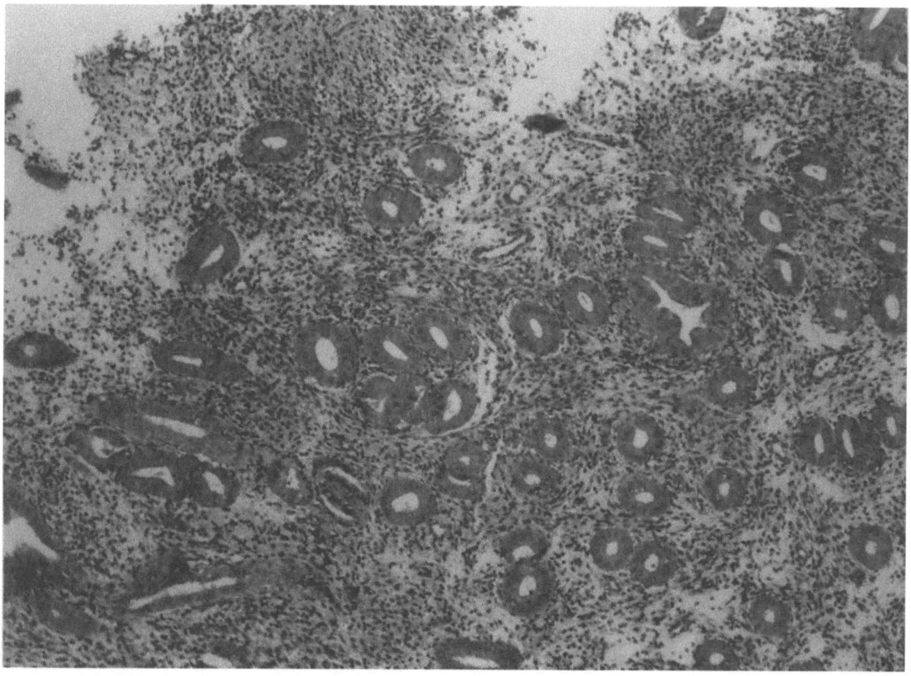

FIGURE 7-3. Mild adenomatous or simple glandular hyperplasia. The pattern is that of an exaggerated proliferative endometrium with occasional crowding of glands. (H&E, × 100.)

endoplasmic reticulum, and mitochondria; the cell is equipped for growth, with well-formed protein-producing and -assembling organelles. The polarity of the cytoplasmic organelles and of the nucleus is preserved. The amount of stroma between the glands is normal.

Moderate Adenomatous Hyperplasia

Moderate adenomatous, or complex glandular, hyperplasia consists of a more pronounced glandular proliferation with outpouching of the glands and tufts of epithelial cells protruding into the lumen, thus creating an irregularly shaped (complex) lumen. Occasional "bridging" with confluence of the epithelial buds is noticed (Fig. 7-4). Individual cells show no marked cytologic atypia. There is crowding of the glands with occasional back-to-back features, but stroma is always present, even if reduced in amount because of the expansion of the glands. The epithelial lining is piled up and nuclei are enlarged but uniform, with no marked atypia. Mitoses are common but not necessarily more numerous than in simple hyperplasia. Ultrastructurally, free ribosomes are very abundant and the cellular structure is somehow more simplified, showing reduced Golgi and vacuolar systems but numerous profiles of rough endoplasmic reticulum. DNA studies of cystic-glandular, mild, and moderate adenomatous hyperplasia reveal a preponderantly diploid-tetraploid range.[14]

Severe Adenomatous Hyperplasia

Severe adenomatous, or atypical, hyperplasia is a potentially precancerous state, often seen in the context of an overt endometrial adenocarcinoma. The glands have an irregular outline and are increased in number while the stroma is reduced or completely absent in areas in which the glands show a back-to-back configuration. Epithelial tufts and projections into the glandular lumen form "bridges" that further subdivide the

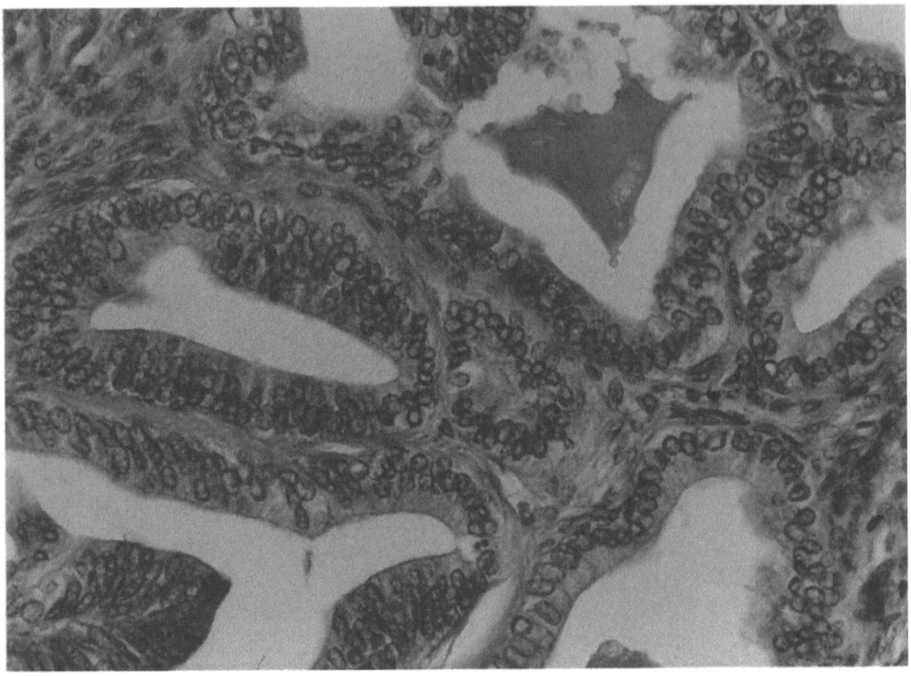

FIGURE 7-4. Moderate adenomatous (complex) endometrial hyperplasia. Glands show crowding, irregular lumen, and epithelial "budding." Stroma is present, although reduced, and epithelial cells are parallel and uniform. (H&E, × 200.)

glandular lumen, resulting in irregular, complex structures. Ciliated cells diminish in number with the increased severity of the hyperplasia. There is true stratification and loss of polarity of the epithelial cells, along with nuclear pleomorphism. Mitotic activity maybe high with atypical, asymmetrical mitoses. Squamous morules are occasionally seen consisting of round structures composed of immature but benign squamous cells (Fig. 7-5). Stromal foam cells are often encountered in adenomatous hyperplasia and in endometrial adenocarcinoma. When necrosis and inflammation are absent, these plump mesenchymal cells with relatively small nuclei and foamy lipid-containing cytoplasm represent a site of storage for lipids (possibly converting them into sex steroids), and are usually seen in a context of hyperestrogenism[15,16] and commonly in obesity (Fig. 7-6). Their precise function is still unknown.

The histologic characteristics of atypical endometrial hyperplasia are often interpreted as endometrial adenocarcinoma. The differential diagnosis is based on the invasion of the stroma[7] and the presence of a cribriform configuration in which glandular spaces are separated only by epithelial tissue in endometrial carcinoma.[12,17] The presence of necrosis and acute inflammatory exudate is also seen more commonly in endometrial adenocarcinoma than in adenomatous hyperplasia. There is, however, great variation in the evaluation of this histologic lesion by individual pathologists. The term "carcinoma in situ" or "endometrial intraepithelial neoplasia" [EIN; in analogy to early intraepithelial neoplasms of the cervix (CIN), vulva (VIN), etc.], applied to lesions characterized by a cribriform glandular pattern, nuclear atypia, and DNA aneuploidy,[9] is not considered suitable for the endometrium[12] and is not included in the World Health Organization classification.[13] For practical purposes, severe adenomatous hyperplasia according to Gusberg,[11] or atypical endometrial hyperplasia,[12] is an indication for hysterectomy, especially in postmenopausal

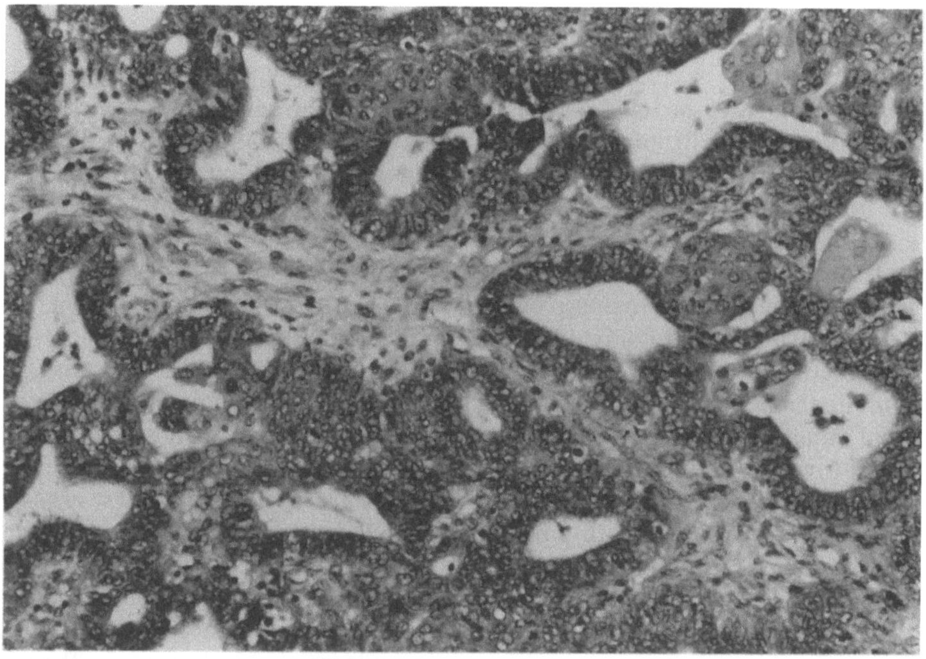

FIGURE 7-5. Severe (atypical) adenomatous hyperplasia. Glands are irregular and very crowded, with only minimal intervening stroma. Note epithelial bridging and squamous "morules". (H&E, × 200.)

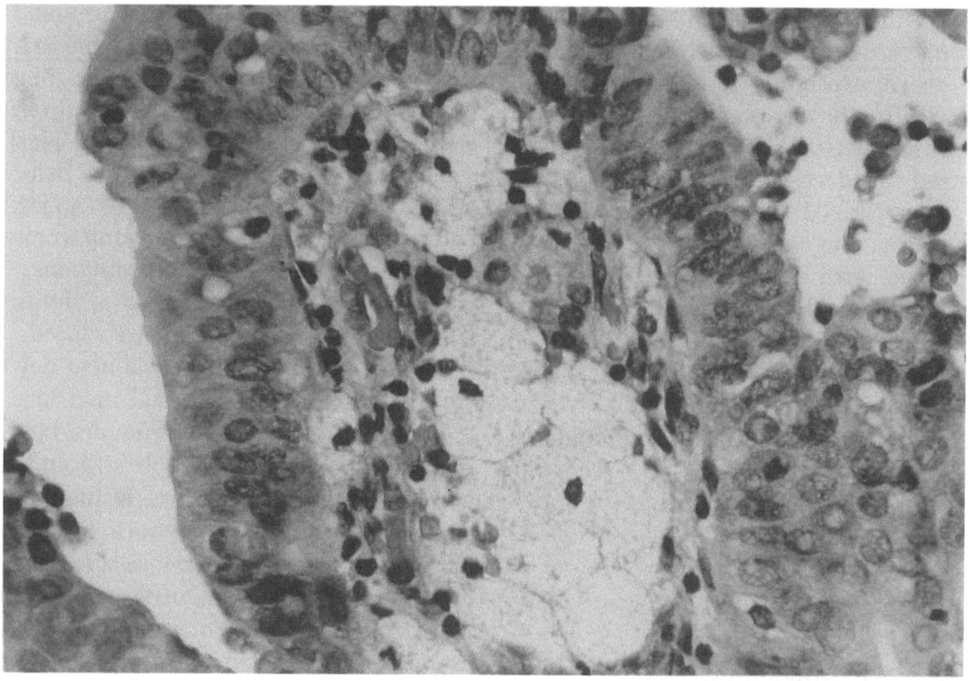

FIGURE 7-6. Severe adenomatous (atypical) hyperplasia with stromal foam cells. Epithelial cells are loosing their polarity and show atypical nuclei. (H&E, × 100.)

patients. The old German saying "Nicht Karzinoma aber besser heraus" has kept its validity.

The progesterone receptor level decreases with the increasing severity of endometrial hyperplasia.[18,19] In atypical endometrial hyperplasia, ultrastructural findings include decrease of cilia and microvilli and bizarre microfilaments with an increased rough endoplasmic reticulum. DNA microspectrophotometry demonstrates aneuploidy in some cells with prolonged S phase, similar to the findings in endometrial adenocarcinoma.[9] Biochemical and ultrastructural studies have pointed toward a spectrum of changes that ranges from proliferative to hyperplastic and neoplastic endometrium in parallel with the progression in severity of the histologic changes.[12,13,18]

Endometrial Polyps

These are common focally hyperplastic processes that often lead to vaginal bleeding in both pre- and postmenopausal women. Grossly, endometrial polyps are ovoid to elongated structures often showing hemorrhagic areas, especially at their distal tip. Histologically, they display a stalk composed of fibrovascular tissue with thickened, often hyalinized blood vessels, surrounded by irregularly arranged glands. The glandular configuration ranges from atrophic cystic to hyperplastic, with a pattern similar to that in the nonpolypoid endometrial lesions. In general, endometrial polyps should not be considered precancerous, except for those showing severe, atypical endometrial hyperplasia. Isolated endometrial carcinomas in polyps are very rare.

Endometrial Carcinoma

General Considerations

Endometrial carcinoma is presently the most common neoplasm of the female pelvis in the United States and in industrialized Western societies and the third most common cancer in women, after breast and colorectal cancer. Its incidence rose dramatically over the past decades, with a somewhat stationary incidence over the past few years. The mortality due to this cancer is slightly decreasing.[19] The risk factors are those of adenomatous hyperplasia—age, nulliparity, obesity (50 pounds of excess weight increases the risk nine times), and estrogen intake[20]—and also include the poorly understood etiopathogenetic factors of heredity, host response, and immunologic incompetence that may lead to cancer in any other site. The American Cancer Society estimated that there were about 40,000 new cases of endometrial carcinoma in 1986, with a mortality of 2,900 per year.[19–21] The peak incidence is between 58 and 60 years of age.

Endometrial adenocarcinoma is generally a neoplasm of elderly women; therefore, the increased life span is an additional risk factor. Young women rarely develop endometrial adenocarcinoma (about 4% of endometrial carcinoma is diagnosed in women under 40 years of age), and cases that histologically resemble endometrial adenocarcinoma may represent reactive hyperplasia due to hyperestrogenism. The histopathologic diagnosis of endometrial adenocarcinoma in young women has to be verified by at least one gynecologic pathology specialist before deciding on a hysterectomy. The majority of endometrial carcinomas in young women are well-differentiated adenocarcinomas, with only a few showing myometrial invasion.[12] Endometrial carcinoma associated with Stein-Leventhal syndrome is also not associated with deep myometrial invasion.[4]

The most common risk factors for endometrial adenocarcinoma, obesity and estrogen therapy, affect a more affluent female population that is also more likely to be under medical care; therefore there is a high rate of early detection. Other sources of hyperestrogenism are functional ovarian tumors, such as thecomas and granulosa cell tumors; ovarian hyperthecosis with stromal luteinization; anovulatory cycles, including those related to PCOD, in premenopausal patients; nulli- and oligoparity; and, rarely, liver diseases. It seems that the natural history of endometrial adenocarcinoma arising in a setting of hyperestrogenism, especially associated with adenomatous hyperplasia, which reflects histologically the patient's hyperestrogenism, is that of a rather low-grade neoplasm[12,19,22–26]

There is a considerable geographic difference in the incidence of endometrial carcinoma between the industrialized nations and the Third World populations. For example, endometrial carcinoma occurs in 33.8:100,000 women in West Germany and 45.8:100,000 in Alameda County, California, as compared to 1.8:100,000 in India and 3.3:100,000 in Brazil.[21] Statistical data from many countries, however, are incomplete and unreliable because of the common use of the term "cancer of uterus," lumping together cervical and corporeal cancer, as cause of death. In fact, an inverse relationship with uterine cervix cancer is seen worldwide. Certain studies[27,28] have pointed out that endometrial adenocarcinoma in multiparous women without risk factors related to hyperestrogenism is a more aggressive cancer. The diagnosis of endometrial adenocarcinoma in patients with no known associated hyperestrogenism is frequently made in more advanced stages of the disease, which are more lethal. According to some of our own investigations the endometrial adenocarcinomas with associated adenomatous hyperplasia are better differentiated and less invasive than those without associated adenomatous hyperplasia.[24] Progesterone receptors are known to be present in higher concentrations in better differentiated endometrial adenocarcinomas,[29] and are also found in considerably higher levels in endometrial adenocarcinoma associated with adenomatous hyperplasia, and therefore in hyperestrogenic patients.[30] Progesterone receptors were either absent or at very low levels in patients with no histologic evidence of hyperestrogenism.[30–32]

These findings suggest that endometrial adenocarcinoma either presents in two different morphologic and biologic forms, one more virulent than the other, or that the different host setting in hyperestrogenic patients implies a different tumoral histology and behavior.[26,27,31,33–36]

Pathologic Findings

Endometrial adenocarcinoma is usually diagnosed on endometrial aspirates or uterine curettage. A friable, sometime malodorous, brownish tan material, the amount of which does not necessarily have to be abundant, is grossly suggestive of endometrial adenocarcinoma. On hysterectomy specimens, the uterus may or may not be enlarged, sounding beyond 8 cm. The surface of the endometrium appears irregular and whitish, usually fungating into the cavity and sometimes infiltrating the myometrium. The entire endometrial cavity may be involved by tumor, which usually does not extend to the endocervical cavity beyond the internal cervical os (Fig. 7-7). In many cases, the endometrial adenocarcinoma involves only focal areas of the endometrial lining, most often the fundus, or is distributed in patchy areas. Not uncommon are the cases in which the endometrial biopsy showed endometrial adenocarcinoma and the hysterectomy specimen failed to reveal any malignancy. In these cases, an extensive sampling of the entire uterine wall with careful study of the entire endometrial lining is recommended and may yield hyperplastic endometrial changes with microscopic foci of malignancy.

Histologic Grading

Histologically, endometrial adenocarcinoma is a glandular cancer in the vast majority of cases. Arising from the endometrial glands, the glandular components of the tumor are somehow reminiscent of the normal proliferative endometrium, especially in the well-differentiated (grade 1) carcinomas in which more than 90% of the tumor is glandular. The endometrial stroma is reduced or absent as a result of the invasion by neoplastic glands, which show a cribriform and back-to-back configuration (Fig. 7-8). In moderately differentiated endometrial adenocarcinomas (grade 2) the glandular pattern is present in 50 to 90% of the specimen, with

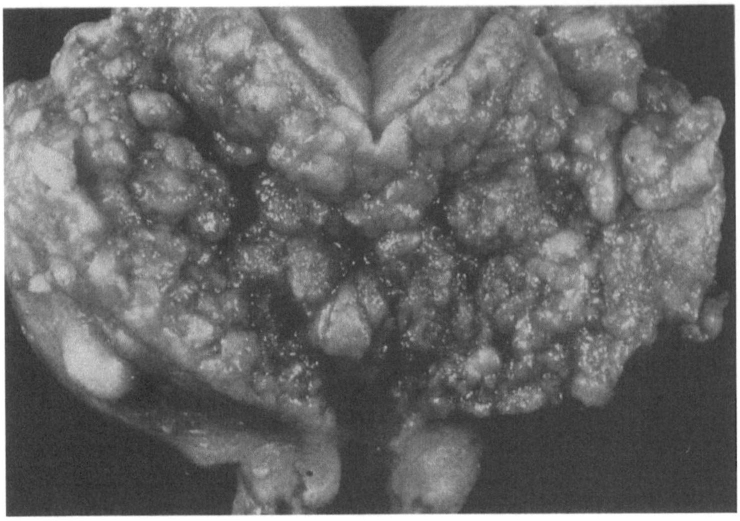

FIGURE 7-7. Endometrial carcinoma involving the entire uterine cavity extending to the internal os of the endocervical canal. Hysterectomy specimen.

solid nests and sheets of tumor cells replacing the glands. The individual tumor cells, which are relatively regular and radially oriented in the well-differentiated tumors, are randomly distributed in the moderate tumors, showing pleomorphic nuclei with intranuclear clearing, coarse clumps of chromatin, and multiple irregular nucleoli (Fig. 7-9). Mitoses are numerous and atypical. The poorly differentiated endometrial adenocarcinoma (grade 3) shows less than 50% glandular differentiation, and consists mostly of solid sheets of tumor cells with a bizarre nuclear configuration, often with multinucle-

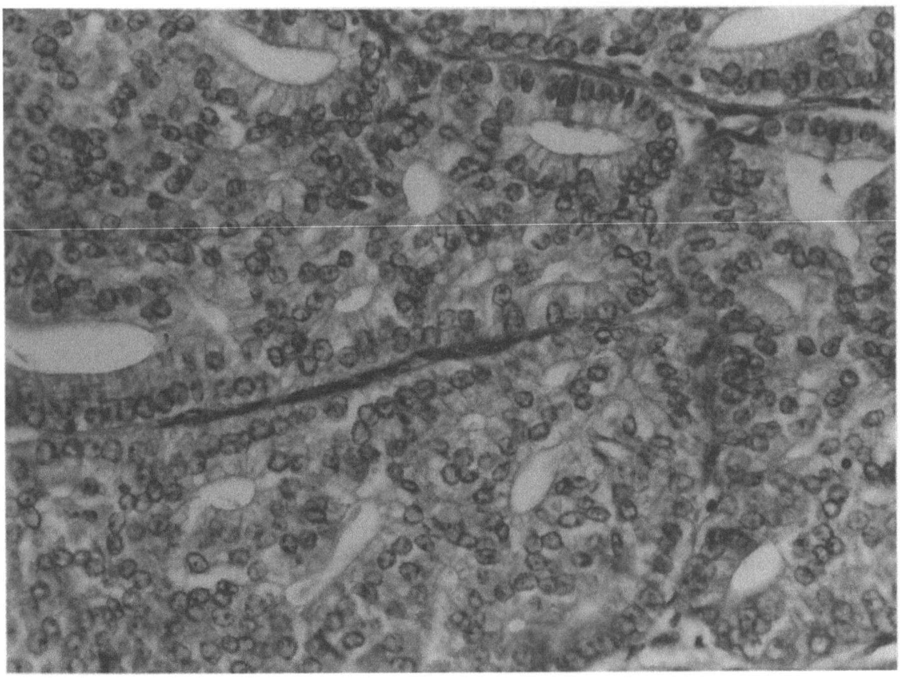

FIGURE 7-8. Well-differentiated adenocarcinoma of the endometrium. Note the cribriform pattern and absence of stroma. (H&E, × 200.)

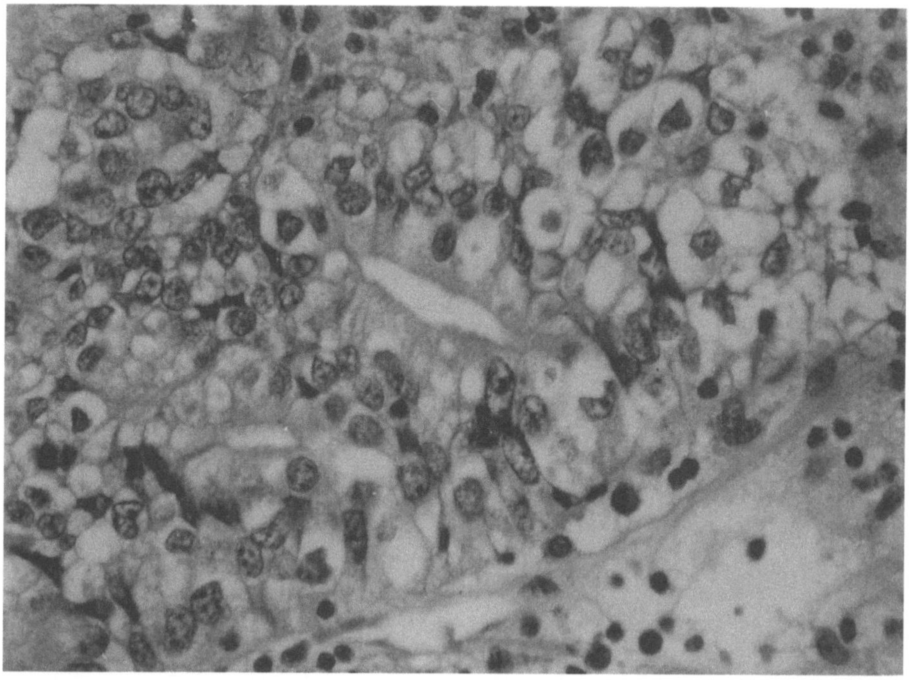

FIGURE 7-9. Moderately differentiated adenocarcinoma of the endometrium. Solid sheets of tumor cells replace glandular structures. Nuclei show clearing, a coarse chromatin pattern, and irregular nucleoli. (H&E, × 400.)

ated giant tumor cells (Fig. 7-10). Most cells have a high nucleus-cytoplasm ratio with very scanty basophilic cytoplasm and marked intranuclear clearing as a result of the segregation of chromatin, characteristic of an intensely proliferating tissue. Anaplastic carcinomas of the endometrium show no glandular differentiation and a very marked degree of cellular anaplasia, with the resemblance to endometrial tissue being difficult to ascertain. The degree of differentiation is closely correlated with the 10-year actuarial survival rate of the patients, ranging from 97% for grade 1 tumors to 71% for grade 3 tumors[23] in stage I endometrial adenocarcinoma.

Histologic Variants

The histologic variants most often encountered in endometrial adenocarcinoma other than glandular are those seen in other organs of the müllerian and extended müllerian system.[37,38] Squamous, mucinous, papillary, and clear cell carcinomas are often intermingled with the glandular elements or may constitute the endometrial carcinoma by themselves. An interesting observation is that the endometrial adenocarcinoma in its pure, glandular form and associated with squamous elements is seen more often in hyperestrogenic patients in whom adenomatous hyperplasia is also present.[24,31] In a number of cases, concomitant ovarian endometrioid carcinoma is found simultaneously with endometrial neoplasm with or without endometriosis, displaying similar histologic characteristics.[39] Papillary, clear cell, and anaplastic endometrial cancers are seen in patients with no associated adenomatous hyperplasia, usually presenting in more advanced stages of the disease and with a more aggressive behavior.

Adenocarcinoma with Squamous Elements

Adenocarcinoma with squamous elements is very common. A subdivision of this group of

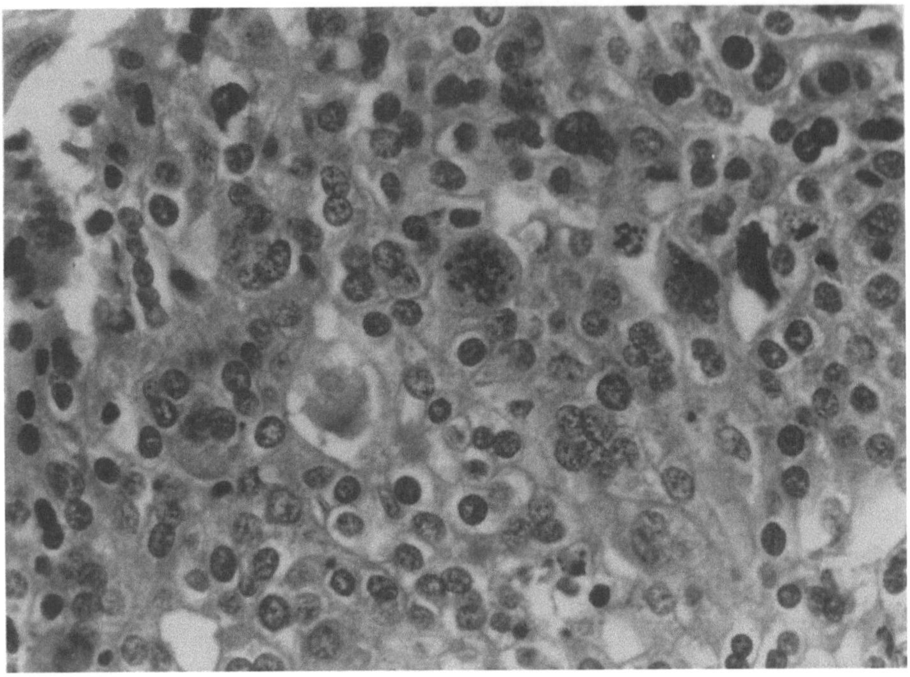

FIGURE 7-10. Poorly differentiated endometrial adenocarcinoma. Glandular differentiation of the tumor is minimal; note bizarre and multinucleated tumor cells with atypical mitotic activity. (H&E, × 400.)

cancers into adenoacanthoma (with benign squamous cells), presumably with a better prognosis, and adenosquamous carcinoma (with malignant squamous cells), with a poorer prognosis, has not been proven to be useful, despite its use by some investigators.[40] The degree of differentiation of the glandular rather than of the squamous components of the tumor is of prognostic significance.[26,41]

Benign squamous elements are encountered in nonmalignant conditions of the endometrium, either as squamous metaplasias in cases of pyometria or in elderly patients with hematometria or, more often, as squamoid "morules" that are globular structures composed of immature squamous epithelium, present also in adenomatous hyperplasia.[42,43] In endometrial adenocarcinoma, the squamous components may present as immature, spindle-shaped cells or as keratinizing lesions, often with parakeratotic pearls. Glandular structures appear intermingled with squamous elements (Fig. 7-11).

Primary pure squamous cell carcinoma of the endometrium is very rare and may occur through squamous metaplasia from subcolumnar reserve cells.[44,45]

Mucinous Endometrial Adenocarcinoma

Mucinous endometrial adenocarcinoma is encountered in about 9% of endometrial adenocarcinoma cases.[37,46] The tumors are glandular and resemble endocervical carcinoma. They are seen along with the "common" endometrial adenocarcinoma and, rarely, in a "pure" form, in which a primary endocervical cancer can only be ruled out in the absence of tumor in the endocervical canal (Fig. 7-12). Well-differentiated adenocarcinomas may contain mucin-secreting tumor glands with no implication of their behavior or prognosis.

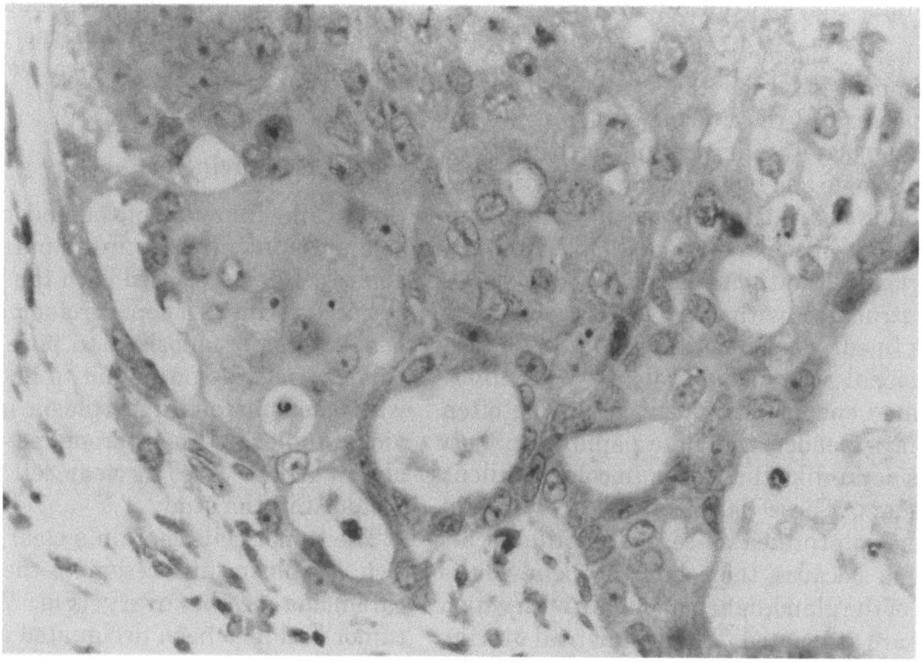

FIGURE 7-11. Glandular tumor of the endometrium intermingled with squamous elements. (H&E, × 400.)

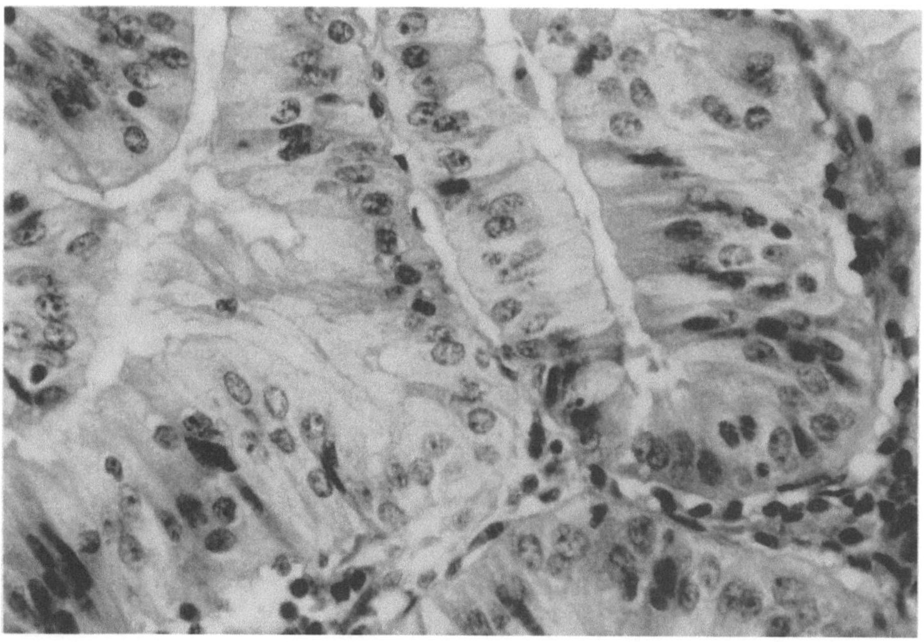

FIGURE 7-12. Mucinous adenocarcinoma of endometrium. The tumor is well differentiated and contains intracellular and intraluminal mucus. (H&E, × 400.)

Papillary Endometrial Adenocarcinoma

Papillary endometrial adenocarcinoma is uncommon in its pure form. Hendrickson et al.[47] described the uterine serous papillary carcinoma as a separate entity with different morphologic and biologic characteristics, resembling ovarian carcinoma in its configuration and mode of spread (Fig. 7-13). Histologically, the tumor is composed of papillary structures with a coarse fibroconnective stalk lined by anaplastic epithelial cells with occasional psammoma bodies.[48] Well-differentiated endometrial adenocarcinoma often displays slender, frondlike papillary projections not unlike those seen in colonic villous adenoma. These papillary features should not be interpreted as serous papillary carcinoma because their biologic behavior is that of the glandular components with which they are associated. The endometrial uterine serous papillary carcinoma often invades the myometrium, spreads to the peritoneum, and requires a therapeutic regimen similar to that for ovarian carcinoma. In our previous studies, it has not been found to be associated with adenomatous hyperplasia, and progesterone receptors were at very low levels.[24,30,31]

Clear Cell Carcinoma

Clear cell carcinoma of the endometrium is also uncommonly seen in its pure form and is most often associated with the less differentiated forms of adenocarcinoma.[26,49,50] It has a worse prognosis than pure adenocarcinoma and was not found to be associated with adenomatous hyperplasia in our previous studies.[24] Progesterone receptors were absent in the cases of clear cell carcinoma included in the study.[30,31]

The tumor cells contain a clear, glycogen-rich cytoplasm and resemble the clear cell carcinoma of the ovary (Fig. 7-14). This tumor had also been designated as mesonephroid because of a superficial resemblance to renal cell carcinoma, but ultrastructural studies have shown marked differences.[51] In the female genital tract, clear cell carcino-

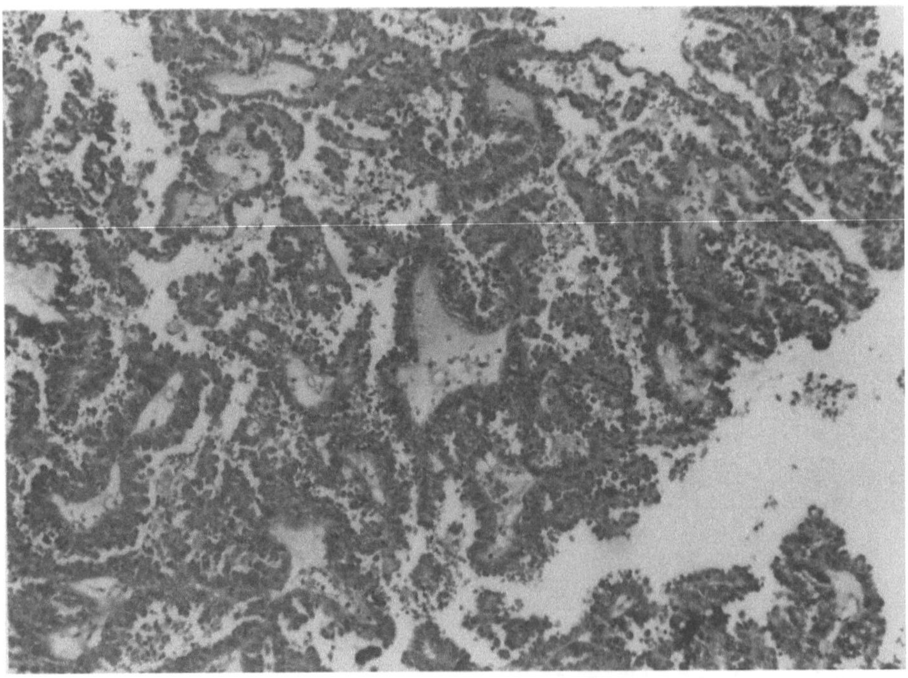

FIGURE 7-13. Uterine serous papillary carcinoma. The histologic pattern is similar to that of ovarian papillary tumors. (H&E, × 100.)

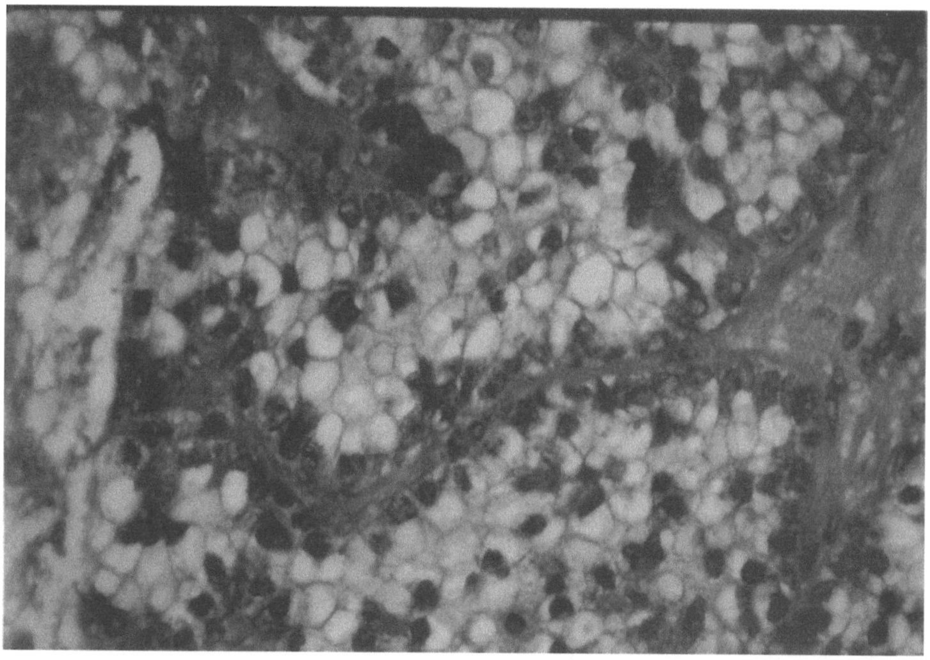

FIGURE 7-14. Clear cell endometrial carcinoma. Tumor cells contain clear cytoplsm, rich in glycogen. (H&E, × 400.)

mas are also seen in offsprings of diethylstilbestrol-exposed women, in association with adenosis, in the vagina and exocervix and in the endocervical canal. A primary origin in the endometrium of the pure clear cell carcinoma can be ascertained only by negative findings in other locations. Microscopically, the tumor cells are arranged in tubules or in solid sheets and occasionally show a hobnail configuration. Secretory carcinoma is an uncommon variant that resembles the secretory endometrium and must be distinguished from clear cell carcinoma because of its more favorable prognosis.[52] It shows relatively regular nuclei with uniform sub- and supranuclear vacuoles, as opposed to the pleomorphic cells in clear cell carcinoma.

Invasion and Spread

Endometrial carcinoma invades the stroma by destroying the glandular basement membrane,[53,54] producing a local immune host response with lymphocytic infiltrates.[55] Stromal foam cells are often seen in well-differentiated adenocarcinoma (Fig. 7-15), as well as in endometrial hyperplasia.[56] Their presence is associated with a high level of progesterone receptors and better histologic differentiation.[31] They are not related to the invasive neoplastic process. Invasion of the myometrium by endometrial adenocarcinoma is diagnosed by the finding of endometrial neoplasm surrounded by smooth muscle; often adenomyosis exhibits a neoplastic transformation of its glandular component, but the presence of surrounding stroma and a round shape of the focus of adenomyosis (Fig. 7-16), rather than the usual irregular configuration of invasive cancer, helps to diagnose a malignancy in adenomyosis. The presence of endometrial carcinoma in adenomyosis located deep in the myometrium may simulate an invasive adenocarcinoma but does not represent a real invasion of the myometrium.[57]

Endometrial adenocarcinoma spreads more often via lymphatics than by a hematogenous route to the pelvic and periaortic

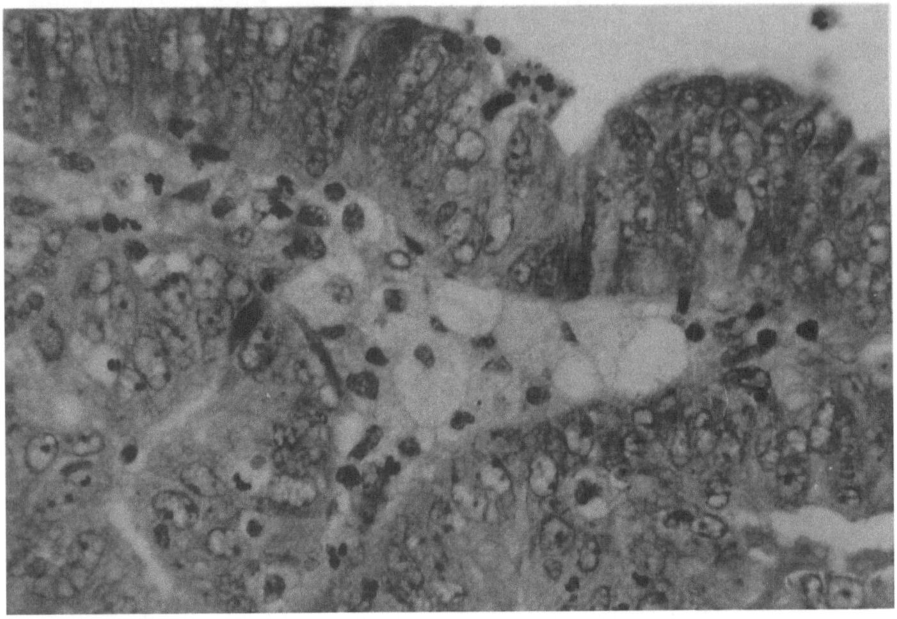

FIGURE 7-15. Endometrial adenocarcinoma with stromal foam cells. Patient was morbidly obese and had adenomatous hyperplasia on previous endometrial biopsies and concomitant with tumor. (H&E, × 40.)

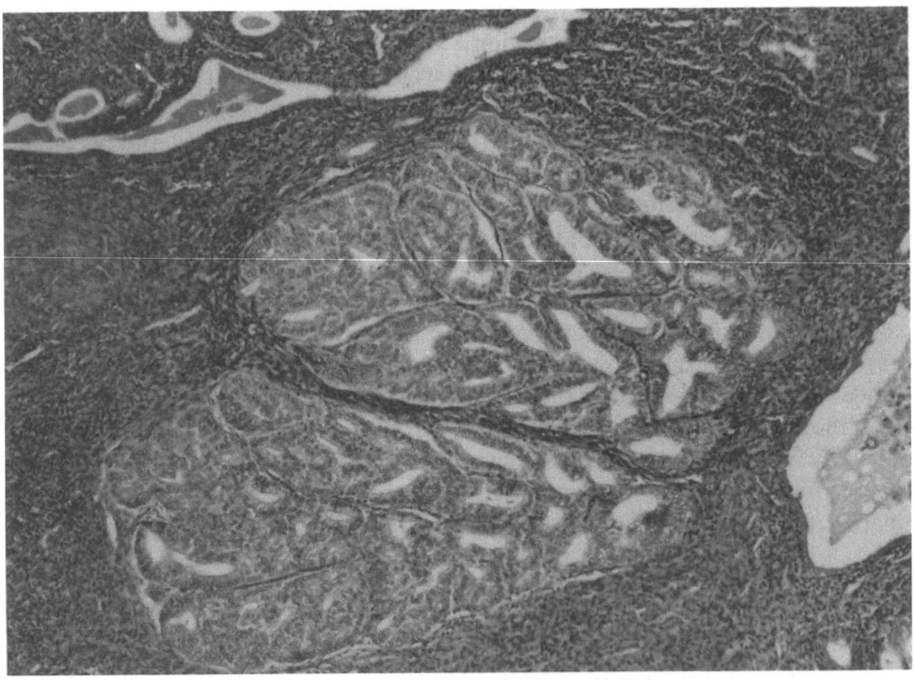

FIGURE 7-16. Well-differentiated adenocarcinoma in adenomyosis. Note benign endometrial glands and stroma at the periphery. (H&E, × 10.)

lymph nodes. Vascular invasion is usually associated with high-grade tumors and deep myometrial invasion. Metastatic endometrial adenocarcinoma involves the rectum, urinary bladder, lungs, brain, and any other distant organ. The overall survival is closely correlated with grading and staging.

According to the FIGO (International Federation of Gynecology and Obstetrics) classification, stage I includes about 80% of patients, and consists of a superficially spreading endometrial carcinoma; myometrial invasion, when present, involves less than the inner half of the thickness of the myometrium in most cases. The endometrial carcinoma is confined to the uterine corpus. In stage II, the endometrial carcinoma involves corpus and cervix. In stage III, the endometrial carcinoma extends outside the uterus but not beyond the pelvis. In stage IV the endometrial carcinoma is outside the pelvis and/or involves the rectum and/or urinary bladder. Details on staging and therapeutic procedures are given in Chapter 26.

Five-year survival is 90% for stage I, 50% for stage II, 20% for stage III, and 8% for stage IV.[26]

In addition to staging and grading, presence of lymph node involvement, peritoneal cytology, and hormone receptor status are helpful for the adequate diagnostic work-up of the patient with endometrial carcinoma.

Tumor antigens in endometrial carcinoma have been studied by immunopathology methods that included the identification of endogenous peroxidase activity,[18] and carcinoembryonic antigens[58]; however their use for diagnostic and follow-up purposes is limited.

Conclusion

Endometrial carcinoma has a high morbidity and a relatively low mortality. When associated with known risk factors leading to hyperestrogenism it is less virulent and has a generally good prognosis. The relatively smaller proportion of patients with no identified risk factors (other than family history of cancer), involving often an older and economically more deprived female population, should be the focus of preventive measures aimed at early diagnosis. These "independent" (from hormones) endometrial carcinomas have a more aggressive biologic behavior and a higher mortality.

References

1. Gurpide E, Tseng L, Gusberg SB: Estrogen metabolism in normal and neoplastic endometrium. Am J Obstet Gynecol 1977; 129:809–816.
2. Gorsky J, Stormshak F, Harris J, et al: Hormone regulation of growth: stimulatory and inhibitory influence of estrogens on DNA synthesis. J Toxicol Environ Health 1977;3:271–272.
3. Muechler EK, Flickinger GL, Mangan CE, et al: Estradiol binding by human endometrial tissue. Gynecol Oncol 1975;3:244.
4. Fechner RE, Kaufman RH: Endometrial adenocarcinoma in Stein-Leventhal syndrome. Cancer 1974;34:444–452.
5. Chanlian DL, Taylor HB: Endometrial hyperplasia in young women. Obstet Gynecol 1970;36:659–666.
6. Jackson RL, Dockerty MB: The Stein-Leventhal syndrome. Analysis of 43 cases with special references to association with endometrial adenocarcinoma. Am J Obstet Gynecol 1957;73:161–173.
7. Louka MH, Ross RD, Lee JH, et al: Endometrial carcinoma in Turner's syndrome. Gynecol Oncol 1978;6:294.
8. Hulka BS, Kaufman DG, Fowler WC, et al: Predominance of early endometrial cancers after long term estrogen use. JAMA 1989; 244:2419–2422.
9. Ferenczy A: Cytodynamics of endometrial hyperplasia and neoplasia. Part II: In vitro DNA histo-auto-radiography. Hum Pathol 1983; 14:77–82.
10. Papadaki L, Beilby JOW, Chowaniec J, et al: Hormone replacement therapy in the menopause; a suitable animal model. J Endocrinol 1979;83:67–77.
11. Gusberg SB: Precursors of corpus carcinoma: estrogens and adenomatous hyperplasia. Am J Obstet Gynecol 1947;54:905–927.
12. Silverberg SG: Hyperplasia and carcinoma of the endometrium. Semin Diagn Pathol 1988; 5:135–153.

13. Kurman RJ, Norris HF: Endometrial hyperplasia and metaplasia, in Kurman RJ (ed): Blaustein's Pathology of the Female Genital Tract, ed. 3. New York, Springer-Verlag, 1987, pp 322–337.
14. Sachs H, Wambach EV, Würthner K: DNA content of normal, hyperplastic and malignant endometrium determined cytophotometrically. Arch Gynaekol 1974;217:349–365.
15. Dallenbach-Hellweg G: Foam cells and estrogen activity of the human endometrium. Arch Gynaek 1974;217:335–341.
16. Fechner RE: Ultrastructure of endometrial foam cells. Am J Clin Pathol 1979;72:628–633.
17. Kurman RJ, Norris HJ: Evaluation of criteria for distinguishing atypical endometrial hyperplasia from well-differentiated carcinoma. Cancer 1982;49:2547–2559.
18. Fenoglio CM, Crum CP, Ferenczy A: Endometrial hyperplasia and carcinoma: are ultrastructural, biochemical and immunocytochemical studies useful in distinguishing between them? Pathol Res Pract 1982;174:257–284.
19. Disaia PJ, Creasman WT: Adenocarcinoma of the uterus, in Clinical Gynecologic Oncology, ed 3. St. Louis: CV Mosby Co, 1989, pp 161–162.
20. Disaia PJ, Creasman WT, Boronow MD, et al: Risk factors and recurrent patterns in stage I endometrial cancer. Am J Obstet Gynecol 1985;151:1009–1015.
21. Mahboudi E, Eyler N, Wynder EL: Epidemiology of cancer of the endometrium. Clin Obstet Gynecol 1982;25:5–17.
22. Beckner ME, Mori T, Silverberg SG: Endometrial carcinoma: non-tumor factors in prognosis. Int J Gynecol Pathol 1980;4:131–145.
23. Christopherson WM, Connelly PJ, Alberhasky RC, et al: Carcinoma of the endometrium V. An analysis of prognosticators in patients with favorable subtypes and stage I disease. Cancer 1983;51:1705–1709.
24. Deligdisch L, Cohen CJ: Histologic correlates and virulence implications of endometrial carcinoma associated with adenomatous hyperplasia. Cancer 1985;56:1452–1455.
25. Gusberg SB: Current concepts in cancer. The changing nature of endometrial cancer. N Engl J Med 1980;302:719–731.
26. Kurman RJ, Norris HJ: Endometrial carcinoma, in Kurman RJ (ed): Blaustein's Pathology of the Female Genital Tract, ed 3. New York, Springer-Verlag 1987, pp 338–372.
27. Bokhman JV: Two pathogenetic types of endometrial carcinoma. Gynecol Oncol 1983; 15:10–17.
28. Silverberg SG, Sasano N, Yajima A: Endometrial adenocarcinoma in Miyagi Prefecture, Japan: Histopathologic analysis of a cancer-based series and comparison with cases in American women. Cancer 1982; 49:1504–1510.
29. Creasman WT, Soper JT, McCarty KS Jr, et al: Influence of cytoplasmic steroid receptor content on prognosis of early stage endometrial carcinoma. Am J Obstet Gynecol 1985; 151:922–932.
30. Deligdisch L, Holinka CF: Progesterone receptors in two groups of endometrial adenocarcinoma. Cancer 1986;57:1385–1388.
31. Deligdisch L, Holinka CF: Endometrial carcinoma: two diseases? Cancer Detection Prevent 1987;10:237–246.
32. Holinka CF, Deligdisch L, Gurpide E: Histologic evaluation of in vitro-responses of endometrial adenocarcinoma to progestins and their relation to progesterone receptor levels. Cancer Res 1984;44:293–296.
33. Gray LA, Christopherson WM, Hoover RN: Estrogens and endometrial carcinoma. Obstet Gynecol 1977;49:385–389.
34. Lucas WE, Yen SC: A study of endocrine and metabolic variables in postmenopausal women with endometrial carcinoma. Am J Obstet Gynecol 1979;134:180–186.
35. Silverberg SG, Muller D, Faraci JA, et al: Endometrial adenocarcinoma: clinical pathologic comparison of cases in postmenopausal women receiving and not receiving estrogens. Cancer 1985;56:1452–1455.
36. Smith M, McCartney AJ: Occult high risk endometrial cancer. Gynecol Oncol 1985; 22:154–161.
37. Hendrickson M, Ross J, Eifel P, et al: Adenocarcinoma of the endometrium: analysis of 256 cases with carcinoma limited to the uterine corpus. Gynecol Oncol 1982;13:373–392.
38. Laughlan SC: The secondary Mullerian system. Obstet Gynecol Surv 1972;27:133–143.
39. Eifel P, Hendrickson M, Ross J, et al: Simultaneous presentation of carcinoma involving the ovary and uterine corpus. Cancer 1982; 50:163–170.
40. Ng ABP, Reagan JW, Storaasli JP, et al:

Mixed adenosquamous carcinoma of the endometrium. Am J Clin Pathol 1973;59:765–781.
41. Zaino RJ, Kurman RJ: Squamous differentiation in carcinoma of the endometrium: a critical appraisal of adeno-acanthoma and adenosquamous carcinoma. Semin Diagn Pathol 1988;5:154–171.
42. Dutra F: Intraglandular morules of the endometrium. Am J Clin Pathol 1959;31:60–65.
43. Tang EY, Bonfiglio TA, Tanga K: Effect of estrogen and progesterone on the development of endometrial hyperplasia in the Fischer rat. Biol Reprod 1984;31:399–413.
44. Kay S: Squamous cell carcinoma of the endometrium. Am J Clin Pathol 1974;61:264–269.
45. Silverberg SG: Significance of squamous elements in carcinoma of the endometrium: a review. Prog Surg Pathol 1981;4:115–119.
46. Ross J, Eifel PH, Cox RS, et al: Primary mucinous adenocarcinoma of the endometrium. A clinico-pathologic and histochemical study. Am J Surg Pathol 1983;7:715–729.
47. Hendrickson M, Ross J, Eifel P, et al: Uterine papillary serous carcinoma. A highly malignant form of endometrial adenocarcinoma. Am J Surg Pathol 1982;6:93–108.
48. Kuebler DL, Nikrui N, Bell DA: Cytologic features of endometrial papillary serous carcinoma. Acta Cytol 1989;33:120–126.
49. Kurman RJ, Scully RE: Clear cell carcinoma of the endometrium. An analysis of 21 cases. Cancer 1976;37:872–882.
50. Silverberg SG, De Georgil LS: Clear cell carcinoma of the endometrium: clinical pathologic and ultrastructural findings. Cancer 1973;31:1127–1140.
51. Nilsson O: Electron microscopy of human endometrial carcinoma. Cancer Res 1962; 22:491–494.
52. Tobon H, Watkins GJ: Secretory adenocarcinoma of the endometrium. Int J Gynecol Pathol 1985;4:328–335.
53. Bulletti C, Galassi A, Jassoni VM, et al: Basement membrane components in normal, hyperplastic and neoplstic endometrium. Cancer 1988;62:142–149.
54. King A, Seraj IM, Wagner RJ: Stromal invasion in endometrial carcinoma. Am J Obstet Gynecol 1984;149:10–14.
55. Deligdisch L: Morphologic correlates of host response in endometrial carcinoma. Am J Reprod Immunol 1982;2:54–57.
56. Dewagne MP, Silverberg SG: Foam cells in endometrial carcinoma. A clinico-pathologic study. Gynecol Oncol 1982;13:67–75.
57. Hernandez E, Woodruff JD: Endometrial adenocarcinoma arising in adenomyosis. Am J Obstet Gynecol 1980;138:827–832.
58. Van Nagel J, Donaldson ES, Hanson MB, et al: The role of carcinoembryonic antigen in the management of patients with gynecologic cancer. Am J Diagn Gynecol Obstet 1979; 1:103–107.

8

Endometrial Response to Hormonal Therapy

LIANE DELIGDISCH

General Considerations

Hormone therapy is used presently for a variety of conditions, ranging from oral contraceptives for family planning to antiestrogens for neoplastic conditions. The histologic changes in the endometrium resulting from these influences are still poorly defined and encompass a wide variety of morphologic features. The endometrium is an extremely sensitive target tissue for steroid sex hormones, able to modify its structural characteristics with promptitude and versatility. In order to comprehend the changes induced by hormone therapy, a brief review of the "natural" hormone-related endometrial changes seems appropriate.

During the reproductive life, in normal cycles the menstrual shedding is followed by endometrial proliferation under estrogenic stimulation. The endometrial thickness increases over tenfold as a result of active growth of glands, stroma, and blood vessels. This proliferative, estrogen-induced phase has a variable length, from 10 to 20 days, with an ideal duration of 14 days. The presence of estrogen receptors in the nucleus of the endometrial cells is responsible for the prompt translation of hormonal impulses into structural changes, including protein synthesis by free ribosomes by the rough endoplasmic reticulum and accumulation of intermediate filaments, mitochondria, Golgi apparatus and lysosomes. A brisk mitotic activity leads to increase in the number of cells, in addition to the increase of the individual cell mass.

After ovulation, the secretion of progesterone inhibits the proliferative activity of the endometrium and induces a complex activity consisting of polarization of the glycogen to a subnuclear location, followed by its transport via microfilaments to the apical region of the cell. An abundant Golgi apparatus "packages" the mucopolysaccharides and glycoproteins. These "packages" eventually, with the help of secretory granules, are expelled into the glandular lumen. The secretory changes take place only in an estrogen-primed endometrium.

The secretory activity found in the second half of the menstrual cycle is characterized by a diversity of structural changes that can be seen on light microscopic examination of the endometrial tissue, showing a different pattern on every single day of the menstrual cycle. "Dating" the endometrium is identifying morphologic changes characteristic for early, middle, and late proliferative endometrium and for each of the 14 days of secretory endometrium.[1–3] While in early proliferation the endometrial glands are tubular and show a moderate number of mitoses, during midproliferation they become coiled, with abundant mitoses; toward the end of the proliferative phase the glands are extremely tortuous, displaying a crowded, pseudostratified epithelium. After ovulation, at only a few hours, ultrastructural examination reveals subnuclear glycogen vacuoles, giant mitochondria, and a nucleolar channel system composed of an interlacing microtubular structure that has a basket-weave appearance on section. Proliferation ceases completely in the glands with the advent of progesterone secretion.

Perhaps the most significant change during the secretory phase in terms of adequacy of the luteal phase is that involving the blood vessels. The thin endometrial arterioles undergo a process of endothelial proliferation, thickening of the wall, and coiling, forming the spiral arterioles on the ninth postovulatory day. These arterioles have a critical role in the process of implantation because of the tropism for arterial blood that characterizes trophoblastic tissue. Their coiling increases the surface to be "tapped" by the implanting trophoblast; therefore the chances for a normal implantation are reduced if the spiral arterioles are not well developed. The biopsy of menstrual endometrium does not offer reliable information on the quality of the luteal phase. The endometrial biopsy, properly performed and evaluated, is one of the most important diagnostic tools in the work-up for infertility because if offers an insight into the tissue that will play host to the implanting conceptus, assessing its adequacy. It is recommended to perform an endometrial biopsy a few days before the onset of menstruation.[3]

After the menopause, the endometrium normally undergoes a gradual atrophy, starting with an inactive phase in which neither proliferation nor secretion are present and ending up in a thin layer, often riddled with cystic cavities, lined by a cuboidal or flat epithelium in which the organelles are pushed to random locations and the stroma becomes fibrotic and collagenized.[4] As discussed in Chapter 7, a source of estrogen, endogenous or exogenous in origin, may produce proliferation and hyperplasia and even be associated with neoplasia in the postmenopausal endometrium.

The endometrial response to hormonal therapy is often unpredictable and varies from patient to patient. In the same endometrial tissue, this response may not be the same in all glands, and a wide variation of changes in glands and stroma in various combinations may result in a diversity of histologic patterns.

Hormone therapy is used at the present time for a wide variety of conditions. The endometrial changes described further will include those related to some of the most commonly used hormonal therapies: (a) hormonal contraceptives, (b) ovulation induction therapy, (c) replacement hormone therapy of the menopause, (d) hormonal therapy of endometriosis, and (e) hormonal therapy used for endometrial neoplasms.

Endometrial Histologic Changes Related to Hormone Therapy

Hormonal Contraceptives

Hormonal contraceptives in use now are "combination" pills containing estrogen and progesterone. They inactivate the naturally secreted estrogens in the first few cycles. The endometrial biopsies show less mitotic activity during the proliferative phase, and less secretion and marked edema during the secretory phase. Later, a progressive stromal hyperplasia develops, often with decidual reaction, that is more marked in its perivascular location, showing granulocytic infiltrate. The glands become inactive and take on a simple tubular configuration, lined by columnar to cuboidal epithelial cells, with oval to round, bland, and basally located nuclei. Often, secretory vacuoles can be seen, with no particular polarity ("abortive" secretion); these sometime may fill the glandular lumen (Fig. 8-1). Prolonged use of oral contraceptives leads to a complete glandular atrophy, the spaces between the glands becoming markedly widened and the stroma markedly hyperplastic. The blood vessels display proliferation of their endothelium and medial muscular layer, resembling the thickened spiral arterioles of pregnancy (Fig. 8-2). After prolonged contraceptive use

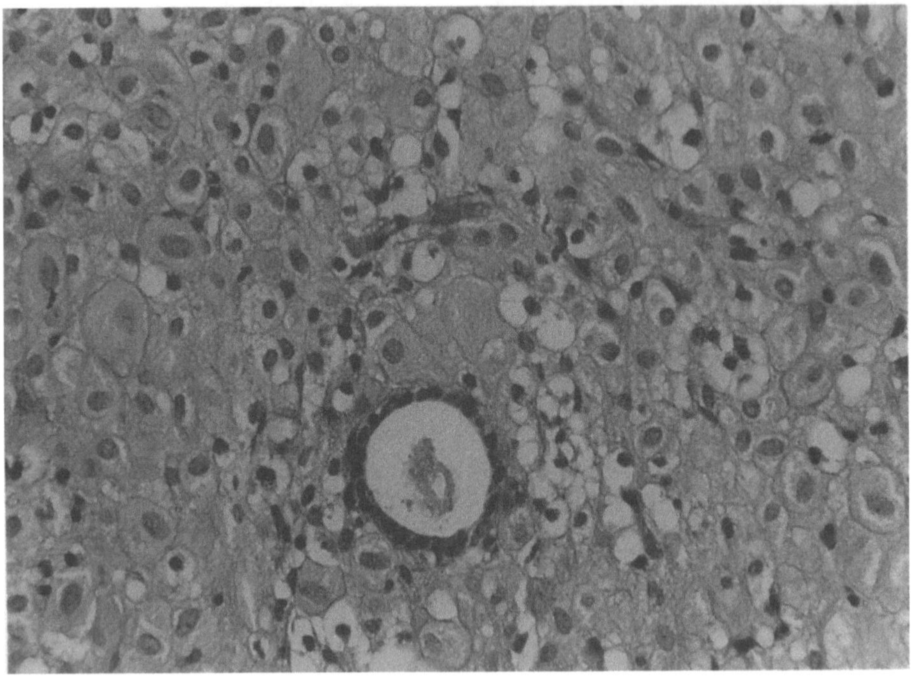

FIGURE 8-1. Endometrial changes in response to combination oral contraceptives: the endometrial gland is atrophic and contains minimal ("abortive") secretion and the stroma is abundant and decidualized. (H&E, × 200.)

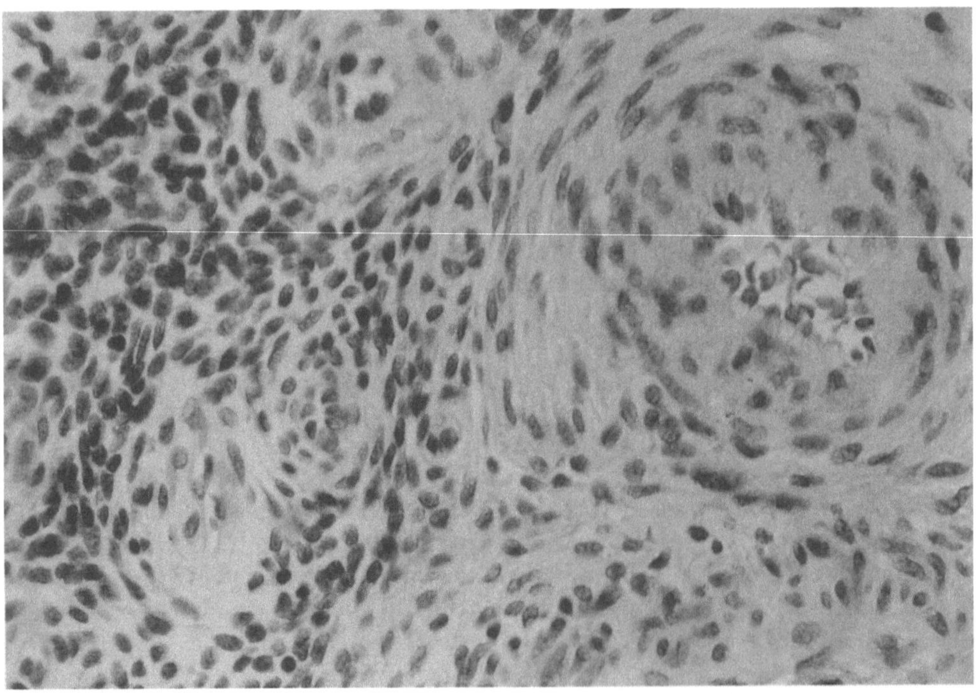

FIGURE 8-2. Pseudosarcomatous nodules and thickened vascular wall in prolonged high-dose oral contraceptive therapy. (H&E, × 100.)

the blood vessels may also become thin and sinusoidal.[5,6]

Pseudosarcomatous nodules formed by spindle cells of the stroma and of smooth muscle fibers have been reported, usually associated with oral contraceptives containing higher doses of hormones.[5,7] Intermittent, intranasal luteinizing hormone–releasing hormone (LH-RH) agonist sequentially combined with an oral progestogen was used as a contraceptive, producing incomplete secretory changes of the endometrium.[8] The use of gonadotropin-releasing hormone (GnRH) superagonist for contraception, followed by inhibition of ovulation, produced a weakly proliferative endometrium.[9]

Sequential contraceptives have not been used in the United States since 1976. They were less reliable as contraceptives and the use of highly potent estrogen alone for 14 to 16 days had been reported to be associated with a higher incidence of endometrial carcinoma.[10,11] The combined contraceptives containing low-dose estrogen and progesterone reduce the risk of endometrial carcinoma[12] and are considered to reduce also the risk of breast and ovarian carcinoma.

Ovulation Induction Therapy

Ovarian failure manifested as infrequent ovulation or chronic anovulation resulting from deficient gonadotropins unable to stimulate follicle maturation can be treated with ovulation induction therapy, and pregnancy can be achieved in many cases. Induction of ovulation with hyperstimulation of the ovaries, resulting in the selection of multiple oocytes, is used for in vitro fertilization.

Hormonal manipulation is aimed at establishing endometrial adequacy for normal implantation. The effect on the histology of the endometrium after using induction of ovulation may vary with the hormones and their combinations. The use of subcutaneous, pulsatile GnRH for induction of ovulation produced luteal-phase defects that persisted despite progesterone and clomiphene citrate therapy. Combination of this therapy with human chorionic gonadotropin could result in normal endometrial maturation.[13]

Clomiphene citrate has been used successfully for therapy of anovulatory cycles and for inadequate luteal phase, producing close to normal cycles.[14] Gonadotropin-releasing hormone is also used for its indirect action via anterior pituitary stimulation, preserving the normal feedback mechanisms between ovarian steroids and gonadotropins. However, it was shown that both clomiphene and GnRH may alter steroid receptor kinetics in the endometrium by lowering estrogen receptors by competition.[15] Luteal-phase defects, known to occur as a result of inadequate estrogenic stimulation, were noted with clomiphene citrate therapy because of its estrogenic and antiestrogenic properties, the latter occurring as a result of the competitive effect on the uterine receptor site. Persistent luteal-phase defects reported after clomiphene treatment was probably due to the inhibition of estrogen-induced progesterone receptors.[16] Luteal-phase defects consisted of a discrepancy of 5 days or more between the chronologic and histologic dating.[15] Other authors, however, have not found specific deleterious effects of clomiphene citrate on the endometrium.[17]

Clomiphene citrate is used in combination with human menopausal gonadotropin (hMG) and/or human chorionic gonadotropin (hCG) to obtain an effective ovulation induction leading to normal cyclic events in the endometrium.[18] This therapy may result in reduction of the binding affinity of the estrogen receptors that correlates better with the endometrial histology than serum hormone levels, according to a recent report.[19]

Synchronization of an otherwise asynchronous endometrial cycle was attempted by therapeutic protocols, including the administration of sequential estrogen and progesterone. The endometrial maturation process was enhanced by accelerating the secretory changes in the stroma, thus creating a disparity in which the glands showed changes consistent with day 17 or 18, and the stroma displayed spiral arterioles and marked edema consistent with day 22 or 23

(Fig. 8-3). Very high nonincremental progesterone (Pl_4) levels abolished the glandular-stromal disparity and were conducive to normal pregnancies.[20]

An optimal follicular development is crucial for the subsequent endometrial maturation, as evaluated by endometrial biopsies.[21] The endometrial biopsy is the ultimate diagnostic tool for assessing the adequacy of hormonal manipulations aimed at obtaining a mature endometrium appropriate for implantation.

Postmenopausal Replacement Hormone Therapy

Patients treated with hormone replacement therapy, now represent a large proportion of the female population. The effects on the endometrium depend on the duration of therapy, the hormones used (estrogens or estrogens and progesterone), and their dosages. Estrogen therapy has an effect on the postmenopausal endometrium that is similar to that of estrogens in normal cycles. Conjugated estrogens alone are known to have a proliferative effect that accumulates and may lead to endometrial hyperplasia. Many estrogen preparations subject the endometrium to a potent stimulus. Scanning electron microscopy studies have shown marked increase in ciliogenesis of the endometrial surface epithelial cells in postmenopausal patients treated with estrogens.[22] Morphometric analysis has shown that glandular epithelial cells and nuclei increase their profile areas by 80% under estrogenic effect.[23] Although complete protection by progesterone has not been achieved, progesterones appear capable of protecting against the development of cancer and hyperplasia.[24,25] Currently, the most commonly used regimen consists of low doses of conjugated estrogens daily, to which progesterone is added sequentially during the last 10 days of the 3-week estrogen therapy.[26]

Following insertion of intrauterine devices releasing progesterone, with no exo-

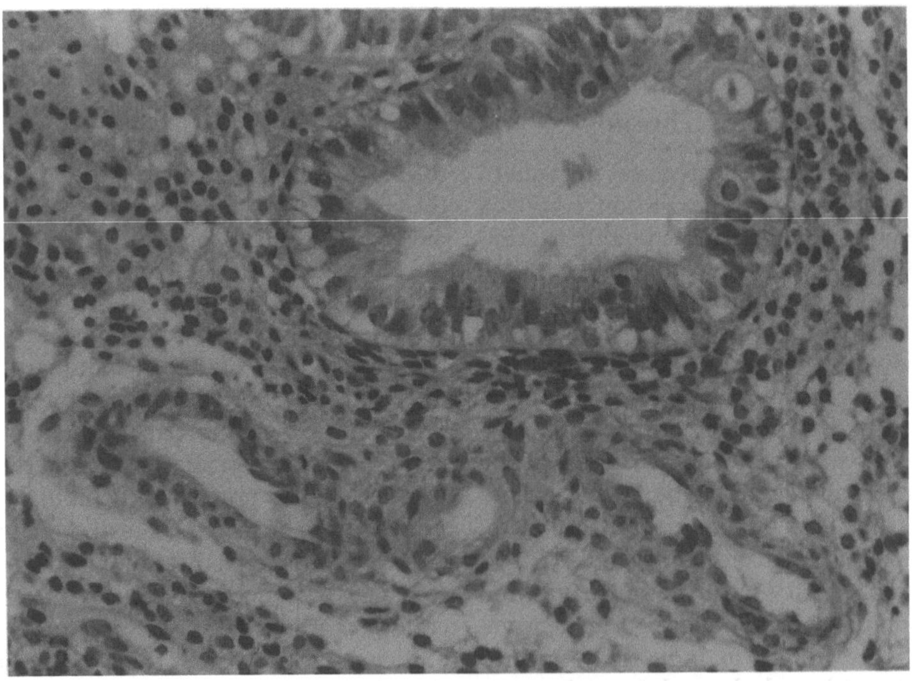

FIGURE 8-3. Discrepancy between glands and stroma with ovulation induction therapy: subnuclear vacuolization corresponds to day 17 and spiral arterioles to day 23. (H&E, × 200.)

genous hormone in the blood, a suppression of the proliferative activity and a marked reduction in alkaline phosphatase and glycuronidase were noticed. Apparently, progesterone arrests the cells in the growth (G_1) phase and prevents DNA synthesis. The antiestrogenic effect is also due to the depression of estrogen receptor levels as opposed to the progesterone receptor levels, which are enhanced by estrogens. Three mechanisms are involved in the antiestrogenic activity of progesterone:

1. Reduction of estrogens in the systemic circulation
2. Inactivation of estradiol by metabolism (i.e., transformation of estradiol to estrone) at the target tissue
3. Lowering of estrogen receptors in the target tissue

The histologic changes of the endometrium in menopausal patients estrogenized by continuous daily conjugated estrogens and treated with progesterone consist of dose-dependent sceretory features.[27] Suppression of DNA synthesis and nuclear estrogen receptor is manifested by an arrest in mitotic activity. A decidualization of the stroma takes place with focal necrosis and inflammatory infiltrates, not unlike that seen in pregnancy. The glands, in the majority of cases, do not show proliferative activity with mitoses and piling up of cells and prominent nucleoli, although the general architectural pattern of crowded glands and scanty interglandular stroma may persist (Fig. 8-4). Secretory changes in the glands are often seen with subnuclear or supranuclear glycogen vacuoles and intraluminal secretion. The epithelium may respond to progestational therapy with a quiescent appearance characteristic of normal secretory endometrium, or may show focal proliferative changes alternating with secretory glands ("mixed endometrium"). Dose-dependent responses to oral progesterone in estrogenized patients are manifested by the extent of secretory changes, including ultrastructural features such as subnuclear vacuoles, nucleolar channel systems, and giant mitochondria.[27]

In a number of cases, despite prolonged progesterone therapy, the endometrium maintains its proliferative and hyperplastic pattern. The stroma may become edematous or decidualized, or keep its compact, immature configuration. In some cases, the glands show minimal ("abortive") secretion and the stroma is markedly hyperplastic (Fig. 8-5).

In practice, endometrial biopsies taken from women on replacement hormone therapy display a wide range of changes, including inactive or weakly proliferative endometrium,[28] focally abortive or diffusely secretory endometrium, mixed proliferative and secretory endometrium with squamoid "morules,"[29] endometrial stromal and/or glandular hyperplasia, and occasional metaplasia mimicking neoplasia (Fig. 8-6). Decidual reaction is often seen, reminiscent of gestational changes. We have observed in one case an extensive decidual transformation not only of the endometrium but of endometriosis involving the abdominal cavity, resulting in a tumor-like proliferation that obstructed the ureter and was complicated by hydronephrosis.[30] The morphologic effects on the endometrium result from the dosage, duration of therapy, endogenic receptors, and metabolic pathways of the patient, and from the combination of these factors.[5,31] For example, continuous conjugated equine estrogens and medroxyprogesterone acetate result in inactive endometrium while the same hormones given in cyclic fashion produce proliferative endometrium.[28] Proliferative changes may alternate with secretory changes in the same endometrium (Fig. 8-7). Periodic endometrial biopsy surveillance is desirable with hormone replacement therapy at least once every 2 years, even in the absence of unscheduled bleeding.

Hormone Therapy of Endometriosis

Historically, the hormonal therapy of this painful condition consisted of synthetic estrogens, then of androgens, and then of pro-

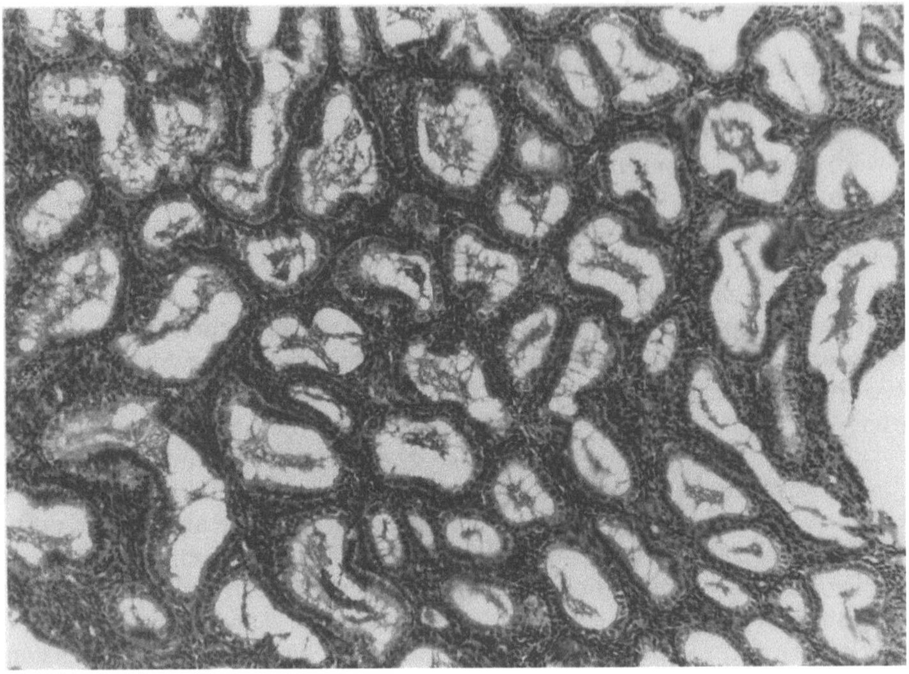

A

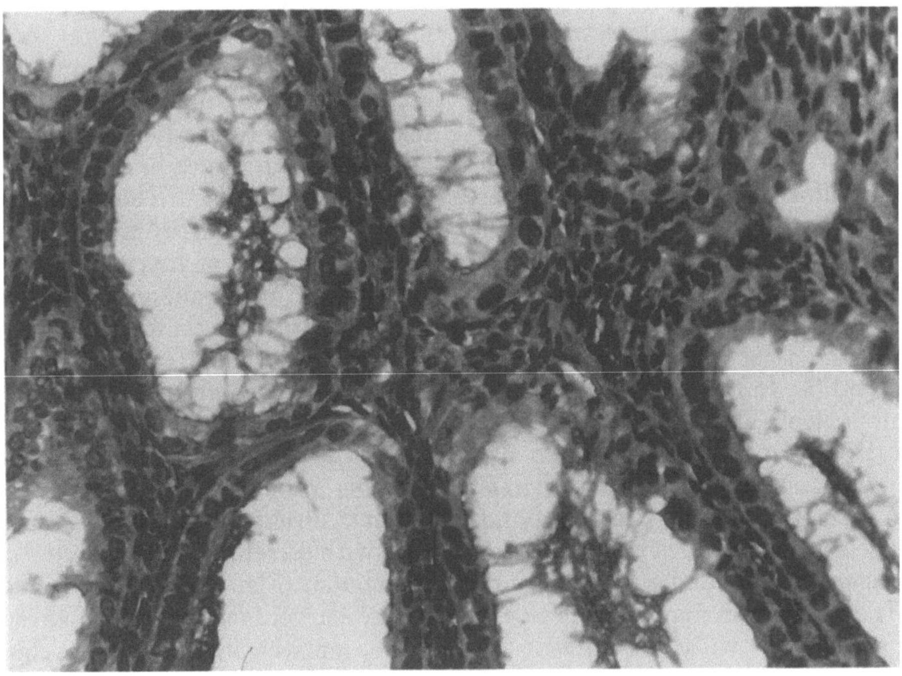

B

FIGURE 8-4. Endometrial biopsy of patient receiving replacement therapy with conjugated estrogens and progesterone. (A) The glands are dilated and crowded; the general architecture is that of endometrial hyperplasia. (H&E, × 40.) (B) Glands are lined by a cuboidal, "quiescent" epithelium and contain secretion in the lumen. (H&E, × 200.)

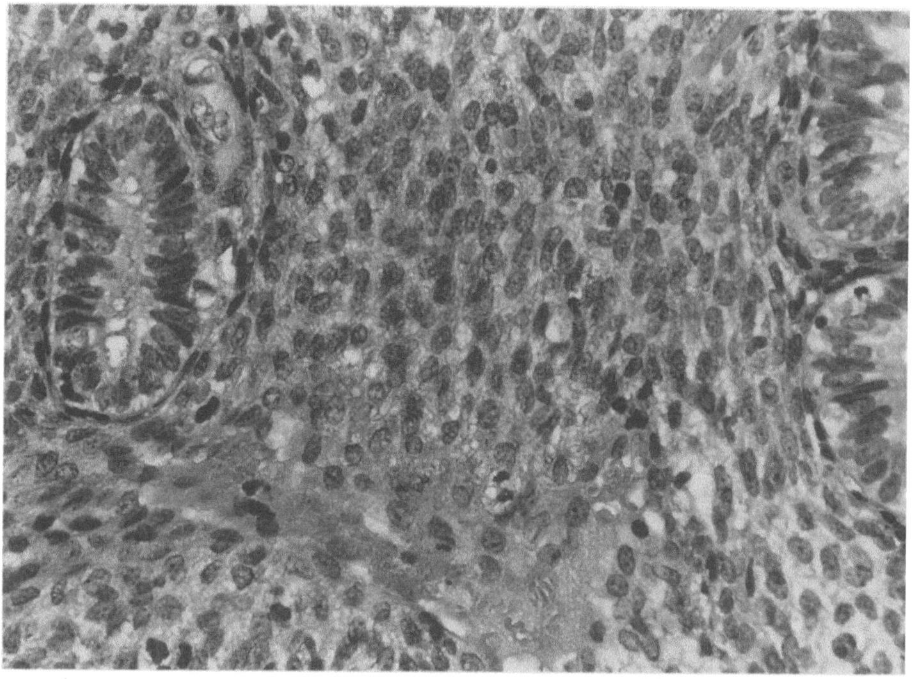

FIGURE 8-5. Endometrial biopsy of patient on estrogen-progesterone replacement therapy: the glands show minimal ("abortive") secretion and the stroma is markedly hyperplastic, displaying mitotic activity. (H&E, × 150.)

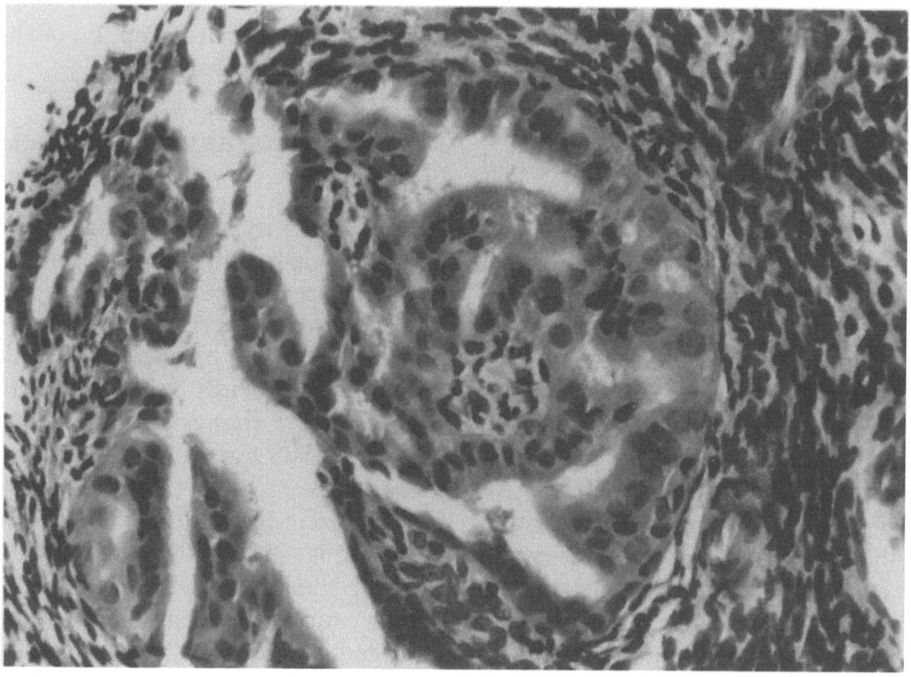

FIGURE 8-6. Endometrial biopsy of patient on estrogen-progesterone replacement therapy: irregular proliferation of glands with papillary and eosinophilic metaplasia, mimicking neoplasia. (H&E, × 200.)

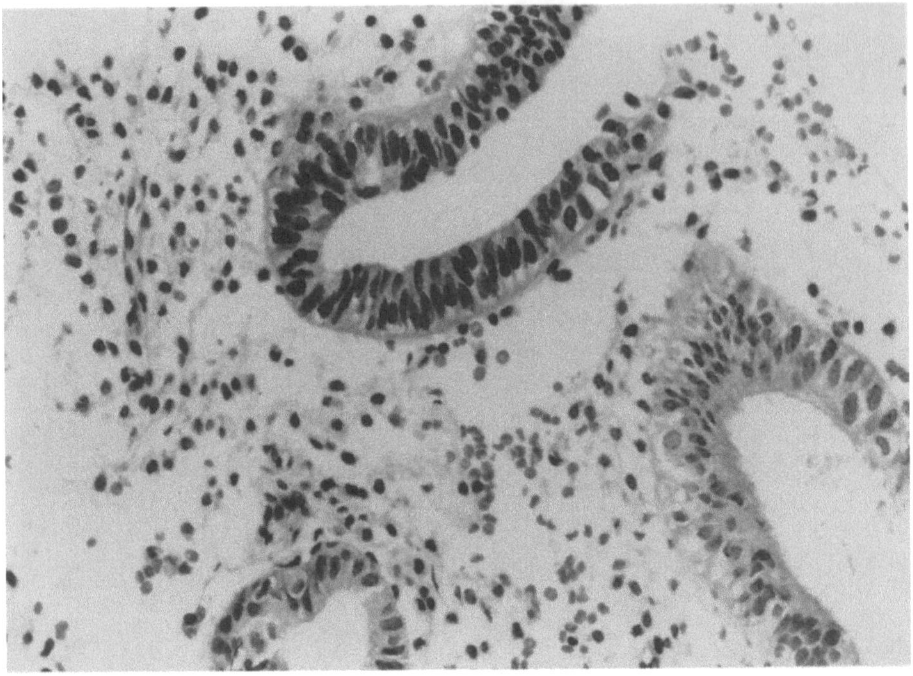

FIGURE 8-7. Endometrial biopsy of patient on estrogen-progesterone therapy: glands show marked proliferation with hyperchromatic nuclei and piling up and display irregular secretion stroma is edematous. (H&E, × 200.)

gestins. Because of undesirable side effects and unsatisfactory results, at present the most widely used nonsurgical therapy is danazol, which has antigonadotropic effects suppressing both luteinizing hormone and follicle-stimulating hormone. Most patients become amenorrheic by a mechanism resembling menopause. Their endometrial biopsy shows inactive and atrophic glands that are straight and tubular in shape, lined by cuboidal or columnar epithelium exhibiting neither proliferation nor secretion and a compact stroma with collagenization of the intercellular matrix and reduction of vascularity (Fig. 8-8). This effect is also obtained at the level of the ectopic endometrium, through interaction with steroid receptors, as shown experimentally.[32] Pseudodecidualization in the endometrium was also described in therapy with progesterone.[33] These changes also involve the ectopic endometrial tissue, and relieve the symptoms due to the cyclic changes taking place in endometriosis. Cessation of therapy is usually followed by prompt resumption of the cyclic changes in the endometrium.

Hormone Therapy of Endometrial Neoplasms

Adenomatous hyperplasia results from prolonged unopposed (by progesterone) estrogen effect. In premenopausal patients, it is seen with anovulatory cycles. The treatment includes, of course, progesterone,[24] although antiestrogenic drugs have also been used.[34] The histologic changes of the endometrium vary with the dosage and duration of the therapy, as well as with the severity of the adenomatous hyperplasia. Under the influence of progesterone therapy, mitoses are inhibited, the glandular proliferation is arrested, and the stroma may remain unchanged.[3] More estrogen has to be counteracted than in normal cycles; therefore the adenomatous hyperplasia may persist during and after therapy in some cases. In most cases of prolonged progesterone

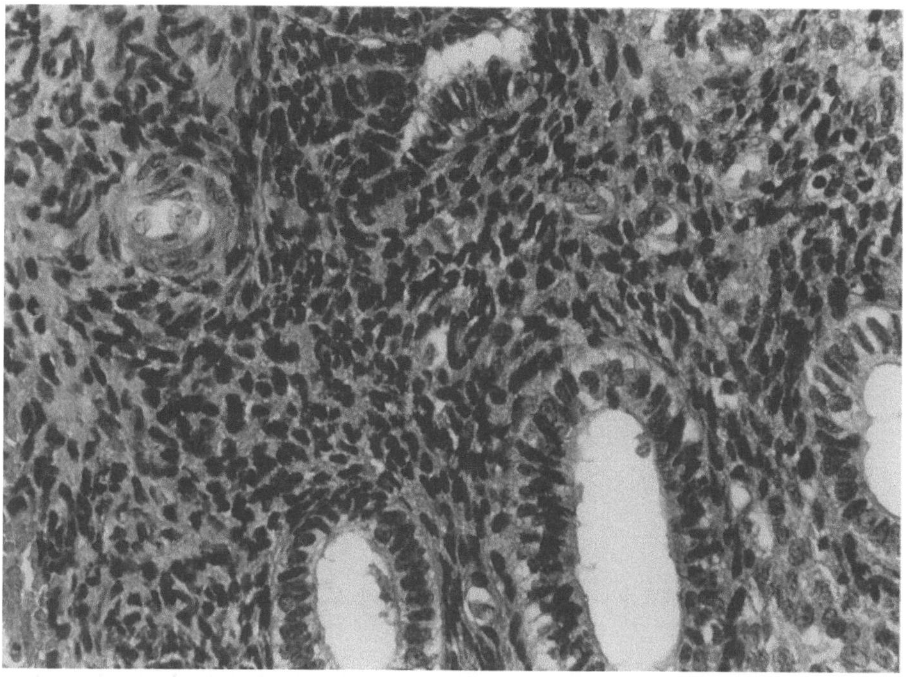

FIGURE 8-8. Endometrial biopsy of patient on danazol therapy for endometriosis: glands are tubular and stroma is compact. (H&E, × 200.)

therapy, however, secretory features are seen in the glands and decidual reaction may take place in the stroma.

Most cases of adenomatous hyperplasia, including the severe, atypical ones, regress with hormone therapy. The risk of a coexisting residual endometrial carcinoma, however, increases with the age of the patient; therefore, hysterectomy is the treatment of choice for women over 50.

The use of progestins for endometrial adenocarcinoma is a controversial issue. The consensus, at the present time, is that progestins should be used as an adjuvant therapy, with surgery and radiotherapy being the major forms of treatment. It has been documented that progestins have the ability to inhibit DNA synthesis and induce regression of endometrial hyperplasia.[24,26] The therapy of endometrial adenocarcinoma should be correlated with the progesterone receptor status.[35,36] The patients who may benefit most are those with a well-differentiated adenocarcinoma of stage I, and with high estradiol and progesterone receptor levels.[35] The histologic effect on the endometrium of progesterone therapy, if used before hysterectomy, consists of the appearance of secretory features in the neoplastic endometrium with a patchy distribution, resembling the early postovulatory endometrium (Fig. 8-9). The glands have a quiescent appearance and mitotic activity is arrested, but the crowding, back-to-back configuration and cribriform pattern are not reversible, as opposed to the effect obtained in adenomatous hyperplasia. Tamoxifen, an estrogen antagonist, has also been used to revert endometrial neoplastic growth.[37] In a growing number of cases, however, tamoxifen therapy for breast carcinoma has been associated with endometrial hyperplasia, polyps, and carcinoma.[38] Tamoxifen used in vitro on human endometrial adenocarcinoma cells (Ishikawa line) have produced proliferation, in even greater proportion than estradiol.[39] It seems, therefore, that this drug, considered an estrogen antagonist, has an agonis-

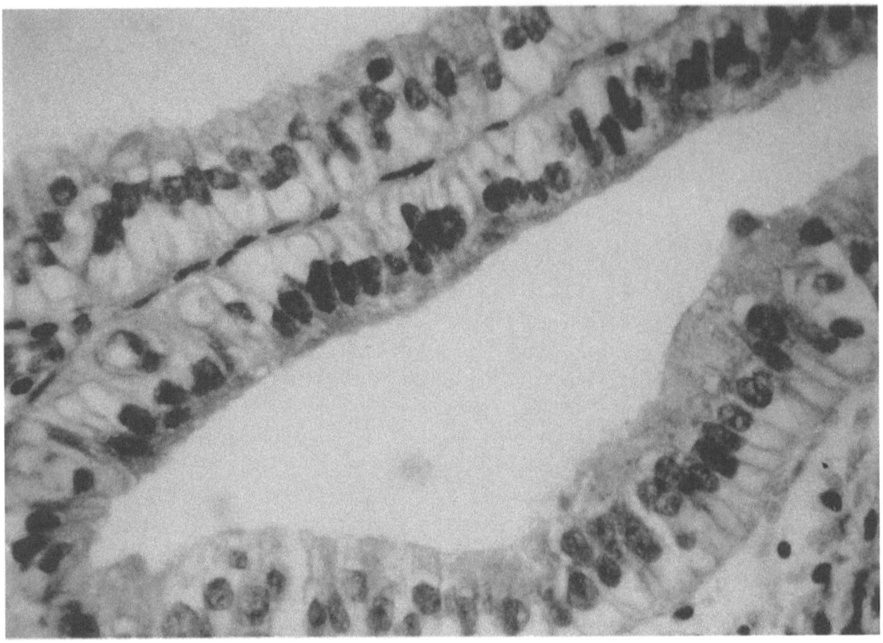

FIGURE 8-9. Endometrial biopsy of patient with endometrial adenocarcinoma, treated by medroxyprogesterone acetate and subsequent hysterectomy. Note back-to-back glands, and well-formed subnuclear secretory vacuoles. (H&E, × 400.)

tic effect in certain circumstances. Danazol, a known antiestrogen, has also been used to reverse hyperplastic and neoplastic endometrium, resulting in fragmentation of the basement membranes.[34]

In young women desirous to maintain their reproductive function, the diagnosis of endometrial adenocarcinoma poses a challenging therapeutic problem. Progesterone therapy alone may be followed by reversal of some neoplastic features.[40] The appearance of areas of secretory endometrium by no means guarantees the reversal of the neoplasm, since secretory endometrium and neoplastic endometrium may coexist for the duration of therapy.

Invasive cancer may follow even after a prolonged progesterone therapy of endometrial adenocarcinoma, probably because endometrial neoplastic tissue, insensitive to hormone effect, continues its aggressive behavior despite the response to therapy in some areas. The complete reversal of neoplastic-appearing endometrium after progesterone therapy, with a total disappearance of malignant characteristics, probably means that the lesion was not cancerous to begin with. It was noted in Chapter 7 that severe, atypical adenomatous hyperplasia can simulate endometrial adenocarcinoma, and the regression of neoplastic features after progesterone therapy confirms the biologic, noncancerous nature of the lesion.

The finding of secretory changes in the endometrial biopsy of a patient with endometrial adenocarcinoma treated with progesterone does not mean that the neoplasm was cured. The absence of endometrial cancer can only be assessed on hysterectomy specimens.

Conclusion

Sex hormone therapy is a vastly expanding field, and new regimens are being implemented. Their effect on the histology of the endometrium is wide ranging. The modalities of response of the endometrial tissue to sex hormone stimulation are mediated by receptors. The target tissue responds by a variety of features ranging from metaplasia[41] to hyperplasia and even neoplasia, often combined in an unpredictable fashion. The interpretation of the morphologic findings should be done in close collaboration between gynecologist and pathologist, with active cooperation of the patient, since an accurate history of drug intake is crucial.

It is therefore advisable to evaluate the endometrial histology during prolonged hormone therapy in order to assess both the therapeutic effect and the potential carcinogenic effect of the hormone(s).

References

1. Noyes RW: Normal phases of the endometrium, in Norris JH, Hertig AT, Abell MR (eds): The Uterus. Baltimore, Williams & Wilkins 1973, pp 110–135.
2. Ferenczy A: Anatomy and histology of the uterine corpus, in Kurman RJ (ed): Blaustein's Pathology of the Female Genital Tract, ed 3. New York, Springer-Verlag, 1987, pp 257–291.
3. Dallenbach-Hellweg G, Poulsen H: Atlas of Endometrial Histopathology. Philadelphia, WB Saunders Company, 1985, pp 10–11.
4. Deligdisch L, Yedvab G, Persitz A, David MP: Ultrastructural features in normal and hyperplastic post-menopausal endometrium. Acta Obstet Gynecol 1978;57:439–452.
5. Ludwig H: The morphologic response of the human endometrium to long-term treatment with pro-gestational agents. Am J Obstet Gynecol 1982;142:796–816.
6. Silverberg SG, Makowski EL: Endometrial carcinoma in young women under 40 years of age. Comparison of cases in oral contraceptive users and non-users. Cancer 1977;39:592–598.
7. Cruz-Aquino M, Shenker L, Blaustein A: Pseudosarcoma of the endometrium. Obstet Gynecol 1967;29:93–96.
8. Lemay A, Jean C, Faure S: Endometrial histology during intermittent intranasal luteinizing hormone-releasing hormone (LH-RH) agonist sequentially combined with an oral progestogenesis an anti-ovulatory contraceptive approach. Fertil Steril 1987;48:775–782.
9. Gudmundsson JA, Lundquist O, Bergquist C, et al: Endometrial morphology after 6 months of continous treatment with a new gonadotropin-releasing hormone superagonist for contraception. Fertil Steril 1987;48:52–56.
10. Hilliard GD, Norris JH: The pathologic effects of oral contraceptives. Recent Results Cancer Res 1979;66:49–71.
11. Kelly HW, Miles PA, Buster JE, et al: Adenocarcinoma of the endometrium in women taking squentially oral contraceptives. Obstet Gynecol 1976;47:200–202.
12. The Cancer and Steroid Hormone Study of the Centers for Disease Control and the National Institute of Child Health and Human Development: Combination oral contraceptive use and the risk of endometrial cancer. JAMA 1987;257:796–800.
13. Campbell BF, Phipps WR, Nagel TC, et al: Endometrial biopsies during treatment with subcutaneous pulsatile gonadotropin-releasing hormone and luteal phase human chorionic gonadotropin. Int J Fertil 1988;33:329–333.
14. Lamb EJ, Colliflower WW, Williams JW: Endometrial histology and conception rates after clomiphene citrate. Obstet Gynecol 1972; 39:389–396.
15. Birkenfeld A, Beier HM, Schenker JG: The effect of clomiphene citrate on early embryonic development, endometrium and implantation. Hum Reprod 1986;1:387–395.
16. Daly DC, Walters CA, Soto-Albors CE, et al: Endometrial biopsy during treatment of luteal phase defects is predictive of therapeutic outcome. Fertil Steril 1983;40:305–310.
17. Thatcher SS, Donachie KM, Glasier A, et al: The effects of clomiphene citrate on the histology of human endometrium in regularly cycling women undergoing in vitro-fertilization. Fertil Steril 1988;49:296–301.
18. Rosenwaks Z: Donor eggs: their application in modern reproductive technologies. Fertil Steril 1987;47:895–909.
19. Seliger E, Schoneich C, Kaltwasser P, et al: Menstrual cycle stimulation in the in vitro fertilization program—effects on estrogen and progesterone receptors of the endometrium. Zentralbl Gynakol 1988;110:1499–1506.
20. Navot D, Anderson TL, Droesch K: Hormonal manipulation of endometrial maturation. J Clin Endocrinol Metab 1989;68:801–807.
21. Graf MJ, Reyniak JV, Battl-Mutter P, et al: Histologic evaluation of the luteal phase in

women following follicle aspiration for oocyte retrieval. Fertil Steril 1988;49:616–619.

22. Thom MH, Davies KJ, Senkus RJ, et al: Scanning electron microscopy of the endometrial cell surface in post-menopausal women receiving estrogen therapy. Br J Obstet Gynecol 1981;88:904–913.
23. Aycock NR, Jollie WP: Ultrastructural effects of estrogen replacement on post-menopausal endometrium. Am J Obstet Gynecol 1979; 135:461–466.
24. Kistner RW: Histological effects of progestins on hyperplasia and carcinoma in situ of the endometrium. Cancer 1959;12:1106–1122.
25. King RJB, Lane G, Siddle N, et al: Assessment of estrogen and progestin effects on epithelium and stroma from pre and post-menopausal endometria. J Steroid Biochem 1981; 15:175–181.
26. Greenblatt RB, Gambrell RD Jr, Stoddard LD: The protective role of progesterone in the prevention of endometrial cancer. Pathol Res Pract 1982;74:297–318.
27. Lane G, Siddle NC, Ryder TA, et al: Dose dependent effects of oral progesterone on the oestrogenised post-menopausal endometrium. Br Med J 1983;287:1241–1245.
28. Prough SG, Aksel S, Wiebe RH, et al: Continous estrogen/progestin therapy in menopause. Am J Obstet Gynecol 1987;157:1449–1453.
29. Dallenbach-Hellweg G, Poulsen H: Iatrogenic changes, in Atlas of Endometrial Histopathology. Copenhagen, Munksgaard, 1985, pp 134–140.
30. Goodman MH, Kredentzer D, Deligdisch L: Post-menopausal endometriosis associated with hormone replacement therapy. J Reprod Med 1989;34:231–233.
31. Whitehead MI, Townsend PT, Pryse-Davies J, et al: Actions of progestins on the morphology and biochemistry of the endometrium of postmenopausal women receiving low-dose estrogen therapy. Am J Obstet Gynecol 1982; 142:791–795.
32. Henig I, Rawlins RG, Weinrib HP, et al: Effects of danazol gonadotropin-releasing hormone agonist and estrogen-progestin combination on experimental endometriosis in the ovariectomized rat. Fertil Steril 1988;49:349–355.
33. Luciano AA, Turksoy RN, Carlos J: Evaluation of oral medioxyprogesterone acetate in the treatment of endometriosis. Obstet Gynecol 1988;72:323–327.
34. Bulletti C, Jasonni VM, Tabanelli S, et al: Danazol reverses hyperplasia to normal endometrium. Acta Eur Fertil 1987;18:185–187.
35. Lellouche D, Gairard B, Renaud R: Hormone dependence and steroid receptors in adenocarcinomas of the endometrium. Rev Fr Gynecol Obstet 1987;82:325–329.
36. Holinka CF, Deligdisch L, Gurpide E: Histological evaluation of in vitro responses of endometrial adenocarcinoma to progestins and their relation to progesterone receptor levels. Cancer Res 1984;44:293–296.
37. Slavik M, Petty WM, Blessing JA, et al: Phase II clinical study of tamoxifen in advanced endometrial adenocarcinoma. Cancer Treat Rep 1984;68:809–811.
38. Nuovo MA, Nuovo GJ, McCafrey RM, et al: Endometrial polyps in postmenopausal patients receiving tamoxifen. Int J Gynecol Pathol 1989;8:125–131.
39. Anzai Y, Holinka CF, Kuramoto H, et al: Stimulatory effects of hydroxytamoxifen on proliferation of human adenocarcinoma cells (Ishikawa line). Cancer Res 1989;49:2362–2365.
40. John HA, Cornes JS, Jackson WD, et al: Effect of a systematically administered progesterone on histopathology of endometrial adenocarcinoma. J Obstet Gynaecol Br Commonw 1974;81:786–790.
41. Hendrickson MR, Kempson RL: Endometrial epithelial metaplasias. Proliferations frequently misdiagnosed as adenocarcinoma. Report of 89 cases and proposed classification. Am J Surg Pathol 1980;4:525–542.

9

Mesenchymal Tumors of the Uterus

MARTIN LEFKOWITZ and HENRY J. NORRIS

The uterus is formed from the paired müllerian ducts, the latter derived from mesenchyme at the eighth to ninth week of gestation. Mesenchyme has the capacity to form connective tissue or collagen, smooth and striated muscle, fat, cartilage, bone, and the epithelium lining the uterus and tubes. The potentiality of the uterine primordium is retained in the cells making up the adult uterus.[1,2] As a result, some neoplasms that subsequently arise in the uterus express the high potentiality of their ancestry by forming a mixture of epithelial and mesodermal components. Given this derivation, numerous mixed overlapping tumor types are common, a feature evident microscopically and in the cross-reactivity of immunocytochemical reactions.

Mesenchymal tumors other than leiomyoma are uncommon, representing only about 3% of uterine malignancies.[3] Despite their rarity, there are many types and variants to challenge the pathologist. There are four major categories of mesenchymal neoplasms (Table 9-1): smooth muscle tumors, endometrial stromal tumors, mixed mesodermal tumors, and closely related tumor-like conditions (Table 9-2), such as adenomyosis, adenomyomas, and adenomatoid mesotheliomas. The microscopic features of mesenchymal tumors of the uterus are greatly dependent on proper formalin fixation. Formalin penetrates solid and bulky structures at a rate of less than 1 cm in 24 hours. Tissue can autolyze in this time and mitotic figures can proceed to completion.[4] Good histology for microscopic diagnosis requires prompt and proper fixation, which in turn depends on early examination of the specimen and exposure of maximum surface area to fixative. Lack of attention to this detail produces uncertainty and delay in arriving at a diagnosis.

Smooth Muscle Tumors

Smooth muscle tumors are the most common tumors of the uterus—about 20% of women 30 to 50 years of age have them.[5] The two key features in the classification of smooth muscle tumors are the degree of cytologic atypia and the degree of mitotic activity.[6,7] Clinical information needed for the proper interpretation of this class of tumors includes the patient's age, oral contraceptive or steroid use, current or recent pregnancy, and the presence of a hormonally active tumor. We recommend that one block of tissue be taken for microscopic examination for each centimeter of tumor diameter from all smooth muscle or stromal tumors, except ordinary leiomyomas. If it can be provided, a rim of normal tissue around a tumor is used to gauge the interface of the tumor with surrounding structures and determine whether the tumor has a pushing or infiltrating margin. However, myoinfiltration by itself is not

---

The opinions and assertions contained herein do not purport to be the views of the Departments of Army or Defense.

TABLE 9-1. Classification of mesenchymal tumors of the uterus

Smooth muscle tumors
Leiomyoma
Leiomyoma with increased mitotic figures
Cellular leiomyoma
Hemorrhagic cellular leiomyoma
Symplastic (atypical) leiomyoma
Epithelioid leiomyoma
Intravenous leiomyomatosis
Leiomyomatosis peritonealis disseminata
Benign metastasizing leiomyoma
Smooth muscle tumors of uncertain malignant potential
Leiomyosarcoma, including myxoid & epithelioid variants
Endometrial stromal tumors
Stromal nodule
Low-grade endometrial stromal sarcoma (endolymphatic stromal myosis)
High-grade endometrial stromal sarcoma
Endometrial stromal variants
Mixed mesodermal tumors
Adenofibroma
Adenosarcoma
Mixed mesodermal tumor (homologous and heterologous)
Related tumor-like conditions (see Table 9-2)

TABLE 9-2. Tumors and tumor-like conditions

Adenomyosis
Adenomyoma, and atypical polypoid adenomyoma
Adenomatoid mesothelioma
Inflammatory pseudotumor of the uterus
Lymphoma
Metastatic tumors to the uterus

evidence of malignancy, because the majority of smooth muscle tumors with infiltration are benign.[6]

Leiomyomas

Leiomyomas occur in the reproductive years. Usually multiple, they vary from microscopic size to 20 cm in diameter, and are hormonally responsive. Symptoms are generally due to size, position in the uterus, and pressure on adjacent structures.[8] They can be submucosal and produce symptoms of a polyp, with bleeding. Usually, intramural leiomyomas attain the greatest size. By expanding subserosally they may become pedunculated and undergo torsion and infarction. Grossly, leiomyomas are spherical and firm to hard, and their cut surface bulges above the surrounding myometrium, showing a characteristic whorled, fibrous, gray-white to tan appearance. Despite their solid nature, leiomyomas are sensitive to vascular impairment, which can result in edema, necrosis, and hemorrhage, all of which may produce a rapid change in size and a yellow to red or even a cystic appearance.

The typical leiomyoma is circumscribed and has fascicles or bundles of uniform, bland spindle cells. The spindle cells have abundant eosinophilic fibrillar cytoplasm and elongated nuclei with blunt or tapered ends. The nuclear chromatin is finely dispersed, similar to that of myocytes in the surrounding myometrium, giving a normochromatic appearance to these cells. The cells have no atypia, and mitotic figures are rare. A leiomyoma differs from the surrounding myometrium by its circumscription, nodularity, greater cellularity, and disorganized fascicular arrangement.

Leiomyomas seldom, if ever, give rise to leiomyosarcomas.[3,9] They tend to involute with age, undergoing fibrosis and hyalinization.[8]

Leiomyoma with High Mitotic Figures

A leiomyoma with a high number of mitotic figures differs from an ordinary leiomyoma only by having 5 to 15 mitotic figures per 10 high-power fields (HPFs). Because of the mitotic activity, these leiomyomas can be confused with leiomyosarcoma. The cells, generally, have little increase in nuclear size over cells in the adjacent myometrium, no cytologic atypia, and no atypical mitotic figures. However, in a series of 10 leiomyomas with high mitotic figures, there was mild cytologic atypia in 7, moderate atypia in 3, and rare abnormal mitotic figures in 5, and fingerlike projections into the adjoining myometrium in 1.[10] In another study of 70 patients with leiomyomas lacking atypia but with 5 to 9 mitotic figures per 10 HPFs followed an average of 5 years after hysterectomy, none of the leiomyomas metastasized. Fourteen of these patients were treated with myomectomy, but only one had persistent tumor after a median follow-up of 96 months.[11] Further studies of the potential of unusual smooth muscle tumors to recur or metastasize after myomectomy are needed.

Cellular Leiomyomas

Cellular leiomyomas are benign variants of ordinary leiomyomas. They present with similar clinical and gross pathologic features. Their importance lies in the fact that their microscopic features may create confusion with a leiomyosarcoma or, when composed of small cells, with an endometrial stromal sarcoma.

Cellular leiomyomas have a greater number of cells per unit area compared to ordinary leiomyomas, have less intervening collagen, and have up to five mitotic figures per 10 HPFs. Close attention to the surrounding myometrium allows comparison of nuclear detail and helps indicate those slides that are thickly cut or heavily stained. Unlike a leiomyosarcoma, neither the cells nor the mitotic figures are atypical, and the number of mitotic figures does not exceed five per 10 HPFs.[3,12,13] Unlike a stromal sarcoma, the cells have eosinophilic cytoplasm, the nuclei are spindle shaped, the cells tend to occur in a fascicular arrangement, and immunocytochemical reactions for smooth muscle, including smooth muscle actin, desmin, and vimentin, are evident. Stromal sarcomas, in contrast, are more infiltrative, with finger-like projections at their margins.

Hemorrhagic Cellular Leiomyomas

Hemorrhagic cellular leiomyomas occur in patients 24 to 48 years of age who are pregnant or who are receiving oral contraceptives.[14,15] The majority are subserosal, and one third erode, rupture, or produce hemoperitoneum.[15] Women present with uterine enlargment, abnormal uterine bleeding, and acute abdominal pain secondary to rupture of the tumor with hemoperitoneum. The rapid increase in size may suggest malignancy. Grossly, there is a knobby, irregular enlargement of the uterus due to multiple leiomyomas. The hemorrhage is often multifocal and is most visible on the cut surface. Hemorrhagic cellular leiomyomas are otherwise pink-gray or tan nodules, ranging in size up to 11 cm.

Microscopically hemorrhagic cellular leiomyomas are circumscribed, densely cellular lesions with oval and spindle-shaped smooth muscle cells surrounding central areas of hemorrhage and edema. The cells have small nuclei that are narrow and elongated, with little or no increase in chromatin, and prominent nucleoli. Atypia is slight to moderate in 20%, and there are two to four mitotic figures per 10 HPFs, most noticeable adjacent to areas of shearing artifact, edema, and hemorrhage.

Differentiation of a hemorrhagic cellular leiomyoma from a leiomyoma with an infarct or hemorrhage is based on the greater cellularity of the former. Ordinary infarction in a leiomyoma is sharply demarcated and does not show an adjoining increase in cellularity, but both may evoke increased mitotic activity. Differentiation of hemorrhagic cellular leiomyoma from leiomyosarcoma is based on the multiple small nodules, dense cellularity, small size of the cells, and the limitation of the high mitotic activity to a narrow zone adjacent to the hemorrhage. Hemorrhagic cellular leiomyomas are more circumscribed than a leiomyosarcoma, have little or no nuclear atypia, and have no abnormal mitotic figures.[14,15]

Symplastic (Atypical) Leiomyoma

Symplastic (atypical) leiomyoma is a designation for a smooth muscle tumor composed of enlarged, multinucleate, hyperchromatic cells with bizarre nuclear forms. The clinical behavior and gross appearance are similar to those of the usual leiomyoma.

The microscopic features of symplastic leiomyoma include cell enlargement with markedly enlarged hyperchromatic nuclei in which there is chromatin clumping and smudging (Fig. 9-1).[13,16] The enlarged multiple nuclei produce bizarre cells that can lead to an erroneous diagnosis of malignancy. Symplastic change is often patchy in distribution within an ordinary leiomyoma. The presence of fewer than five mitotic figures per 10 HPFs, the lack of atypia in the underlying leiomyoma, and the young age of the patient point to its benign nature.

Symplastic leiomyomas do not recur or metastasize, but symplastic nuclear forms may occur in leiomyosarcomas, making their interpretation more difficult. Leiomyosarcomas, however, have higher mitotic activity (five or more mitotic figures per 10 HPFs) and abnormal mitotic figures.

Epithelioid Leiomyomas

Epithelioid leiomyomas (clear cell and plexiform leiomyoma, leiomyoblastoma) produce abnormal bleeding, pelvic pain, and abdominal enlargement of usually short duration.[17,18] They tend to be solitary, but otherwise are not significantly different in their gross appearance from the usual leiomyoma.

Epithelioid leiomyomas have a distinctive microscopic appearance in that they have either pure or mixed patterns of epithelioid, clear cell, and plexiform differentiation.[17,19] Rather than spindle shaped, as in the usual leiomyoma, the cells are round or polygonal with centrally placed round nuclei, and are often arranged in cordlike or clustered patterns. The cells are often compartmentalized by hyalin material. Usually there is transition to ordinary smooth muscle at the periphery or within the neoplasm. There are fewer than five mitotic figures per 10 HPFs, and little or no atypia. Clear cell leiomyomas have polygonal cells with abundant clear cytoplasm. The cytoplasm contains glycogen, sparse lipid, and no mucin. Signet ring forms may appear, but do not usually present a diagnostic problem since muscle differentiation is usually apparent, and mucin and cytokeratin reactions are negative. Plexiform tumors are rare. They are small epithelioid leiomyomas of the leiomyoblastoma type arranged in a striking cordlike arrangement. None have metastasized.

Epithelioid and leiomyoblastic leiomyomas (Fig. 9-2) are composed of polygonal cells that often show little cohesion, a feature that may explain why epithelioid leiom-

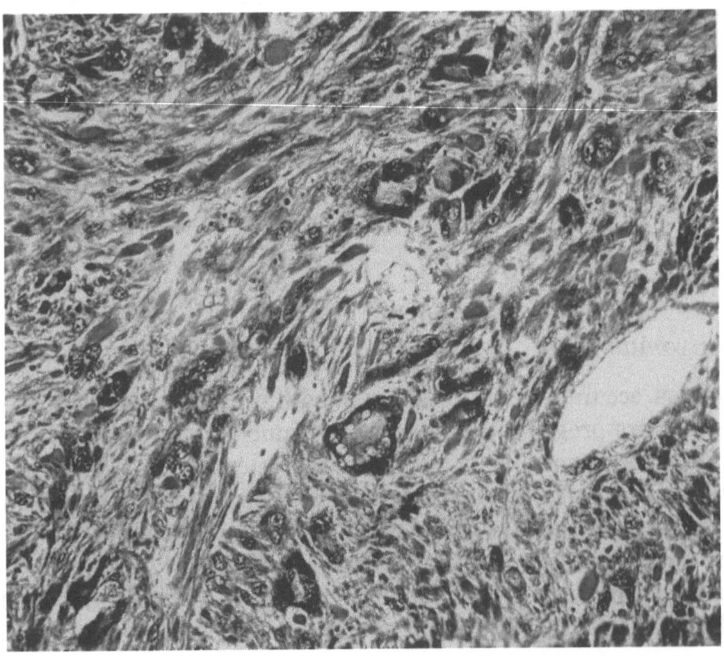

FIGURE 9-1. Symplastic leiomyoma. Despite the hyperchromatic and bizarre multinucleate cells, mitotic activity is very low. (H&E ×180.)

FIGURE 9-2. Epithelioid leiomyoma, leiomyoblastoma pattern. (H&E, × 390.)

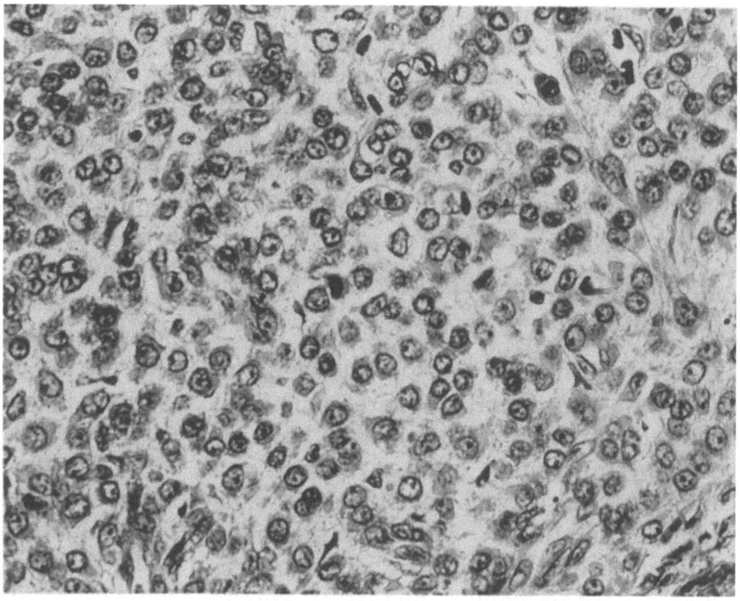

yomas are not as predictable in their behavior as other leiomyomas. A few have metastasized despite low degrees of mitotic activity and atypia. Necrosis and a size over 5 cm predispose to metastasis. Thus, hysterectomy and follow-up are recommended for patients with an epithelioid leiomyoma. Epithelioid features may also be found in tumors of uncertain malignant potential and in leiomyosarcomas.

Intravenous Leiomyomatosis

Intravenous leiomyomatosis is a very rare tumor seen in women ranging in age from 36 to 70 years who present with abnormal bleeding, pelvic pain, discomfort or pressure, and a pelvic mass.[20,21] In intravenous leiomyomatosis the leiomyomas appear as coiled or nodular, well-demarcated masses in the myometrium with convoluted, often mobile, wormlike growths of smooth muscle into the uterine veins within the broad ligament and beyond. Extension or metastasis of intravenous leiomyomatosis to the right atrium occurs, but seldom does it reach the lungs. Intravenous leiomyomatosis arises from two sources: by direct vascular invasion by a leiomyoma or by origin from smooth muscle in the walls of veins. Intrusion or invasion of a leiomyoma into veins by itself is not diagnostic of malignancy. It occurs in as many as one fifth of leiomyomas and probably is of no clinical significance.[11,13] The majority of cases have visibly evident masses within vascular channels of the myometrium. The cut surface of these wormlike growths is pinkish white or gray, and the consistency varies from soft and spongy to rubbery and firm.

Intravenous leiomyomatosis is usually composed of spindled-shaped smooth muscle cells arranged in a whorled leiomyoma pattern, but other forms are relatively acellular and unusually vascular. Still others are collagenous with few smooth muscle cells. Intravenous leiomyomatosis usually has less than two mitotic figures per 10 HPFs and usually no atypia.[20] In addition to venous channels, lymphatics are also involved, but no arteries. The rare occurrence of endometrial-type glands within intravenous leiomyomatosis, reflecting the pluripotential nature of müllerian derivatives, should not preclude the diagnosis. The distinction from endometrial stromal sarcoma is usually not a problem microscopically because the cell types are markedly different.

Treatment of intravenous leiomyomatosis with total abdominal hysterectomy and bilateral salpingo-oophorectomy has had a favorable prognosis. Three of four patients with recurrence had retained ovaries, suggesting that estrogen stimulation might play a role in recurrence.[21]

Leiomyomatosis Peritonealis Disseminata

Leiomyomatosis peritonealis disseminata (LPD) is a rare condition. Three fourths of cases are patients who are or who have recently been pregnant, or have received oral contraceptives, or who have an estrinizing granulosa cell tumor.[22] The finding of many small surface nodules throughout the peritoneal cavity resembling metastatic spread of a leiomyosarcoma in a woman at the time of cesarean section is characteristic.

Leiomyomatosis peritonealis disseminata is composed of bland spindle-shaped smooth muscle cells arranged in whorls and accompanied by fibroblasts and decidual cells. The clinical setting, the numerous small (less than 1 cm) nodules, and the lack of atypia and mitotic activity help identify LPD and distinguish it from leiomyosarcoma, which tends to have fewer but larger nodules. Leiomyosarcoma also occurs in older women and seldom is seen in a hormonal milieu. The cells of origin are the pluripotential subperitoneal mesenchymal cells that demonstrate their potential to differentiate into fibroblasts, myofibroblasts, smooth muscle cells, and endometrial stromal cells and glands. The demonstration of estrogen and progesterone receptors on these myocytic cells suggests a physiologic mechanism for their hormonal responsiveness.[23] With rare exceptions, LPD involutes when the hormonal stimulus is removed.[24]

Benign Metastasizing Leiomyoma

Benign metastasizing leiomyomas (BMLs) are rare lesions that have been encountered in patients with solitary or multiple lung, peritoneal, and occasionally lymph node smooth muscle proliferations, usually found following a hysterectomy or myomectomy for leiomyomas. These have been noted in the third or fourth decade, often in black women. The diagnosis of BML is one of exclusion. Included in the designation BML have been benign leiomyomas arising in the lung, fibroleiomyomatous hamartomas or hormonally induced hyperplasias resembling them, and metastatic leiomyosarcomas.[25] Most reports discuss the roentgenographic appearance in the lungs, but not the gross morphologic features. In the lung BMLs are generally bilateral and range in size from 0.2 to 8 cm. The microscopic appearance is entirely benign, because nuclear enlargement, hyperchromatism, necrosis, mitotic figures, and atypia are absent. This has given rise to the theory that BMLs represent local growth of fragments of smooth muscle cells embolized at the time of surgery. Leiomyomas metastasize very rarely and grow very slowly. There are so many leiomyomas of the uterus that if one in a million metastasized, a few would do so each year. Despite their bland appearance, many lesions that have been called BMLs are in fact metastatic low-grade leiomyosarcomas that resemble leiomyomas. Most earlier reports describe uterine tumors that were not sampled thoroughly, and a sarcoma may have been overlooked. Leiomyosarcomas occur in the uterus in an older age group, grow slowly, multiply, and, when they metastasize to the lung, are fatal, unlike BMLs.

Smooth Muscle Tumors of Uncertain Malignant Potential

Among smooth muscle tumors, there are a small number that fall into a gray zone between benign and malignant (Table 9-3). These smooth muscle tumors are similar in clinical presentation and gross appearance to usual leiomyomas. A diagnosis of uncertain malignant potential is used when smooth muscle tumors fail to satisfy all the requirements for leiomyosacoma. For example, a smooth muscle tumor with no cellular atypia but with a mitotic figure count

TABLE 9-3. Smooth Muscle Tumors (The gray zone)

Diagnosis	Atypia	Mitoses	Metastatic potential
Leiomyoma with increased mitoses	0	5–9	0
Atypical leiomyoma	Moderate to marked	<5	0
Uncertain malignant potential	Equivocal to minimal	5–9	Unknown
Leiomyosarcoma low grade	1+	5–9	Yes

of 5 to 9 per 10 HPF has no metastatic potential. A neoplasm with grade one atypia and the same mitotic range (5 to 9 per 10 HPF) is a low grade leiomyosarcoma, but if the atypia is equivocal, the behavior is unknown.[11] Thus, smooth muscle tumors of uncertain malignant potential are those with mitotic activity of 5 to 9 mitotic figures per 10 HPF and with equivocal or minimal nuclear atypia. If there is moderate to marked atypia and fewer than five mitotic figures per 10 HPFs the metastatic rate is very low to nonexistent and the lesion is designated an atypical leiomyoma.[6,11]

A diagnosis of smooth muscle tumor of uncertain malignant potential becomes a special concern in myomectomy specimens because all of the clinicopathologic studies and prognosis are based on follow-up after hysterectomy. Studies are needed to determine the rate of metastasis from smooth muscle tumors removed at myomectomy.

Leiomyosarcoma

Leiomyosarcomas are relatively rare, occurring about once for every 800 leiomyomas. Found predominantly in older women (median age 51 years), a rapid increase in the size of what is suspected to be a leiomyoma is a common presentation.[3] The majority are identified by the pathologist after hysterectomy, and are solitary rather than multiple. There is little evidence that ordinary leiomyomas transform into leiomyosarcomas.[9] As noted above, leiomyomas are often multiple, yet patients with multiple leiomyomas are not at greater risk to develop leiomyosarcomas. Predisposing factors such as pelvic radiation have not been demonstrated for leiomyosarcoma, as they have for other uterine sarcomas.[9]

Gross examination of the cut surface of a leiomyosarcoma typically shows an ill-defined, soft, fleshy mass with hemorrhagic or necrotic gray or yellow areas, unlike the ordinary leiomyoma with its circumscribed, whorled, firm, gray-white cut surface (Fig. 9-3). Leiomyosarcomas tend to have a larger median diameter (about 9 cm) compared to leiomyomas.[9] They tend not to impinge upon the endometrial cavity or invade blood vessels as stromal sarcomas do (Table 9-4).

Microscopically leiomyosarcomas are spindle cell sarcomas that vary in their composition and differentiation from recognizable smooth muscle cells containing abundant eosinophilic fibrillar cytoplasm and ple-

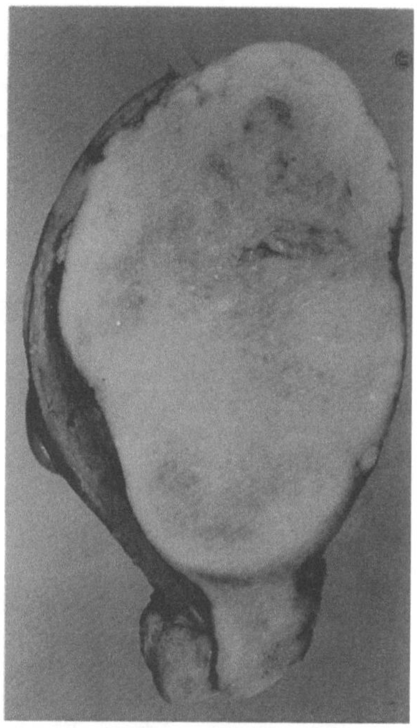

FIGURE 9-3. Leiomyosarcoma. (Reprinted from ref. 9, with permission.)

TABLE 9-4. Comparison of the gross pathology of leiomyoma and leiomyosarcoma

Leiomyoma	Leiomyosarcoma
Usually multiple	Often solitary
Variable size (3–5 cm)	Large (>10 cm)
Firm, whorled cut surface	Soft, fleshy cut surface
White	Yellow or tan
Hemorrhage & necrosis infrequent	Hemorrhage & necrosis frequent

Reprinted from ref. 7, with permission.

omorphic fusiform nuclei to poorly differentiated round cell tumors resembling stromal sarcomas. In the presence of marked atypia or poor differentiation, immunocytochemical reactions for desmin and actin may demonstrate smooth muscle differentiation. In tumors with fusiform cells containing eosinophilic cytoplasm (Fig. 9-4), the diagnostic microscopic findings include increased cellularity, high mitotic activity, nuclear hyperchromasia, and prominent nucleoli.[9,13] Infiltrating margins are noted in 25% and vascular invasion in about 10%. Microscopically, a uterine leiomyosarcoma always has a degree of atypia and usually has more than 10 mitotic figures per 10 HPFs. Some have moderate to marked atypia and more than five mitotic figures per 10 HPFs.[13]

The distinction of leiomyosarcoma from cellular leiomyoma and leiomyoma with increased mitotic figures is based on the absence of cytologic atypia in the latter. Symplastic leiomyomas have atypia and multinucleated cells with enlarged smudged nuclei, but their mitotic rate does not exceed two per 10 HPFs and there is often a background of a usual leiomyoma. Epithelioid leiomyomas, including plexiform, clear cell, and leiomyoblastoma types, have characteristic polygonal cells with eosinophilic cytoplasm and fewer than five mitotic figures per 10 HPFs.[17] Intravenous leiomyomatosis differs from leiomyosarcoma by having a characteristic wormlike intravascular growth, less than two mitotic figures per 10 HPFs, and little or no atypia.[20] Leiomyomatosis peritonealis disseminata ordinarily has a

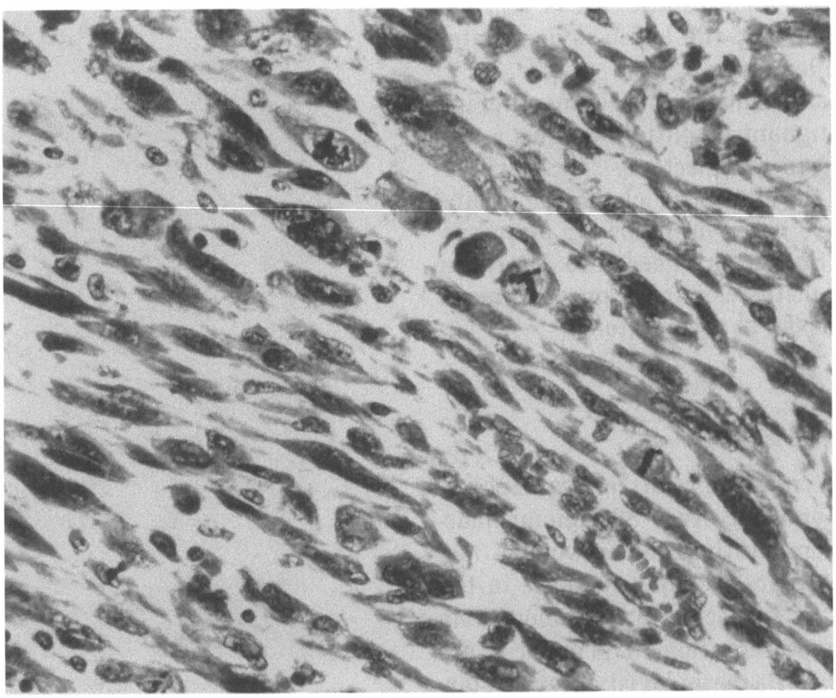

FIGURE 9-4. Leiomyosarcoma. (H&E, × 400.) (Reprinted from ref. 13, with permission.)

hormonal background of pregnancy or history of oral contraceptive use. In LPD multiple nodules less than 1 cm in size form that contain no atypia and have very little mitotic activity, unlike leiomyosarcoma.[9]

Hysterectomy for stage I and II leiomyosarcomas has produced a 40 to 50% 5-year survival rate, but the overall prognosis is poor, with a 20% 5-year survival.[3,7]

Myxoid Leiomyosarcoma

Myxoid leiomyosarcoma is a very rare variant of leiomyosarcoma that tends to occur in women 47 to 68 years of age, with a median age of 56 years.[26,27] Myxoid leiomyosarcoma has a deceptively bland appearance, hence the need to recognize it as a separate specific variant. Myxoid leiomyosarcoma produces symptoms similar to leiomyosarcoma, including vaginal bleeding and a pelvic mass. It ranges from 5 to 17 cm in diameter and arises in the myometrium as a circumscribed mass with a gelatinous cut surface.

Microscopically, the tumor cells are spindle shaped but widely spaced in a myxoid matrix. The cells are occasionally arranged in small fascicles. Cytoplasm is scant, nuclei are small and pale, and nucleoli inconspicuous (Fig. 9-5). Myxoid leiomyosarcomas usually have less than two mitotic figures per 10 HPFs. However, if a correction is applied for the wide separation of cells, a higher count would emerge, underlining the malignant nature of this tumor. The prognosis with myxoid leiomyosarcoma is as poor as with ordinary leiomyosarcoma. Metastatic sarcoma is found in pelvic peritoneal surfaces, liver, and lung.

Epithelioid Leiomyosarcoma

Epithelioid leiomyosarcoma is another very rare variant of leiomyosarcoma.[17] Two of three patients are over the age of 50 years. The tumors range from 6 to 16 cm in diameter. Microscopically, epithelioid leiomyosarcoma has an admixture of rounded cells with either clear or eosinophilic cytoplasm

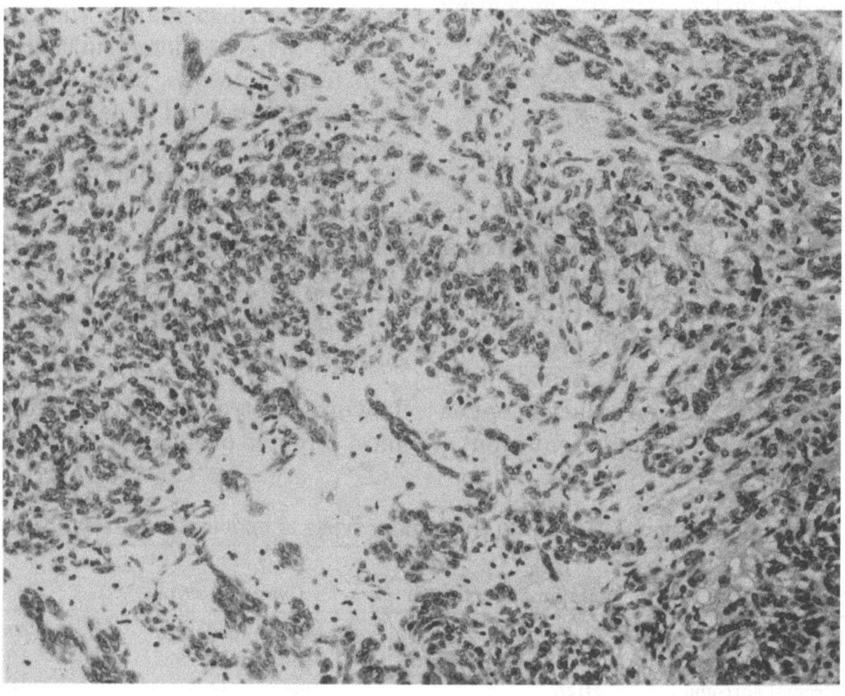

FIGURE 9-5. Myxoid leiomyosarcoma. (H&E, × 60.)

or some mixture of them and elongated spindle-shaped smooth muscle cells. Cells are arranged in clustered or cordlike patterns. Nuclei are central or paracentral. Features associated with metastatic potential include an infiltrating margin, extensive necrosis, and variable mitotic activity. Features that do not have prognostic significance include nuclear atypicality and vascular invasion, because these may be present in both malignant and benign neoplasms. Primary therapy is hysterectomy.[17]

Endometrial Stromal Tumors

There are three types of endometrial stromal tumors: stromal nodule and low- and high-grade endometrial stromal sarcoma (Table 9-5). Stromal nodule and low-grade stromal sarcoma resemble the stroma of proliferative endometrium, whereas high-grade stromal sarcoma tends to be less differentiated and less easily recognizable as being of endometrial stromal origin.[13,28,29] Flow cytometric analysis of DNA content and rate of cell turnover, as demonstrated by the percentage of cells in S phase, has been recommended as an additional estimate of prognosis.[30]

Stromal Nodules

Stromal nodules are benign. Unfortunately, less than a quarter of endometrial stromal tumors are stromal nodules. Patients range in age from 23 to 75 years, with a median age of 47 years, and two thirds present with abnormal bleeding. Pelvic and abdominal pain are common.[28] Uterine enlargement occurs in more than half of patients, but a stromal nodule is an incidental finding in 12%. More than half involve the myometrium alone, while the remainder involve endometrium or both endometrium and myometrium. Nodules often protrude into the endometrial cavity as a polyp. Typically they are oval masses ranging in size from 0.8 to 15 cm. The cut surface bulges above the myometrium and is fleshy, gray white to orange, and well circumscribed (Fig. 9-6).

The cells comprising stromal nodules are identical to those found in midproliferative endometrial stroma. There is no cytologic atypia and generally few mitotic figures, although between 5 and 15 mitotic figures per 10 HPFs are found in 7%.[28] Characteristic features are the circumscription, highly vascular supporting stroma, and lack of vascular invasion. Glandlike structures, including plexiform, trabecular, and tubular forms, may be present, reflecting the pluripotential nature of the endometrial stromal cell.

Stromal nodules do not recur or metastasize regardless of the extent of surgery. Although similar cytologically to low-grade endometrial stromal sarcoma, a stromal nodule differs from low-grade stromal sarcoma by its circumscribed margin and lack of infiltration. When endometrial curettings contain fragments of a stromal tumor, a hysterectomy is needed to determine the tumor margin, because this is the prime criterion for separating a stromal nodule from the more aggressive low-grade endometrial stromal sarcoma.[28]

Low-Grade Endometrial Stromal Sarcoma

Low-grade stromal sarcoma (LGSS) is the most common of endometrial stromal tumors

TABLE 9-5. Histologic characteristics of the three main types of endometrial stromal tumors

Tumor	Malignant potential	Cytologic atypia	Mitoses/ 10 HPFs
Stromal nodule	None	Mild to moderate	Usually 0–3
Low-grade stromal sarcoma	Low to intermediate	Mild to moderate	<10, usually 1–3
High-grade stromal sarcoma	High	Moderate to marked	>10

Reprinted from ref. 13, with permission.

FIGURE 9-6. Endometrial stromal nodule. Myometrium (top) is sharply demarcated from the endometrial stromal tumor (bottom). (H&E, × 100.)

and was formerly known as endolymphatic stromal myosis and stromatosis. It is formed by small, oval, nondescript stromal cells that comprise most of the endometrial stroma. Half of patients present with stage I disease, while the other half present with extension to parametrium or with metastases. The age range of patients with LGSS varies widely, from 20 to 80 years, but more than half are premenopausal. Menorrhagia, cramping pain, and uterine enlargement are the main symptoms. Anemia is common from excessive bleeding. The cut surface characteristically has yellow to pink wormlike plugs infiltrating the myometrium and projecting into vascular spaces. Often a polypoid mass of similar tissue protrudes into the endometrial cavity.

Microscopically, a monotonous pattern of small ovoid or fusiform cells with scant cytoplasm is punctuated by a relatively large number of arterioles and capillaries (Fig. 9-7). At the periphery, LGSS infiltrates the adjacent myometrium with finger-like projections. Usually there are only three or fewer mitotic figures per 10 HPFs. Epithelial, trabecular, and cordlike differentiation may be found, but this does not alter their behavior. The main differential diagnostic problem arises with smooth muscle neoplasms. If the usual cytologic features do not allow for differentiation, immunocytochemistry may demonstrate a positive reaction for vimentin in LGSS and positive reactions for desmin and muscle-specific actin in smooth muscle tumors.[31]

Recurrences develop in stage I tumors in about a quarter of cases, most commonly when tumor has extended to the uterine serosa above. It may be 20 years or more before the recurrence, which tends to involve pelvic structures first, then spreads to involve the periotneum and omentum, and perhaps the lung. Isolated metastases to the ovaries or peritoneum also may be the first sign of metastasis, and in some instances are responsible for the presenting complaints. Some patients have metastases discovered years

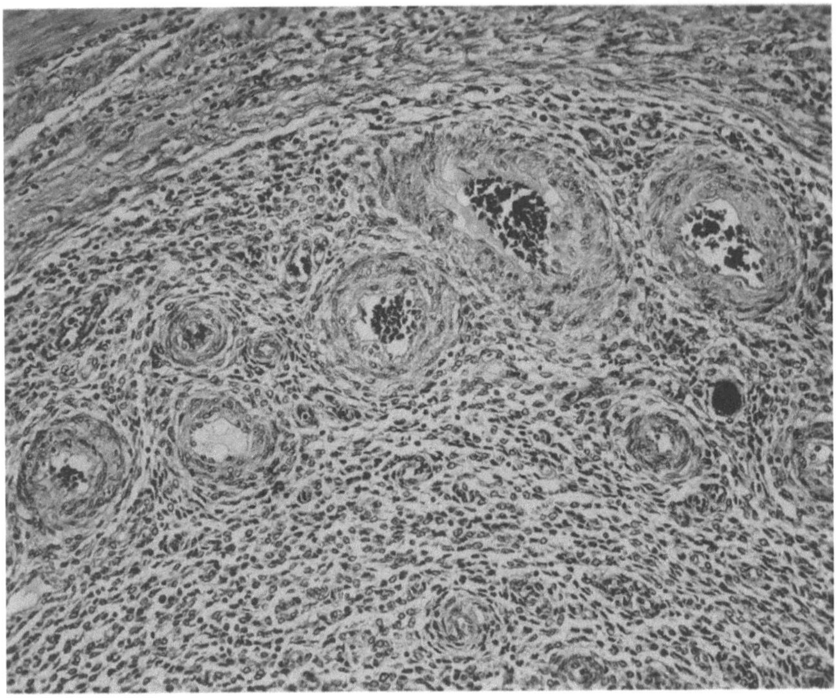

FIGURE 9-7. Low-grade endometrial stromal sarcoma. Note the prominent small muscular arteries. (H&E, × 160.) (Reprinted from ref. 13, with permission.)

before the primary site in the uterus is discovered.

High-Grade Endometrial Stromal Sarcoma

Patients with high-grade endometrial stromal sarcoma (HGSS) are generally over 50 years of age. They present with abdominal pain and abnormal vaginal bleeding that gradually becomes more severe. Anemia may be present from blood loss. The uterus is enlarged and has a soft, fleshy, tan to white polypoid mass that fills the endometrial cavity and invades the underlying myometrium (Fig. 9-8).[29,32] The microscopic appearance is that of an endometrial tumor with destructive infiltrative growth and necrosis in the myometrium (Fig. 9-9). The cellular composition is often without distinctive features except for larger, more pleomorphic and hyperchromatic cells than are found in a LGSS, and much more numerous mitotic figures, often between 10 and 20 per 10 HPFs.[29,32] Poorly differentiated forms are difficult to recognize as being of endometrial stromal origin, and most reports contain undifferentiated carcinomas or unclassified sarcomas reported as HGSS (Fig. 9-10). The

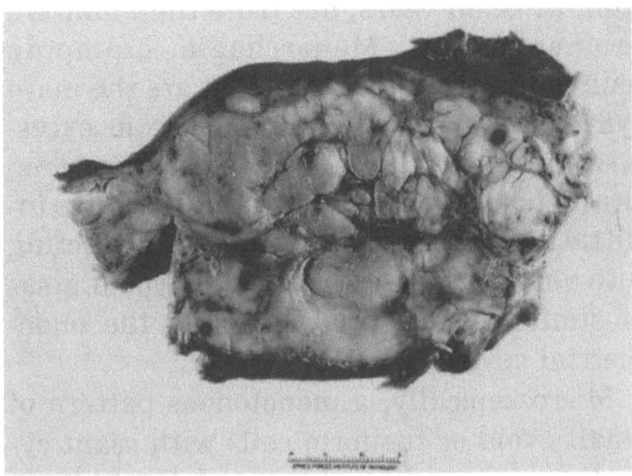

FIGURE 9-8. High-grade endometrial stromal sarcoma. Nearly all of the myometrium is replaced by tumor.

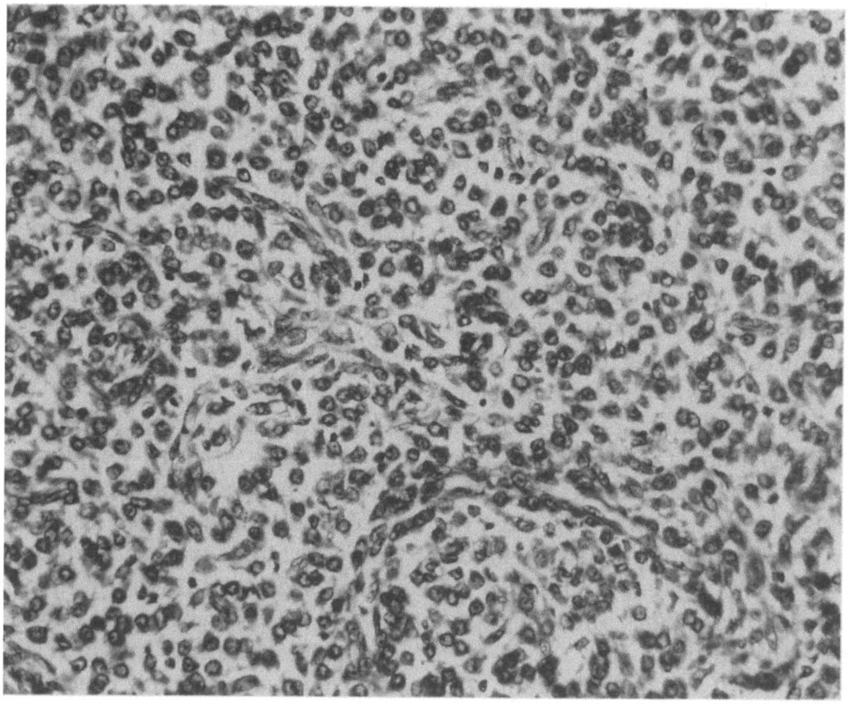

FIGURE 9-9. High-grade stromal sarcoma. The underlying vascularity is evident. (H&E, × 250.)

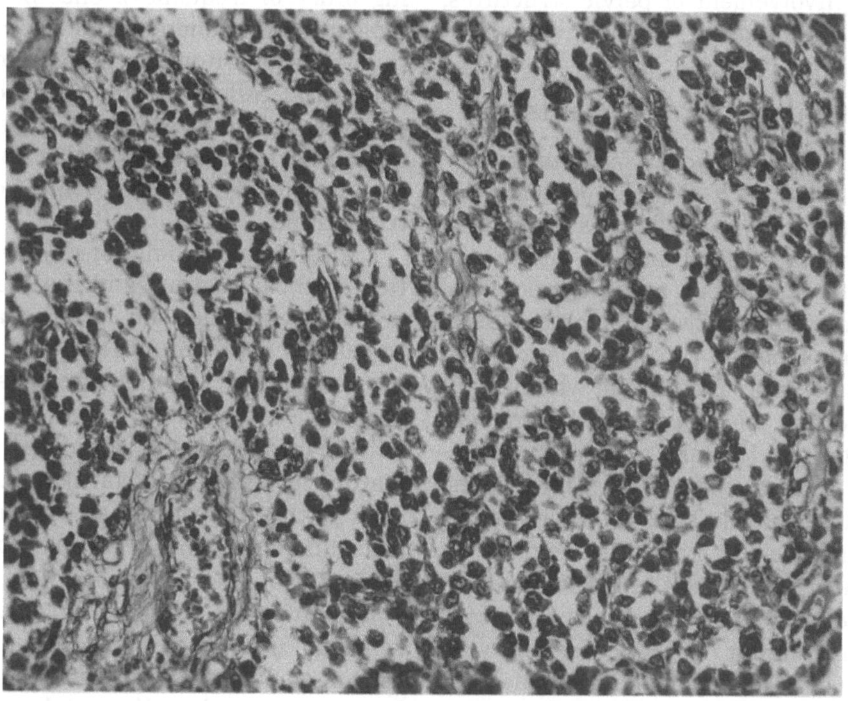

FIGURE 9-10. Undifferentiated sarcoma (sarcoma unclassified). The small cells resemble endometrial stromal sarcoma more than any other kind of sarcoma.(H&E, × 250.)

differential diagnosis involves metastatic undifferentiated carcinoma, which usually shows larger, less cohesive cells and positive reactions for cytokeratin and mucins. Leiomyosarcoma and HGSS share morphologic features such as pleomorphism and atypia. Immunocytochemical reactions are often necessary to demonstrate muscle differentiation by means of desmin or smooth muscle actin reactions. Electron microscopy may be necessary in some cases. Lymphoma has a predilection for the cervix, tends to have a sharp peripheral margin, has a more monotonous cell population with scanty cytoplasm, and has a positive reaction for leukocyte common antigen and a negative one for vimentin. Melanoma can be distinguished in most instances by positive immunohistochemical reactions for S-100 protein and HMB 45 (melanoma-specific antigen), and a negative reaction for cytokeratin.

Recurrences of HGSS are frequent and usually occur within 2 years.[6,33–35] Rather than the isolated metastases initially seen when LGSS metastasizes, there is commonly combined involvement of pelvic structures, abdomen, and lungs.

Endometrial Stromal Variants

Combined smooth muscle–stromal tumor (stromomyoma) is a form of endometrial stromal tumor.[36,37] Resembling a leiomyoma grossly, it has areas of smooth muscle admixed with stroma, or areas composed of cells intermediate between smooth muscle and stroma. This is a rare tumor, occurring in patients of reproductive age who present with abnormal bleeding and uterine enlargement. It is thought to arise from endometrial stromal cells.[37] Very few metastasize. Those that do have more characteristics of endometrial stromal sarcoma, such as infiltrating margins, than of leiomyoma. *Uterine tumors resembling ovarian sex cord tumors* are similar to LGSS in behavior. Most have areas of endometrial stromal sarcoma, either low grade or high grade.

Other variants of endometrial stromal tumors contain areas of epithelioid differentiation with trabeculae, tubules, or cords that arise from stromal cells, as in stromal nodules and LGSS, but not in HGSS. The cells are small and have scant cytoplasm. They have hyperchromatic nuclei, small nucleoli, and rare mitotic figures (Fig. 9-11). This variant might be confused with metastatic adenocarcinoma or mixed mesodermal tumor, homologous type, but the absence of cytologic atypia, pleomorphism, or high mitotic activity should make the distinction clear.[29,36]

When the sex cord pattern is a minor constituent in a stromal neoplasm, pelvic recurrence occurs in about half. When it predominates, there are few recurrences.[32]

Low-grade *extrauterine endometrial stromal sarcoma* occurs as a primary neoplasm, usually arising in endometriosis.[38] There are instances in which careful study has excluded a primary uterine endometrial stromal sarcoma as the source of extrauterine stromal sarcoma. Sources of this occurrence other than metastasis are (in decreasing order of frequency): endometriosis, ectopic growth of endometrial stroma present in the retroperitoneum; and from the pluripotential cells of pelvic mesothelium and submesothelial mesenchyme (the "secondary müllerian system"), which have the inherent capacity to differentiate to endometrial stroma.

Extrauterine stromal sarcomas are identical to those that arise in the uterus. They are both cystic and solid, have a chalky white to yellow-brown cut surface, and, like uterine endometrial stromal sarcomas, are composed of a sea of small, bland, fusiform cells with scant cytoplasm. The usual associated features of an endometrial stromal sarcoma, such as the characteristic vascularity, foam cells, and eosinophilic hyaline globules, are present.[38]

Extrauterine stromal sarcoma is often confused with leiomyosarcoma, malignant schwannoma, spindle cell mesothelioma, fibrosarcoma, and malignant hemangiopericytoma.[38] Leiomyosarcoma, in contrast to

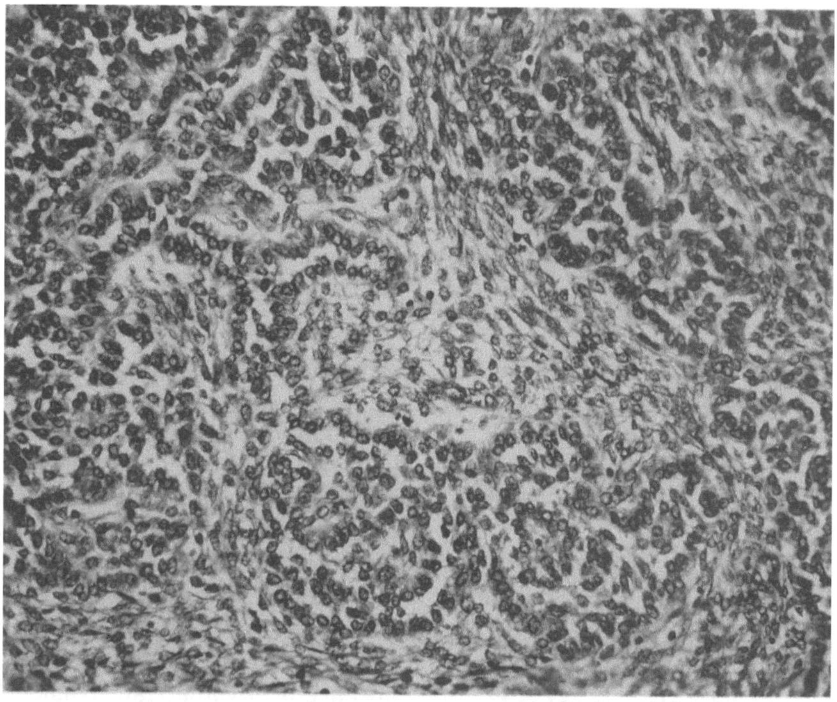

FIGURE 9-11. Low-grade endometrial stromal sarcoma variant with cordlike arrangement as in the "uterine tumor resembling sex cord stroma." (H&E, × 250.)

those, usually has more atypia, more mitotic activity, and larger, more pleomorphic cells with fibrillar cytoplasm, and may have positive reactions for desmin and actin.[31] Malignant schwannoma is usually found in patients with neurofibromatosis. It usually is associated with a plexiform neurofibroma and has plump, spindle-shaped cells with hyperchromatic nuclei having a wavy outline that are arranged in whorls and interlacing fascicles. Stains for S-100 protein are usually positive. Fibrous mesothelioma, like fibrosarcoma, has spindle cells arranged in interlacing bundles with prominent mitotic activity. Hemangiopericytoma has variably sized gaping sinusoids or slitlike capillary channels surrounded by plump, spindle-shaped pericytes with scant cytoplasm, indistinct cell borders, and features of smooth muscle cells. Endometrial stromal sarcoma lacks the prominent staghorn-shaped sinusoids of hemangiopericytoma. Malignant lymphoma has round to oval cells, but not usually a spindle cell proliferation, and does not have the regular structured arteriolar component. Vimentin is demonstrated in endometrial stromal sarcoma, but little or no desmin or muscle-specific actin, and no lymphoma markers such as leukocyte common antigen.[31]

Mixed Mesodermal Tumors

Mixed mesodermal tumors are a spectrum of tumors in which the pluripotential nature of müllerian-derived structures reaches its greatest expression (Table 9-6). Mixed mesodermal tumors consist of both epithelial and mesenchymal elements that in adenofibroma, the low end of the spectrum, are both benign; in adenosarcoma the epithelial element is benign and the stromal element is malignant, and in mixed müllerian tu-

TABLE 9-6. Classification of mixed mesodermal tumors

Tumor	Malignant potential	Epithelial component	Mesenchymal component
Adenofibroma	None	Benign	Benign, homologous
Adenosarcoma	Low to intermediate	Benign	Sarcoma, homologous or heterologous
MMT: Homologous	High	Carcinoma	Sarcoma, homologous
MMT: Heterologous	High	Carcinoma	Sarcoma, heterologous

Reprinted from ref. 7, with permission.
MMT = mixed mesodermal tumor.

mors both epithelium and stroma are malignant. Their relative frequency is set forth in Table 9-7.

Adenofibroma

Adenofibroma is a benign tumor of menopausal and postmenopausal women. Patients have a median age of 68 years and usually present with vaginal bleeding and abdominal pain and enlargement.[39,40] Adenofibroma constitutes a very small proportion of mixed mesodermal tumors. Curettings contain large polypoid fragments. Adenofibromas are lobulated or papillary tumors that range up to 10 cm. They are tan-brown and vary from soft to rubbery in consistency. Adenofibroma is superficial and does not involve the myometrium. The stromal cells either resemble endometrial stroma or are fibroblastic in type, and show no atypia, although three or fewer mitotic figures are found in 10 HPFs.[40,41] The stroma does not condense about glands, as in adenosarcoma.

TABLE 9-7. Relative frequency of uterine sarcomas

Type	% of uterine sarcomas
Mixed müllerian tumor (homologous or heterologous)	30
Leiomyosarcoma	27
Endometrial stromal sarcoma (LGSS & HGSS)	26
Adenosarcoma	8
Unclassified sarcoma	6
Miscellaneous sarcomas	3

Reprinted from ref. 7 with permission.

The main histologic differential diagnoses are with endometrial polyps and adenosarcoma. Polyps have more rounded contours without papillary processes; more glands, usually with a more hyperplastic lining; and a prominent vascular center. Differentiation from adenosarcoma is based on the absence of mitotic figures, atypia of stromal cells and absence of myometrial invasion.[41] Whereas both epithelium and stroma in adenofibroma are benign, in adenosarcoma only the epithelium appears benign. Endometrial stromal sarcomas do not show the architectural complexity of glands or papillary epithelial-lined processes found in adenofibroma and adenosarcoma. They invade the myometrium with finger-like projections. Extrauterine adenofibromas may arise in the ovary or in the pelvic peritoneum, and should not be mistaken for a metastasis from an endometrial adenofibroma.[41]

Although none of the patients with adenofibroma of the uterus have had a recurrence after hysterectomy, hysterectomy is recommended because curettings alone do not provide for complete excision with the opportunity to exclude an adenosarcoma.

Adenosarcoma

Adenosarcoma is a malignant tumor occurring in women with a history of vaginal bleeding and ranging from 14 to 79 years of age, with a median age of 57 years.[41–43] Adenosarcoma has more exuberant stroma than adenofibroma, and may be much larger, measuring up to 12 cm in size. Adenosarcoma is typically polypoid, soft to firm in consistency, and tan, brown, or gray, with focal necrosis and hemorrhage (Fig. 9-12).

Papillary or polypoid fronds consisting of either endometrial stromal or fibroblastic cells project into cystic spaces (Fig. 9-13) lined by a variety of benign epithelia, mostly columnar epithelium (Fig. 9-14). The stromal cells have four or more mitotic figures per 10 HPFs, and may have mild, moderate, or severe cytologic atypia. A heterologous element such as striated muscle, cartilage, or fat appears in about a quarter of cases. The epithelium and glands appear benign, but usually have a condensation of stroma about them (Fig. 9-15).

The potential for metastasis is correlated with overgrowth by sarcoma,[43] invasion of the myometrium, the number of mitotic figures, and the degree of cytologic atypia. The usual treatment is hysterectomy and bilateral salpingo-oophorectomy. Although adenosarcomas are usually superficial, they may recur in as many as 30% of cases over a median period of 5 years, generally within the pelvis or vagina. In the differential diagnosis, mixed mesodermal tumors differ from adenosarcoma in that they have malignant epithelium as well as stroma, and as long as both elements are sampled the distinction from adenosarcoma should not present a problem. The malignant stroma, however, may be identical in both lesions. Extrauterine adenosarcomas may arise elsewhere in the pelvis or ovary, and should not be mistaken for a metastasis from the endometrium.[41]

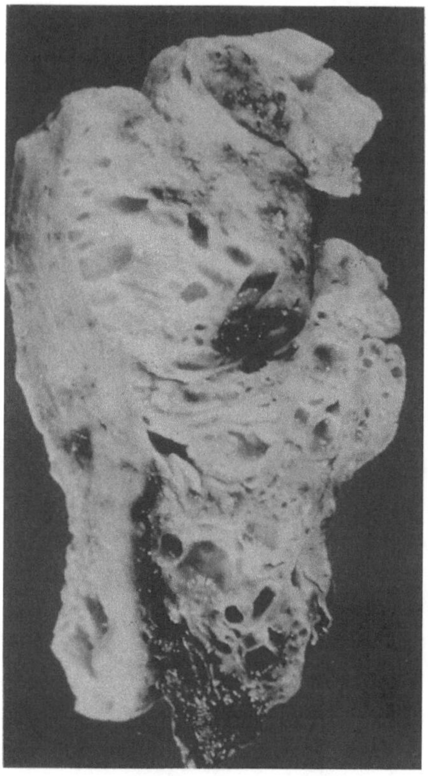

FIGURE 9-12. Adenosarcoma occupying most of the endometrial cavity. (Reprinted from ref. 7, with permission.)

Homologous and Heterologous Mixed Mesodermal Tumors

Homologous (carcinosarcoma) and Heterologous ("mixed mesodermal tumor") subtypes of mixed mesodermal tumors are equally common and similar in all characteristics but the presence of a heterologous component in the heterologous form. Mixed mesodermal tumors occur in older women, ranging from 43 to 85 years of age, with a median age of 68 years. Patients with mixed mesodermal tumors tend to present with vaginal bleeding, an abdominal mass, and pain.[33,35,44] About 10% of patients have a history of prior pelvic radiation. Those with this history tend to develop mixed mesodermal tumors at an earlier age than those without it. Mixed mesodermal tumors are more common than leiomyosarcomas. They arise in the endometrium more often than in the endocervix, and are bulky, polypoid masses that range from 1 to 20 cm and often fill the endometrial cavity (Fig. 9-16). They are typically tan and firm, with focal hemorrhage and necrosis. Both epithelium and the stroma are malignant (Fig. 9-17), with the latter comprising the majority of the tumor. Because of this, curettings may miss

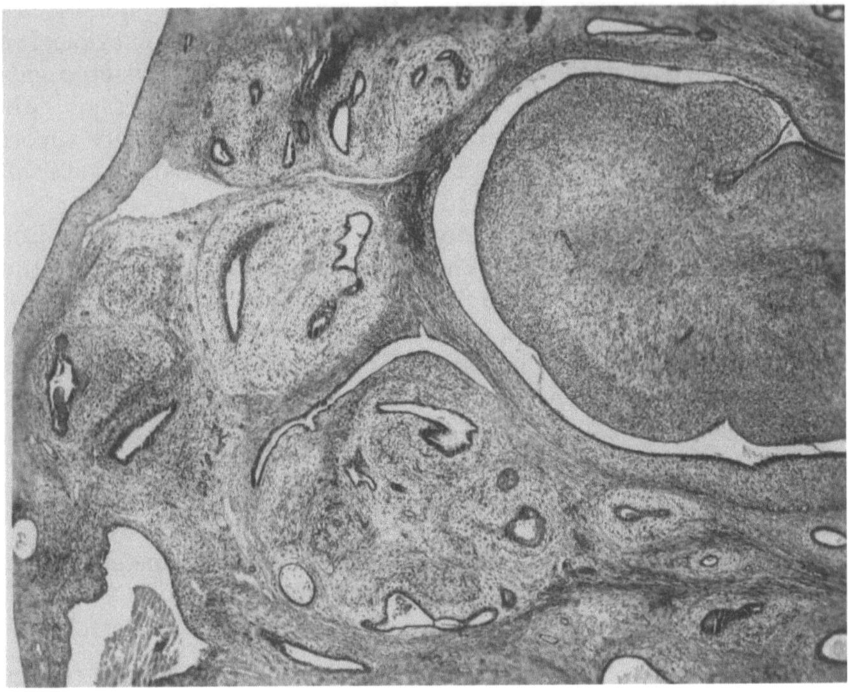

FIGURE 9-13. Adenosarcoma. The stroma shows more variation than in an adenofibroma and denser cellularity. (H&E, ×75.) (Reprinted from ref. 41, with permission.)

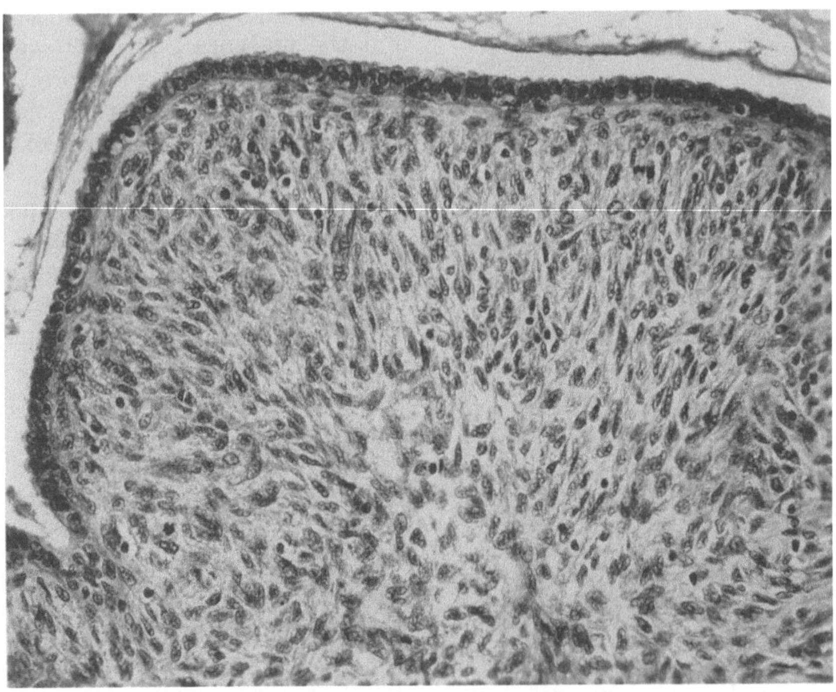

FIGURE 9-14. Adenosarcoma. (H&E, × 250.)

FIGURE 9-15. Adenosarcoma. Note concentric arrangement of the cells. (H&E, × 250.)

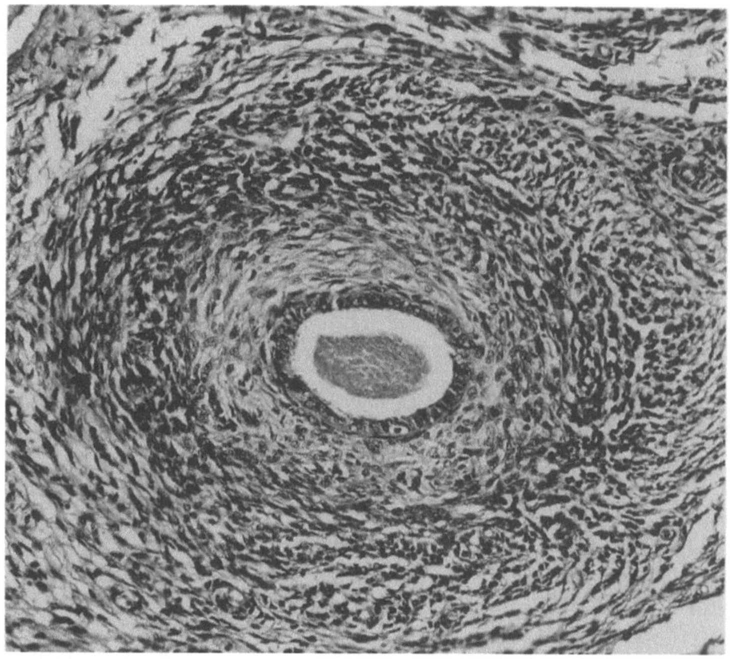

the epithelial component, which is only discovered by examination of the uterus proper.

The microscopic findings are different in each case. The epithelial component is mostly adenocarcinoma, but there is a squamous component. The sarcoma varies from well differentiated to high grade and may include heterologous elements. Tumor stage is the most significant factor in prognosis, but morphologic features also play a role.[44] Homologous mixed mesodermal tumor is made up of elements common to the uterus (fibrosarcoma, endometrial stromal sarcoma, and leiomyosarcoma) and tends to be less bulky than heterologous mixed mesodermal tumor. About a third of patients with the homologous type are in stage III or IV at discovery, with a 5-year survival of 36%, whereas patients with heterologous mixed mesodermal tumor with rhabdomyoblastic or chondrosarcomatous elements (Fig. 9-18) are more often in stage III or IV, with a 5-year survival of 14%.[44] Precise identification of these elements often requires immunocytochemical study. Mixed mesodermal tumors are rapidly growing, invasive tumors. Apart from the stage, the prognosis is most dependent on the depth of myometrial invasion. When invasion is more than halfway through the myometrium, pelvic lymph node metastasis is found in 30% of patients.[45,46] Five-year survival figures range from 20 to 40%. Patients with tumor confined to an endometrial polyp tend to do well.

Differentiation of mixed mesodermal tumor from primary or metastatic carcinoma and from metastatic melanoma, lymphoma, and metastatic sarcoma is based on clinical presentation, use of immunocytochemistry for identification of otherwise obscure heterologous elements, and careful search of multiple sections to find the characteristic biphasic pattern. Grossly, none of the above except mixed mesodermal tumors forms a bulky endometrial mass.

Extrauterine mixed mesodermal tumors are rare, but arise from the ovary and fallopian tube as well as from the pelvic peritoneum, usually near or in the region of the adnexal structures. Origin outside of the uterus, ovary, or fallopian tube is very rare, and a reflection of the pluripotential nature of the peritoneum and subperitoneal mesenchyme.[47,48]

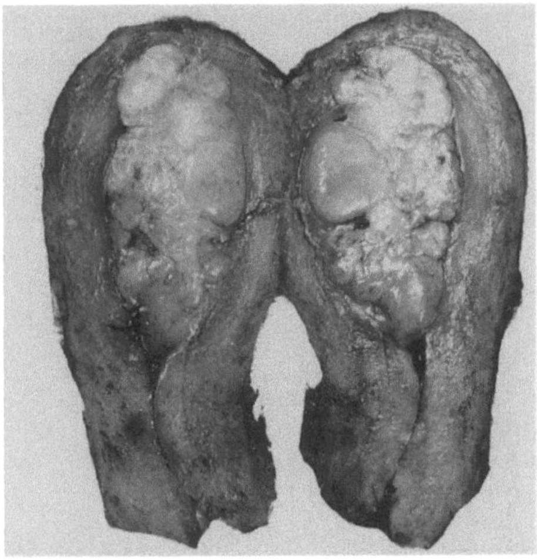

FIGURE 9-16. Carcinosarcoma with the typical polypoid broad-based configuration. (From Norris HJ, Taylor HB: Mesenchymal tumors of the uterus. III. A clinical and pathologic study of 31 carcinosarcomas. Cancer 1966;19:1459–1465. Reprinted with permission.)

Tumor-like Conditions

Adenomyosis

Adenomyosis occurs in 15 to 20% of women of reproductive or perimemopausal age.[49] The clinical presentation is usually a combination of dysmenorrhea, abnormal bleeding, and enlargement of the uterus. The uterus may be two or three times its normal size, often in globoid or eccentric fashion. The cut surface has hemorrhage and a spongy or trabeculated appearance in areas of adenomyosis, but no discrete tumor mass. This distortion is partly a result of hyperplasia of the myometrium adjacent to adenomyosis.

The endometrial-myometrial interface is often irregular. A minimum criterion for the diagnosis requires that the interface extend more than half a low-power field (about 2 mm) from the basalis layer of the endometrium. Theories of pathogenesis of endometriosis also apply to adenomyosis, with local extension of glands the most likely event, perhaps favored by local structural changes in the myometrium induced by leiomyomas. Adenomyosis coexists with extrauterine endometriosis in 15% of instances and with leiomyomas in half of cases.[50] Unlike endometriosis, the glands of adenomyosis usually do not show cyclic menstrual changes, probably because of a low level of progesterone receptor, as in the basalis.[51]

Adenomyoma

Adenomyoma is a circumscribed mass of smooth muscle, endometrial stroma, and glands. Usually found protruding from the endometrium as a form of polyp, it seldom presents a problem in diagnosis. The *atypical polypoid adenomyoma of Mazur* (APAM) is a rare variant of an adenomyoma that can be a diagnostic problem for the pathologist. It occurs in premenopausal women with menometrorrhagia, 75% of whom have endometrial hyperplasia.[52]

Atypical polypoid adenomyoma of Mazur is usually formed by an intimate admixture of atypical endometrial glands without stroma abutting on fibrous tissue and smooth muscle (Fig. 9-19), suggesting an invasive process. The epithelial component includes proliferative glands, usually with prominent squamous metaplasia. The epithelial cells have nuclear enlargement, hyperchromasia, and prominent nucleoli (Fig. 9-20). When APAM is encountered in curettings, the diagnosis is more difficult than in a hysterectomy specimen, where its polypoid circumscribed nature and lack of myometrial invasion is apparent. In curettings, the whorled density of the muscle containing the epithelium suggests a benign diagnosis. Carcinoma cannot always be ruled out in the curettings, and since some of these tumors have persisted for up to 4 years with repeated curettage, hysterectomy may be necessary for therapeutic reasons if not for diagnostic reasons.[53] Differentiating APAM from adenomyosis is not difficult since adenomyosis has glands that are not atypical, has little or no squamous differentiation, and has endometrial stroma surrounding the glands. Adenosarcoma seldom contains areas of smooth muscle, but it does have a

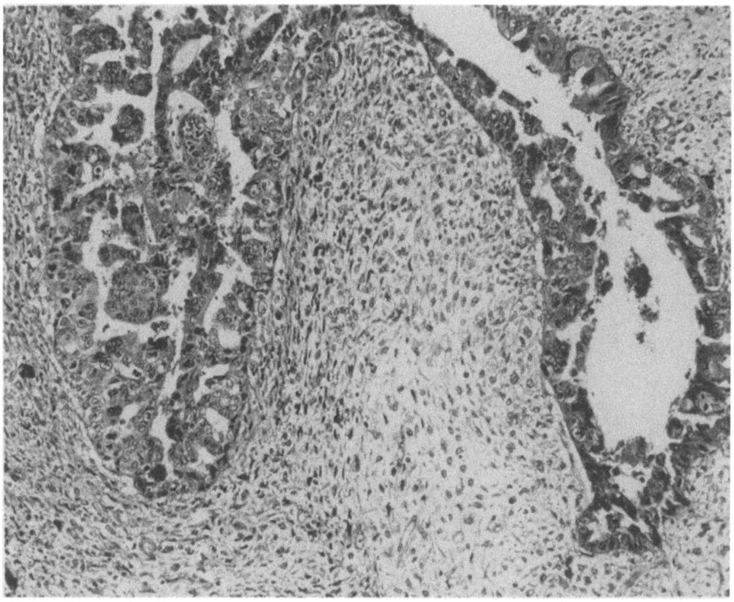

FIGURE 9-17. Carcinosarcoma with blending of the cells with epithelial differentiation (center) from the surrounding malignant spindle cells. (H&E, × 200.)

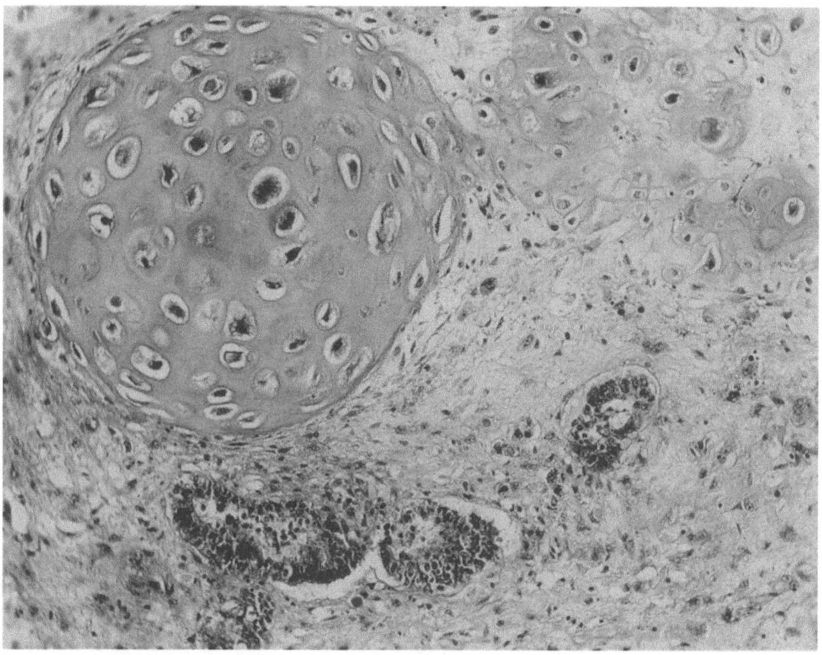

FIGURE 9-18. Mixed mesodermal tumor. Malignant cartilage (left) with epithelium (below). (H&E, × 115.)

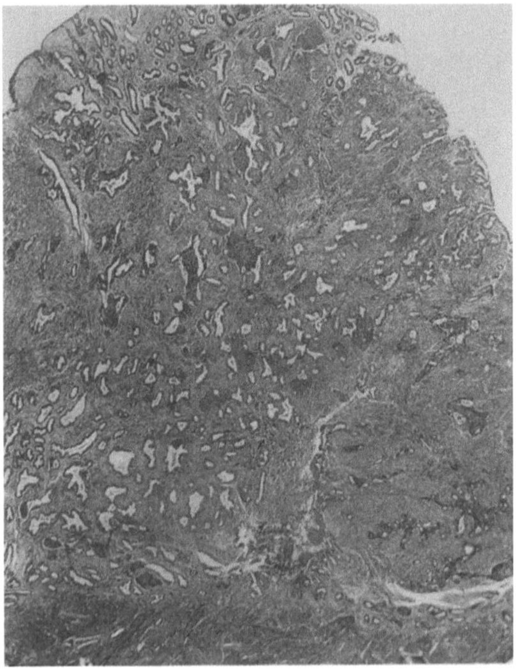

FIGURE 9-19. Atypical adenomyoma seen in low magnification with the endometrial surface at the top, myometrium at the bottom. (H&E, × 10.)

mitotically active and atypical fibroblastic or endometrial-type stroma. The epithelium in an adenosarcoma is usually less hyperplastic than in an atypical adenomyoma and seldom has squamous metaplasia.

Adenomatoid Mesotheliomas

Adenomatoid mesotheliomas are benign tumors found incidentally in about 1% of uteri of women of reproductive age. Patients have a median age of 42 years. Similar lesions are found less frequently in the fallopian tube, ovary, and omentum. They are usually solitary and asymptomatic, resembling leiomyomas occurring in a subserosal location, usually near the cornu. Adenomatoid mesotheliomas are found and rubbery and usually very small, but may measure up to 4 cm in diameter.[54] Usually found in hysterectomy specimens, they may be confused with a metastatic carcinoma because of the infiltrating margin. Because of their peculiar intermingling of mesothelial-lined, capillary-like spaces and smooth muscle, adenomatoid

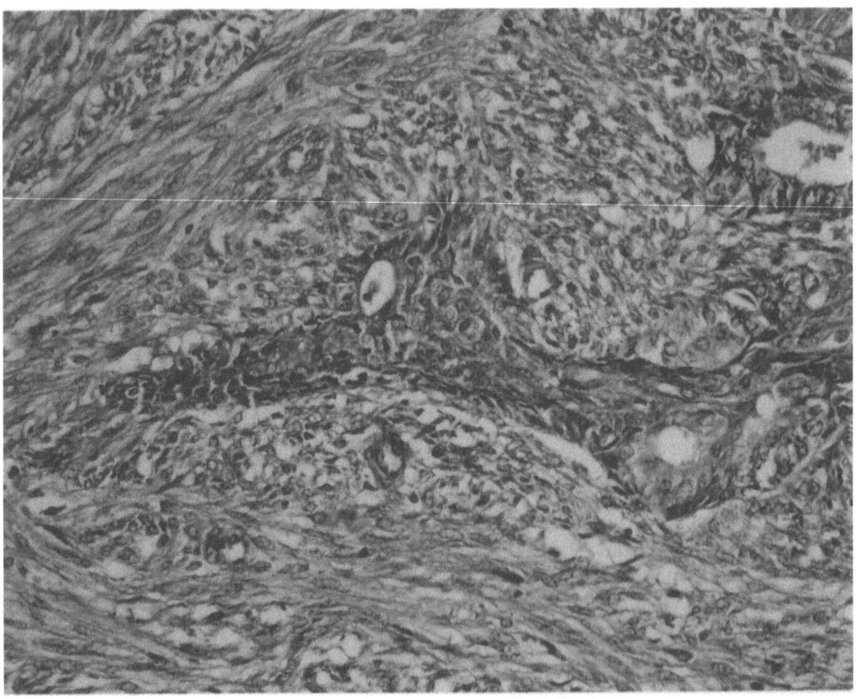

FIGURE 9-20. Atypical adenomyoma with dense squamous metaplasia giving the appearance of myometrial invasion. Higher view of same sample as in Figure 9-19. (H&E, × 220.)

mesotheliomas may be confused with signet cell carcinoma and some sarcomas.[55]

The microscopic appearance is variable and consists of haphazardly placed spaces lined by flattened cells seemingly infiltrating nodules of smooth muscle. Grossly they usually resemble a yellowish leiomyoma. The mesothelial cells lining these spaces vary from a flattened, endothelial-like lining to columnar epithelium. The absence of mitotic figures and atypia favor a benign process. The mesothelial derivation is proved by the presence of hyaluronic acid and the absence of epithelial mucin, with immunocytochemical reactions positive for cytokeratin, but not for carcinoembryonic antigen or factor VIII.[56]

Inflammatory Pseudotumor of the Uterus

Inflammatory pseudotumor of the uterus is a very rare, benign, reactive process similar to that seen in the lung, mesentery, liver, stomach, pancreas, spleen, retroperitonoum, renal pelvis, kidney, pelvic cavity, and bladder.[57] One of the two patients reported to have this condition presented at age 6 with abdominal pain and distention, while the other was 30 years of age and asymptomatic. Inflammatory pseudotumor appears as a solitary, circumscribed, spherical leiomyoma-like mass in the myometrium ranging from 4 to 12 cm. It is firm, with a homogeneous tan and fleshy cut surface without hemorrhage or necrosis.

Microscopically inflammatory pseudotumor consists of myofibroblastic spindle cells positive for actin and arranged in fascicles. These have oval to spindle-shaped nuclei with fine to coarse chromatin and occasional nucleoli. Cytoplasm is finely fibrillar and eosinophilic. There is no atypia, and mitotic figures are rare. Numerous plasma cells are present, accompanied by a polymorphous inflammatory infiltrate and variable amounts of collagen with hyalinization.

The differential diagnosis of inflammatory pseudotumor includes the closely related postoperative spindle cell pseudotumor, which tends to involve the vagina and appear after trauma or a surgical procedure such as an episiotomy or cervical biopsy and has high mitotic activity. Both are self-limited spindle cell proliferations, lacking atypia or atypical mitotic figures. Inflammatory pseudotumor can be separated from a leiomyoma on the basis of its plasmacytic infiltrate, and from leiomyosarcoma by the young age of the patient, low mitotic activity, and lack of atypia. The absence of a storiform pattern and histiocytic proliferation helps separate it from fibrous histiocytoma. It is differentiated from plasmacytoma by its polymorphous composition, by the benign-appearing plasma cell population, and by the absence of other manifestations of plasmacytoma.

Lymphoma

Lymphoma as an initial manifestation in the uterus is rare (estimated as less than 1 of 730 women with lymphoma), compared with its occurrence in 10 to 15% of women with systemic disease.[58,59] Granulocytic sarcoma as an initial manifestation in the uterus is even more rare. Patients with granulocytic sarcoma range in age from 20 to 88 years and present with abnormal vaginal bleeding and abdominal pain.[58,60] Lymphoma tends to involve the cervix, and may produce a barrel-shaped cervix. The endometrium is involved less frequently, and the myometrium only rarely. Lymphomas range from 1.7 to 18 cm in size, and appear grossly as firm rubbery nodules with a homogeneous white or gray-white fish flesh appearance on cut surface, sharply demarcated from surrounding tissue.[58,60]

Lymphomas involving the uterus tend to be either the diffuse, large-cell lymphocytic type or a follicular, small, cleaved lymphoma. Hodgkin's disease is very rare if it occurs.[60] The malignant infiltrate spares glands, but infiltrates vessel walls. Lymphomas lack cell cohesion, and have cells with scant cytoplasm and nuclei that are often angulated and cleaved, with prominent nucleoli.

The major diagnostic considerations are chronic and follicular cervicitis and small-cell carcinoma. Features in lymphoma that may be helpful in differentiating it from reactive processes include deep stromal invasion, monomorphous cells, close packing of follicles, presence of cleaved lymphocytes between follicles, the absence of plasma cells, and lack of a starry sky pattern. Small-cell carcinoma resembles oat cell carcinoma of the lung, may have a spindle cell component, and may show cohesion of cells and a nesting pattern. Neuroendocrine granules are often present, identified by argyrophilic stains or immunohistochemical studies. An immunocytochemical reaction for leukocyte common antigen and a negative one for cytokeratin may be necessary to differentiate lymphoma from small-cell carcinoma.

The prognosis is closely correlated with the stage of disease and histologic type. Patients with a low-grade diffuse lymphoma may have localized disease and can survive with treatment, while those with high-grade disease do poorly.[60]

Metastatic Tumors to the Uterus

Metastatic tumors to the uterine corpus from extragenital primary sites are more common than many of the disease entities discussed above. About 50% of metastatic growths are of gastrointestinal origin and 35% are from the breast.[61] The myometrium is involved in nearly all cases of metastatic carcinoma, whereas combined myometrial and endometrial involvement is seen in one third and endometrial involvement alone in less than 4%.[61,62] When endometrial involvement is the first indication of a tumor, the main symptoms are vaginal bleeding or a pelvic mass.

Inasmuch as the majority of metastatic tumors are carcinomas, their microscopic features often allow for differentiation from primary genital tract neoplasms. The prognosis for these patients is poor since there is usually widespread metastatic disease.[61]

References

1. Ober WB: Uterine sarcoma: histogenesis and taxonomy. Ann NY Acad Sci 1959;75:568–585.
2. Gruenwald P: Developmental basis of regenerative and pathological growth in the uterus. Arch Pathol 1943;35:53–65.
3. Christopherson WM, Williamson EO, Gray LA: Leiomyosarcoma of the uterus. Cancer 1972;29:1512–1517.
4. Donhuijsen K, Schmidt U, Hirche H et al: Changes in mitotic rate and cell cycle fractions caused by delayed fixation. Hum Pathol 1990;21:709–714.
5. Buttran VC, Reiter RC: Uterine leiomyomata: etiology, symptomatology, and management. Fertil Steril 1981;36:433–445.
6. Kempson RL, Hendrickson MR: Pure mesenchymal neoplasms of the uterine corpus: selected problems. Semin Diagn Pathol 1988; 5:172–198.
7. Zaloudek C, Norris HJ: Mesenchymal tumors of the uterus, in Kurman RJ (ed): Blaustein's Pathology of the Female Genital Tract. New York, Springer-Verlag, 1987, pp 373–408.
8. Persand V, Arjoon PD: Uterine leiomyoma. Incidence of degenerative change and a correlation of associated symptoms. Obstet Gynecol 1970;35:432–436.
9. Taylor HB, Norris HJ: Mesenchymal tumors of the uterus. IV. Diagnosis and prognosis of leiomyosarcomas. Arch Pathol 1966;82:40–44.
10. Perrone T, Behner LP: Prognostically favorable "mitotically active" smooth muscle tumors of the uterus. A clinicopathologic study of ten cases. Am J Surg Pathol 1988;12:1–8.
11. O'Connor DM, Norris HJ: Leiomyomas with increased mitotic figures treated by myomectomy. Hum Pathol 1990;21:223–227.
12. Burns B, Curry RH, Bell MEA: Morphologic features of prognostic significance in uterine smooth muscle tumors: A review of 84 cases. Am J Obstet Gynecol 1979;135:109–114.
13. Zaloudek CJ, Norris HJ: Mesenchymal tumors of the uterus. In: Fenoglio CM, Wolff M (eds): Progress in Surgical Pathology, vol III. New York, Masson Publishing, 1981, pp 1–35.
14. Myles JL, Hart WR: Apoplectic leiomyomas of the uterus. A clinicopathologic study of five distinctive hemorrhagic leiomyomas associated with oral contraceptive useage. Am J Surg Pathol 1985;9:798–805.
15. Norris HJ, Hilliard GD, Irey NS: Hemorrhagic cellular leiomyomas ("apoplectic leiomyoma") of the uterus associated with pregnancy and oral contraceptives. Int J Gynecol Pathol 1988;7:212–224.

16. Fechner RE: Atypical leiomyomas and synthetic progestin therapy. Am J Clin Pathol 1968;49:697–704.
17. Kurman RJ, Norris HJ: Messenchymal tumors of the uterus. VI. Epithelioid smooth muscle tumors including leiomyoblastoma and clear cell leiomyoma. A clinical and pathologic analysis of 26 cases. Cancer 1976; 37:1853–1865.
18. Rywlin AM, Pecher L, Benson J: Clear cell leiomyoma of the uterus. Report of 2 cases of a previously undescribed entity. Cancer 1964; 17:100–104.
19. Kaminski PF, Tavassoli FA: Plexiform tumorlet: a clinical and pathologic study of 15 cases with ultrastructural observations. Int J Gynecol Pathol 1984;3:124–134.
20. Norris HJ, Parmley T: Mesenchymal tumors of the uterus. V. Intravenous leiomyomatosis. A clinical and pathologic study of 14 cases. Cancer 1975;36:2164–2178.
21. Evans AT, Symmonds RE, Gaffey TA: Recurrent pelvic intravenous leiomyomatosis. Obstet Gynecol 1981;57:260–264.
22. Williams LJ Jr, Pavlick FJ: Leiomyomatosis peritonealis disseminata. Two case reports and a review of the medical literature. Cancer 1980;45:1726–1733.
23. Due W, Pickartz H: Immunohistologic detection of estrogen and progesterone receptors in disseminated peritoneal leiomyomatosis. Int J Gynecol Pathol 1989;8:46–53.
24. Tavassoli FA, Norris HJ: Peritoneal leiomyomatosis (leiomyomatosis peritonealis disseminata): a clinicopathologic study of 20 cases with ultrastructural observations. Int J Gynecol Pathol 1982;1:59–74.
25. Gal AA, Brooks JSJ, Pietra GG: Leiomyomatous neoplasms of the lung: a clinical, histologic, and immunohistochemical study. Mod Pathol 1989;2:209–216.
26. King ME, Dickersin GR, Scully RE: Myxoid leiomyosarcoma of the uterus. A report of six cases. Am J Surg Pathol 1982;6:589–598.
27. Chen KTK: Myxoid leiomyosarcoma of the uterus. Int J Gynecol Pathol 1984;3:389–392.
28. Tavassoli FA, Norris HJ: Mesenchymal tumors of the uterus. VII. A clinicopathologic study of 60 endometrial stromal nodules. Histopathology 1981;5:1–10.
29. Fekete PS, Vellios F: The clinical and histologic spectrum of endometrial stromal neoplasms: a report of 41 cases. Int J Gynecol Pathol 1984;3:198–212.
30. August CZ, Bauer KD, Lurain J, et al: Neoplasms of endometrial stroma: histopathologic and flow cytometric analysis with clinical correlation. Hum Pathol 1989;20:232–237.
31. Lifschitz-Mercer B, Czernobilsky B, Dgani R, et al: Immunocytochemical study of an endometrial diffuse clear cell stromal sarcoma and other endometrial stromal sarcomas. Cancer 1987;59:1494–1499.
32. Norris HJ, Taylor HB: Mesenchymal tumors of the uterus: a clinical and pathological study of 53 endometrial stromal tumors. Cancer 1966;19:755–766.
33. Marchese MJ, Liskow AS, Crum CP, et al: Uterine sarcomas: a clinicopathologic study, 1965–1981. Gynecol Oncol 1984;18:299–312.
34. Schwartz Z, Dgani R, Lancet M, et al: Uterine sarcoma in Israel: a study of 104 cases. Gynecol Oncol 1985;20:354–363.
35. Wheelock JB, Krebs H, Schneider V, et al: Uterine sarcoma: analysis of prognostic variables in 71 cases. Am J Obstet Gynecol 1985; 151:1016–1022.
36. Clement PB, Scully RE: Uterine tumors resembling ovarian sex cord tumors. A clinicopathologic analysis of fourteen cases. Am J Clin Pathol 1976;66:512–525.
37. Tang C, Toker C, Ances IG: Stromomyoma of the uterus. Cancer 1979;43:308–316.
38. Ulbright TM, Kraus FT: Endometrial stromal tumors of extrauterine tissue. Am J Clin Pathol 1981;76:371–377.
39. Abell MR: Papillary adenofibroma of the uterine cervix. Am J Obstet Gynecol 1971; 110:990–993.
40. Vellios F, Ng A, Reagan JW: Papillary adenofibroma of the uterus: a benign mesodermal mixed tumor of mullerian origin. Am J Clin Pathol 1973;60:543–551.
41. Zaloudek CJ, Norris HJ: Adenofibroma and adenosarcoma of the uterus. A clinicopathologic study of 35 cases. Cancer 1981;48:354–366.
42. Clement PB, Scully RE: Mullerian adenosarcoma of the uterus. A clinicopathologic analysis of ten cases of a distinctive type of mullerian mixed tumor. Cancer 1974;34:1138–1149.
43. Clement PB: Mullerian adenosarcomas of the uterus with sarcomatous overgrowth. A clinicopathological analysis of 10 cases. Am J Surg Pathol 1989;13:28–38.
44. Barwick KW, LiVolsi VA: Malignant mixed mullerian tumors of the uterus. A clinicopathologic assessment of 34 cases. Am J Surg Pathol 1979;3:125–135.

45. DiSaia PJ, Morrow CP, Boronow R, et al: Endometrial sarcoma: lymphatic spread pattern. Am J Obstet Gynecol 1978;130:104–105.
46. Peters PJ, Kumar NB, Fleming WP, et al: Prognostic features of sarcomas and mixed tumors of the endometrium. Obstet Gynecol 1984;63:550–556.
47. Deligdisch L, Plaxe S, Cohen CJ: Extrauterine pelvic malignant mixed mesodermal tumors. A study of 10 cases with immunohistochemistry. Int J Gynecol Pathol 1988; 7:361–372.
48. Kao GF, Norris HJ: Benign and low grade variants of mixed mesodermal tumor (adenosarcoma) of the ovary and adnexal region. Cancer 1978;42:1314–1324.
49. Tiltman AJ: Adenomatoid tumors of the uterus. Histopathology 1980;4:437–443.
50. Weed JC, Geary WL, Holland JB: Adenomyosis of the uterus. Clin Obstet Gynecol 1966; 9:412–421.
51. Tamaya T, Motoyama T, Ohono Y, et al: Steroid receptor levels and histology of endometriosis and adenomyosis. Fertil Steril 1979; 31:396–400.
52. Mazur MT: Atypical polypoid adenomyomas of the endometrium. Am J Surg Pathol 1981; 5:473–482.
53. Young RH, Tregert T, Scully RE: Atypical polypoid adenomyoma of the uterus. Am J Clin Pathol 1986;86:139–145.
54. Quigley JC, Hart WR: Adenomatoid tumors of the uterus. Am J Clin Pathol 1981;76:627–635.
55. Carlier MT, Dardick I, Lagace AF, et al: Adenomatoid tumor of uterus: presentation in endometrial curettings. Int J Gynecol Pathol 1986;5:69–74.
56. Said JW, Nash G, Lee M: Immunoperoxidase localization of keratin proteins, carcinoembryonic antigen, and factor VIII in adenomatoid tumors: Evidence for a mesothelial derivation. Hum Pathol 1982;13:1106–1108.
57. Gilks GB, Taylor GP, Clement PB: Inflammatory pseudotumor of the uterus. Int J Gynecol Pathol 1987;6:275–286.
58. Chorlton I, Karnei RF, King FM, et al: Primary malignant reticuloendothelial disease involving the vagina, cervix, and corpus uteri. Obstet Gynecol 1974;44:735–748.
59. Lathrop JC: Malignant pelvic lymphomas. Obstet Gynecol 1967;30:137–145.
60. Harris NL, Scully RE: Malignant lymphoma and granulocytic sarcoma of the uterus and vagina. A clinicopathologic analysis of 27 cases. Cancer 1984;53:2530–2545.
61. Kumar NB, Hart WR: Metastases to the uterine corpus from extragenital cancers. a clinicopathologic study of 63 cases. Cancer 1982; 50:2163–2169.
62. Mazur MT, Hsueh S, Gersell DJ: Metastases to the female genital tract. Analysis of 325 cases. Cancer 1984;53:1978–1984.

10

Colposcopy of the Cervix

LOUIS BURKE

Colposcopy was first developed by Hans Hinselmann in Germany in 1925. Investigators were curious, as they are today, with regard to the origin of cervical cancer. Hinselmann assumed that the primary focus of cervical cancer must occur as a minute ulceration that, although undetectible to the naked eye, might be appreciated by suitable low-power magnification and illumination. He designed an instrument that directed sharply focused light on the cervix and the image of this focused light was viewed through binocular magnification. Thus, a new field of clinical investigation, known as colposcopy, was started. Colposcopy, therefore, may be defined as the magnification of the gross appearance of the epithelium of the cervix, vagina, and vulva by means of an instrument that is essentially a stereoscopic binocular microscope of low magnification, usually 10 to 40 times, and a strong light.

Every colposcopic picture is a counterpart of a specific tissue pattern. Each tissue pattern, in turn, is determined by the nature of its surface epithelium and associated connective tissue stroma. The colposcope produces its effect by illuminating both surface epithelium and underlying stroma. When a sheet of epithelium composed of a particular cell population is interposed between the colposcopic light source and subjacent stroma, a characteristic visual impression is created. The visual image observed is a reflection of epithelial cell number, organization, and morphology. This image is also influenced by the vascular arrangement of the underlying stroma. Each of the many variations can therefore be modified by a number of possible changes in stromal vascular architecture and epithelial cellular morphology.

Instrumentation and Technique

Each colposcope consists of a binocular low-powered microscope, with centered illumination that may be of incandescent or fiberoptic source (Fig. 10-1). It is usually mounted on the extension arm of an adjustable stand. Free movement of the focusing elements is possible. A green filter is inserted between the light source and the tissue to accentuate the color tone differences between normal and abnormal patterns as well as to enhance the vascular pattern. By absorbing the red light, the vascular elements appear as black lines. Care should always be taken to place the colposcope in a location such that the light beam strikes the observed tissue at right angles.

The best focal lens for a working colposcope is between 12 and 25 cm. With a distance of 20 cm between the objective lens and the field of examination, punch biopsies and treatments under visual guidance of the colposcope can be carried out. The diameter of a visualized field varies in different models between 20 and 23 mm, with a magnification

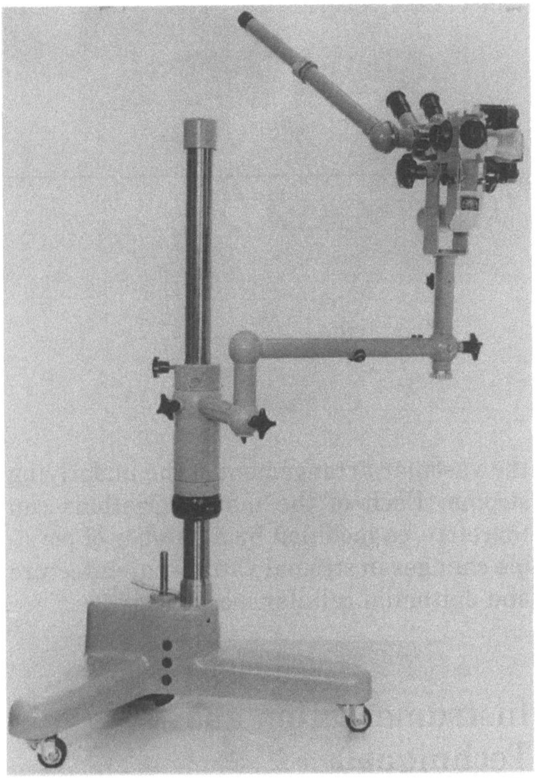

FIGURE 10-1. Zeiss colposcope, Model 1 on rolling base. A 35-mm camera and monocular teaching tube are attached. (Courtesy Carl Zeiss Co, New York.)

of 10 times. The binocular eye pieces are easily manipulated, may be changed from 12.5 to 20 magnifications, and may be adjusted for individual intraocular distances.

Many colposcopes have the ability to alter the magnification by means of a series of lenses inserted between the ocular and the objective lenses. Some colposcopes only have one to three magnifications, such as 6, 10, and 20 times, and others may have as many as five, providing a magnification of 6, 10, 16, 25, and 40 times. Magnification is changed by the simple turning of a knob without the need for refocusing. Most examinations, however, are adequately performed with a magnification of 12.5 to 13.5 times. Most single-magnification colposcopes come with a 25-mm objective lens and a magnification of 12.5 to 13.5 times.

For colposcopic examinations the vagina and cervix are exposed as usual for a gynecologic examination. Good colposcopic technique demands correct positioning of the patient and instrument. The patient must assume the dorsal-lithotomy position. Inasmuch as she may have to maintain this position for as long as 25 to 30 minutes, the examination table should offer maximal patient comfort. Special knee and calf support for the patient are useful. The usual stationary gynecologic table with conventional stirrups is entirely acceptable for colposcopic examination. However, such tables may need to be elevated as much as 2 inches on wooden blocks in order to perform the colposcopic examination without examiner neck or back strain. Tables equipped with an automatic foot control for adjusting heights are ideal.

A warm vaginal speculum without lubricant should be introduced slowly into the vagina; the blades of the speculum must be partially separated soon after entry into the vagina to avoid traumatizing the exposed cervix. If necessary, a Papanicolaou smear can be taken concurrently with the colposcopic examination; however, because as this may initiate bleeding and obliterate features of colposcopic interest it is best if the smears are taken prior to the examination with the colposcope.

The cervix should be inspected first with the surface moistened with normal saline; a dry epithelial surface is nontransparent and gives a poor view of vascular patterns. Use of the green filter provides the best colposcopic impression of vascular patterns. Inspection of the unprepared cervix should be followed immediately by scrutiny of the same areas treated with acetic acid. Acetic acid solution (3 to 5%) is applied to the cervix with moistened cotton balls. For reasons that remain obscure, acetic acid solution shrinks blood vessels, dissolves mucus, and causes an osmolar change in the intercellular space so that certain cells swell. Three percent solution is the usual acceptable strength of the acetic acid. Weaker dilution requires a longer waiting time to allow for the appearance of the various lesions.

Stronger solutions delineate the lesions more rapidly but are quite irritating to mucous membranes, especially after repeated application, and complaints from the patients can be expected.

Native Epithelia

The visible portion of the female genital tract is covered by stratified squamous epithelium, which lines the vagina and exocervix and is designated colposcopically as the original or native squamous epithelium. The endocervical canal is lined by a simple columnar epithelium that is called the original or native columnar epithelium. The junction of these two tissue types is called the squamocolumnar junction and can appear on various locations of the cervix depending upon the age and the estrogen activity of the woman in question (Fig. 10-2).

Zinser and Rosenbauer, in 1960, filled the vascular tree of the cervix with white latex and found that the terminal vessels that supplied oxygen to the epithelium might be divided into four zones. The first zone is a plexus of relatively large, freely anastamosing vessels deep in the stroma. These continue into the second zone of palisade-like vessels running perpendicular or obliquely to the surface. These branch into the third zone of small vessels running parallel to the surface. This is called the basal network. From the basal network the fourth zone of terminal capillaries emerge. Depending on the thickness of the surface epithelium, the basal network and the subepithelial terminal capillaries can be observed. The terminal capillaries can be divided into two types. The most common type are the network capillaries, which form a dense and fairly regular network of fine capillaries. These have an intercapillary distance of about 100 μm and usually less than 200 μm. A second type, usually best seen around the os in the postmenopausal woman, are the hairpin capillaries. They have an ascending and descending branch of fine-caliber vessels that are close together and form a loop at their interchange. Only the crests of the loops are visible, appearing as a fine, regular, punctate pattern. This pattern is very similar to that seen in the presence of a trichomonas infestation.

Native squamous epithelium shows little variation colposcopically from subject to subject. Under colposcopic illumination, it is uniformly pale pink and transluscent, and exhibits a feathery, vascular arrangement. Beneath the normal squamous epithelium there is a flat capillary network in the area between the lamina propria and the epithelium (Fig. 10-3).

Native columnar epithelium is unmistakable colposcopically. It is readily identified by its intense red hue and is thrown up into

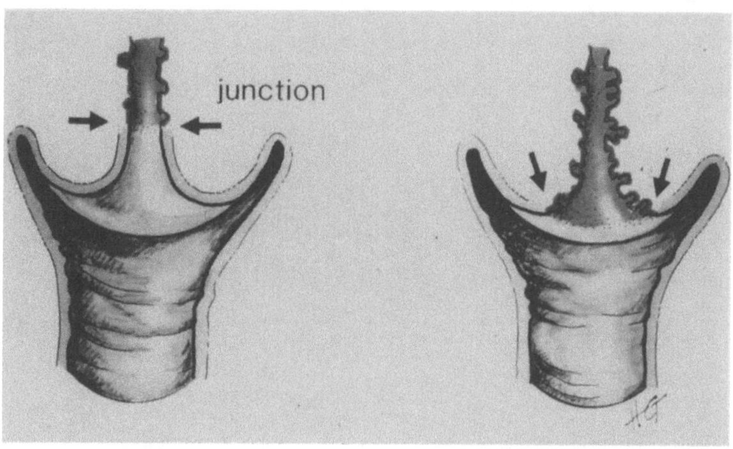

FIGURE 10-2. Interface of the original epithelia at the squamocolumnar junction. This varies at different ages.

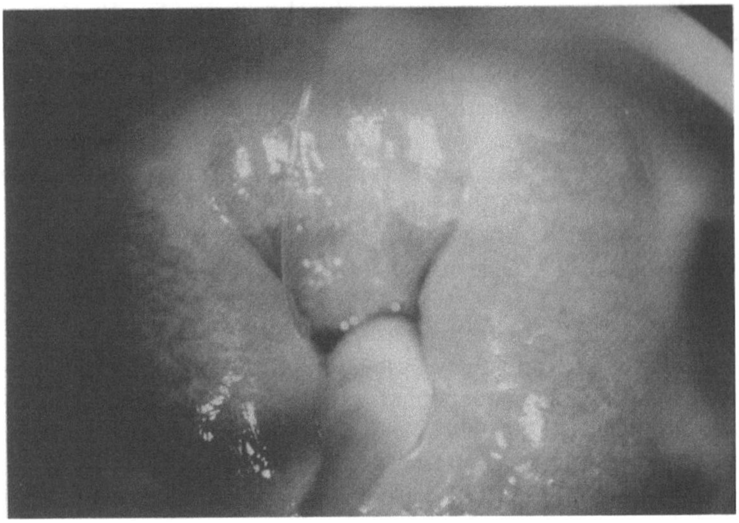

FIGURE 10-3. Colpophotograph of a normal cervix. Note the feathery terminal capillaries of the original stratified squamous epithelium. The cervix is being everted with a cotton-tipped applicator to reveal the squamocolumnar junction.

many papillae and folds. From many of these folds smaller folds are formed. This is especially prominent as one approaches the external os of the cervical canal (Fig. 10-4). The vascular network of the columnar epithelium is very complex. There is a network of uniform coiled vessels consisting of an afferent and efferent capillary. Within each papillus these two capillaries intertwine into a multichannel network. Each papillus is separated from the next by an interpapillary or intervillous crypt. The terminal capillary loops in each papillae are easily seen colposcopically. After being washed with dilute acetic acid, the columnar epithelium takes on a characteristic grapelike appearance (Figs. 10-5 and 10-6).

The colposcopic examination of the cervix is directed primarily toward investigating the tissue in the area where the original squamous and the original columnar epithelia come together. At this squamocolumnar interface, columnar epithelium is gradually transformed into squamous epithelium by the process of metaplasia. This dynamic area of change is known as the "transformation

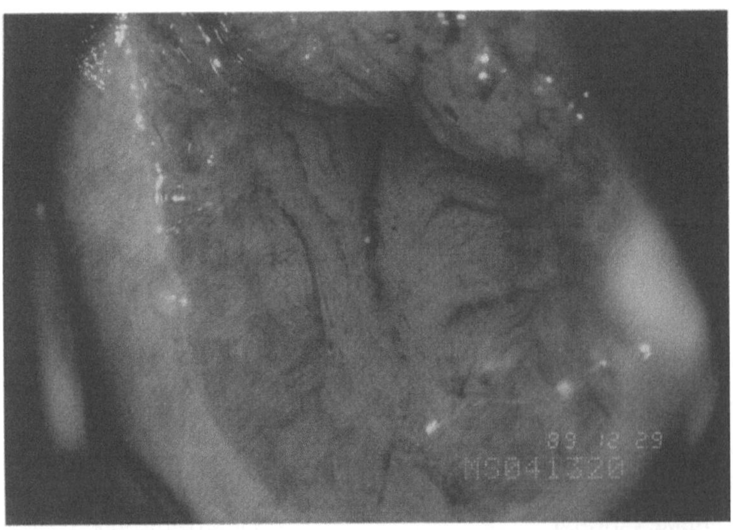

FIGURE 10-4. Colpophotograph of the posterior wall of the endocervical canal prior to the application of acetic acid. Note the many folds and crypts.

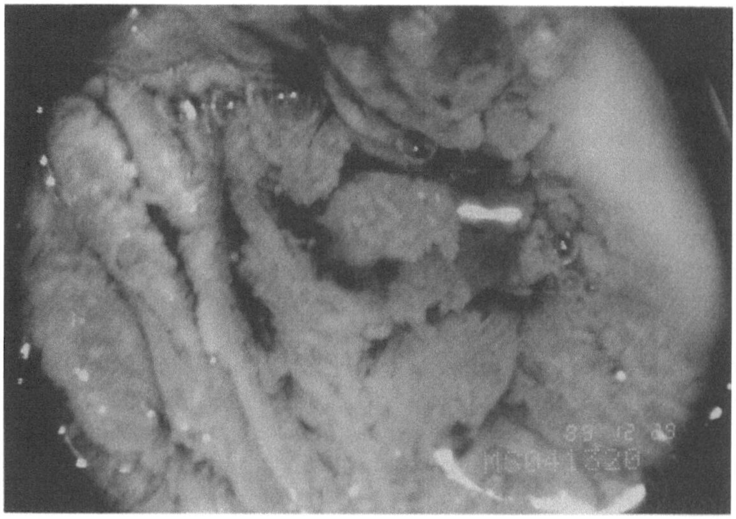

FIGURE 10-5. Colpophotograph of cervix shown in Figure 10-4 after washing with 3% acetic acid.

zone." A clear understanding of the transformation zone is vital not only to colposcopy but also in comprehending the origin and development of cervical neoplasia.

Typical Transformation Zone

The interface of the two original epithelia is called the squamocolumnar junction. As generally described in gynecology texts, it is said to be located at the anatomic external os of the cervix. However, examination of the cervices of women at different ages of their life shows that the area of the squamocolumnar junction seems to move in relationship to the presence or absence of estrogen. In neonates the squamocolumnar junction is located more toward the portio of the cervix than within the canal because of the influence of intrauterine maternal estrogen. In the premenarchal stage, the squamocolumnar junction is more likely to be at the anatomic external os of the cervix. At menarche and first pregnancy, both of which are characterized by high estrogen levels, the

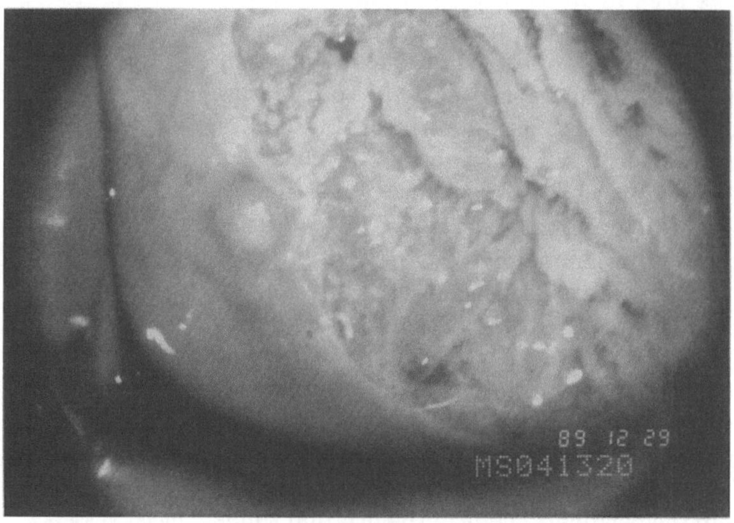

FIGURE 10-6. Columnar epithelium of the cervix after application of acetic acid. Note grapelike appearance at 11 o'clock.

squamocolumnar junction is usually on the portio of the cervix. It will stay in this area throughout the woman's menstruating lifetime and then appear to recede up the canal in the postmenopausal state. This deceptive movement of the squamocolumnar junction really does not occur. Instead, what the high estrogen levels do is to cause a rearrangement of the elements within the connective tissue of the cervix. This results in an eversion of the cervix so that the columnar epithelium is brought into contact with the environment of the vagina. With the withdrawal of this hormonal stimulus, the cervix inverts; the columnar epithelium and squamocolumnar junction recede into the endocervical canal.

Usually accompanying the high estrogen level is a change in acidity of the vagina due to the emergence of lactobacilli. The original columnar epithelium and its subjacent connective tissue stroma, when exposed to an acid medium in a background of high estrogen content, will produce a new cell, called the reserve cell, just beneath the basement membrane. The reserve cell will then divide horizontally, lifting off the original columnar epithelium and replacing it with a multilayered epithelium. This multilayered epithelium will ultimately mature into a palisade-like epithelium, acquire glycogen, and for all intents and purposes be identical to the original stratified squamous epithelium. The area in which this metaplastic process occurs is called the transformation zone.

This process of epithelial succession from columnar to metaplastic epithelium to mature squamous epithelium probably occurs throughout a female's lifetime. However, it is most active in three phases of her life: (a) fetal existence, (b) menarche, and (c) during the first pregnancy.

Squamous metaplasia occurs in sharply defined areas within the columnar epithelium. These areas can be of varying extension and they can lie in the midst of normal columnar epithelium. The various islands of these squamous cells overlying columnar epithelium broaden, coalesce, and eventually join the peripheral edge of the original squamous epithelium. The coalescence of the papillae is seldom complete. Some islands of columnar epithelium often remain surrounded by metaplastic squamous epithelium. Sometimes the columnar epithelium persists in the deeper clefts in the stroma below the metaplastic squamous epithelium. This columnar epithelium has an outlet to the surface from which mucus can be expelled through small channels that persist in the metaplastic epithelium. These are called gland openings. If there is no outlet to the surface, retention cysts and nabothian cysts will develop (Fig. 10-7). The typical transformation zone can be identified by localizing the remnants of the original columnar epithelium, that is, islands of columnar epithelium that were bypassed in the process of metaplasia, gland openings, and nabothian cysts.

The Atypical Transformation Zone

Under certain circumstances, the cause of which is yet unknown, the ectoptic columnar epithelium undergoing metaplasia absorbs some oncogenic factor. This oncogenic factor may be viral in origin (herpes, human papillomavirus) or may be from the sperm

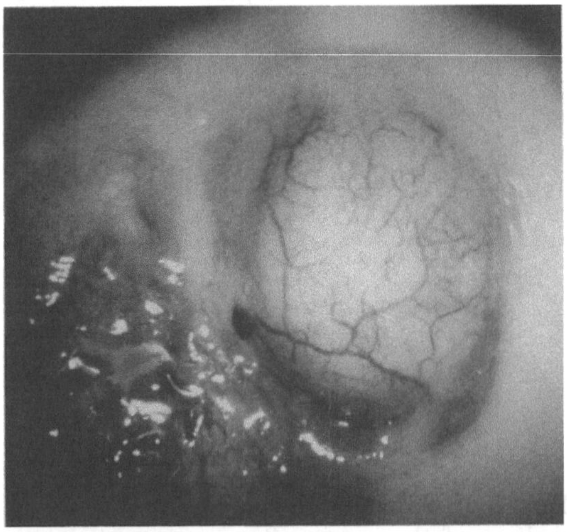

FIGURE 10-7. Large nabothian cyst on anterior lip of cervix. Note characteristic, normal, arborizing blood vessels.

(sperm DNA, histones) or some other unknown factor. Suffice it to say that when the metaplasing cells incorporate these oncogenic factors they will start to grow in a different fashion from that seen with the typical transformation zone. The individual papillae of the columnar epithelium do not coalesce or fuse. The atypical metaplastic epithelium completely fills the clefts and folds of the previous columnar epithelium. The central vascular network, under the influence of tumor angiogenesis factor (TAF), proliferates and remains in the thick stromal papillae, which are surrounded by metaplastic epithelium.

The subepithelial vascular network undergoes profound alterations. The flat capillary network usually found beneath normal cervical epithelium becomes tortuous and compressed vertically by neoplastic epithelium and with extension close to the surface produces the atypical features that are called punctation and mosaicism. Because of severe compression some of the capillaries eventually disappear, which may result in an increase in the intercapillary distance. If the dividing cells produce keratin over the epithelium a leukoplakic lesion will develop. If there is epithelial proliferation with an increase nuclear density, but the vasculature does not penetrate epithelium, the colposcopic picture is one of a white epithelium. Because this can be enhanced with the use of acetic acid it is called aceto-white epithelium. Aceto-white epithelium is usually focal and can be seen only after the application of acetic acid and not with the naked eye. It is a transient phenomenon that is seen in the area of increased nuclear density. If neoangiogenesis has occurred these new capillaries will be seen running parallel underneath the surface epithelium, and are called atypical blood vessels. They appear as irregular vessels with abrupt courses, and various names such as commas, corkscrews, spaghetti, sausages, or large nondividing vessels have been ascribed to them.

In summary, the transformation zone is said to be abnormal if any of the following are seen: (a) aceto-white epithelium, (b) punctation, (c) mosaicism, (d) leukoplakia, or (e) abnormal blood vessels. These atypical colposcopic tissue patterns may occur singly or in combination; they may be unifocal or multifocal and almost invariably are sharply delineated from surrounding normal tissue. Their lateral margins rarely, if ever, extend beyond the original squamocolumnar junction onto the native squamous epithelium.

It should be emphasized that the colposcopic variations of atypia are found primarily within the transformation zone but can also be present in the vagina and on the vulva. Moreover, the same colposcopic findings that often signal a neoplastic change may occur in nonneoplastic conditions such as metaplasia, infection, inflammation, regeneration, and repair following trauma, cautery, or cryosurgery or laser surgery. The presence of an abnormal transformation zone, although highly suggestive, does not prove that neoplasia exists. Thus colposcopy cannot be done in a vacuum but must be done in concert with cytology; if the cytologic examination has suggested neoplasia then these atypical findings take on greater significance.

Colposcopic Atypicalities

Aceto-White Epithelium

The aceto-white epithelium constitues the most common appearance of the atypical transformation zone. Whiteness of the abnormal epithelium presumably is related to the nuclear predominance of atypical cells. Usually aceto-white epithelium is level with the surrounding tissue and no vascular changes are evident (Figs. 10-8 and 10-9). It cannot be distinguished from normal tissue without colposcopic magnification aided by either saline application and screening with a green filter or the use of acetic acid. The speed of change, duration of whiteness, and intensity of whiteness as well as the sharpness of the borders are related to number, size, and concentration of the underlying abnormal cells.

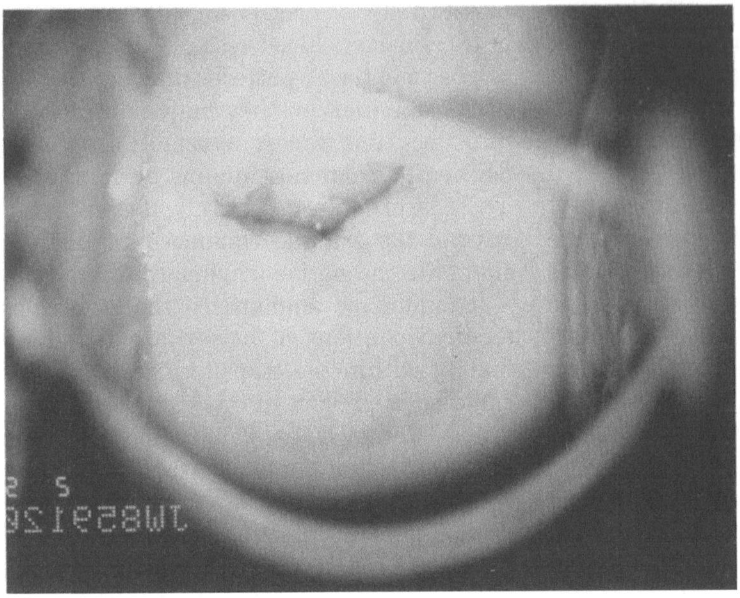

FIGURE 10-8. Colpophotograph of cervix after application of acetic acid. Note that there are two lesions: at 12 o'clock a sharply delineated aceto-white change, and at 4 to 8 o'clock a less definitive one. The squamocolumnar junction is visible. Biopsy at 12 o'clock revealed cervical intraepithelial neoplasia (CIN) II; biopsy at 4 to 8 o'clock, CIN I with condylomatous features.

Punctation and Mosaicism

As columnar epithelium undergoes atypical metaplasia, as the result of the influence of TAF produced by the abnormal cells, changes in the capillaries of the columnar epithelium papillae take place. Thus, in conjunction with the development of blocks of abnormal cells, the vascular network persists and proliferates, and punctation and mosaic structure develop (Fig. 10-10). Since this process is basically similar for both punctation and mosaicism, they are both frequently found in the same lesion. The vessels

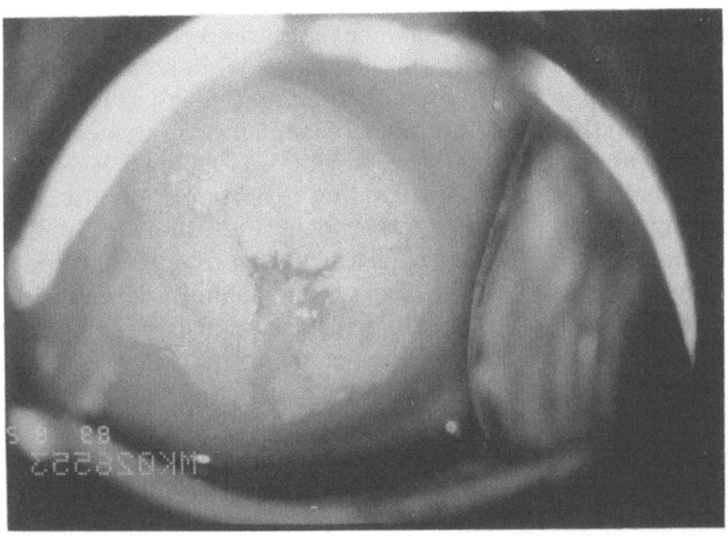

FIGURE 10-9. Colpophotograph of cervix after application of 3% acetic acid. Note the wide, atypical aceto-white transformation zone. Biopsy revealed CIN II with condylomatous features.

FIGURE 10-10. (A) Cervix after application of 3% acetic acid. Note punctation at 4 o'clock and mosaic at 7 o'clock. (B) Higher magnification of cervix shown in A. Note the increased intercapillary distances between the punctate vessels. The vessels protrude above the surface, producing the appearance called "coarse punctation." Biopsy revealed CIN III.

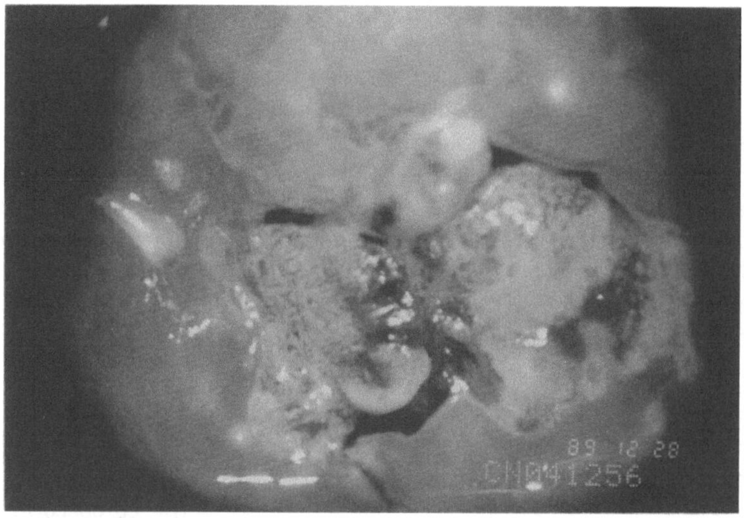

A

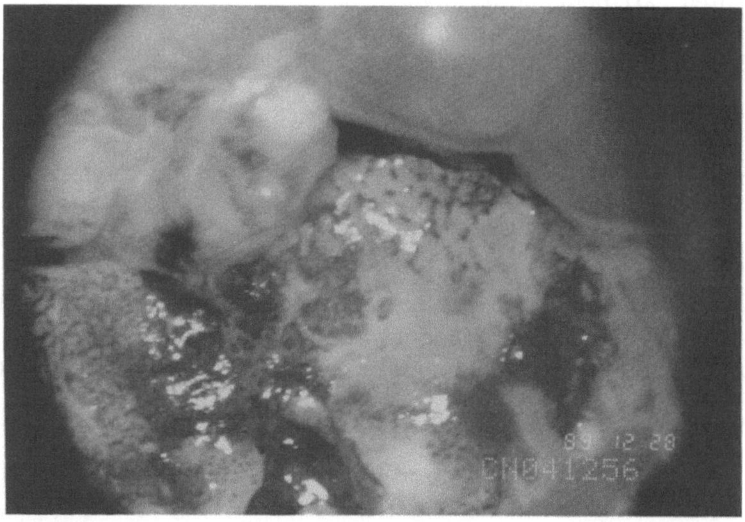

B

may show wide variation in size, shape, arrangement, and intercapillary distances. The greater the variation, the more severe the abnormal histopathology. Punctation and mosaicism may represent an innocuous change of the epithelium. This can be suspected if the surface is smooth and level with the adjacent normal squamous epithelium, and if the epithelium under consideration does not turn white with the application of acetic acid. Significant punctation and mosaicism are only seen in a field of aceto-white epithelium.

Leukoplakia

Leukoplakia, unlike the other abnormalities of the transformation zone, can be diagnosed without colposcopy. It is seen prior to the application of acetic acid and without the aid of low-power magnification. Leukoplakia, colposcopically, appears as a clearly demarcated, white area with an irregular border and raised surface. Because of its considerable opacity, leukoplakia obscures the underlying vascular structure, and prevents analysis of the underlying tissue structure. Thus it can cover areas of pathologic epithelium. Localization of leukoplakia is important. When it overlies normal squamous epithelium it is usually of no great significance. When present within the transformation zone, especially when surrounded by white epithelium, mosaicism, or punctation, it is disturbing. Although leukoplakia is more often found in connection with benign lesions, the possibility of an underlying precancerous lesion or even well-differentiated carcinoma must always be kept in mind. For these reasons all keratotic areas require biopsy.

Abnormal Blood Vessels

At some stage in severe abnormal epithelium, by causes not well understood, a new release of TAF occurs. New capillaries will develop from the adjacent punctate or mosaic vessels. These usually come off in a perpendicular manner so that they are parallel to and close to the surface (Fig. 10-11). They are usually nonarborizing and fragile and have abnormal appearances. Because they are not typical of normal stratified squamous epithelium, they are called "atypical" blood vessels. The presence of atypical blood vessels should always suggest that, until proven otherwise, microinvasive or invasive disease is present.

Grading of Colposcopic Lesions

All things being equal, the more abnormal the colposcopic appearance the greater the underlying histopathologic abnormality. By utilizing the characteristics of the lesion—color, surface contour, speed and duration of aceto-white change, sharpness of the margin of the lesion, vascular patterns, intercapillary distances, and the presence or absence of atypical blood vessels—it is possible for the colposcopist with experience to differentiate the various forms of cervical intraepithelial neoplasia (CIN) from each other and from invasive cancer. However, gynecologists practicing clinical colposcopy should guard against the temptation to make histologic diagnosis and should re-

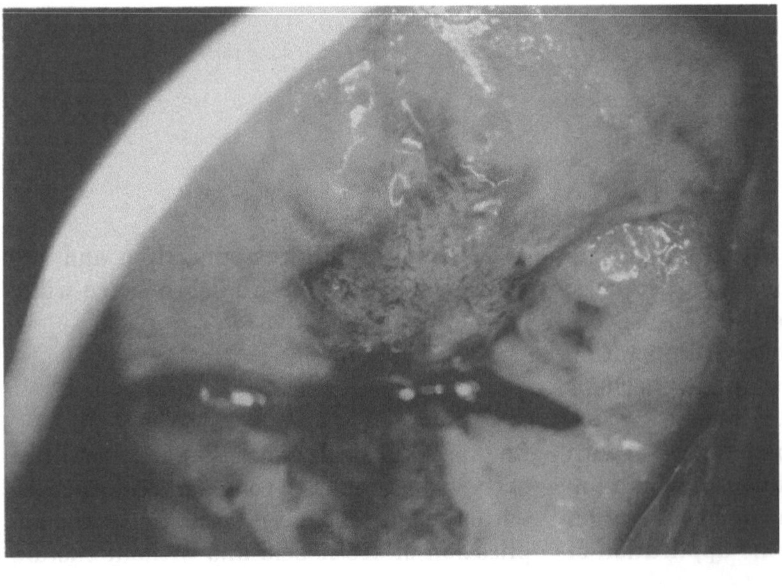

FIGURE 10-11. Anterior lip of the cervix with atypical blood vessels. Biopsy revealed invasive squamous cell carcinoma.

member that aceto-white epithelium, punctation and mosaic structures, and areas of abnormal vasculature require biopsy for definitive confirmation of the true nature of the lesion.

Biopsy

Colposcopy establishes the location and extent of foci of abnormal epithelium. Tissue sampling of specific abnormal areas can be performed easily under colposcopic guidance. A biopsy taken in the area of greatest colposcopic abnormality will yield the diagnostic information necessary for implementation of appropriate therapy.

Carefully excised, well-preserved, well-oriented, and rapidly fixed specimens afford the best opportunity for accurate diagnosis. A variety of instruments may be used for punching out small sections of epithelium. The common instruments used for removing these bits of tissue include the Eppendorfer, Kevorkian, Tischler, and Burke biopsy forceps. The Eppendorfer and Kevorkian forceps take small, relatively superficial samples. Unless the cervix is stablized, they tend to slip off and the resultant biopsy is unsatisfactory. These instruments not infrequently crush rather than cut through the epithelium, producing an unsatisfactory specimen. The Tischler and Burke biopsy forceps have projections on each blade that act as stablizing teeth so that additional immobilization of the cervix is hardly ever necessary. The Tischler head is 5 × 5 × 4 mm. It takes a longitudinal sample; however, there is a tendency to take too deep a biopsy with resultant excessive bleeding. The Burke biopsy forceps, whose head measures 5 × 4 × 3 mm, was designed to obviate the problem of depth (Fig. 10-12). It takes a biopsy that extends only 2 mm in depth, which is adequate to obtain not only epithelium but underlying stroma. Forceps should not be autoclaved because this procedure will blunt the cutting edge of the biopsy forceps. They should either be soaked in Cidex or sterilized with gas. If they become dulled, they may be sent back to the manufacturer for resharpening, or the clinician may sharpen them using a jeweler's file, with the colposcope as a magnifier.

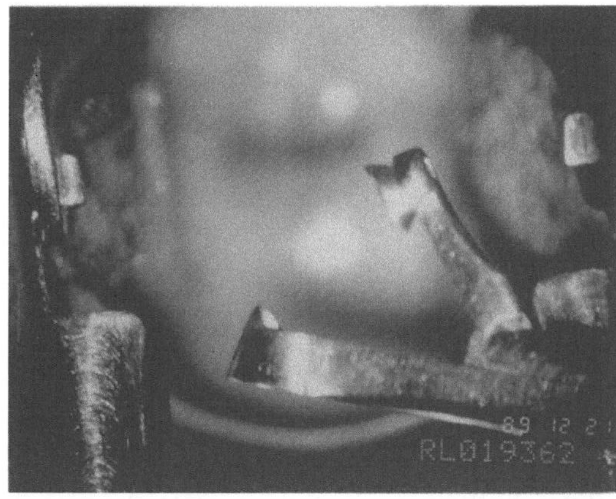

FIGURE 10-12. Head of Burke biopsy forceps.

Obtaining biopsy specimens can be facilitated by stabilizing the cervix using a fine hook such as the Iris or Burke hook. The hook holds the cervix firmly so that the biopsy forceps can be applied perpendicular to the epithelium and an adequate specimen taken. The biopsy should be taken to a depth sufficient to obtain enough stroma for differential diagnosis.

Biopsies can be submitted to the pathologist floating free in multiple-labeled jars of formalin. In order to conserve specimen jars, several specimens may be placed within the same container if each specimen is individually wrapped and labeled. A technique utilized and developed by the author involves placing the specimen between two pieces of thin styrofoam, which are then held together with paper clips or staples (Fig. 10-13). The circles are labeled with the hours of the clock to designate the area from which the biopsy was taken. A special pen whose ink will not wash off in the fixative must be used.

Hemostasis after biopsy is obtained by applying Monsel's solution (ferric subsulfate) followed by gentle pressure of cotton-tipped applicators. The Monsel's solution should be allowed to remain in a wide-necked jar open to the air until its consistency is that of mus-

A

B

FIGURE 10-13. (A) Biopsy of cervix on styrofoam circle. (B) Styrofoam circle, noting that the biopsy was taken at 11 o'clock, added to circle shown in A and attached with paper staples.

tard. At the time its efficiency as a hemostatic agent will be markedly improved. Silver nitrate, either on sticks or in solution, may also be used for hemostasis; however, it has the disadvantage of leaving small deposits of silver that may interfere with subsequent colposcopic evaluation of the area. Cautery, which has been used at the base of the biopsy site, destroys the surrounding tissue and renders it unsuitable for later histologic or colposcopic examination.

Endocervical Curettage

As part of the colposcopic method, endocervical curettage may have to be performed. This is indicated if a lesion extends up the canal, if its upper edge cannot be biopsied, if the colposcopic examination is unsatisfactory, or if the cytology and biopsy specimens have a two-grade discrepancy, indicating that a more severe lesion is present than was determined by the directed biopsy. The routine use of endocervical curettage following a satisfactory colposcopic examination, as advocated by some,* will produce an excessive number of false-positive results, resulting in needless excisional conization.

To perform the endocervical curettage, the usual instrument is either the Kevorkian endocervical currette or one of the newer, smaller curettes. The latter have been devised to prevent sampling of the lesion on the portio as the curette is brought down from the canal. The canal should be sampled without withdrawing the curette to minimize the possibility of contaminating the sample with strips of abnormal epithelium from the portio of the cervix. The specimen is collected on pieces of filter paper or Telfa gauze, including all the blood and mucus from the endocervical canal. These are all placed in formalin.

Unsatisfactory Colposcopy

If the transformation zone and the squamocolumnar junction are located within the endocervical canal and cannot be adequately visualized, the colposcopic evaluation is termed unsatisfactory. This is most likely to occur in the older postmenopausal woman; it also occurs after conization, cryocautery, or laser stenosis. If this occurs in conjunction with an abnormal cytologic exam, endocervical curettage is mandatory.

* *Editorial Comment:* Some advise routine endocervical curettage at the time of cervix biopsy.

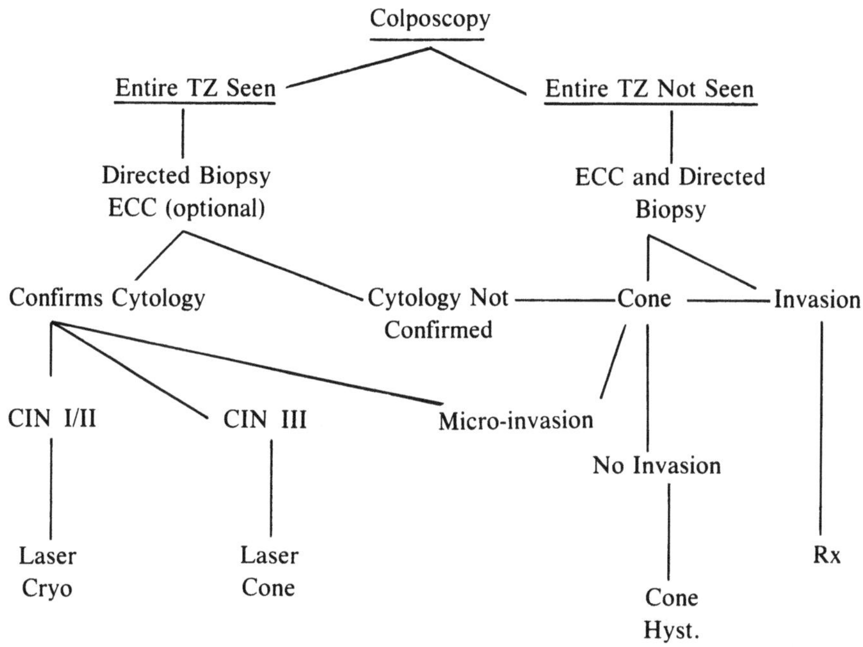

FIGURE 10-14. Algorithm for colposcopy of the cervix. TZ = transformation zone, ECC = endocervical curettage.

Indications for Colposcopy of the Cervix

Colposcopy of the cervix is indicated for the following reasons: (a) abnormal cytologic examination, (b) to direct the cervical biopsy, (c) prior to excisional conization, (d) prior to hysterectomy for CIN III, (e) evaluation of the female offspring of mothers who took diethylstilbestrol (DES) during pregnancy,(f) follow-up of patients who were treated with radiation for carcinoma of the cervix, (g) follow-up of patients who were treated for CIN, and (h) follow-up of the DES-exposed patient. Figure 10-14 provides a suggested method of using colposcopy following an abnormal cytologic smear.

The ability of the gynecologist to utilize this technique depends on many factors, the most important of which is the physician's dedication to increasing his or her expertise. This depends on the liberal use of biopsies and correlating the colposcopic findings with cytohistopathology. Colposcopy must be learned with an in-depth understanding of the underlying cytopathology. It must explain the cytologic abnormality. It is only by constant review of the cytology, colposcopy, and pathology can the individual gynecologist incorporate this technique into his or her armamentarium.

Bibliography

Burke L, Mathews BE: Colposcopy in Clinical Practice. Philadelphia, FA Davis and Co, 1977.

Burke L, Antonioli DA, Ducatman BS: Colposcopy: Text and Atlas. Norwalk, CT, Appleton and Lange, 1990.

Coppleson M, Pixley E, Reid B: Colposcopy, ed 2. Springfield, IL, Charles C Thomas, 1978.

Coppleson M: The colposcopic terminology. J Reprod Med 1976;16:214–219.

Ferenczy A: The ultrastructural dynamics of normal and neoplastic cervical squamous epithelium. The Colposcopist, December 1982, pp 1–6.

Folkman J: Tumor angiogenesis: therapeutic implications. N Engl J Med 1971;285:1182.

Heinzl S: The relevance of various elements contributing to colposcopic diagnosis. The Cervix 1989;7:93–99.

Kolstad P: Colposcopic diagnosis of dysplasia, CIS and early invasive cancer of the cervix. Acta Obstet Gynecol Scand 1964;43(suppl)7:105–108.

Kolstad P: The development of the vascular bed in tumors as seen in the squamous cell carcinoma of the cervix uteri. Br J Radiol 1965; 38:216.

Kolstad P: Colposcopic nomenclature and grading. The Colposcopist, September 1979, pp 1–3.

Moseley KR, Tung VD, Hannigan EV, et al: Necessity for endocervical curettage in colposcopy. Am J Obstet Gynecol 1986;154:992.

Oliveira A, Keppler M, Luisi A, et al: Comparative evaluation of abnormal cytology, colposcopy and histopathology in preclinical cervical malignancy during pregnancy. Acta Cytol 1982;26:636.

Pixley E: Basic morphology of the prepubertal and youthful cervix. J Reprod Med 1976; 16:221–230.

Sakuma T, Hasegawa T, Tsutsui F, Kurihara S: Quantitative analysis of the whiteness of the atypical cervical transformation zone. J Reprod Med 1985;30:773.

Sillman F, Boyce J, Fuchter R: The significance of atypical vessels and neovascularization in cervical neoplasia. Am J Obstet Gynecol 1981; 139:154.

Singer A: The uterine cervix from adolescence to the menopause. Br J Obstet Gynaecol 1975; 32:81–99.

Stafl A, Mattingly RF: Angiogenesis of cervical neoplasia. Am J Obstet Gynecol 1975;121:845.

Toplis PJ, Casemore V, Hallam M, Charnock M: Evaluation of colposcopy in the postmenopausal woman. Br J Obstet Gynaecol 1986;93:843.

Wetrich D: An analysis of the factors involved in the colposcopic evaluation of 2194 patients with abnormal Papanicolaou smears. Am J Obstet Gynecol 1986;154:1339.

Zinser HK, Rosenbauer KA: Utersuchungen uber die angioarchitecktonic der normalen und pathologisch veranderten cervix uteri. Archiv Gynakol 1960;194:79–112.

11

Endometrial Sampling Techniques

FREDERICK FRIEDMAN, JR. and MICHAEL L. BRODMAN

Abnormal uterine bleeding is one of the most frequent presenting gynecologic complaints. The relative accessibility of the uterine lining via a transcervical route makes it an ideal organ in which to diagnose pathology. Such sampling usually enables the gynecologist to determine the etiology of the bleeding—whether neoplastic, structural, or hormonal in origin.

Because endometrial cancer is the most common genital tract malignancy, uterine bleeding in the perimenopausal or postmenopausal woman may trigger anxiety in both the patient and her physician. The availability today of outpatient endometrial sampling techniques, in concert with rapid pathologic diagnostic services, provides a diagnosis within days; this curtails the waiting and alleviates much of the anxiety.

Similarly, advances in techniques of uterine sampling have also facilitated the work-up of infertility. The ability to provide accurate dating of the endometrium in an office setting and with minimal patient discomfort has resulted in a more rapid, less invasive, and less costly approach.

In this chapter we will discuss the different techniques of endometrial sampling. Specifically, we will review the indications and contraindications of the operative techniques, as well as complications and the respective limitations of the various procedures.

History

Endometrial sampling has been practiced with varying techniques since at least 1843, when Recamier designed a curette to scrape off uterine "fungosities."[1] Sims advocated the use of curettage for a variety of disorders.[2] Such usage was not only for abnormalities of bleeding (oligomenorrhea or menorrhagia), but also for dysmenorrhea. At times sponge tents were used to dilate the cervix, affording subsequent digital palpation of the uterine cavity itself.[2] However, sampling or scraping the endometrium was not uniformly embraced as an appropriate surgical procedure. Indeed, many believed it was a barbaric and unscientific endeavor.[1,3] For this reason, the technique fell into disuse until several years later.

Nonetheless, such early pioneers led to the eventual description of the cyclic nature of changes in the endometrium by Hitschmann and Adler in 1908.[4] By the time Bell wrote his controversial treatise on gynecology in 1910, uterine curettage had a variety of diagnostic and therapeutic indications.[4] Included in this original list of treatable abnormalities was "congenital anteflexion of the uterus."[4] Simultaneous flushing of the uterus during curettage was also advocated by some.

Kelly in 1925 advocated the use of outpatient endometrial curettage without anesthesia.[6,7] This advanced concept did not become routinely accepted until approximately 50 years later. In the interim others, such as Novak[8] and Randall[9] in 1935, had described techniques of endometrial aspiration biopsy. These devices were, until recently, the most frequently used instruments for suction biopsy. The use of sponge biopsy was advocated briefly by Chatfield and Watson.[10] With this technique, a V-shaped sponge was introduced into the uterine cavity and, with a brushing motion, a specimen was obtained.

Palmer in 1950 reported great success using aspiration with curettage for the diagnosis of endometrial cancer.[11] Nonetheless, formal dilation with curettage (D&C) under general anaesthesia remained the favored technique for endometrial sampling. Perhaps the most important step in universalizing outpatient sampling came as a result of work by Jensen and Jensen in 1968.[12] They implemented vacuum uterine sampling in an attempt to detect early uterine carcinoma. Modifications of their device led to the Vabra aspirator (Berkeley Medevices, Inc., Berkeley, CA), and later the Tis-u-Trap (Milex Products, Inc., Chicago, IL). Numerous reports in the literature have been published that favorably testify to the accuracy of aspiration endometrial sampling.[1,7,11–18] When compared to conventional D&C outpatient aspiration biopsy is an accurate, less invasive, less costly, and safer technique.

More recently, there has been further progress in the development of instruments for outpatient endometrial sampling. The most recently introduced devices are narrow cylinders with self-contained internal pistons used to generate suction. Examples of these include the Pipelle (Unimar, Inc., Wilton, CT) and the Z-Sampler (Zinnanti Surgical Instruments, Inc., Chatsworth, CA). The principal indication for use of these devices has been for the evaluation of infertility. These instruments enable the operator to obtain a reasonable specimen for endometrial dating[19–22] as well as for other purposes.

Indications

Uterine curettage may be performed for therapeutic or diagnostic reasons (Table 11-1). In this chapter we shall concentrate on diagnostic indications, but a brief review of the former is appropriate.

Therapeutic

The principal conditions for which curettage serves a therapeutic purpose are incomplete abortion and postpartum retained products of conception. In these settings, bleeding may be massive and rapid evacuation of endometrial contents is desired. Suction curettage with large-bore catheters is often used in concert with conventional sharp curettage.

Dysfunctional uterine bleeding (DUB), the main cause of which is anovulation due to a disorder of the hypothalmic-pituitary-ovarian axis, is best managed with hormonal therapy. However, with severe bleeding, sharp curettage may be warranted to acutely stop the hemorrhage. Indeed, because of its rapid efficacy, it is the treatment of choice in women with DUB who are hypovolemic.[23]

While some physicians have advocated the use of sharp curettage as a treatment for irregular or heavy bleeding, there does not appear to be any literature to support this.[1] In fact, in a controlled study by Haynes and colleagues, reduction of blood loss lasted only

TABLE 11-1. Indications for endometrial sampling

Therapeutic
Incomplete abortion
Retained products of conception
Dysfunctional uterine bleeding
Diagnostic
Abnormal bleeding in patient ≥35 years old
Postmenopausal bleeding
Infertility
Chronic uterine infection
?Patients on estrogen replacement therapy (see text)
(Chorionic villus sampling)

one month following D&C; there was actually an exacerbation of bleeding in subsequent months.[24]

Diagnostic

The majority of patients who require diagnostic endometrial sampling have abnormalities of bleeding (pre- or postmenopausal) or infertility. With the widespread administration of estrogens beginning in the climacteric, the need for sampling women taking hormone replacements must also be addressed. Confirmation of a chronic pelvic infection is a fourth consideration. A recently introduced additional indication for endometrial sampling is the chorionic villus biopsy. However, this is a diagnostic modality for determination of fetal karyotype and not of endometrial histopathology itself; a discussion of this technique is therefore beyond the scope of this chapter.

Premenopausal Abnormal Bleeding

Patients in the reproductive years experiencing abnormal bleeding—either metrorrhagia or menorrhagia (or both)—after the age of 35 years should have sampling of their endometrium. Although adenocarcinoma of the endometrium is uncommon in women under 40 years of age,[1,25–27] we prefer to sample the uterus in patients with abnormal bleeding patterns beginning at age 35. Others have supported this.[28] We perform this in the office or clinic using aspiration curettage (see "Techniques" below) at the time of presentation, following a careful history and physical examination. If there is any question as to pregnancy or other contraindication (see below), the procedure is deferred.

Postmenopausal Bleeding

Although the most frequent cause of postmenopausal uterine bleeding is not endometrial carcinoma, this remains the working diagnosis until proven otherwise. Therefore, all such patients must undergo biopsy of the endometrium. Outpatient sampling provides a rapid and accurate diagnosis,[1,7,11–18] and is possible in the vast majority of patients.

Patients Receiving Estrogen Replacement Therapy

There is controversy as to when, if ever, sampling of the uterus is necessary for patients receiving estrogen replacement therapy. There is strong evidence to support pretreatment and then subsequent annual biopsies for the patients with an intact uterus receiving unopposed estrogen therapy.[29–31] However, for the woman receiving combination estrogen/progestin therapy with predictable bleeding patterns, routine sampling does not appear necessary.[32,33] If any atypical or unscheduled bleeding arises, sampling at that time is strongly indicated.[32,34]

Infertility

Endometrial sampling with accurate dating of the endometrium is an integral component of the investigation of the infertile couple.[35] For accurate interpretation of biopsy results, one must have the cooperation of the patient, clinician, and pathologist. To diagnose luteal-phase deficiencies,[36] for example, one needs to know the timing of the subsequent menstruation or, better, the date of ovulation relative to the biopsy. The Pipelle and similar devices have been shown to be accurate for dating of the endometrium.[19,21,37] Because one study showed a discrepancy in results with different biopsy techniques, as well as for other practical reasons, it is generally recommended that at least two equivalent determinations be obtained before assigning the diagnosis of luteal-phase inadequacy.[38,39] A detailed discussion of this process is beyond the scope of this chapter; briefly, one looks for a "lag" in development of the endometrium by more than 2 days (secretory development of the endometrium compared to postovulation chronology).

Infection

Endometrial biopsy is useful for diagnosis and documentation of chronic uterine infections. For pelvic tuberculosis, it provides the pathologist with a specimen in which giant cells and granulomata may be found. Herpes virus, cytomegalovirus, and human papillomavirus, which are the only viruses presently known to cause endometritis,[40] can also be diagnosed in this fashion.

Recently, Martens et al. reported the efficacy of endometrial aspiration biopsy for cases of postpartum endometritis.[41] This technique has also been successfully applied to patients with nonpuerperal endometritis/salpingitis.[42] It appears to be a useful method to determine the spectrum of pathogenic microorganisms involved.

Contraindications

There are relatively few contraindications to endometrial sampling. Foremost of these is pregnancy. A careful history must be obtained; if there is any question, it is best to defer the procedure until the nongravid state can be confirmed.

Purulent cervicitis is a relative contraindication. One would in theory prefer to delay sampling until the risk of infection is minimal. However, occasionally endometrial biopsy may be used for documentation or clarification of an infectious process (see above). In these settings only, biopsy may proceed.

For the unanesthetized patient in the office setting or clinic two additional contraindications arise. These are stenosis of the cervix and poor patient tolerance. These circumstances, along with the uncooperative patient, are best handled by conventional D&C in the operating room, on an ambulatory basis. The vast majority of patients are suited to office sampling.[13,15,16,28,20,21]

Another consideration is the patient on anticoagulant therapy. Naturally, decisions about technique must be individualized. Some patients require preoperative hospitalization awaiting reversal of anticoagulation, followed by subsequent postsurgical hospitalization to reestablish the desired effects of oral agents.

It is paramount for all operators to be familiar with the equipment, patient, risks, and limitations of the different techniques before subjecting the patient to the procedure. Lack of such knowledge is a contraindication in and of itself.

Limitations

While office endometrial sampling is a convenient and reliable method to evaluate bleeding abnormalities, certain limitations must be understood. For the patient with a large uterus, greater than 8 to 10 weeks in size, biopsy with Pipelle-type devices may not provide sufficient sampling to be considered representative of the entire uterus. Similarly, with either the Vabra-type or Pipelle-type devices, the presence of endometrial polyps could easily be missed. In addition, the therapeutic effects of both of these are limited. One should not expect to successfully utilize these techniques for incomplete abortions, retained products of conception, or the hemorrhaging DUB patient.

Techniques

Prior to undertaking endometrial biopsy, as with all operating techniques, a careful history and physical exam must first be performed. The size and degree of flexion and version of the uterus must be noted to limit complications. Informed consent must be obtained.

Dilation and Curettage

Dilation and currettage is usually performed in an operating room on an ambulatory basis. Generally anesthesia or sedation with a paracervical block is used. After induction of anesthesia (or sedation), the patient is placed in lithotomy position and a thorough examination of pelvic anatomy is conducted. A sterile prep of the vagina and labia/perineum/mons is then performed. Sterile drapes are used to isolate the operative area.

Trendelenburg position is offered—care must be taken with the compromised patient. Using a weighted or Sims posterior retractor, the cervix is visualized and grasped on its anterior lip with a single toothed-tenaculum. At times, an anterior retractor will be needed to visualize the cervix. Gentle traction is then placed on the tenaculum to straighten the cervical canal.

Endocervical curettings are then obtained and placed in fixative as a separate specimen. The uterus is then sounded to determine cavity size and direction. Gentle serial dilatation with cervical dilators is performed to approximately 8 mm. Gauze is then placed in the posterior vaginal fornix to trap the tissue and blood that may subsequently be expressed. Sharp curettage is then performed in a systematic and serial fashion—anterior, right lateral, posterior, left lateral—in such a way that the entire endometrium is sampled. Curettage should involve multiple gentle but firm long strokes to minimize trauma to the specimen. The tissue should be placed with forceps into fixative. During curettage, an evaluation of the cavity's contour should be made.

An ovum or ureteral stone forceps is then gently introduced into the uterine cavity and moved systematically throughout the uterus in an opening-and-closing fashion. This is to grasp any polyps that may have escaped detection with the curette. Any polyps so obtained should be sent as a separate specimen. Random cervical biopsies should not be taken because the yield is low with such a technique.

Aspiration Curettage

Aspiration curettage is most often performed in an office or clinic setting without anesthesia. We usually administer ibuprofen 800 mg orally, 30 minutes before the procedure. We have not found the need for paracervical block, because dilatation is not necessary. The cannulas range from 2 to 4 mm in diameter.

Following history and pelvic examination, with the patient is lithotomy position, a bivalve speculum is placed in the vagina for visualization of the cervix. The cervix and vaginal walls are cleansed with antiseptic and the anterior lip of the cervix is grasped with a sterile single-toothed tenaculum. The metal (or plastic) cannula is then inserted through the cervical os and into the uterine cavity while gentle traction on the tenaculyum straightens the canal. Once the tip of the cannula is at the fundus, suction is begun to approximately 60 cm of water pressure. The cannula is rotated 360° three times while gentle forward-and-backward motions are used to yield representative sampling of the entire endometrial cavity. During the biopsy procedure, the operator's finger must cover the small holes in the cannula to complete the delivery of suction.

By capping the specimen trap and adding fixative to it, there is little manipulation of the tissue. (An example of this device is depicted in Figure 11-1.)

Hand-Held Aspiration Curettage

Hand-held aspiration curettage is also performed in the office and without anesthesia. In most cases, no analgesic is required either. The technique is similar to that for the Vabra aspirator above except that no ad-

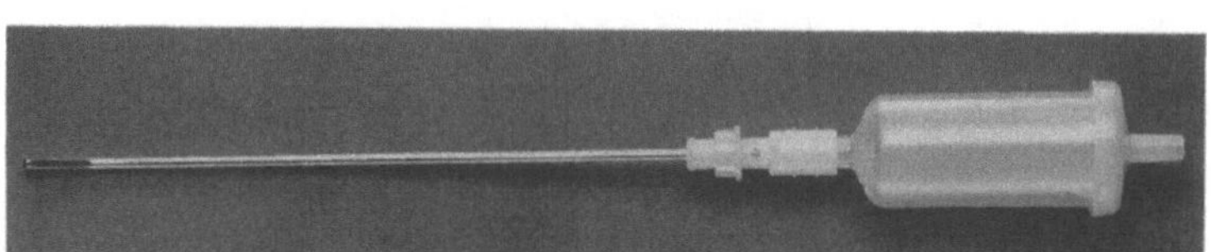

FIGURE 11.1. The Vabra aspirator. (Courtesy of Berkeley Medevices, Inc., Berkeley CA.)

ditional devices are necessary. After history, physical, and cervicovaginal preparation, the 3-mm cannula is inserted through the cervical canal to the fundus. A tenaculum may not be needed in multiparous patients. Suction is generated by pulling the internal piston backward to its maximal length. Three or four gentle strokes forward and backward are preformed while rotating the device. The cannula is then removed. The tip is then cut off proximal to the aspiration port and the specimen is released into fixative. (An example of this type of device is depicted in Figure 11-2.)

Comments

Hysteroscopic-guided uterine curettage has also been recently advocated.[43,44] A full discussion of this technique, however, is beyond the scope of this chapter.

As a routine, we have not administered antibiotics during endometrial sampling. However, for the patient with valvular disease or other medical conditions that warrant antibiotic prophylaxis, it should be administered in the usual fashion.

Complications

Complications following endometrial sampling are rare, with estimates of 1 to 2% for patients undergoing D&C[45,46] and less than 1% for aspiration biopsies.

Knowledge of the approximate size and flexion/version of the uterus prior to beginning sampling will decrease the risk of perforation. It is well recognized, however, that in patients with cervical stenosis, an atrophied or severely anteflexed or retroflexed uterus, or malignancy perforation is more likely. Additional caution must be exercised in such patients. When using office biopsy techniques, the cannula itself should not be used as a sound. Care must also be taken to avoid creation of a false channel with the rigid cannula.

If perforation does occur during dilation of the cervix, the procedure may need to be completed under laparoxopic guidance. Perforation during curettage itself is usually associated with more extensive injury. The patient should be admitted and observed. If extensive damage is suspected, laparoscopy or laparotomy is indicated. Antibiotics should

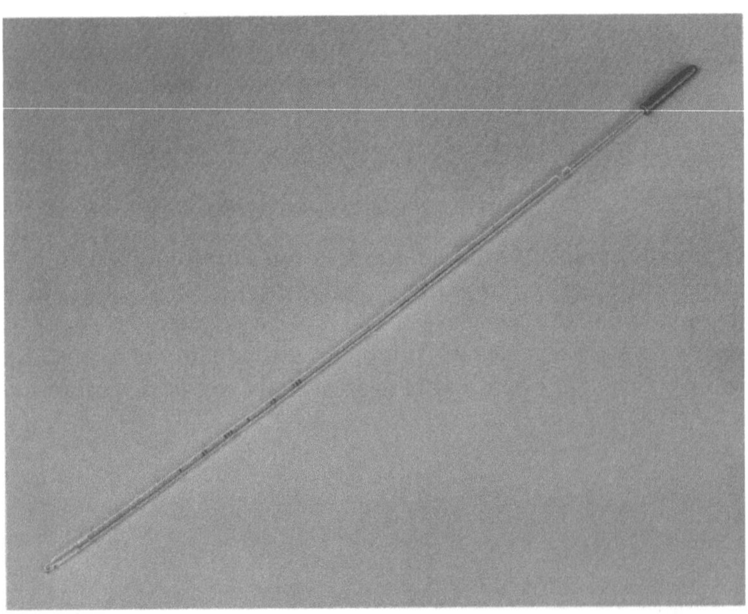

FIGURE 11.2. The Z-Sampler. (Courtesy of Zinnanti Surgical Instruments, Chatsworth, CA.)

be administered to limit the risk of developing a pelvic abscess.

Hemorrhage is a very unusual complication of these procedures. As stated above, certain patients on anticoagulant therapy require hospitalization for proper management.

McElin and colleagues reported a post-D&C febrile morbidity of 0.5%.[46] Following office biopsies the rate is probably lower; we thus do not administer prophylaxis routinely. A study by Livengood and colleagues, however, did report bacteremia in a significant number of premenopausal patients following endometrial biopsy.[47] For patients at risk for endocarditis, antibiotic prophylaxis is mandatory.

Care must also be exercised in compromised patients requiring endometrial biopsy. Patients with medical conditions such as unstable angina should have their health optimized prior to sampling. Most patients would benefit from intraoperative monitoring not available for office procedures.

References

1. Grimes DA: Diagnostic dilation and curettage: a reappraisal. Am J Obstet Gynecol 1982;142:1–6.
2. Sims JM: Clinical Notes on Uterine Surgery with Special Reference to The Management of the Sterile Condition. New York, Samuel and William Wood, 1858.
3. Speert H: Obstetric and Gynecologic Milestones. New York, The Macmillan Co, 1958.
4. Hitschmann F, Adler L: Der Bau der Uterusschleimhaut des geschlechtsreifen Weibes besonderer Berücksichtigung der menstruation. Monatsschr Geburt Gynäkol 1908;27: 1–82.
5. Bell WB: Principles of Gynaecology. London, Longmans, Green and Co, 1910.
6. Kelly HA: Curettage without anesthesia on the office table. Am J Obstet Gynecol 1925; 9:78.
7. Cohen CJ, Gusberg SB, Koffler D: Histologic screening for endometrial cancer. Gynecol Oncol 1974;2:279–286.
8. Novak E: A suction curette apparatus for endometrial biopsy. JAMA 1935;104:1497.
9. Randall LM: Endometrial biopsy. Proc Staff Meet Mayo Clin 1935;10:143.
10. Chatfield WR, Watson AA: Diagnosis of intrauterine disease by sponge biopsy technique. Lancet 1970;1:21.
11. Palmer JP, Kneer WJ, Eccleston HH: Endometrial biopsy—comparison of aspiration curettage with conventional dilatation and curettage. Am J Obstet Gynecol 1950;60:671.
12. Jensen JA, Jensen JG: Abragio mucosae uteri e aspiratione. Ugeskr Laeger 1968;130:2124.
13. Lutz MH, Underwood PB Jr, Kreutner A, et al: Vacuum aspiration: an efficient outpatient screening technique for endometrial disease. South Med J 1977;70:393–395.
14. Saunders P, Rowland R: Vacuum curettage of the uterus without anaesthesia: a comparison with conventional curettage. J Obstet Gynaecol Br Commonw 1972;79:168–174.
15. Barnett JM: Suction curettage on unanesthetized outpatients. Obstet Gynecol 1973; 42:672–674.
16. Webb MJ, Gaffey TA: Outpatient diagnostic aspiration curettage. Obstet Gynecol 1976; 47:239–242.
17. Alberico S, Elia A, Dal-Corso L, et al: Diagnostic validity of the Vabra curettage. Compared study on 172 patients who underwent Vabra curettage and the fractional curettage of the uterine cavity. Eur J Gynaecol Oncol 1986;7:135–138.
18. Walters D, Robinson D, Park RC, et al: Diagnostic outpatient aspiration curettage. Obsete Gynecol 1975;46:160–164.
19. Hill GA, Herbert CM III, Parker RA, et al: Comparison of late luteal phase endometrial biopsies using the Novak curette or Pipelle endometrial suction curette. Obstet Gynecol 1989;73:443–445.
20. Kaunitz AM, Masciello A, Ostrowski M, et al: Comparison of endometrial biopsy with the endometrial Pipelle and Vabra aspirator. J Reprod Med 1988;33:427–431.
21. Check JH, Chase JS, Nowroozi K, et al: Clinical evaluation of the Pipelle endometrial suction curette for timed endometrial biopsies. J Reprod Med 1989;34:218–220.
22. Henig I, Chan P, Tredway DR, et al: Evaluation of the Pipelle curette for endometrial biopsy. J Reprod Med 1989;34:786–789.
23. Mishell Jr DR: Abnormal uterine bleeding, in Droegenmueller W, Herbst AL, Mishell Jr DR, et al (eds): Comprehensive Gynecology.

St. Louis, CV Mosby Company, 1987, pp 953–964.
24. Haynes PJ, Hodgson H, Anderson ABM, et al: Measurement of menstrual blood loss in patients complaining of menorrhagia. Br J Obstet Gynaecol 1977;84:763–766.
25. Deligdisch L, Holinka CF: Endometrial carcinoma: two diseases? Cancer Detection and Prevention 1987;10:237–246.
26. Sutton GP, Geisler HE, Stehman FB, et al: Features associated with survival and disease-free survival in early endometrial cancer. Am J Obstet Gynecol 1989;160:1385–1393.
27. Connelly PJ, Alberhasky RC, Christopherson WM: Carcinoma of the endometrium III. Analysis of 865 cases of adenocarcinoma and adenoacanthoma. Obstet Gynecol 1982; 59:569–575.
28. Stovall TG, Solomon SK, Ling FW: Endometrial sampling prior to hysterectomy. Obstet Gynecol 1989;73:405–409.
29. Whitehead MI: Prevention of endometrial abnormalities. Acta Obstet Gynecol Scand [Suppl] 1986;134:81–91.
30. Weiss NS, Szekely R, Austin DF: Increasing incidence of endometrial cancer in the United States. N Engl J Med 1976;294:1259–1262.
31. Shapiro S, Kelly JP, Rosenberg L, et al: Risk of localized and widespread endometrial cancer in relation to recent and discontinued use of conjugated estrogens. N Engl J Med 1985; 313:969–972.
32. Koss LG, Schreiber K, Oberlander SG, Moussouris SG, et al: Detection of endometrial carcinoma and hyperplasia in asymptomatic women. Obstet Gynecol 1984;64:1–11.
33. Padwick ML, Pryse-Davies J, Whitehead MI: A simple method for determining the optimal dosage of progestin in postmenopausal women receiving estrogens. N Engl J Med 1986; 315:930–934.
34. Ettinger B: Optimal use of postmenopausal hormone replacement. Obstet Gynecol 1988; 72(suppl):31–36.
35. Rosenfeld DL, Chodow S, Bronson RA: Diagnosis of luteal phase inadequacy. Obstet Gynecol 1980;56:193–196.
36. Jones GS: Some newer aspects of the management of infertility. JAMA 1949;141:1123–1129.
37. Witten BI, Martin SA: The endometrial biopsy as a guide to the management of luteal phase defect. Fertil Steril 1985;44:460–465.
38. Honore LH, Cumming DC, Fahmy N: Significant difference in the frequency of out-of-phase endometrial biopsies depending on the use of the Novak curette or the flexible polypropylene endometrial biopsy cannula ('Pipelle'). Gynecol Obstet Invest 1988;26:338–340.
39. Balasch J, Vanrell JA, Creus M, et al: The endometrial biopsy for diagnosis of luteal phase deficiency. Fertil Steril 1985;44:699–701.
40. Kurman RJ, Mazur MT: Benign diseases of the endometrium, in Kurman RJ (ed): Blaustein's Pathology of the Female Genital Tract, ed 3. New York, Springer-Verlag, 1987, p 297.
41. Martens MG, Faro S, Hammill HA, et al: Transcervical uterine cultures with a new endometrial suction curette: a comparison of three sampling methods in postpartum endometritis. Obstet Gynecol 1989;74:273–276.
42. Paavonen J, Aine R, Teisala K, et al: Comparison of endometrial biopsy and peritoneal fluid cytologic testing with laparoscopy in the diagnosis of acute pelvic inflammatory disease. Am J Obstet Gynecol 1985;151:645–650.
43. Walton SM, Macphail S: The value of hysteroscopy in postmenopausal and perimenopausal uterine bleeding. J Obstet Gynaecol 1988; 8:332–336.
44. Mencaglia L, Perino A, Hamou J: Hysteroscopy in perimenopausal and postmenopausal women with abnormal uterine bleeding. J Reprod Med 1987;32:577–582.
45. Mackenzie IZ, Bibby JG: Critical assessment of dilatation and curettage in 1029 women. Lancet 1987;2:566.
46. McElin TW, Bird CC, Reeves BD, et al: Diagnostic dilatation and curettage. Obstet Gynecol 1969;33:807.
47. Livengood CH III, Land MR, Addison WA: Endometrial biopsy, bacteremia, and endocarditis risk. Obstet Gynecol 1985;65:678–681.

12

Abdominal Approach to Pelvic Sonography

NEELA LAMKI and PATRICIA A. ATHEY

Sonography is an imaging modality that utilizes high-frequency sound waves. The sound waves are produced by a crystal within the transducer, a hand-held device that is placed in contact with the area of the body to be scanned. Gel is placed on the patient's skin to ensure good transmission of sound into the body. Each time the sound beam strikes tissue interfaces, an echo will be produced and displayed on a monitor as a dot (either black or white), the intensity of which corresponds to its brightness. The combination of innumerable echoes gives a textural and anatomic representation of the structures scanned.

The ultrasound beam cannot penetrate a soft tissue–air interface. Consequently, all pelvic transabdominal scans utilize a distended urinary bladder to push air-filled loops of bowel out of the way. This also serves to displace the uterus posteriorly and out of its normally anteverted position. Thus we can view the uterus from its anterior to its posterior aspect rather than from fundus to cervix.

Pelvic scans are generally performed with 3.5- or 5-MHz frequency transducers according to the patient's body habitus. The higher the frequency the greater the resolution achieved, but the less the depth penetration. Transvaginal transducers (to be discussed in Chap. 13) are able to utilize higher frequency transducers because they are positioned in close proximity to the pelvic organs. Both sagittal and transverse images of the uterus are obtained.

In addition to its applications in the pelvis, ultrasound is used in the abdomen to examine the liver, spleen, gallbladder, pancreas, biliary tract, and kidneys. It is also utilized to assess for the presence of abscess, hematoma, ascites, adenopathy, and other masses. Further applications include imaging of small parts (i.e., breast, thyroid, scrotum) as well as the extremities and the various applications of Doppler ultrasound in the evaluation of the carotid arteries and peripheral and deep abdominal arteries and veins. In the evaluation of a patient with known or suspected pelvic neoplasm, ultrasound of the abdomen is helpful in assessing for liver metastases, adenopathy, ascites, and hydronephrosis.

Most sonographers receive on-the-job training in a hospital setting. Those individuals with a paramedical background and at least 1 year of hospital-based scanning experience are allowed to take a registry examination. They must pass a physics/instrumentation section and at least one subspecialty exam (e.g., abdomen, obstetrics) to become a Registered Diagnostic Medical Sonographer. Radiologists, either in a hospital-based or outpatient setting, conduct a wide variety of sonographic examinations. Some obstetricians/gynecologists have obtained machines for their private offices. Extensive training in this modality is advised.

The Normal Uterus

The normal uterus is pear-shaped and has a smooth contour (Fig. 12-1). Normally the uterus is slightly anteverted, with a visible angulation at the junction of the cervix and uterine body. The retrodisplaced uterus is sometimes mistaken clinically for a cul-de-sac mass. Sonography will demonstrate the posterior angulation of the fundus. The myometrium is homogeneous and composed of low- to medium-level echoes. The appearance of the endometrial cavity varies depending on the stage of the menstrual cycle, which will be discussed later.

Uterine Abnormalities

Uterine abnormalities can be divided into two broad categories: abnormalities of size, shape, position, contour, and texture; and abnormalities of the endometrial cavity.

Abnormalities of Shape, Size, Position, Contour, and Texture

The size of the uterus varies according to pubertal status, age, and parity. Before puberty, the uterus usually measures 3 × 1 × 1 cm (Fig. 12-2). Following puberty, it usually measures 7 × 4 × 5 cm. In a multiparous patient, the uterus increases 1 to 2 cm in all directions. In the postmenopausal patient the uterus atrophies, measuring 3 to 7 cm in length and 1 to 2 cm in thickness.[1]

Retrodisplaced Uterus

On transabdominal scanning, the retrodisplaced uterus appears lobular and the fundus appears hypoechoic as a result of its increased distance from the transducer, simulating a leiomyoma (Fig. 12-3). Because the endometrium is not perpendicular to the sound beam, it is also difficult to visualize. Thus, detection of leiomyomata in a retrodisplaced uterus can be very difficult. However, transvaginal sonography is excellent for the assessment of the retrodisplaced uterus.[2]

Congenital Anomalies

Congenital anomalies such as bicornuate uterus, uterus didelphys, and uterus bicornis bicollis can be demonstrated sonographically. The most frequently detected anomaly is bicornuate uterus.[3] Sonographically, the transverse dimension of the uterus is increased and may appear lobulated (Fig. 12-4). The identification of two endometrial cavities is crucial, so as not to mistake one horn of the uterus for a leiomyoma or a solid adnexal mass. The identification of other

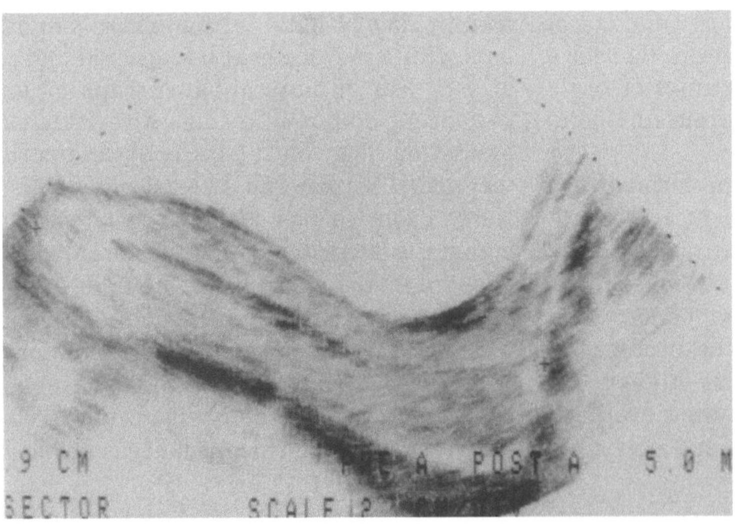

FIGURE 12-1. Sagittal scan of a slightly elongated but otherwise normal-appearing anteverted uterus. The endometrial cavity is seen as a linear echogenicity centrally located.

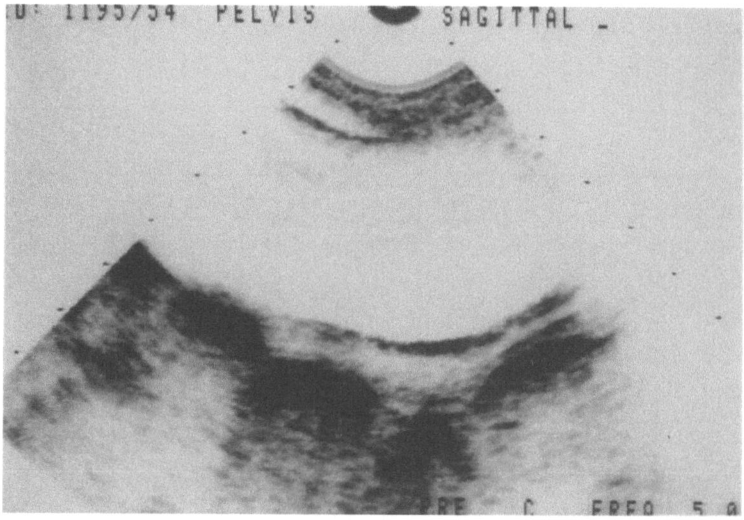

FIGURE 12-2. Sagittal scan of an infantile uterus. Note that the cervix is larger than the body and fundus.

anomalies requires demonstration of two separate uterine bodies. More subtle anomalies, such as uterine arcuatus and those anomalies with a septum, are more difficult to recognize. When a uterine anomaly is identified, the kidneys must also be examined for associated anomalies.[3]

Leiomyomata

The sonographic appearance of leiomyomata varies with the location and the presence or absence of secondary changes. Depending on its location, a fibroid may distort the endometrial cavity, appear as a discrete mass within the myometrium (Fig. 12-5), distort the contour of the uterus (Fig. 12-6), or be attached by a pedicle of varying dimensions (Fig. 12-7). Usually leiomyomata are hypoechoic, with relatively poor transmission of sound. The echogenicity of leiomyomata depends on the relative ratio of smooth muscle to fibrous tissue and on the presence and type of degeneration.[3,4] Leiomyomata also may cause contour deformation and/or enlargement of uterus. If the fibroid is calci-

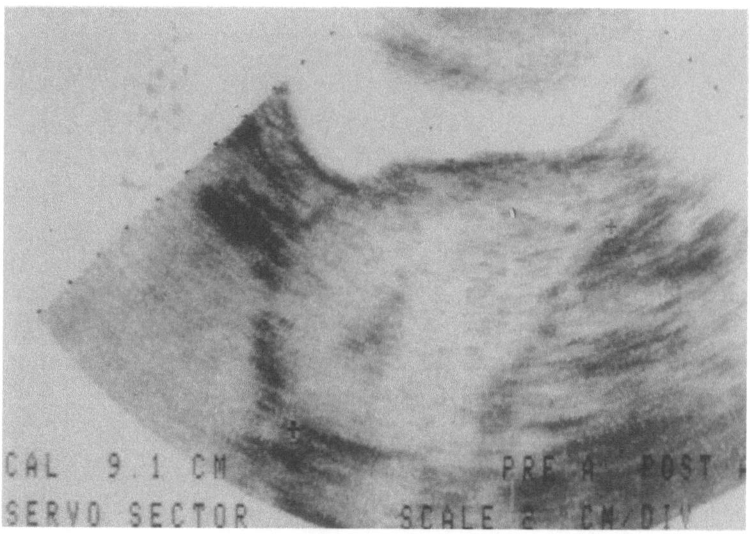

FIGURE 12-3. Sagittal scan of a retrodisplaced uterus. The fundus appears bulky and the endometrial cavity is indistinct.

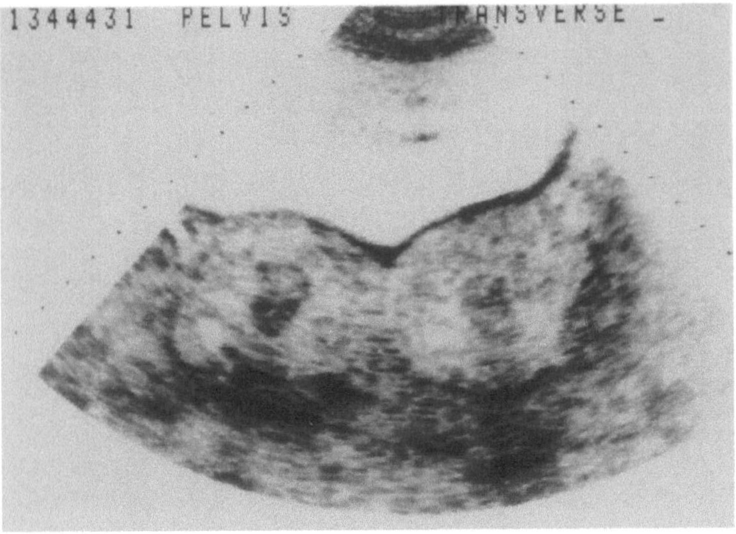

FIGURE 12-4. Transverse scan of a bicornuate uterus. Both cornua are visualized and each contains prominent endometrial cavity echoes.

fied, focal areas of increased echogenicity with acoustic shadowing are present. If degeneration and necrosis have taken place within the fibroid, the echogenicity is decreased with increased through-transmission of sound. If fatty degeneration has taken place in the fibroid, the echogenicity of the fibroid is homogeneously increased compared to the myometrium. In a postmenopausal patient with fibroids, increase of uterine or fibroid size may be secondary to degeneration, but sarcomatous change, although rare, should be suspected.[1,3] The differential diagnosis of leiomyomata includes solid masses adjacent to the uterus, which may be due to solid ovarian neoplasm or metastases in the cul-de-sac.

Adenomyosis

Adenomyosis has no specific sonographic appearance other than smooth generalized uterine enlargement (Fig. 12-8). If there is a focal adenomyoma, it may be difficult to differentiate from a leiomyoma. If the glandular tissue is functioning, irregular ane-

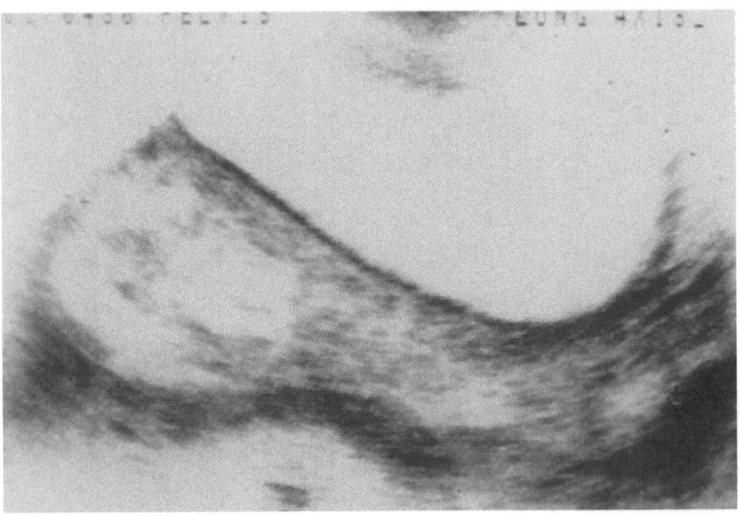

FIGURE 12-5. Sagittal scan of a uterus containing several hypoechoic submucosal and myometrial fibroids in the fundus and body.

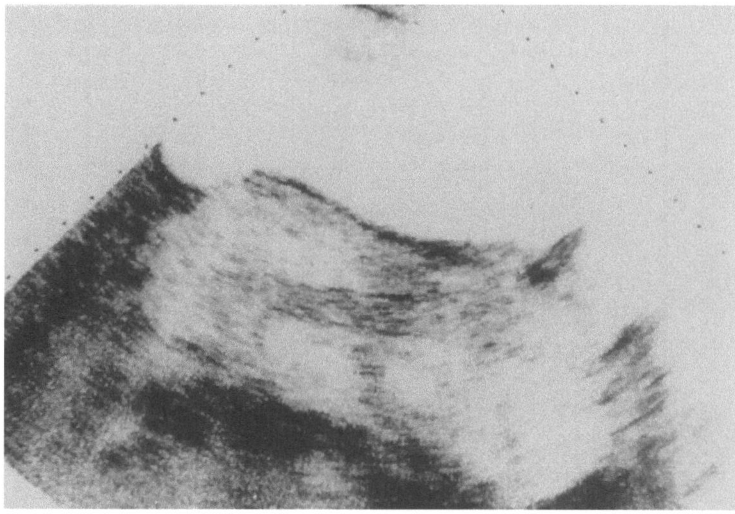

FIGURE 12-6. Sagittal scan of a uterus containing multiple hypoechoic fibroids, the largest of which are in the region of the cervix and distort the contour of the uterus.

choic spaces may be present in the myometrium.[5] Adenomyosis is usually more prevalent in the posterior wall, and acoustic enhancement posterior to the uterus is present, rather than the attenuation often seen with fibroids.[3] Other causes of generalized uterine enlargement include recent delivery or abortion, multiparity, and idiopathic hypertrophy.

Sarcoma

Sonographically, leiomyosarcomas are indistinguishable from leiomyoma, unless invasion or distant metastases are present. They are usually hypoechoic and may contain areas of cystic degeneration. Mixed müllerian sarcoma produces generalized uterine enlargement. The texture sonographically is heterogeneous, and extensive myometrial invasion may be present.

Cervical Carcinoma

Later stages of cervical carcinoma manifest sonographically as enlargement and lobularity of the cervix (Fig. 12-9), which can be confused with leiomyomata. Increased echo-

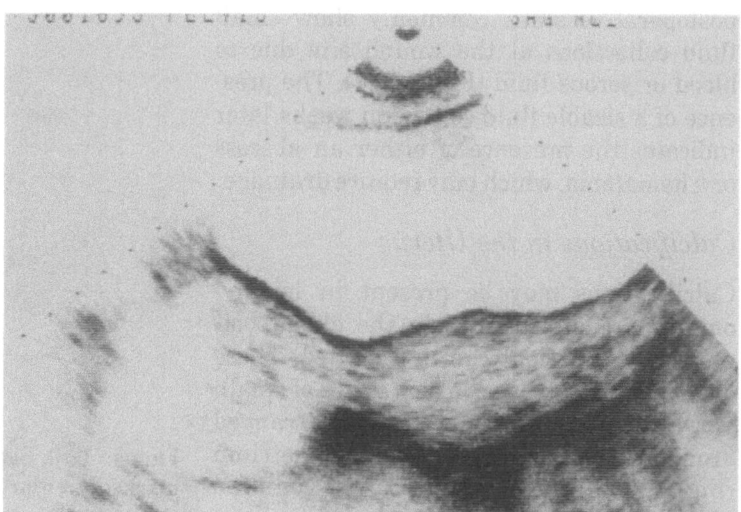

FIGURE 12-7. Sagittal scan of a uterus with a pedunculated fibroid arising from the fundus. The sound beam is markedly attenuated by this large fibroid.

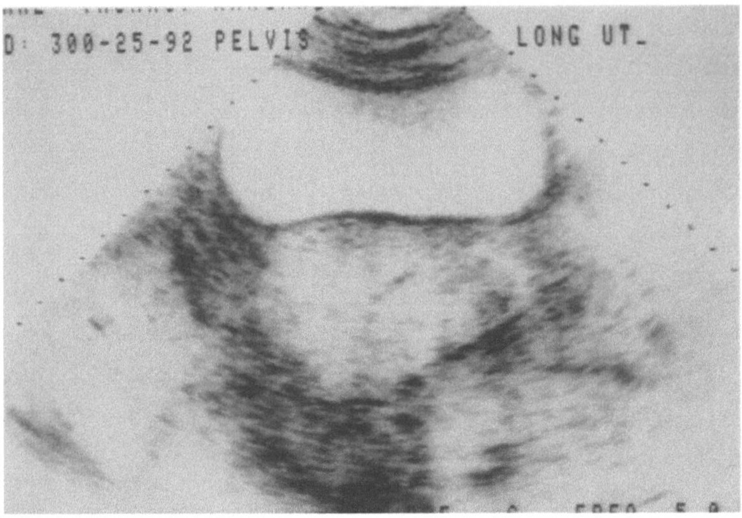

FIGURE 12-8. Sagittal scan of a uterus involved with adenomyosis. The uterus is generally enlarged and inhomogeneous in texture.

genicity with dirty shadowing or reverberation as a result of air in necrotic tissue is occasionally seen within the cervical mass (Fig. 12-10). If there is cervical stenosis resulting from neoplasm, endometrial fluid is present and cannot be differentiated from other causes of hematocolpos or pyometra. Computed tomography[4] and magnetic resonance imaging[6] are preferable for staging.

Postoperative Changes following Cesarean Section

Postoperative changes at the wound site of a cesarean section can cause a contour defect on the anterior surface of the uterus. Early postoperative scans commonly show small fluid collections at the wound site due to blood or serous fluid (Fig. 12-11). The presence of a sizable fluid collection weeks later indicates the presence of either an abscess or a hematoma, which may require drainage.

Calcifications in the Uterus

Calcifications may be present in leiomyomata. Arcuate arteries in the older postmenopausal patient can be calcified. These are sonographically seen as a series of small, very echogenic foci, regularly arranged around the periphery of the myometrium (Fig. 12-12). These should not be confused with fibroid calcifications.

Abnormalities of the Endometrial Cavity

The endometrial cavity is a centrally located, highly echogenic structure seen on both sagittal and transverse scans. It is difficult to identify in a retrodisplaced uterus by transabdominal scanning alone and may be distorted by fibroids. It serves to identify the site of early pregnancy, decidual reaction, and endometrial abnormalities. If a mass touches the endometrial echo complex,

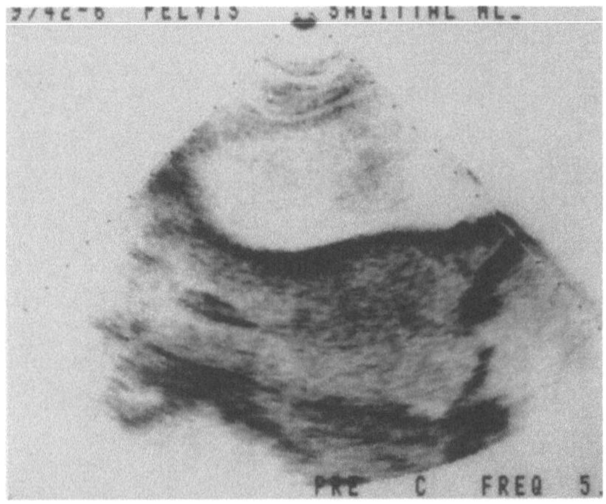

FIGURE 12-9. Sagittal scan of a uterus with a markedly enlarged and lobulated cervix due to cervical carcinoma.

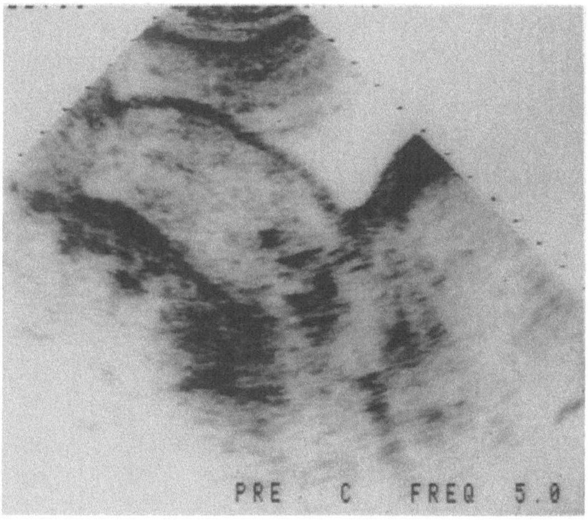

FIGURE 12-10. Sagittal scan of the uterus in another patient with cervical carcinoma. Note the echogenicities in the cervix, which exhibits reverberation and dirty shadowing. These represent air in the necrotic tissue.

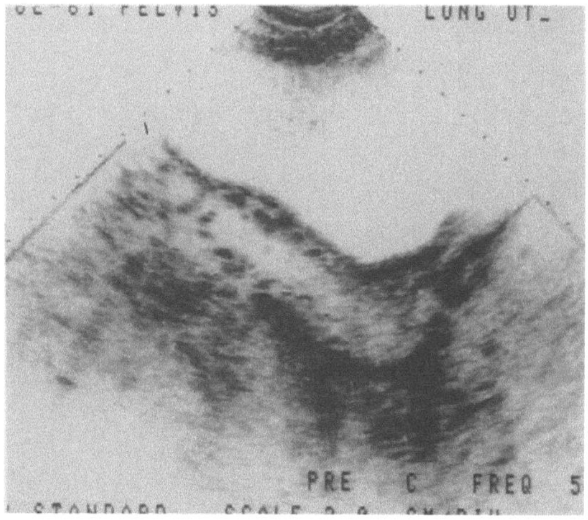

FIGURE 12-12. Sagittal scan of the uterus in a postmenopausal patient. The echogenic foci regularly arranged around the periphery of the myometrium represent arcuate artery calcifications.

it is assumed to be of uterine origin, usually a fibroid.

The sonographic appearance of the endometrium varies depending on the phase of the menstrual cycle as well as the level of ovarian activity.[7,8] Alteration of the normal endometrial echo complex can be divided into two main categories: fluid collections within the endometrial cavity and increase in the echogenicity in the endometrial cavity. Clinical history is helpful in differentiating between these conditions.

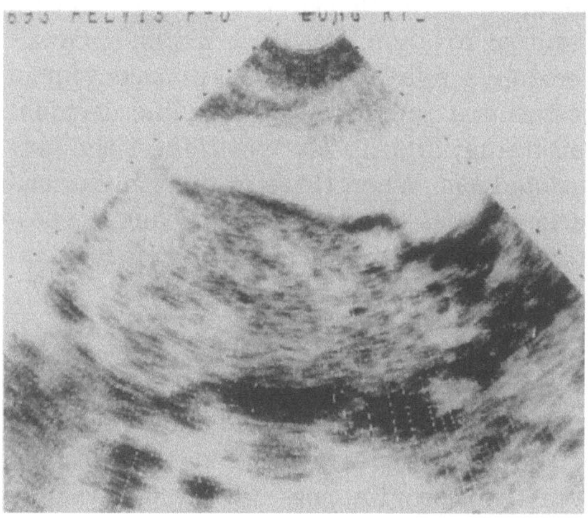

FIGURE 12-11. Sagittal scan of an enlarged postpartum uterus. There is a small fluid collection at the site of the cesarean section wound, causing an anterior contour defect. This is compatible with a small amount of serous fluid or blood.

Fluid Collections within the Endometrial Cavity

Menstruation

During menstruation, the endometrial cavity echo may be separated by blood and sloughed tissue. Sonographically, this is seen as a relatively anechoic space.

Normal Pregnancy

An early intrauterine pregnancy (approximately 5 menstrual weeks) appears sonographically as an endometrial fluid collection surroundedby an echogenic rim (Fig. 12-13). Stimulation of the uterine lining with hormones produced by an ectopic pregnancy can result in changes that are similar in appearance to an early gestational sac. This is referred to as the pseudogestational sac of ectopic pregnancy, and is seen in 10 to 20%

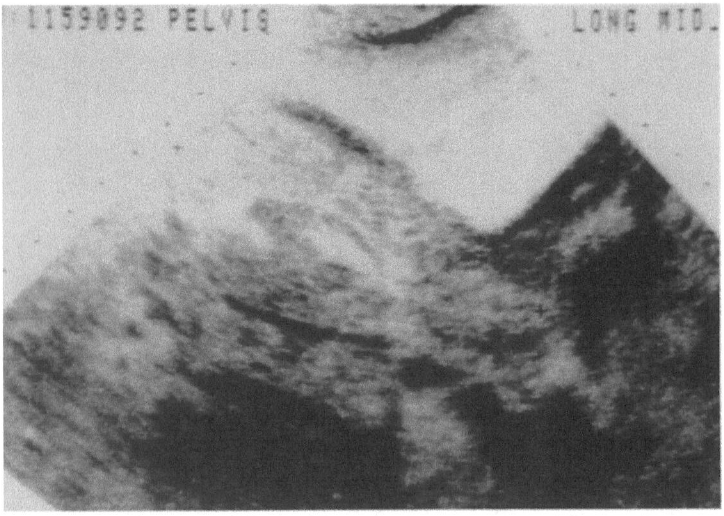

FIGURE 12-13. Transverse scan demonstrating a normal early intrauterine pregnancy. This is a fluid collection surrounded by echogenic trophoblastic tissue. The fetus is also visible.

of cases[9] (Fig. 12-14). More specifically, the early intrauterine gestational sac is surrounded by two concentric echogenic rings thought to represent the decidua parietalis adjacent to the decidua capsularis, referred to as the double decidual sac.[9,10] The pseudogestational sac of ectopic pregnancy is composed of a single decidual layer surrounding an endometrial fluid collection.[9,10] The double decidual sac sign is highly reliable, but does not absolutely exclude pseudogestational sac, nor does it necessarily confirm normal intrauterine pregnancy.[9] Therefore, a follow-up examination is helpful when no embryonic structures can be seen within the gestational sac. Embryonic structures can be detected when the gestational sac is 10 mm or greater, but only consistently when the sac is 15 mm or greater.[11] The presence of embryonic structures, especially cardiac activity, confirms an intrauterine pregnancy. With the advent of transvaginal transducers, intrauterine pregnancy can be diagnosed earlier with marked increase in the confidence of the diagnosis.

Abnormal Intrauterine Pregnancy

Incomplete abortion and blighted ovum also present as intrauterine fluid collections (Fig. 12-15). Difficulty arises in differentiating normal intrauterine pregnancy versus abnormal pregnancy when a normal gestational sac is visualized, but no embryo is discerned.

Major and minor sonographic criteria have been described for abnormal intrauterine pregnancies. The major criteria include: (a) a gestational sac that is small for menstrual dates, and (b) the appearance of the gestational sac—gestational sac 25 mm or greater without embryo, or gestational sac 20 mm or greater without yolk sac. The minor sonographic criteria include: (a) thin decidual reaction, less than 2 mm; (b) weak decidual amplitude; (c) irregular contour of gestational sac; (d) absent double decidual sac lining; and (e) low position of the gestational sac. When three or more minor criteria are present, the specificity and the positive predictivity of an abnormal outcome increases to 100%.[11,12]

Inflammation

Endometritis has a variable and nonspecific sonographic appearance that includes endometrial prominence and/or collection of fluid in the endometrial cavity with smooth or irregular margins. Endometritis can be confused with early intrauterine pregnancy or its complications, ectopic pregnancy, and, in the postpartum period, retained products of conception.

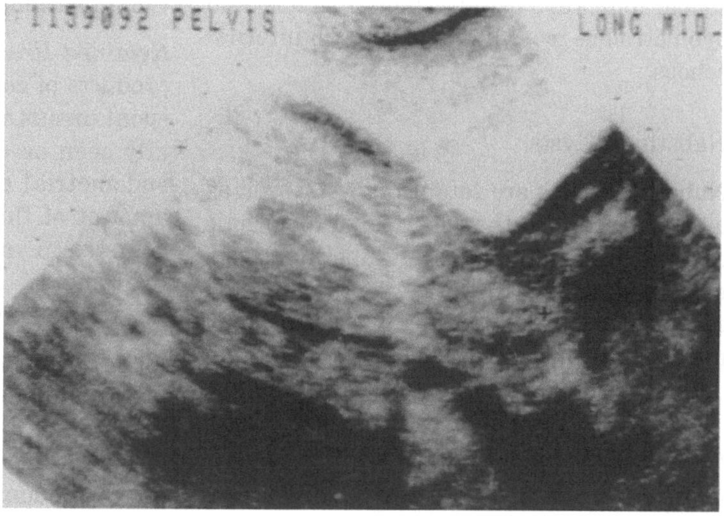

FIGURE 12-14. Sagittal scan of the uterus in a patient with an ectopic pregnancy. The fluid collection in the endometrial cavity, with an echogenic rim, represents a pseudogestational sac.

Hydrometra, Pyometra, and Hematometra

In the young patient, obstruction of the genital tract can be due to congenital causes: imperforate hymen, vaginal membrane, or absence or atresia of the vagina. In the older age group, uterine or cervical malignancy, trauma, radiation, surgery, or inflammation may lead to cervical stenosis.

The sonographic features of hydrometrocolpos or hematocolpos include distention of the endometrial cavity and of the vagina, with the latter demonstrated as a tubular hypoechoic or anechoic structure. The uterus may be enlarged. There may be layering of echogenic material within the hypoechoic or anechoic space, exhibiting a fluid-debris level, or the fluid may contain low-level echoes. Retained products of conception is one of the differential considerations, especially under the appropriate clinical circumstances.

Differentiation between hydrometra, hematometra, and pyometra cannot be made

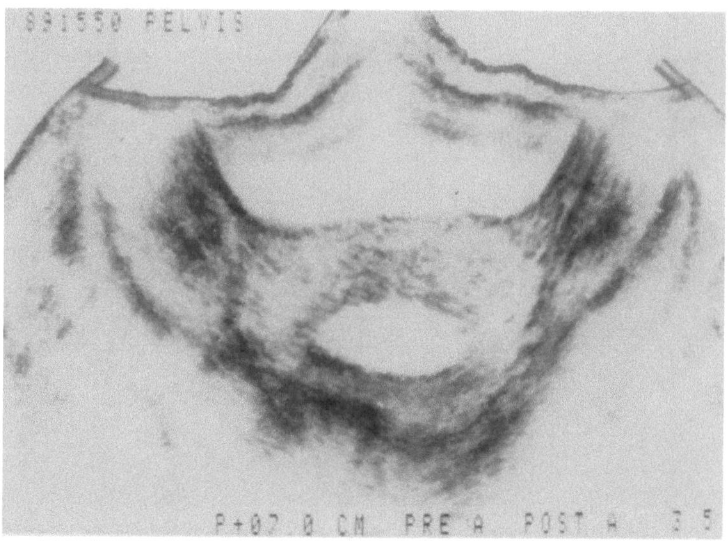

FIGURE 12-15. Transverse scan demonstrates an empty gestational sac compatible with a blighted ovum.

sonographically; usually hematometra and pyometra are more likely to contain internal echoes.

Nabothian Cysts

Nabothian cysts are retention cysts, seen as anechoic areas in the cervix, due to chronic cervicitis in healing phases. Nabothian cysts may be eccentric to the cervical canal and can measure 6 to 20 mm (Fig. 12-16). They may be multiple and are usually within 1 cm of the external os, but may be located higher.

Increased Echogenicity in the Endometrium

Menstrual Cycle/Blood Clots

In the secretory phase of the menstrual cycle, the endometrium attains maximum thickness and echogenicity and measures 5 to 8 mm. This is because of the distended and tortuous glands that contain echogenic secretions (mucus and glycogen).[13] In the late secretory phase the endometrium decreases in thickness, and increased echogenicity persists until menses. In the menstrual phase, the endometrium appears thin and slightly irregular and has an echogenic interface. Blood clots can also cause an increase in the echogenicity of the endometrial cavity, regardless of the cause of uterine bleeding.

Complications of Pregnancy

Retained Products of Conception. Retained products of conception following delivery or spontaneous or therapeutic abortion are usually seen as echogenic material within the endometrial cavity (Fig. 12-17). A variable amount of fluid may be present in the endometrium and may mimic hematometra, hydrometra, or pyometra when fluid is present in the endometrial canal.

Hydatidiform Mole/Choriocarcinoma. The ultrasound appearance of hydatidiform mole is variable in the first trimester. Early mole may present as a small echogenic mass filling the endometrial cavity without the characteristic vesicular appearance. It would be very difficult to distinguish it from a missed abortion in which hydropic degeneration is present. Combined with a high index of suspicion and correlation with the level of serum beta human chorionic gonadotropin, the diagnosis can be suspected, and confirmed pathologically. Classically, the sonographic features of hydatidiform mole include a large, moderately echogenic soft tissue mass filling the endometrial cavity, with numerous small anechoic spaces scattered in the echogenic mass (Fig. 12-18). Choriocarcinoma has a similar appearance. With the advent of transvaginal scanning, myometrial invasion may be detected with more

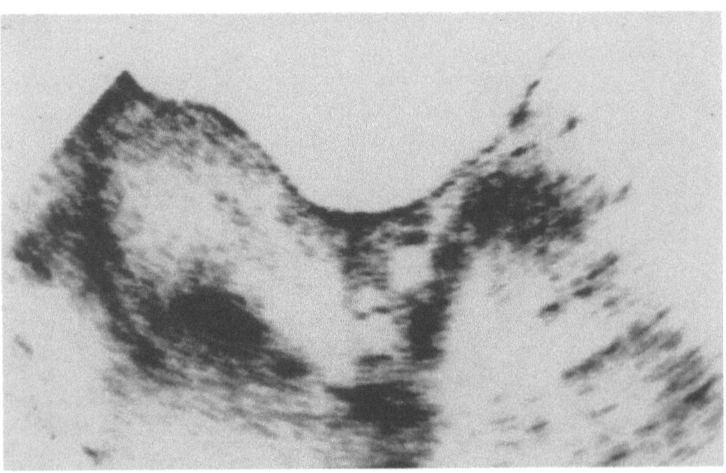

FIGURE 12-16. Sagittal scan of the uterus demonstrates at least three small structures in the cervix compatible with nabothian cysts.

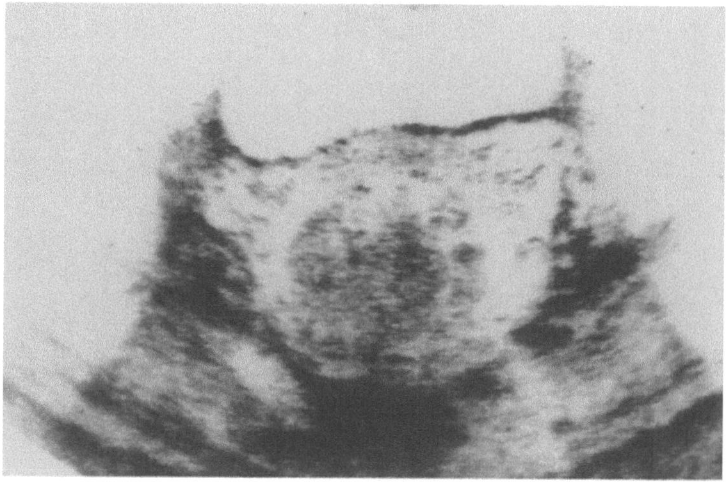

FIGURE 12-17. Transverse scan of the uterus in a patient who had retained products of conception. Sonographically this is seen as an area of increased echogenic material with some adjacent fluid within the endometrial cavity.

success. Theca-lutein cysts, if present, are helpful in the diagnosis of molar pregnancy. These are multiple, large, thin-walled cysts usually occurring bilaterally. The differential diagnosis includes leiomyoma with cystic degeneration.

Ossification or Calcification. Causes of ossification or calcification in the endometrial cavity can be variable, but usually there is a history of instrumentation for therapeutic or spontaneous abortion resulting in retention of the tissues or implantation of fetal tissues, which undergo ossification. Sonographically, calcification or ossification presents as areas of increased echogenicity with irregular margins with posterior acoustic shadowing (Fig. 12-19). The differential diagnosis includes intrauterine contraceptive device (IUD), air, retained products of conception, and calcified fibroid; these can usually be differentiated.

Endometrial Abnormalities

Increased echogenicity of the endometrium can be seen with endometrial hyperplasia, endometrial polyps, and endometrial carcinoma. These cannot be differentiated sonographically.

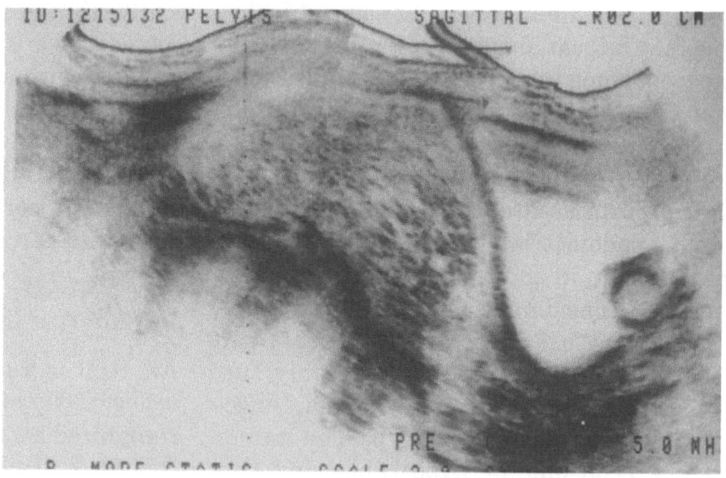

FIGURE 12-18. Sagittal scan of a uterus containing a hydatidiform mole, which is a large echogenic mass filled with many small cystic spaces representing the vesicular material of a molar pregnancy.

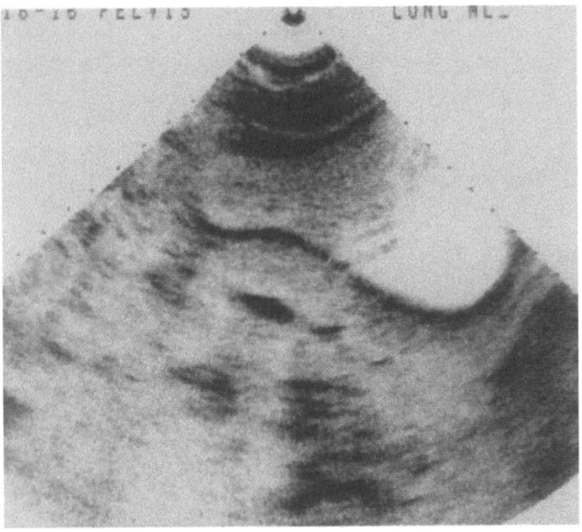

FIGURE 12-19. Sagittal scan of the uterus containing some very echogenic material with faint shadowing. This represented an area of ossification presumably as a sequela of incomplete abortion followed by curettage.

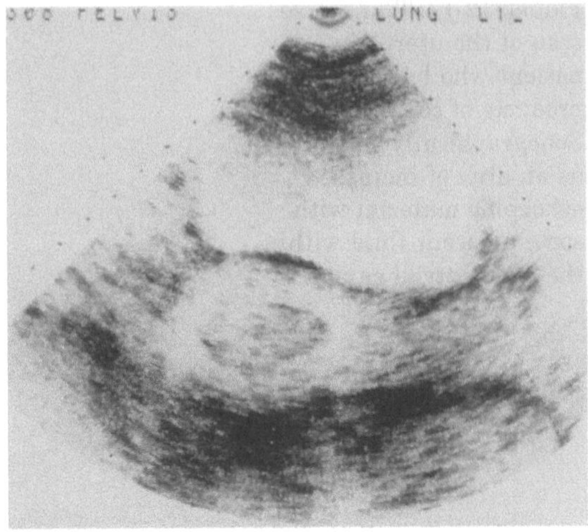

FIGURE 12-20. Sagittal scan of the uterus containing an increased amount of echogenic material within the endometrial cavity. This proved to be cystic hyperplasia of the endometrium.

Endometrial Polyps. Sonographically, endometrial polyp presents as a prominent endometrial echo complex, or it may be seen as a discrete echogenic mass. Uterine enlargement due to endometrial polyps is uncommon.

Endometrial Hyperplasia. Sonographically, the endometrial central echo complex is very prominent and cannot be differentiated from endometrial polyp or endometrial carcinoma (Fig. 12-20).

Endometrial Carcinoma. Sonographically, endometrial carcinoma presents as thickened endometrium, which is usually more than 6 to 8 mm.[14] The endometrial interface may also be irregular and it can be another cause of prominent endometrial echo complex, or increased echogenicity of endometrium. Endometrial echo complex greater than 5 mm in a postmenopausal woman should be considered suspicious for a pathologic process[14]; if the outline of the enlarged endometrium is irregular, it is highly suspicious for endometrial carcinoma, endometrial hyperplasia, or polyp. The tumor may be echogenic or hypoechoic, and this may be related to the grade of the tumor. Usually the more differentiated the tumor the more echogenic it appears to be; anaplastic tumors appear to be relatively hypoechoic.[14] In a later stage, sonographically the uterus is enlarged, with irregular areas of low-level echoes and focal areas of increased echogenicity (Fig. 12-21). Ultrasound also helps in staging the carcinoma. Endometrial carcinoma may obstruct the endometrial cavity, and there may be evidence of fluid collection in the cavity. Sonographically, this is seen as an anechoic or hypoechoic space within the endometrial cavity with or without echoes, depending on the amount of debris present. Distinction between endometrial carcinoma and leiomyomas can be difficult, especially because a significant percentage of patients with endometrial carcinoma have leiomyomas.

Intrauterine Contraceptive Device

An IUD is yet another cause of increased echogenicity of the endometrium and can be recognized by the separate entrance and exit echoes. With the Copper 7 and Copper T, in

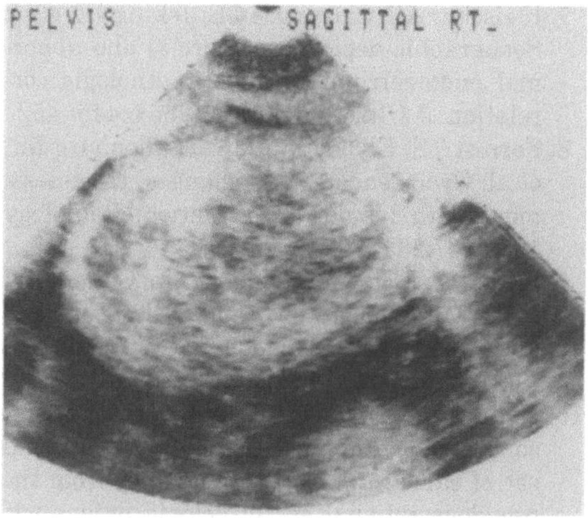

FIGURE 12-21. Sagittal scan of a globally enlarged uterus with a large, inhomogeneous, echogenic endometrial mass that pathologically represented adenocarcinoma of the uterus.

the longitudinal view, strong linear echoes are present in the endometrial cavity and there may be posterior reverberations (Fig. 12-22). The Lippes Loop produces increased echoes in the endometrial cavity, but they have a steplike appearance and acoustic shadowing is seen posterior to the central echoes (Fig. 12-23).

Miscellaneous Causes of Increased Echogenicity

Foreign Bodies. Foreign bodies can simulate an IUD by causing increased echogenicity in the endometrium; hence, history is crucial in the diagnosis. Cerclage sutures are seen as hyperechoic linear structures in the cervical region and are best seen on sagittal scans. Acoustic shadowing may be seen posteriorly depending on the type of material used in the suture.

Air. Air in the endometrial cavity, whether it is intrauterine or in the cervical region, can also present as increased echogenicity of endometrium with acoustic shadowing. Air may be introduced into the endometrium iatrogenically following D&C or endometrial biopsy. Air can also be seen in the endometrial cavity in patients with endometrial carcinoma, presumably resulting from necrosis

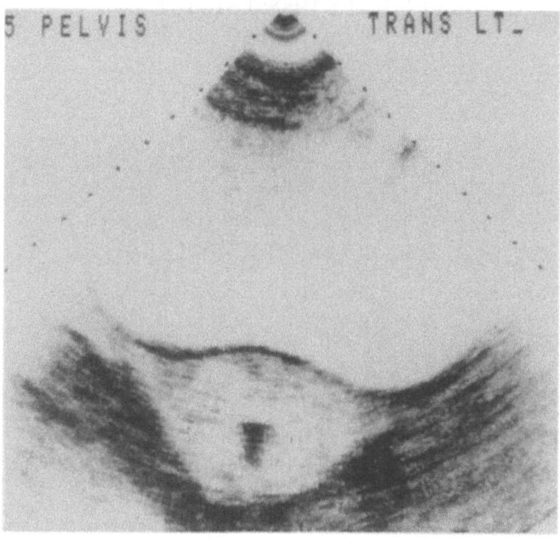

FIGURE 12-22. Transverse scan of a uterus containing a Copper 7 IUD. Note the reverberations from a portion of one limb of the IUD.

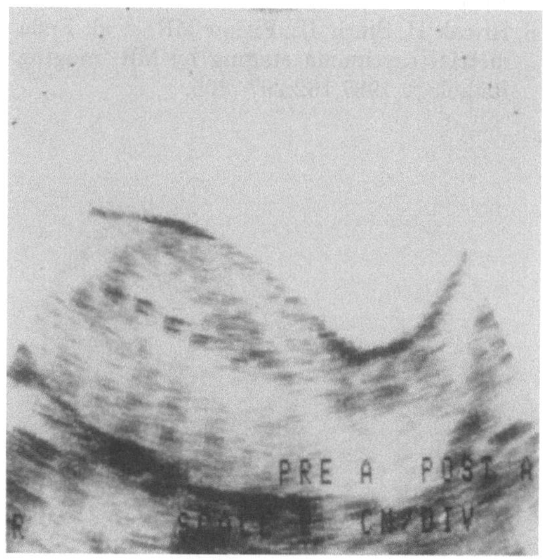

FIGURE 12-23. Sagittal scan of a uterus containing a Lippes Loop. Note the entrance and exit reflections of each coil of the IUD. Also note the acoustic shadowing posterior to each coil.

or infection, or secondary to a surgical procedure.

Acknowledgment. We wish to thank Mrs. Becky Baxter for her excellent secretarial assistance in the preparation of this chapter.

References

1. Cooperberg PL, Kidney MR: Ultrasound evaluation of the uterus, in Callen PW (ed): Ultrasonography in Obstetrics and Gynecology, ed 2. Philadelphia, WB Saunders Co, 1988, pp 393–411.
2. Mendelson EB, Bohm-Velez M, Neiman HL, et al: Transvaginal sonography in gynecologic imaging. Semin Ultrasound, CT, MR 1988; 9(2):102–121.
3. Athey PA: Uterus: Abnormalities of size, shape, contour and texture, in Athey PA, Hadlock FP (eds): Ultrasound in Obstetrics and Gynecology, ed 2. St. Louis, CV Mosby Co, 1985, pp 167–193.
4. Walsh JW, Brewer WH, Schneider V: Ultrasound diagnoses in diseases of the uterine corpus and cervix. Semin Ultrasound 1980;1:30–40.
5. Walsh JW, Taylor KJW, Rosenfield AT: Gray scale ultrasonography in the diagnosis of endometriosis and adenomyosis. AJR 1979; 132:87–90.
6. Hricak H, Stern JL, Fisher MR, et al: Endometrial carcinoma staging by MR imaging. Radiology 1987;162:297–305.
7. Fleischer AC, Calamaris GE, Jeking JE, et al: Sonographic depiction of normal and abnormal endometrium with histopathologic correlation. J Ultrasound Med 1986;5:445–452.
8. Forrest TS, Elyaderani MK, Muilenburg MI, et al: Cyclic endometrial changes: US assessment with histologic correlation. Radiology 1988;167:233–237.
9. Filly RA: Ectopic pregnancy, in Callen PW (ed): Ultrasonography in Obstetrics and Gynecology, ed 2. Philadelphia, WB Saunders Co, 1988, pp 447–466.
10. Nyburg PA, Laing FC, Filly RA, et al: Ultrasonographic differentiation of the gestational sac of early intrauterine pregnancy from the pseudogestational sac of ectopic pregnancy. Radiology 1983;146:755–759.
11. Filly RA: The first trimester, in Callen PW (ed): Ultrasonography in Obstetrics and Gynecology, ed 2. Philadelphia, WB Saunders Co, 1988, pp 19–46.
12. Nyburg DA, Laing FC, Filly RA: Threatened abortion: sonographic distinction of normal and abnormal gestation sacs. Radiology 1986; 158:397–400.
13. Fleischer AC, Entman SS, Kalemeris GE: Sonographic depiction of normal cyclical changes of endometrium. Ultrasound Med Biol 1986;12:271–277.
14. Fleischer AC, Mendelson EB, Bohm-Velez EM, et al: Transvaginal and transabdominal sonography of the endometrium. Semin Ultrasound CT MR 1988;9:81–101.

13

High-Frequency Transvaginal Scanning of the Uterus

ETAN Z. ZIMMER and ILAN E. TIMOR-TRITSCH

Transvaginal sonography (TVS) is a relatively new sonographic technique to image pelvic structures. The transvaginal approach permits the use of probes operating at higher frequencies, which can be placed close to the pelvic organs. The axial and lateral resolutions and the determination of image quality are improved compared to the resolution obtained by conventional 3.5-MHz abdominal transducers. The higher resolution is obtained somewhat at the expense of depth of field. However, most of the relevant anatomy for transvaginal imaging is within 7 to 9 cm of the vaginal fornices. It is sometimes possible to increase the transducer frequency to 7 MHz while attenuation and depth of field are still acceptable in order to image somewhat closer structures.[1]

Magnification does not alter the resolution of high-frequency probes, and it is important to use the largest possible magnification that still enables orientation as well as recognition of the organs or the pathology.

Orientation of transvaginal images is somewhat different from that of images obtained transabdominally. Scanning can be done using the usual sagittal, coronal, and axial views; however, using the vaginal probe an "organ-oriented" approach should be used, wherein the different planes of imaging are related to the organ itself. In this case the scanning planes are related to the longitudinal axis of the uterus itself. The endometrium, which is an echogenic structure, becomes a useful landmark to depict the long axis of the uterus.

The uterus may be anteverted or retroverted. Usually when the uterus is anteverted the probe slips naturally into the anterior fornix. If the uterus is retroverted the probe slips into the posterior fornix. This becomes important in identifying the position of the uterus.

This review summarizes the available literature as well as our experience in the transvaginal sonographic evaluation of the uterus.

The Cervix

The cervix should be scanned at the very beginning of the pelvic evaluation or as a part of scanning of the uterus. The examination is performed as the probe penetrates halfway or two thirds of the way into the vagina and about 2 to 3 cm before the tip of the probe reaches the cervix itself. Scanning of the cervix is done through both longitudinal and horizontal planes. On the longitudinal sagittal plane the length of the cervical canal, its shape, and its relation to the uterine cavity and to the endometrium can be followed. On the horizontal plane the thickness or any dilatation of the cervix can be imaged. Nabothian cysts usually appear as sonolucent cystic structures close to the cervical lip; their walls are extremely thin. Cystic structures closer to the endocervical canal may represent dilatation of the cervical glands. The mucus within the endocervical canal usually appears as an echogenic interface. This may become hypoechoic as the cervical canal becomes somewhat distended and fluid filled in mid-cycle, because the cervical mucus has a higher fluid content.

Brown et al.[2] were the first to use the vaginal approach to evaluate the cervix and lower uterine segment in pregnancy. Kushnir et al.[3] established nomograms of cervical length throughout pregnancy. According to their study, cervical length achieved its maximum at 20 to 25 weeks' gestation. An accurate measurement of cervical length in pregnancy may prove to be of importance because it could improve the risk assessment for preterm labor and preterm delivery. Andersen et al.[4] compared cervical length measurements performed by transabdominal sonography (TAS), by TVS, and by manual examination. They found that TVS measurements are an efficient method for determining risk of preterm delivery in the high-risk population as well as in the low-risk population.

The position of a cervical cerclage can be determined on the sagittal section by the echogenicity of the suture material. (Fig. 13-1). The bulging of the amniotic membranes into the dilated cervix can sometimes be seen in cases in which treatment failed and patients went into labor.

Examination of the cervix in a pregnant patient with vaginal bleeding is controversial. However, the careful use of TVS is possible because examination is performed while the tip of the probe is 2 to 3 cm away from the external os. Localization of the placenta and definition of placenta previa in particular is sometimes difficult and inaccurate with TAS, whereas with TVS it could be easily done.[5]

Although very rare, the possibility of cervical pregnancy should always be kept in mind while scanning a patient with vaginal bleeding or suspected ectopic pregnancy.

About 15% of all cancers in women occur in the uterine cervix. Accurate clinical staging becomes of importance when comparing results of treatment. Unfortunately, to date there are no studies in the English literature concerning the TVS features of this malignancy. It is our belief that TVS might be useful in early recognition and staging of cervical malignancy.

The Uterine Body

Scanning in the longitudinal and transverse uterine planes reveals the size of the uterus, its position, and myometrium, and the endometrial lining. The endometrial interface, which is typically echogenic, is a useful landmark to depict the long axis of the uterus.

In cases in which it is technically difficult to depict an enlarged uterus, the widest possible scanning angle should be used. Another possibility is to use the "split-screen" technique. This technique consists of displaying sequential images of sagittal sections of the corpus and cervical regions on split screens. By adding up the measurements on the two screens the total size of the uterus is obtained. However, if the uterus seems to be excessively enlarged TAS will enable a better examination.

The normal myometrium is usually uniformly echogenic and contains at its outer third the sonolucent arcuate vessels, which sometimes are prominent (Fig. 13-2).

A rather interesting and intriguing property of the myometrium is its ability to exhibit contractions propagating in different directions. These contractions of the inner myometrium were observed to increase in frequency toward the proliferative phase and to slightly decrease in the secretory phase. Lyons et al.[6] studied three groups of women for the presence, frequency, symmetry, and distance traveled by the contraction of the inner myometrium. Group 1 included fertile controls, group 2 oral contraceptive users and women who use intrauterine contraceptive devices, and group 3 infertile women. Significant differences were found in the above-mentioned variables between the three groups. Patients

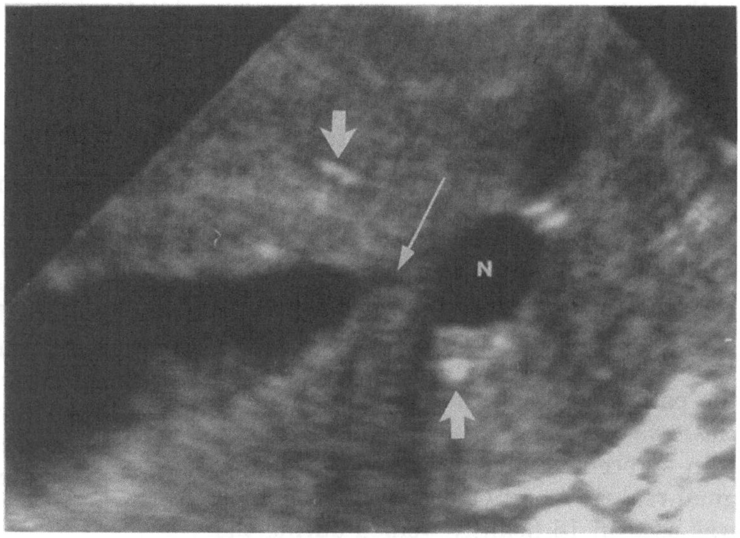

FIGURE 13-1. The sagittal section of the lower uterine segment and cervix. The small arrow point to the cross-section of the suture material. The internal os is imaged (arrow), and a nabothian cyst can be seen (N).

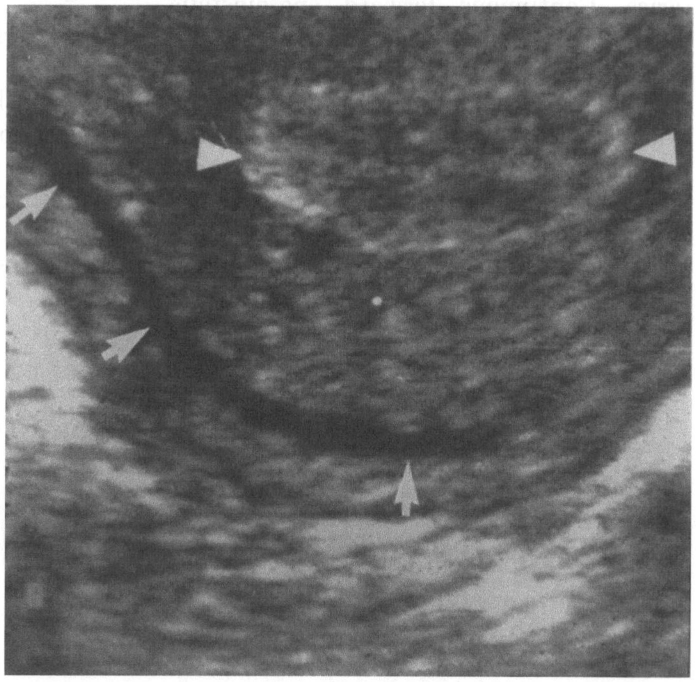

FIGURE 13-2. Transverse scan of the uterus revealing the prominent arcuate vein in the outer third of the myometrium (small arrows). The endometrium is marked by arrowheads.

with contraceptive devices had uncoordinated asymmetric contractions throughout the cycles. The infertile patients also had disturbed contraction patterns. We believe more studies are needed in order to arrive at more significant results as well as duplicating results of previous studies. If the results are determined to be accurate and reproducible, the observation of these contraction waves on a speeded up replay of the video recordings could prove important in understanding cases of infertility and maybe even the pathogenesis of some ectopic pregnancies.

In a recent study[7] a combined transvaginal real-time and pulsed Doppler method was used in 16 healthy women during the normal menstrual cycle to record flow velocity waveforms in the uterine and ovarian arteries. Continuous forward end-diastolic flow velocities were documented in 74% of the ovarian and 96.5% of the uterine artery flow velocity waveforms. In contrast to the ovarian artery, the pulsatility index of the uterine artery seems to be involved only marginally in the observed impedance changes during the luteal phase of the menstrual cycle. The advanced technique of transvaginal Doppler color flow mapping of bilateral uterine arteries and branches has also been reported.[8]

The Postmenopausal Uterus

In the postmenopause period the uterus becomes gradually but significantly smaller and usually looses its anteversion and anteflexion. It has a uniform echogenicity with an extremely thin endometrial lining that is hard to differentiate from the myometrium. In some patients a sonolucent area can be seen in the uterine cavity. It is possible that in these cases uterine secretions accumulated as a result of variable degrees of cervical stenosis (Fig. 13-3).

Congenital Anomalies

Transvaginal sonography can easily detect congenital anomalies such as septate, didelphys, and bicornuate uteri (Fig. 13-4). The examination usually relies on the shape of the uterus on the transverse scan, as well as the presence of a horizontal figure-8–shaped endometrium or two distinct and separate endometrial echoes, to indicate a bicornuate uterus.

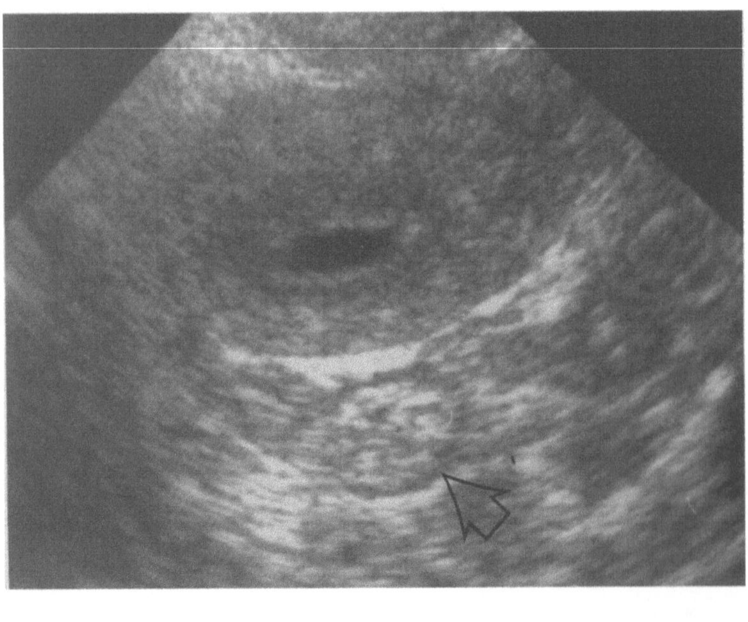

FIGURE 13-3. The uterus of a postmenopausal patient. The endometrium is thin and there is a small collection of fluid inside the uterine cavity. The cross-section of the rectum is also imaged (open arrow).

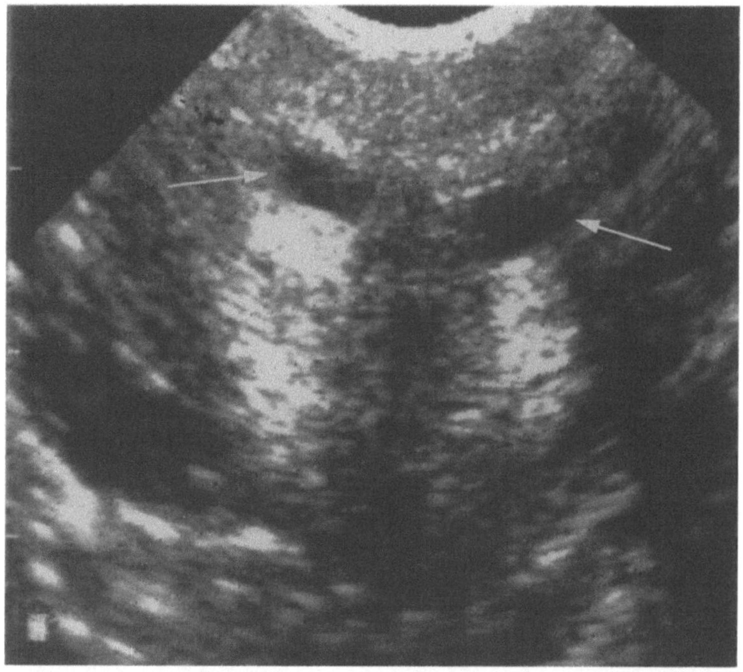

FIGURE 13-4. A patient with a double uterus. The double cervical canals (arrows) are seen on the same scanning plane.

Adenomyosis

At the present time there are no documented data of a reliable evaluation of this disorder by TVS.

Fibroids

Uterine fibroids are the most common tumors encountered in almost every busy ultrasound laboratory. Their location as well as their relationship to the serosa, the endometrium, the cervix, and the vessels can readily be determined (Fig. 13-5). Fibroids as small as 0.5 cm can be seen.

The sonographic appearance of fibroids consists of a variable echogenic pattern with bundles of "turbulent" structures interspersed with acoustic shadows generated distally to them by dense fibroid tissue. Central necrosis will result in "cavity formation" (Fig. 13-6). These cavities contain a low-level sonolucent material and can be potentially misdiagnosed as gestational sacs. Sometimes their size may be extremely large. In this case a differential diagnosis of an ovarian cyst may arise.

Localization of submucous fibroids and determination of the extent of distortion of the uterine cavity are possible. In cases of pelvic adhesions or when a differentiation between a uterine fibroid and an adnexal mass or ovary arises, the diagnosis can be done by using the "sliding organs sign." The maneuver is performed by pointing the transducer's tip, under vision, toward the organ or structure in question and applying a gentle pressure to repeatedly move it. One then appreciates its relative sliding motion compared to that of a "fixed" structure or the pelvic wall. If their movements are synchronous, they are probably part of the same organ or they are adhered to each other. If they move in opposite directions, they are (in the majority of cases) separate or nonadherent structures.

The Endometrium

The endometrium has a variety of appearances depending on its stage of development. Fleischer et al.[9] studied these changes in

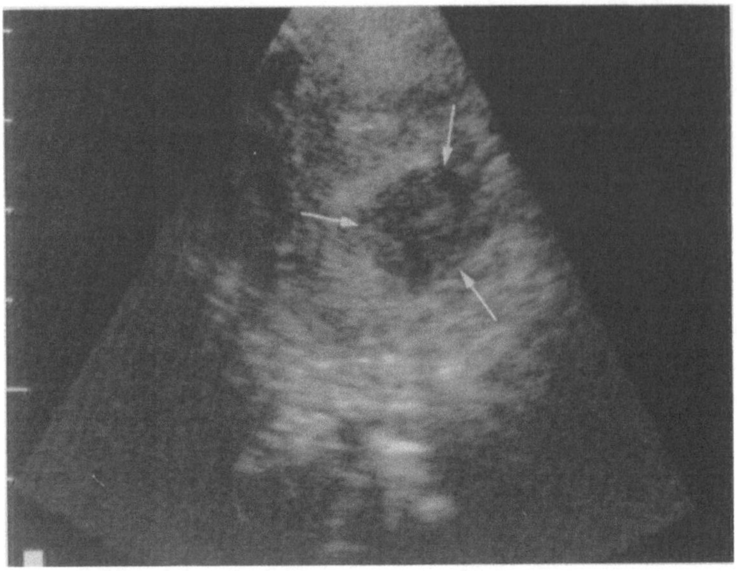

FIGURE 13-5. Longitudinal scan of the uterus. The small but well-defined intramural myoma is marked by arrows.

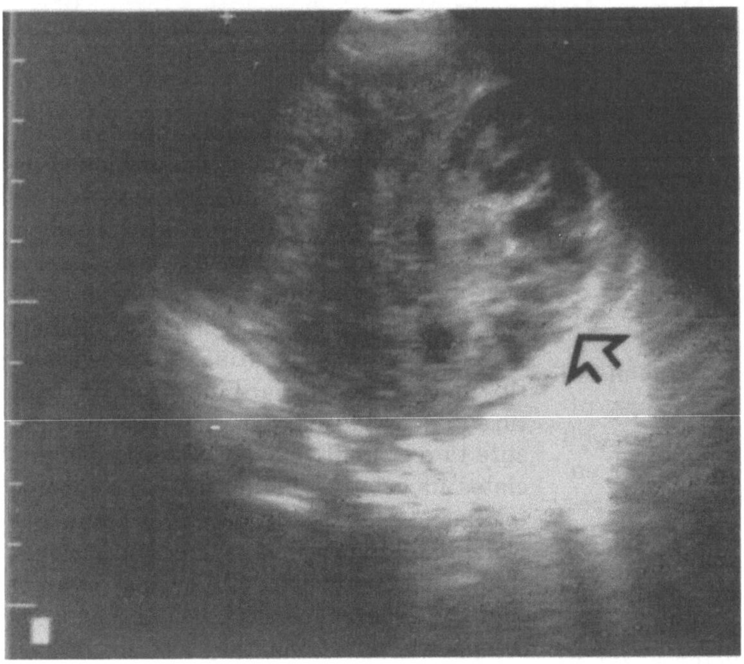

FIGURE 13-6. Degenerating myoma. A longitudinal scan showing the fundal region containing the lesion (open arrow). Sonolucent cavities and hyperechogenic strands of tissue are scattered within the myoma.

normal cycles. During the menstrual phase the endometrium appears as an echogenic, interrupted interface of 1 to 4 mm in anteroposterior width (Fig. 13-7). A hypoechoic outer layer can be identified surrounding the endometrium, corresponding to the inner layer of the myometrium. In the proliferative phase the endometrium thickens to 4 to 8 mm and has an isoechoic or slightly hyperechoic texture relative to the surrounding myometrium. In the late proliferative or periovulatory phase a multilayered endometrium can be seen. In the secretory phase the endometrium achieves a width of 8 to 16 mm and becomes echogenic, probably as a result of the increased mucus and glycogen within the glands.

Endometrial Response to Ovarian Hyperstimulation

Various sonographic parameters of the endometrium have been evaluated as possible predictors of implantation in in vitro fertilization. The endometrial response to ovarian hyperstimulation can be seen on sonographic examination as a quantitative change in thickness and a qualitative change in appearance or reflectivity.

Welker et al.,[10] who examined 190 in vitro fertilization patients, could not find a relationship between the thickness of the endometrium, hormonal values, and the likelihood of implantation. Even diameters as low as 6 mm coincided with pregnancies. However, three endometrial patterns were detected after hormonal stimulation:

Pattern A: homogeneously hyperechogenic
Pattern B: an outer hyperechogenic endometrial layer and an inner hypoechogenic layer
Pattern C: uterine cavity distended by a certain amount of fluid

Pattern B was the one that correlated in a positive fashion with subsequent implantation. The authors suggested that the inner layer in this pattern could be caused by the stretched glands and vessels.

Gonen et al.[11,12] performed two consecutive studies. In the first study[11] on 108 patients, the endometrium was significantly thicker on the day before ovum retrieval in the pregnant than in the nonpregnant

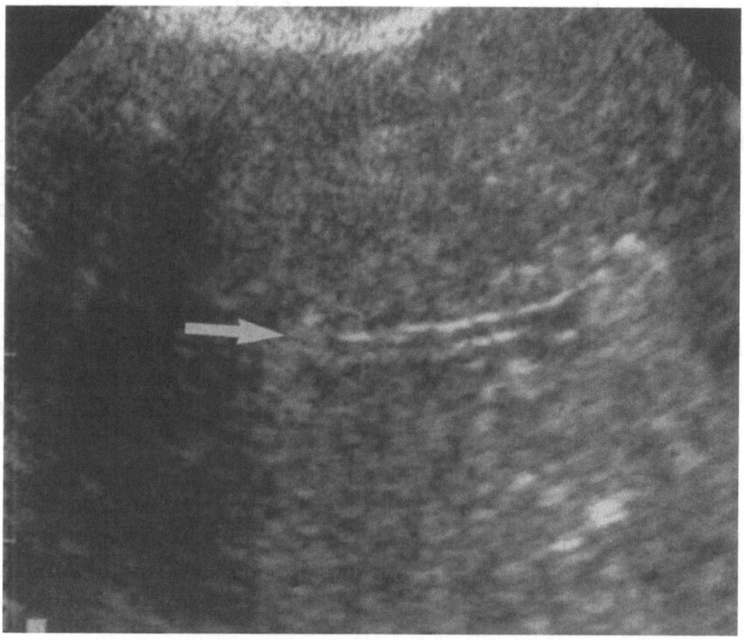

FIGURE 13-7. The thin endometrial line (arrow) of a woman during a day of her menstrual flow.

women, whereas the mean estradiol levels did not differ between the groups. In the second study[12] endometrial texture was graded by comparing reflectivity or gray scale appearance of the endometrium to that of the surrounding myometrium. The three types of endometrium have been defined:

Type A: an entirely homogeneous hyperechogenic endometrium
Type B: an intermediate isoechogenic pattern
Type C: a multilayered endometrium consisting of prominent outer and central hyperechogenic lines with inner hypoechogenic or sonolucent regions

Significantly more patients had multilayered endometrium (type C) in the group who conceived than in the group who did not.

Again, the thickness of the endometrium was found to be a reliable predictive parameter for implantation. No pregnancy occurred when endometrial thickness was less than 6 mm the day before ovum retrieval. Fleischer et al.[9] noted that patients who had a multilayered periovulatory endometrium achieved pregnancy more frequently than did the group that did not.

Sterzik et al.[13] studied the hemodynamic parameters in women in an in vitro fertilization program. Using vaginal Doppler flow measurements they found that the vascular resistance of uterine arteries on the day of follicular aspiration was significantly lower in the women who became pregnant when compared to women who did not become pregnant.

Benign Endometrial Disorders

Endometrial Polyps

Endometrial polyps appear as hyperechogenic structures that are surrounded by a very thin sonolucent fluid interface. If uterine bleeding also occurs, the outline of an intracavital structure becomes more obvious.

Asherman's Syndrome

Serpiginous irregularities or bridges between the two sides of the uterine wall can be seen in Asherman's syndrome.[14]

Pyometra

Pyometra is recognized by the homogenous low echogenic content of the cavity (Fig. 13-8).

Intrauterine Contraceptive Device

A high proportion of women who have an intrauterine device have various symptoms attributed to the device. It is possible to locate the device (Fig. 13-9) and to indicate whether the device is in the uterine cavity, has moved into the region of the lower uterus and upper cervix, or is embedded in the myometrium.[15]

Malignant Endometrial Disorders

A thick endometrium in postmenopausal patients is a cause for concern (Fig. 13-10). Osmers et al.[16] performed vaginosonographic measurements of the endometrium in 155 patients without abnormal gynecologic findings and free of any clinical symptoms. An endometrial thickness of equal to or greater than 4 mm was clarified histologically by means of curettage. Endometrial carcinomas were found in 7 cases, cervical carcinoma in 1, endometrial polyps in 10, and glandular cystic hyperplasia in 3 cases.

Goldstein et al.[17] performed a prospective study on 30 women with postmenopausal bleeding. Of these 18 were on hormone replacement therapy and 12 were not. In two patients no endometrial echo was imaged because of associated myomas. Eleven patients demonstrated a thin linear endometrial echo, and the maximal endometrial thickness was 5 mm or less. In all of these patients the endometrial sampling yielded amounts that were insufficient for diagnosis or contained "scant cellular material." Seventeen patients had an echogenic endometrium. In

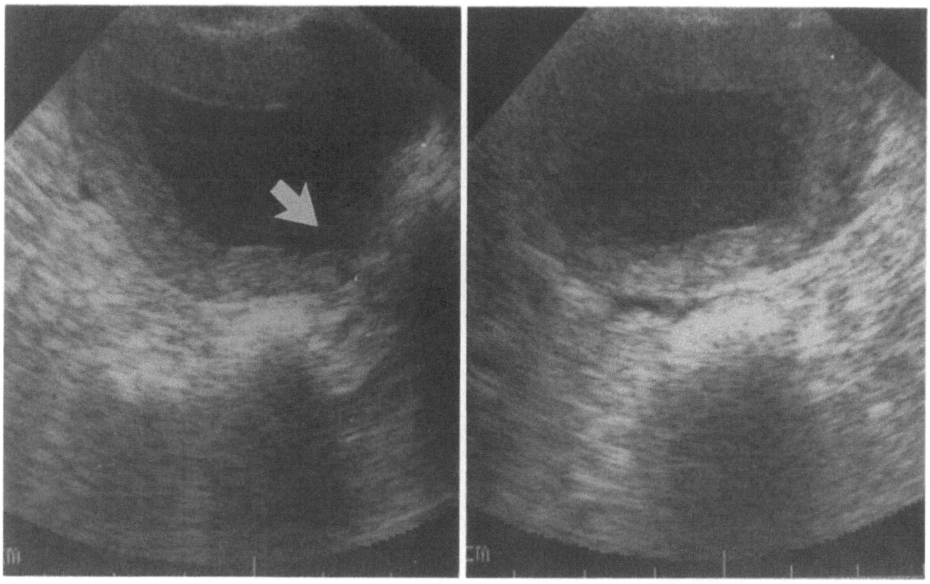

FIGURE 13-8. Pyometra. The uterine cavity is distended and filled with fluid of low-level echogenicity. The arrow on the left-side scan points to an area of myometrial invasion by tumor.

some cases the endometrium was homogeneous and symmetrically enlarged, and in others it was heterogeneous and irregular. The only consistent finding in this second group was that of endometrial thickness of 6 mm or more. In two of these patients there was not enough tissue for diagnosis at the sampling. The pathologic examination in the other cases showed six proliferative, three secretory, and three hyperplastic types of endometrium, two interauterine polyps, and one case of malignancy. The conclusion of

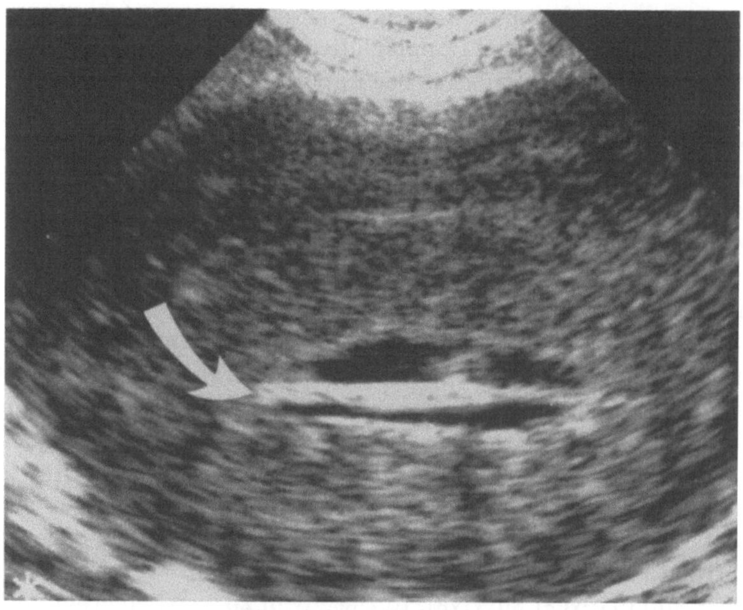

FIGURE 13-9. The intrauterine contraceptive device presents as a well-defined hyperechogenic structure inside the uterine cavity (arrow).

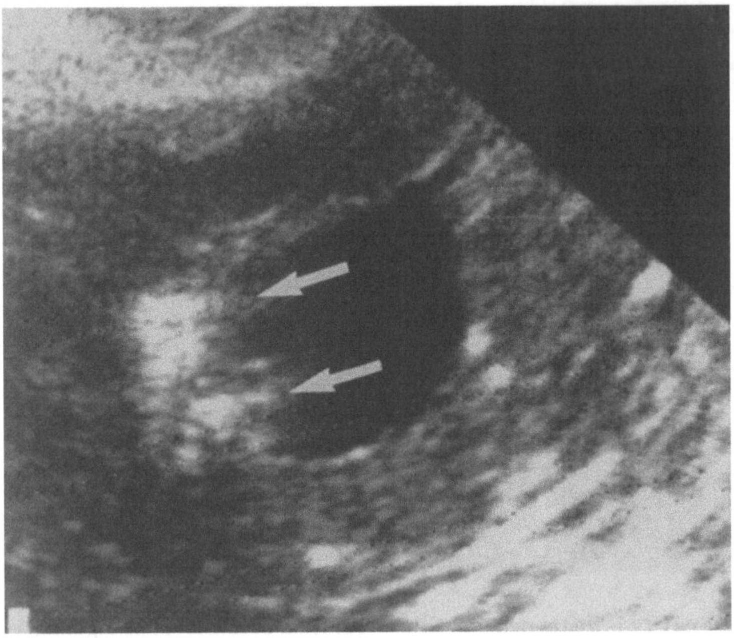

FIGURE 13-10. Papillary endometrial carcinoma. The longitudinal scan of the uterus shows fluid in the uterine cavity with a cluster of echogenic areas (arrows) projecting into both the cavity and the myometrium.

this study was that in postmenopausal women with an endometrial thickness of 6 mm or more a histologic diagnosis is imperative. However, at present it is impossible to differentiate benign from malignant endometrial disorders based on their transvaginal sonographic appearance.

Uterine morphology and blood flow in postmenopausal women have been assessed by transvaginal pulsed Doppler ultrasonography.[18] The technique has been applied to 12 patients with confirmed endometrial cancer, 20 apparently healthy women, and 6 women with postmenopausal bleeding with normal histology on endometrial biopsy. All women with endometrial cancer had a significantly thicker endometrium than either set of control subjects. In addition all patients with uterine cancer were found to have a markedly lower pulsatility index, suggesting decreased impedance of blood flow through the uterine arteries.

An accurate preoperative assessment of myometrial invasion, cervical involvement, and adnexal spread is of importance since it correlates to the risk of lymph node metastasis and to 5-year survival. In a study of 25 patients with histologically proved endometrial carcinoma Fleischer and Gordon[19] correctly identified the extent of myometrial invasion in 21 patients. In two cases an overestimation occurred. Both were exophytic tumors that thinned the myometrium. Two underestimations of invasion occurred in tumors with microscopic invasion. A good correlation between ultrasound and myometrial invasion was reported by Cruickshank et al.[20] in five patients with endometrial neoplasia [International Federation of Gynecology and Obstetrics (FIGO) stage I]. Hata et al.[21] were able to delineate cervical invasion of an endometrial cancer.

Pregnancy

One of the most valuable applications of TVS is in early identification of a normal or abnormal pregnancy. On the average TVS can detect embryonic or fetal structures 1 to 3 weeks earlier than TAS.

Normal Uterine Pregnancy

When pregnancy is achieved there is a thickening of the echogenic postovulatory endometrium and an increase in the prominence of the arcuate vessels in the myometrium. Timor-Tritsch et al.[22] and Warren et al.[23] documented the early stages of embryonic development. At 4 weeks a gestational sac can be seen. It consists of a central sonolucent chorionic sac that is surrounded by an echogenic rim.

At about 5 menstrual weeks the yolk sac is visible. Lacunar structures with visible blood flow can be seen within the endometrium, predominantly on one side of the gestational sac. These lacunar structures are probably the forerunners of the placental circulation. At around 6 weeks the fetal crown-rump length and heartbeat were identified. Other milestones in embryonic development are as follows. The central nervous system can be seen as early as the seventh week with the appearance of the unpartitioned single ventricle. The spine can be reliably seen from week 9. The lower limb buds appear at 7 weeks and precede the appearance of the upper limb buds. At 10 to 11 weeks digits are identified. The physiologic process of the midgut herniation can be followed starting at 8 weeks until its complete disappearance during the 12th week.

Thaler et al.[24] studied the changes in uterine blood flow during normal human pregnancy. Using a transvaginal duplex Doppler ultrasonography system they were able to show an increase in the diameter of the ascending uterine artery and a decrease in the resistance to flow throughout pregnancy. These changes were visible as early as 6 to 8 menstrual weeks.

Abnormal Early Pregnancy

The diagnosis of very early abnormal pregnancy is at times difficult. Keeping in mind the temporal appearance of embryonic and extraembryonic structures, one could evaluate the presence or absence of an abnormal pregnancy if the correct dating of pregnancy itself is known. If any doubt concerning the dating exists, serial ultrasound scans are appropriate for clinical follow-up. It is agreed by almost all that a 1-cm or greater chorionic sac without an apparent yolk sac is a sign of abnormal gestation. The presence or absence of the fetal heartbeat may also indicate a normally or abnormally developing embryo since a well-dated 6½-week embryo should have an apparent heartbeat.[22]

Previously, "blighted ovum" and missed abortion were considered as separate entities. Actually they may represent the same process occurring at different times in the first trimester. Pirrone et al.[25] were able to demonstrate embryonic or extraembryonic structures in 20 out of 22 cases previously judged by abdominal sonography to be blighted ova. The mechanism for this presumed "embryonic resorption" is probably also involved in the "vanishing twin" phenomenon in which a spontaneous reduction in the number of chorionic sacs or embryos is noted.

Other signs of abnormal pregnancy are an irregular shape of the decidua and subchorionic bleeding.[1] This bleeding is identified as a sonolucent collection outside the gestational sac.

Ectopic Pregnancies

The tubal changes and the sonographic findings of the ectopic embryo or fetus are not the subject of this chapter. We would like to describe the pertinent intrauterine findings of this entity. The endometrium responds to the hormonal stimulation of the extrauterine gestation and appears as a thick and echogenic structure. Hemorrhage and necrosis may occur later in the decidua. This blood collection is depicted as an echo-free area and is usually centrally placed in the uterine cavity, assuming a longitudinal shape.[26] This finding corresponds to the "pseudogestational sac" seen with TAS. It differs from the normal gestational sac, which is asymmetrically positioned in the decidua somewhat distant from the visible endometrial canal.

Hydatidiform Mole

The sonographic picture of a hydatidiform mole obtained by TVS shows the multitude

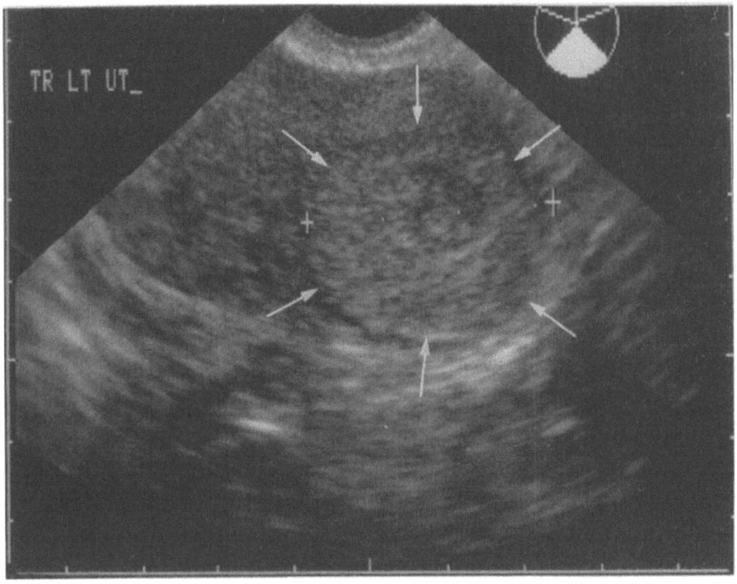

FIGURE 13-11. Choriocarcinoma. The longitudinal image of the uterus reveals a round structure surrounded by a sonolucent line of demarcation (arrows). The lesion contains mixed echogenicity as well as tiny sonolucent structures. Total myometrial invasion is evident.

of small sonolucent structures having various sizes with high fidelity. If suspected, choriocarcinoma can also be identified (Fig. 13-11). The possibility of invasion of the myometrium in this disease is a cause of great concern. Schneider et al.[27] were able to demonstrate an intramyometrial nodule of invasion and to monitor its disappearance during chemotherapy.

Summary

Transvaginal scanning opened important avenues to better imaging of the uterus and, hence, to improved patient care through a more accurate and faster diagnostic process. The ultrasonographic evaluation of the uterus is an eloquent example of how the two different scanning routes (TVS and TAS) obtain different and important data specific to each of the two modalities, to be incorporated into a truly combined report rendering an improved diagnosis. It is our belief that the main advantage of ultrasonography will be in patients with uterine malignancy. We anticipate that ultrasonography will be used both to evaluate the extent of spread of tumors and to detect the disease in asymptomatic women.

References

1. Timor-Tritsch IE, Rottem S, Thaler I: Review of transvaginal ultrasonography: a description with clinical application. Ultrasound Q 1988;6:1–34.
2. Brown JE, Thieme GA, Shah DM, et al: Transabdominal and transvaginal endosonography: evaluation of the cervix and lower uterine segment in pregnancy. Am J Obstet Gynecol 1986;155:721–726.
3. Kushnir O, Vigil DA, Izquierdo L, et al: Vaginal ultrasonographic assessment of cervical length changes during normal pregnancy. Am J Obstet Gynecol 1990;162:991–993.
4. Andersen FH, Nugent CE, Wanty SD, et al: Prediction of preterm delivery by ultrasonographic measurements of cervical length. Am J Obstet Gynecol 1990;163:859–867.
5. Farine D, Peisner DB, Timor-Tritsch IE: Placenta previa—is the traditional diagnostic approach satisfactory? J Clin Ultrasound 1990;18:328–330.
6. Lyons EA, Ballard G, Zheng X-H, et al: Contraction of the inner myometrium in fertility (abstract). Radiology 1989;173:114.
7. Scholtes MCV, Wladimiroff JW, van Rijen

HJM, et al: Uterine and ovarian flow velocity waveforms in the normal menstrual cycle: a transvaginal Doppler study. Fertil Steril 1989;52:981–985.
8. Hata T, Hata K, Daisaku S, et al: Transvaginal Doppler color flow mapping. Gynecol Obstet Invest 1989;27:217–218.
9. Fleischer AC, Gordon AN, Entman SS, et al: Transvaginal sonography (TVS) of the endometrium: current and potential clinical applications. Crit Rev Diagn Imaging 1990; 30:85–110.
10. Welker BG, Gembruch U, Deidrich K, et al: Transvaginal sonography of the endometrium during ovum pickup in stimulated cycles for in vitro fertilization. J Ultrasound Med 1989; 5:549–553.
11. Gonen Y, Casper RF, Jacobson W, et al: Endometrial thickness and growth during ovarian stimulation: a possible predictor of implantation in in vitro fertilization. Fertil Steril 1989;52:446–450.
12. Gonen Y, Casper RF: Prediction of implantation by the sonographic appearance of the endometrium during controlled ovarian stimulation for in vitro fertilization (IVF). J In Vitro Fert Embryo Transf 1990;7:146–152.
13. Sterzik K, Grab D, Sasse V, et al: Doppler sonographic findings and their correlation with implantation in an in vitro fertilization program. Fertil Steril 1989;52:825–828.
14. Mendelson EB, Bohm-Velez M, Joseph N, et al: Endometrial abnormalities: evaluation with transvaginal sonography. Am J Roentgenol 1988;150:139–142.
15. Parsons J, McEwan JA: Transvaginal ultrasonography to ascertain the position of an IUD. N Engl J Med 1989;321:546–547.
16. Osmers R, Volksen M, Roth W, et al: Vaginosonographische messungen des postmenopausale endometriums zur fruherkennung des endometriumkarzinoms. Geburtshilfe Frauenheilkd 1989;49:262–265.
17. Goldstein SR, Nachtigall M, Snyder JR, et al: Endometrial assessment by vaginal ultrasonography before endometrial sampling in patients with postmenopausal bleeding. Am J Obstet Gynecol 1990;163:119–123.
18. Bourne TH, Campbell S, Steer C, et al: Pulsed Doppler transvaginal ultrasonography—a novel technique for the early diagnosis of uterine cancer (abstract). J Ultrasound Med 1990;9:540.
19. Fleischer AC, Gordon AN: Assessment of myometrial invasion by transvaginal sonography (abstract). Radiology 1989;173:114.
20. Cruickshank DJ, Randall JM, Miller JD: Vaginal endosonography in endometrial cancer. Lancet 1989;1:445–446.
21. Hata K, Hata T, Aoki S, et al: Efficiency of transvaginal scanning in obstetric and gynecologic fields. Gynecol Obstet Invest 1990; 29:41–46.
22. Timor-Tritsch IE, Farine D, Rosen MG: A close look at early embryonic development with the high frequency transvaginal transducer. Am J Obstet Gynecol 1988;159:676–681.
23. Warren WB, Timor-Tritsch I, Peisner D, et al: Dating the early pregnancy by sequential appearance of embronic structures. Am J Obstet Gynecol 1989;161:747–753.
24. Thaler I, Manor D, Itskovitz J, et al: Changes in uterine blood flow during human pregnancy. Am J Obstet Gynecol 1990;162:121–125.
25. Pirrone E, Monteagudo A, Timor-Tritsch IE: Does blighted ovum really exist? (abstract). J Ultrasound Med 1990;91:S41.
26. Rempen A: Vaginal sonography in ectopic pregnancy. A prospective evaluation. J Ultrasound Med 1988;7:381–387.
27. Schneider DF, Bukovsky I, Wienraub Z, et al: Transvaginal ultrasound diagnosis and treatment follow up of invasive gestational trophoblastic disease. J Clin Ultrasound 1990; 18:110–113.

14

Hysterosalpingography

CLAUDE BLOCH and SAUL J. DAN

Indications

There has been a marked increase in the demand for accurate evaluation of the causes of sterility. This has been brought about by the increased incidence of pelvic inflammatory disease, and by the recent advent of new medical and surgical techniques for the treatment of sterility. Hysterosalpingography plays a pivotal role in determining the causes of sterility, and thus is being ordered with increased frequency. It is also used in the differential diagnosis of abnormal uterine bleeding, and when congenital anomalies of the genitourinary tract are suspected. The status of the postoperative uterus and the success of tubal surgery can be assessed by hysterosalpingography. It is also of use in the diagnosis of Asherman's syndrome and the evaluation of uterine abnormalities in patients who have been exposed to maternal diethylstilbestrol (DES).

Technique

When performed with care, hysterosalpingography is well tolerated. Premedication is almost never required if the procedure is fully explained to the patient beforehand. In some cases, tranquilizers and/or analgesics can be prescribed to allay apprehensions and to avoid pain. The examination is best performed after menstruation has ceased and before ovulation has occurred (days 7 to 14 of the cycle). After emptying her bladder, the patient is placed on the fluoroscopy table in the lithotomy position. Using sterile technique, the cervix is visualized and cleansed with Betadine before the instruments are inserted. The patient is then positioned under the image-amplified fluoroscope, her legs extended and slightly apart to allow the instruments to remain in place. A preliminary spot film of the pelvis is taken prior to the fractionated injection of contrast medium with fluoroscopic monitoring. Other spot-films are obtained in the anteroposterior and both oblique projections as required.

There are numerous methods used to fill the uterine cavity and the fallopian tubes by preventing reflux of the contrast agent into the vagina. One simple and reliable technique uses a small rubber balloon to occlude the endocervical canal. Several different catheters are available in various sizes, and some require the additional use of a tenaculum on the anterior lip of the cervix. The tenaculum placement may cause slight momentary pain, but this allows traction upon the cervix to straighten an otherwise flexed uterus. Other common techniques include a vacuum suction device placed against the cervix, and a metal cannula with a rubber or plastic tip that is held firmly against the external os with the aid of a tenaculum.

It is important to instill the liquid gently and slowly, using an amount sufficient to fill but not overdistend the uterine cavity and to outline the fallopian tubes. In this manner, the patient will not experience much discomfort and a good-quality hysterosalpingogram will be obtained. As a rule, one needs only 8 to 10 ml of contrast medium to fill the uterine cavity and the tubes, but on occasion an enlarged cavity or an imperfect seal requires more contrast material to fill adequately. A delayed film is important to evaluate the spread of the contrast medium within the pelvic cavity.

The modern contrast media are almost all water soluble. In most instances one can employ the same agents that are used for intravenous pyelography. The new low-osmolar contrast materials may decrease the pain that occurs with peritoneal spillage. Only rarely is oil-soluble contrast medium needed. In most cases, this is only necessary in patients who have undergone tuboplastic or reconstructive procedures, when a 24-hour delayed film may be important to properly evaluate the distribution of the agent outside the tubes. Contrary to previous belief, the posthysterosalpingogram fertility rate and the patient's discomfort during and after hysterosalpingography are similar using water-soluble or oily contrast medium.[1]

Following the removal of the instruments, it is important to allow the patient to remain on the x-ray table for about 10 minutes to avoid any vasovagal symptoms, such as diz-

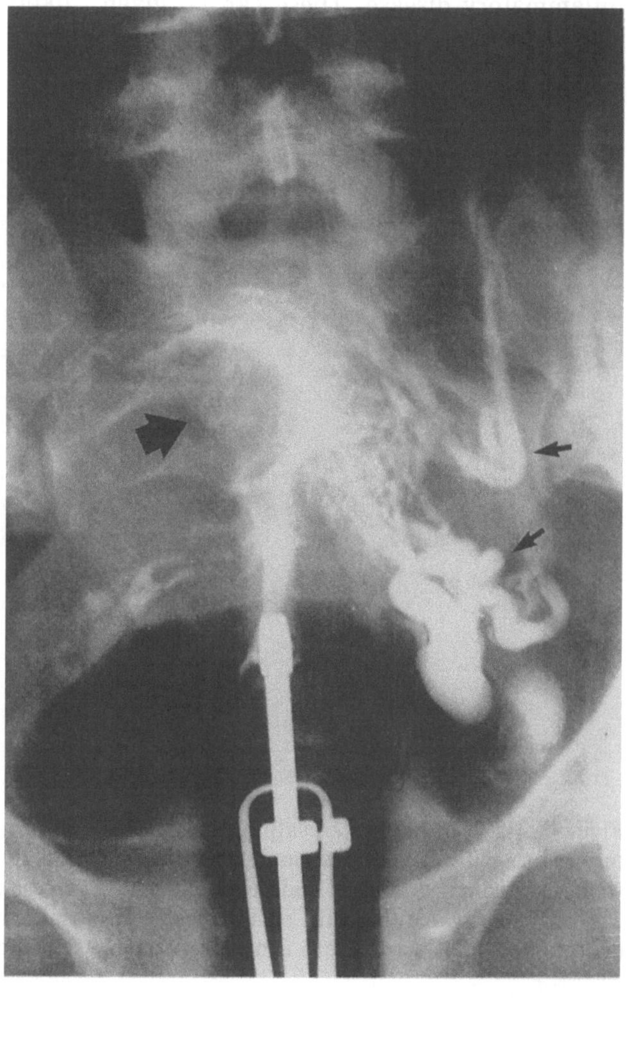

FIGURE 14-1. Intravasation. Anteroposterior film shows the uterine cavity to be flattened and distorted by a large filling defect (large arrow) projecting from the right lateral wall; this represents a fibroid that is partially intramural and partially submucosal in position. There is extensive intramural intravasation of contrast medium and accompanying filling of large pelvic veins (small arrows).

ziness, and also to wait for any pelvic pain to subside.

Complications

One of the most serious complications is the development of acute pelvic inflammatory disease. According to the literature, this occurs in 0.3 to 3% of cases.[2] In our personal experience with over 15,000 cases (C.B.), it is less than 0.2%, and with rare exceptions is only seen in patients with chronic pelvic inflammatory disease and with dilated fallopian tubes. When these conditions are suspected, broad-spectrum antibiotics should be used to prevent an exacerbation of the pelvic inflammatory disease. These can be given either 1 day before or just after the hysterosalpingography is performed.

Procedural pain is the most frequent side effect of hysterosalpingography, but it is transient and rarely of major proportion. Intravasation of contrast medium into the pelvic venous circulation is seen only when excessive pressure is used to instill the contrast material, usually in patients with closed fallopian tubes (Fig. 14-1). If an oil-soluble contrast medium is used, side effects of embolization are potentially serious. Allergic reaction to the contrast medium is almost never a problem, but given a history of severe prior reaction to intravenous contrast media, proper precautions must be taken. Uterine and tubal perforation and posthysterosalpingography hemorrhage are very rare reported complications, and are unlikely to occur if proper technique is used.[3]

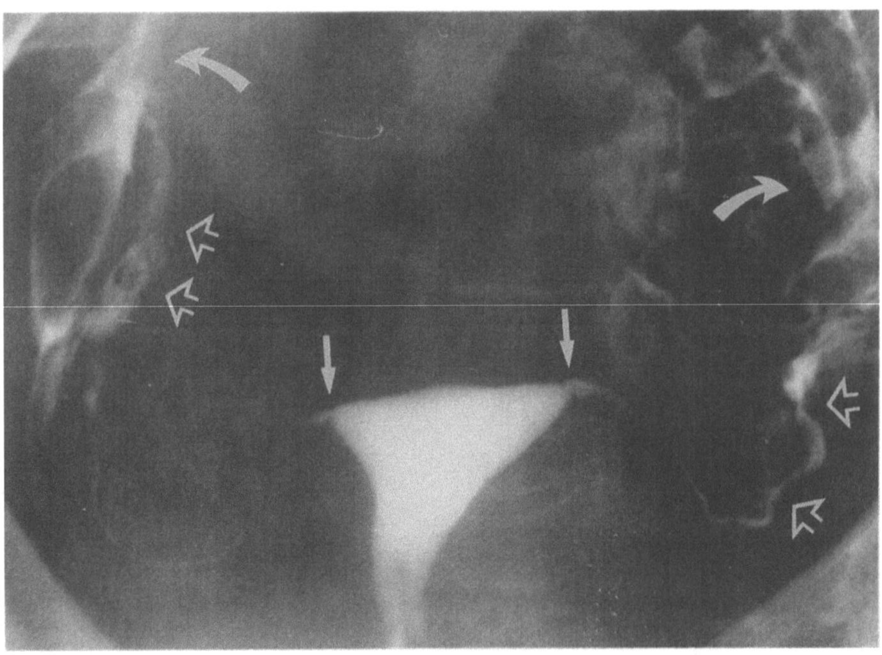

FIGURE 14-2. Normal hysterosalpingogram. Anteroposterior film shows a normal cervical canal with serrated appearance of endocervical glands. The cavity is triangular with smooth walls. Indentations (small arrows) are a normal variant of cornual anatomy. Proximal portions of the fallopian tubes are thin and they gradually widen to the ampullary portion (open arrows). Contrast spills into the peritoneal cavity at the fimbriated ends (curved arrows).

FIGURE 14-3. Leiomyoma. (A) Anteroposterior film demonstrates a round, smoothly marginated filling defect enlarging and distorting the uterine cavity. (B) Anteroposterior film shows a large filling defect arising from the right lateral uterine wall, causing marked distortion, flattening, and enlargement of the cavity. (*continued*)

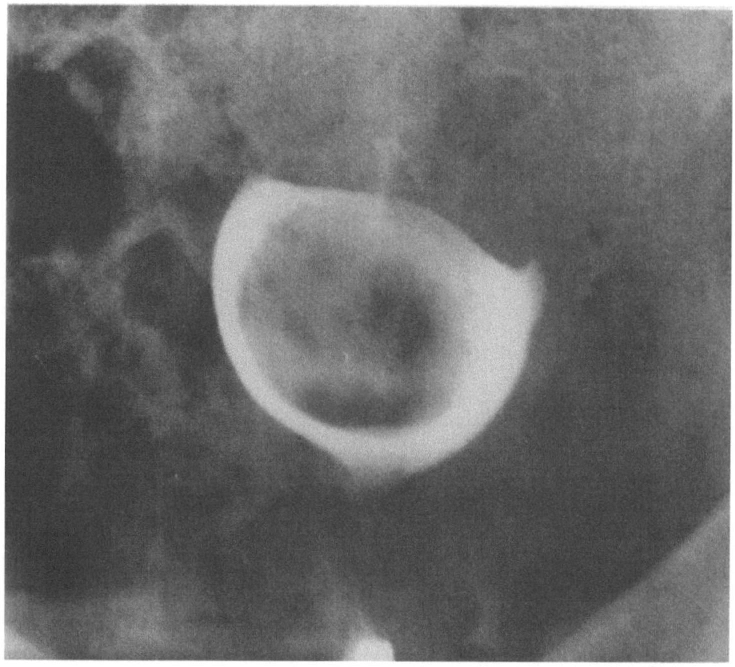

A

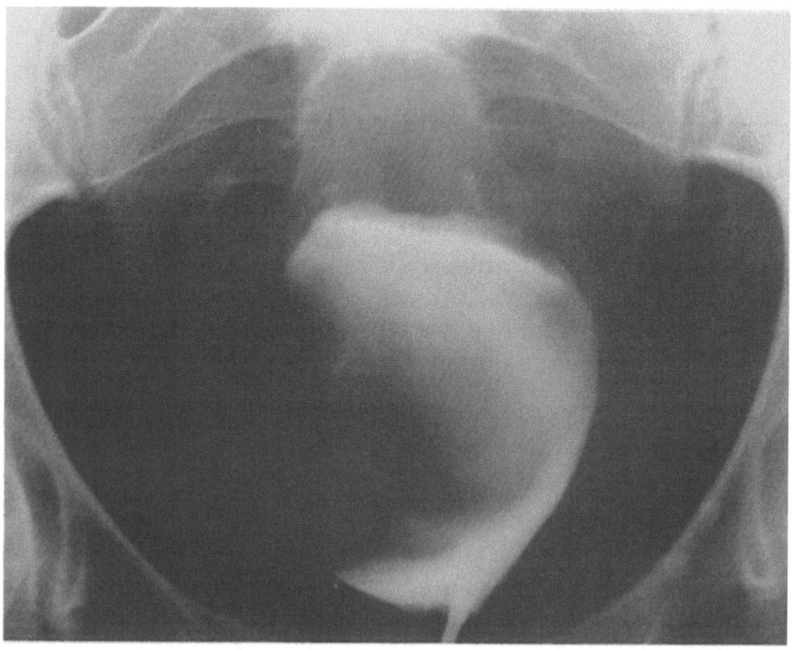

B

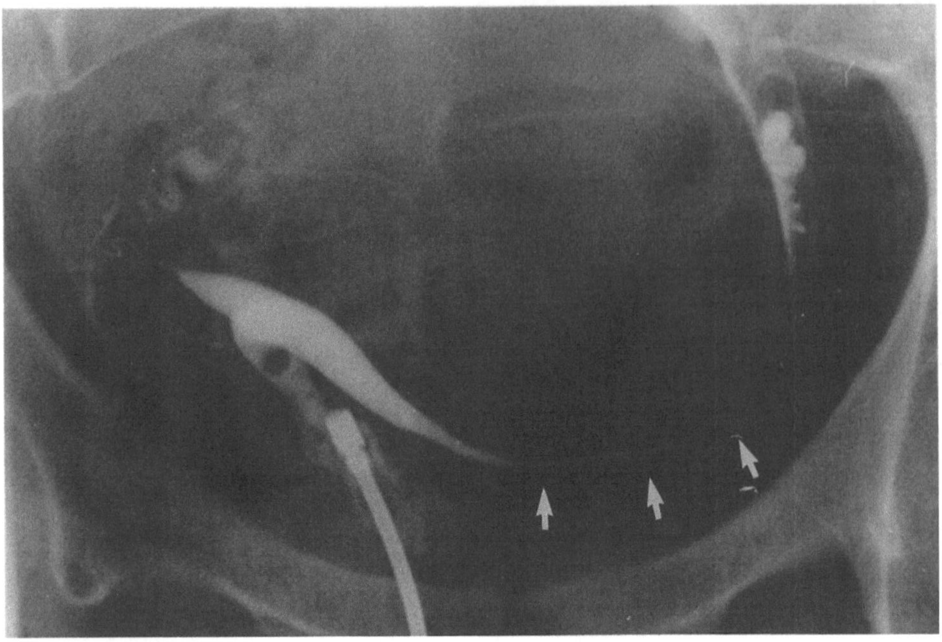

C

FIGURE 14-3. (*continued*) (C) Oblique film shows the uterine cavity to be displaced to the right and posteriorly by a large intraligamentous fibroid. Left isthmic fallopian tube is stretched and deviated downward (arrows).

Contraindications

The presence of clinical pelvic inflammatory disease is a definite contraindication to hysterosalpingography. Active uterine bleeding and intrauterine pregnancy are other conditions that preclude performing hysterosalpingography.

Normal Anatomy

The uterine cavity is triangular in shape, with smooth, regular, symmetrical borders. In nullipara, the endocervical canal is relatively long with serrated borders, termed "arborization" (see Fig. 14-7). The fallopian tubes arise from the uterine cornua and extend laterally toward the pelvic walls. The interstitial segment lies within the myometrium and is only 1 to 2 cm long. The isthmic portion is thin and tortuous, ending distally in a flaired fashion at the ampulla, which contains longitudinal mucosal folds. From the fimbriated ends, the contrast medium then spills into the peritoneal cavity and outlines the pelvic bowel loops (Fig. 14-2; see also Figs. 14-5 and 14-10).

Hysterosalpingographic Findings

Acquired Uterine Conditions

Leiomyomas

Leiomyomas are either single or multiple and are benign growths arising from the myometrium. When they are submucous in position, they present as intraluminal filling defects, usually with a concomitant contour abnormality. Small submucous fibroids cannot be differentiated from large endometrial polyps, although typically polyps appear to have sharper and better etched outlines than submucous fibroids (Fig. 14-3A). Large submucous fibroids cause marked enlargement, distortion, and contour abnormalities of the uterine cavity (Figs. 14-1 and 14-3B). Intraligamentous fibroids may cause marked extrinsic pressure and displacement of the uterine cavity and the fallopian tubes (Fig. 14-3C). Some leiomyomas contain mottled, coarse calcifications, which are recognizable on the preliminary film of the pelvis.

Polyps

Endometrial polyps are pedunculated or sessile benign growths projecting as small, well-circumscribed, rounded filling defects within the uterine cavity. They may be single or multiple and as a rule do not distort the outline of the uterine cavity (Fig. 14-4A; see also Fig. 14-9). In rare cases, one has to differentiate large polyps from retained products of conception.

In endometrial hyperplasia, the cavity is usually slightly enlarged. Its borders are irregular and serrated with wavy margins. This appearance is due to numerous small intraluminal filling defects related to the abnormally prominent mucosal folds (Fig. 14-4B).

Adenomyosis

Often this diagnosis cannot be made by hysterosalpingography. However, the finding of an enlarged uterus with an irregular spiculated border may suggest it, especially if one can outline contrast medium within tiny

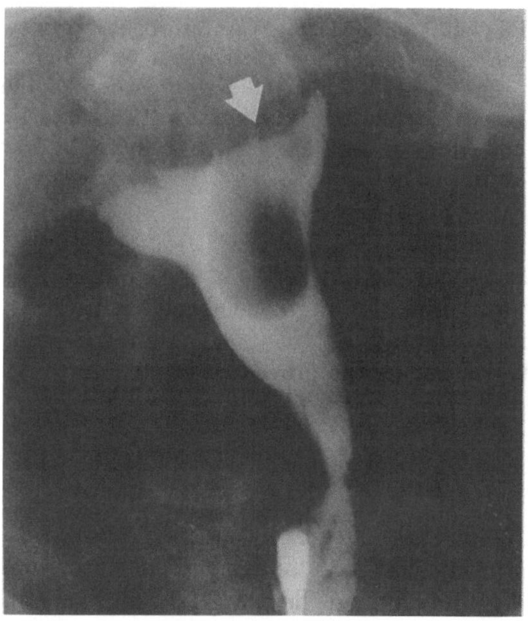

A

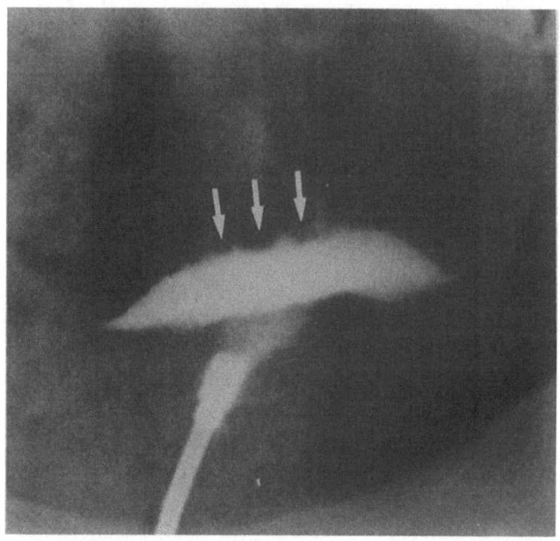

B

FIGURE 14-4. (A) Endometrial polyp. Anteroposterior view reveals a large sessile intraluminal filling defect arising from the uterine fundus (arrow). It has smooth margins and does not distort the uterine cavity. (B) Endometrial hyperplasia. Oblique film demonstrates a slightly enlarged uterine cavity with an irregular wavy contour. The contrast medium outlines innumerable nodular defects along the endometrial lining (arrows).

lollipop-like outpouchings within the myometrium (Fig. 14-5).

Synechiae

Intrauterine adhesions usually occur after dilation and curettage or other endometrial trauma. When small, they cause linear, angular or rectangular intraluminal filling defects unassociated with any other distortion of the uterine cavity, and they are usually asymptomatic (Fig. 14-6). However, when the synechiae are extensive they can obliterate the entire uterine cavity (Asherman's syndrome) and cause amenorrhea. In these cases it is impossible to instill contrast medium into the uterine cavity past the internal os (Fig. 14-7). One has to be certain that the endocervical canal is properly occluded so as to avoid any reflux of the contrast material into the vagina. Often this requires repositioning of the instruments several times during the procedure.

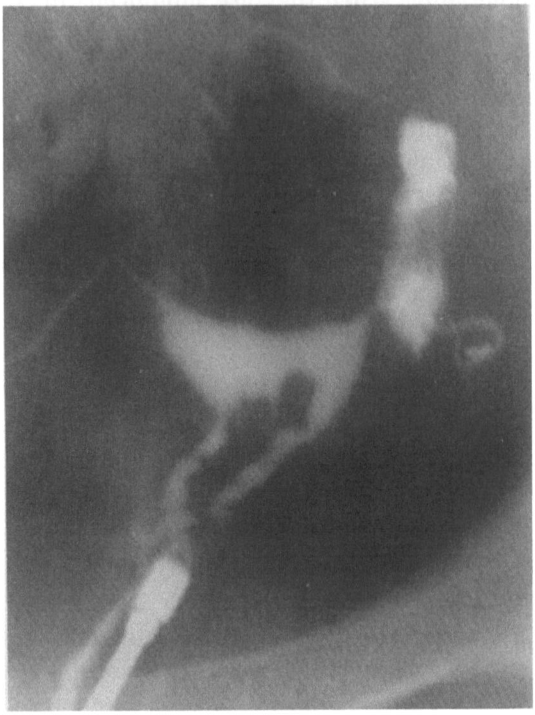

FIGURE 14-6. Synechia. Anteroposterior view shows an angular and rectangular intraluminal filling defect within the lower portion of the uterus, encroaching on the cavity. The anterior and posterior walls are partially adhesed by scar tissue.

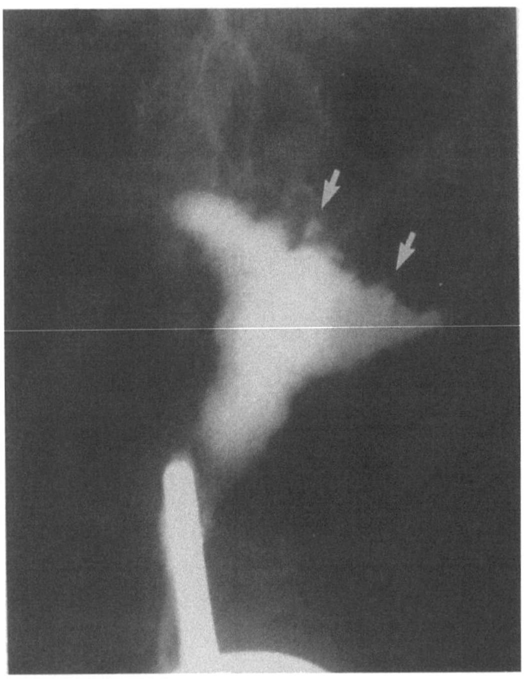

FIGURE 14-5. Adenomyosis. Anteroposterior film demonstrates an enlarged uterine cavity with a spiculated irregular border caused by contrast medium entering glands embedded within the myometrium (arrows).

Intrauterine Device

Hysterosalpingography is the preferred diagnostic test to establish the exact position of an intrauterine device (IUD). This is often requested by the clinician when the IUD string is no longer visible, and when several unsuccessful attempts have been made to remove the IUD by instrumentation. A plain film of the abdomen must be taken to identify the presence of a radiopaque IUD. After instillation of the contrast agent, films are

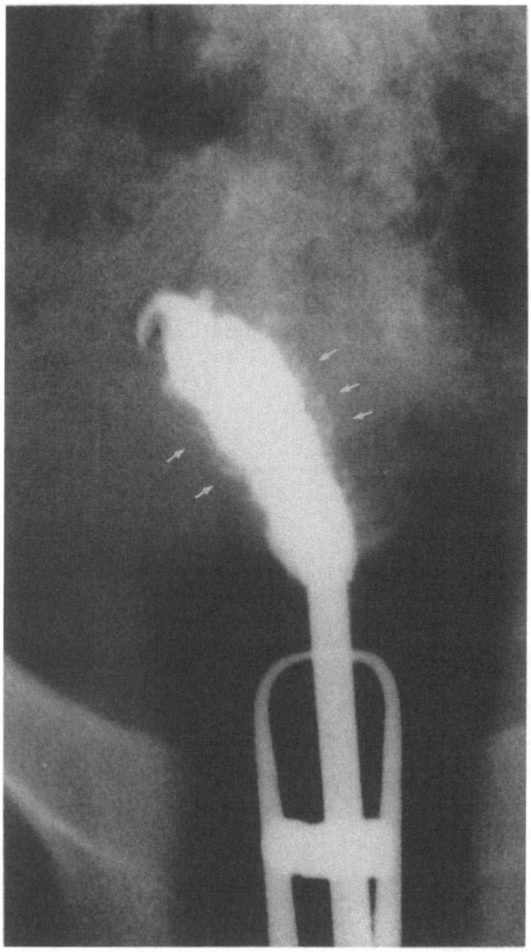

FIGURE 14-7. Asherman's syndrome. Anteroposterior view shows good filling of the endocervical canal, but no contrast material enters the uterine cavity, which has been totally obliterated by scarring. The normal endocervical gland filling causes a serrated border, termed "aborization" (arrows).

taken at numerous angles, including the lateral projection, to rule out a partial or complete perforation (Fig. 14-8).

Cesarean Section Deformity

A typical angular, diverticulum-like outpouching can be seen on the anterior wall of the lower uterine cavity after cesarean section. This can vary in appearance from a thin linear depression to a deep outpouching (Fig. 14-9).

Incompetent Cervix

The evaluation of the uterine cavity by hysterosalpingography in patients with a history of repeated miscarriages is controversial and inconsistent. The external os must be occluded without the use of an endocervical instrument. If the diameter of the isthmus is greater than 1 cm, cervical incompetence is suggested (Fig. 14-10).

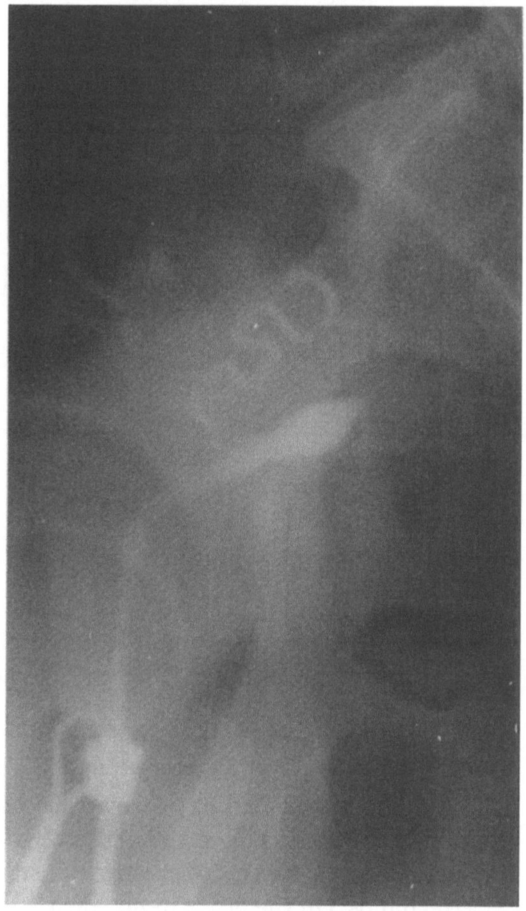

FIGURE 14-8. Perforation by an intrauterine device. Lateral view shows contrast material filling a retroverted uterine cavity. A radiopaque Lippes Loop is noted to be at a considerable distance anterior to the cavity, proving a perforation by the IUD.

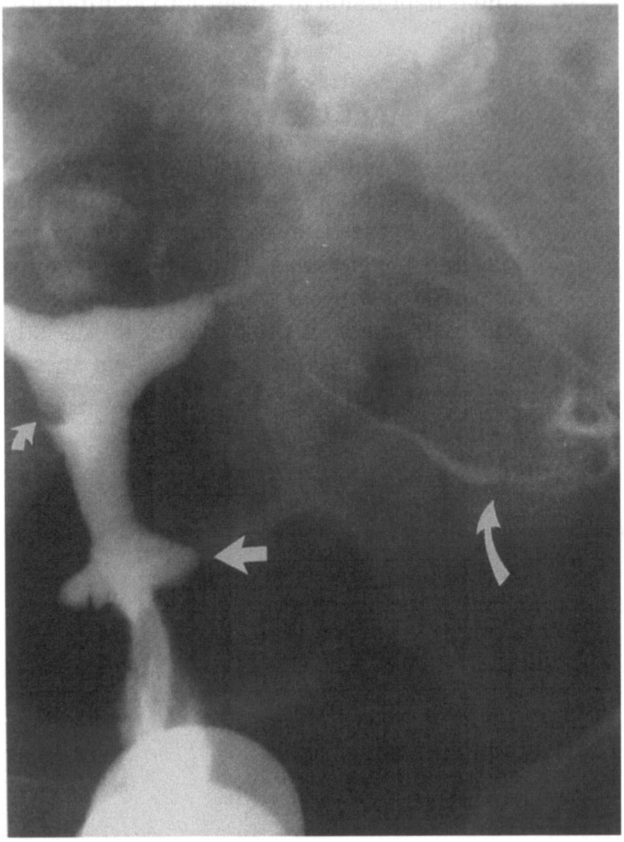

FIGURE 14-9. Cesarean section deformity. Anteroposterior view demonstrates a typical angular deformity within the lower uterine segment (arrow) at the site of a previous cesarean section. Also note that the distal portion of the fallopian tube has been reanastomosed following previous tubal ligation (large curved arrow) and demonstrates tubal patency. There is also a small endometrial polyp (small curved arrow).

Congenital Anomalies

With uterus didelphys, there is a duplex uterus with a vaginal septum as a result of the total failure of the müllerian ducts to fuse embryologically (Fig. 14-11A).

A bicornuate uterus with a single vagina is due to the partial failure of müllerian duct fusion. There may be an asymmetry in the size of the two halves. The two leaves may join just above the cervix, resulting in a single external os. On the other hand, each leaf may have its own cervix opening (Fig. 14-11B).

In septate uterus, a vertically oriented intrauterine septum is present as a result of the failure of normal embryologic resorption of the median septum (Fig. 14-11C). In unicornuate uterus there is aplasia of one the uterine horns.

When a congenital uterine anomaly is suspected, correlation with a thorough physical examination of the pelvis is mandatory in order to be certain that the entire uterine anatomy has been outlined with contrast agent, thus fully evaluating the anomaly. Laparoscopy may be necessary to help differentiate between the various types of congenital anomalies.

Diethylstilbestrol Exposure

In women with DES exposure the uterine cavity may be T-shaped or small or both, and often it has symmetrical constrictions in its

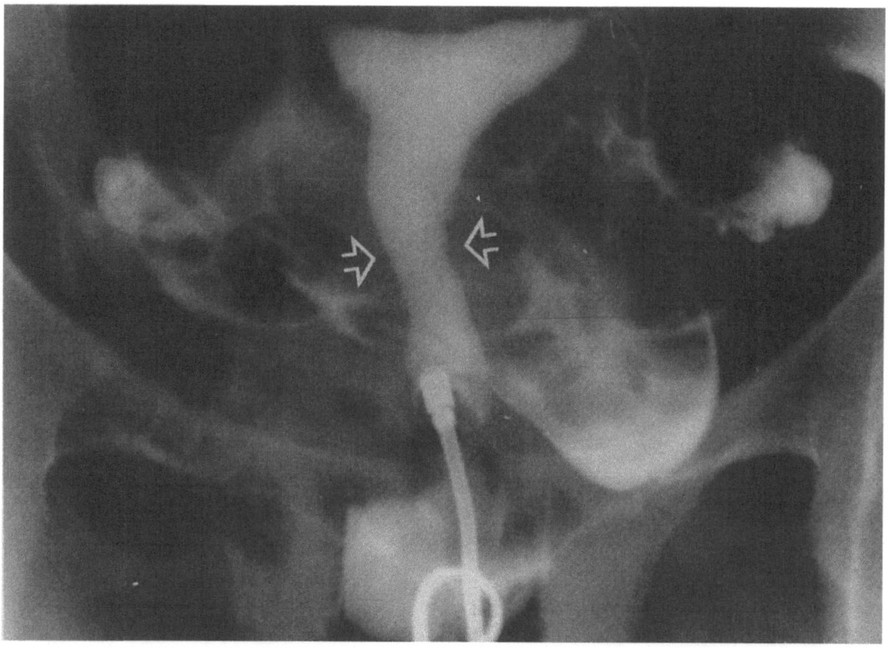

FIGURE 14-10. Cervical incompetence. Anteroposterior view of the filled uterine cavity, without any instruments within the endocervical canal or endometrial cavity, shows that the diameter of the internal os is 11 mm; it should normally be smaller than 1 cm. There is contrast material in the peritoneal cavity.

lower portion, causing a waisting deformity in the inferior third of the cavity. The endocervical canal is usually narrowed and elongated, whereas the superior segment of the cavity is relatively more normal, but is often somewhat shallow (Fig. 14-12). Irregular diverticula can arise from various portions of the uterine cavity. The fallopian tubes are unaffected as a rule.

Diseases of the Fallopian Tubes

Inflammatory disease of the fallopian tubes causes either unilateral or bilateral abnormalities, usually ending in occlusion. Cornual spasm must be distinguished from pathologic occlusion, and enough contrast medium must be instilled to fill the tubes. An injection system with an adequate seal at the cervix allows the exertion of steady pressure during the injection procedure. Intravenous glucagon or sublingual nitroglycerine can also help overcome any spasm encountered. The technique of selective salpingography, using a guidewire and catheter threaded into the proximal portion of the fallopian tube, is a promising new development. When proximal occlusion is due to intraluminal debris, passing a fine flexible guidewire and a No 3 French catheter into the tube may result in recanalization.[4]

In most cases of tubal occlusion secondary to pelvic inflammatory disease, there is dilatation of the ampullary portion of the tube with a clubbing deformity of the fimbriated end, causing a hydrosalpinx and resulting in tubal occlusion (Fig. 14-13).

Peritubal adhesions may not be visualized

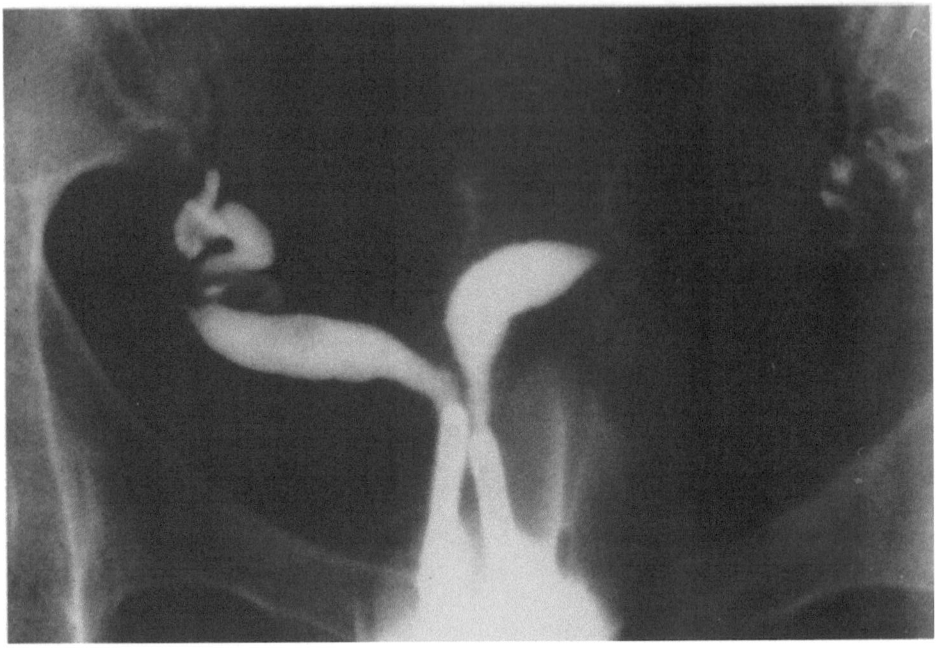

A

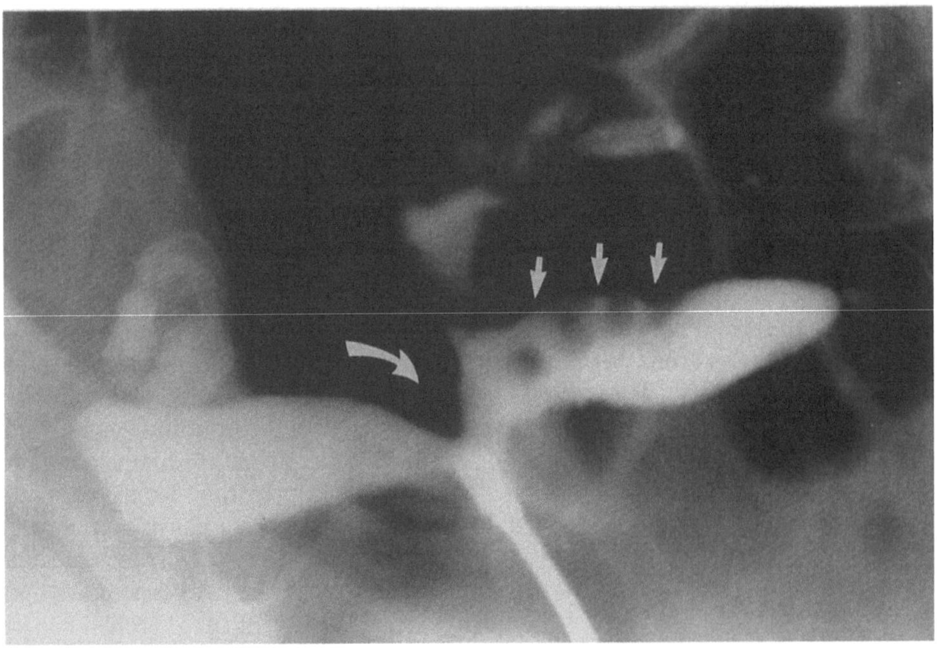

B

FIGURE 14-11. (A) Uterus didelphys. Anteroposterior view of the pelvis reveals two sets of cannulas outlining double cervical canals and separate uterine cornuae and fallopian tubes, which appear normal. (B) Bicornuate uterus. Anteroposterior film shows a single cervical canal and double uterine cornua that have a relatively obtuse angle separating them (curved arrow) and normal fallopian tubes. There are air bubbles within the left cornu (arrows). *(continued)*

FIGURE 14-11. (*continued*) (C) Anteroposterior film shows a cannula in a single cervical canal and double uterine cornua with an acute angle separating the two segments (arrow); this favors a septum rather than a bicornuate uterus.

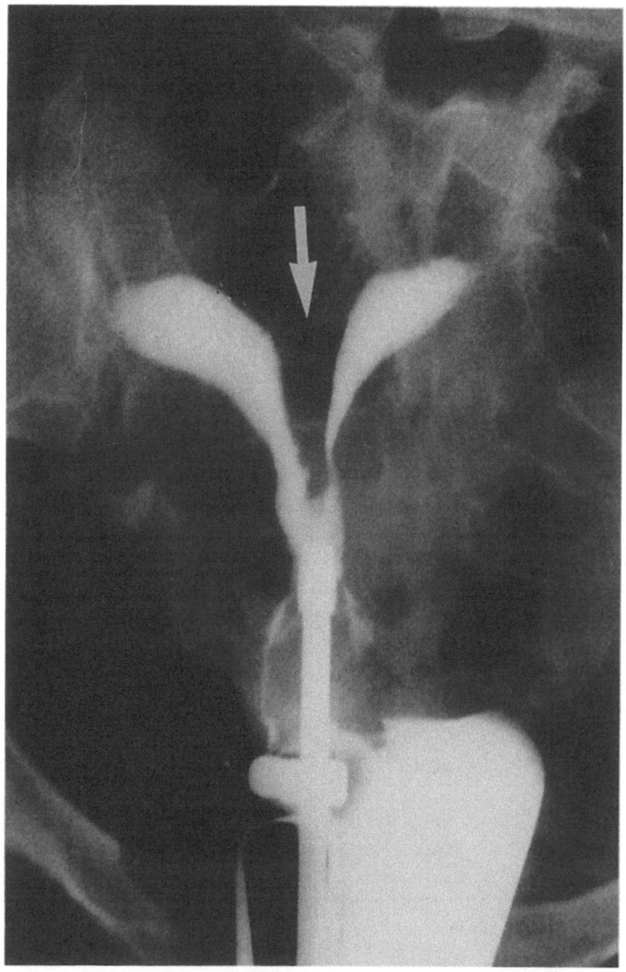

C

by hysterosalpingography. When the contrast medium leaving the fimbriated portion of the fallopian tube appears to pool within the peritubal regions both during the examination and on a delayed film of the pelvis, adhesions should be suspected (Fig. 14-14). There maybe a concomitant abnormality of the fimbria with slight clubbing or an abnormal vertical position of the distal segment of the tube. When compared with laparoscopy, there is a 75% correlation in the accuracy of this diagnosis by hysterosalpingography.[5]

Salpingitis isthmica nodosa is a condition limited to the isthmic portion of the fallopian tubes and characterized by small, fine, bead-like outpouchings adjacent to the lumen of the tube. They can be arranged in a distinct cluster along the linear portion of the tube, or may sometimes be sparse and subtle in appearance (Fig. 14-15). Salpingitis isthmica nodosa may be unilateral or bilateral, and may be seen in association with tubal occlusion with or without concomitant hydrosalpinx. This condition is usually thought to be a sequela of pelvic inflammatory disease and is associated with a significant increase in ectopic pregnancies.

Other conditions that can affect the fallopian tubes are endometriosis, tuberculosis,

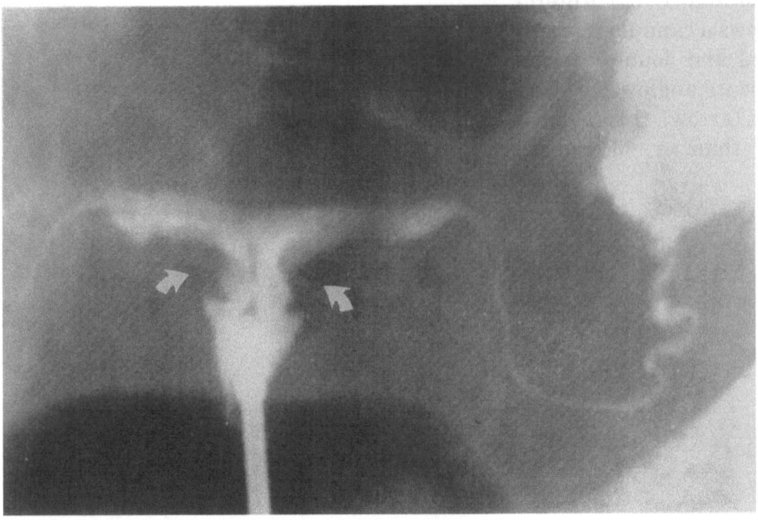

FIGURE 14-12. Diethylstilbestrol exposure. Anteroposterior film demonstrates a small T-shaped uterine cavity, with symmetrical constrictions laterally (arrows) within the upper uterine body and cornua.

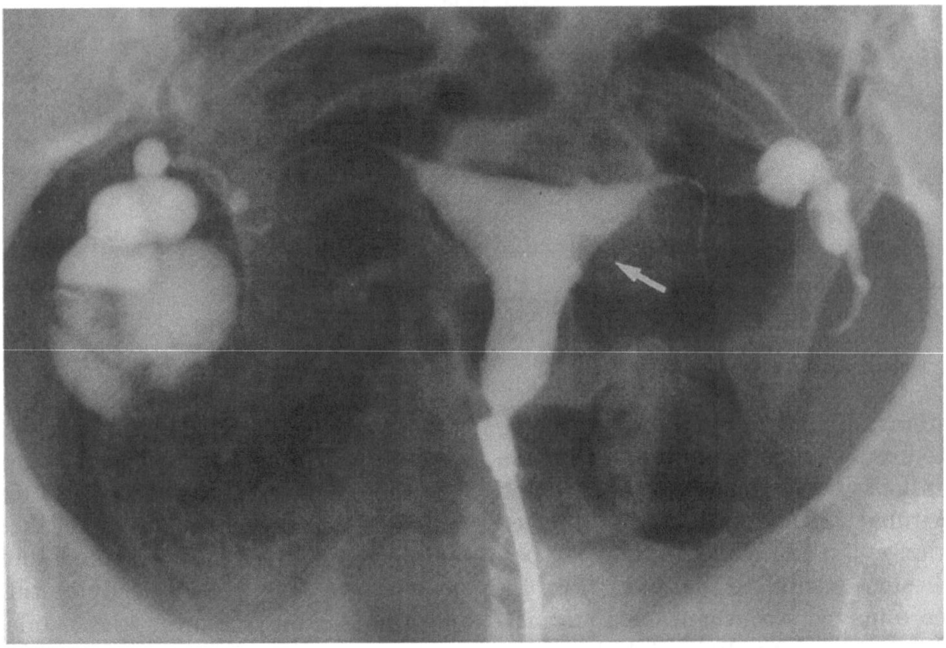

FIGURE 14-13. Hydrosalpinx. Anteroposterior film shows markedly dilated right and moderately dilated left fallopian tubes at their fimbriated ends with bilateral distal tubal occlusion. There is a small endometrial polyp (arrow).

FIGURE 14-14. Peritubal adhesions. Anteroposterior film shows the ampullary portion of the right fallopian tube to be normal (straight arrows). Contrast material spills out from the fimbria into a well-marginated peritubal collection as a result of adhesions (curved arrows). There is a small polyp in the lower uterus (open arrow).

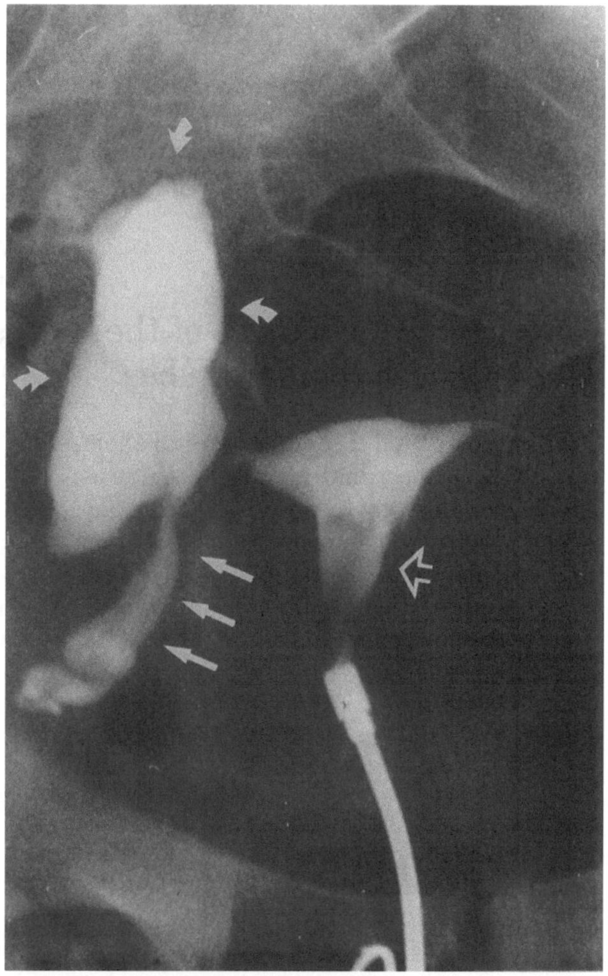

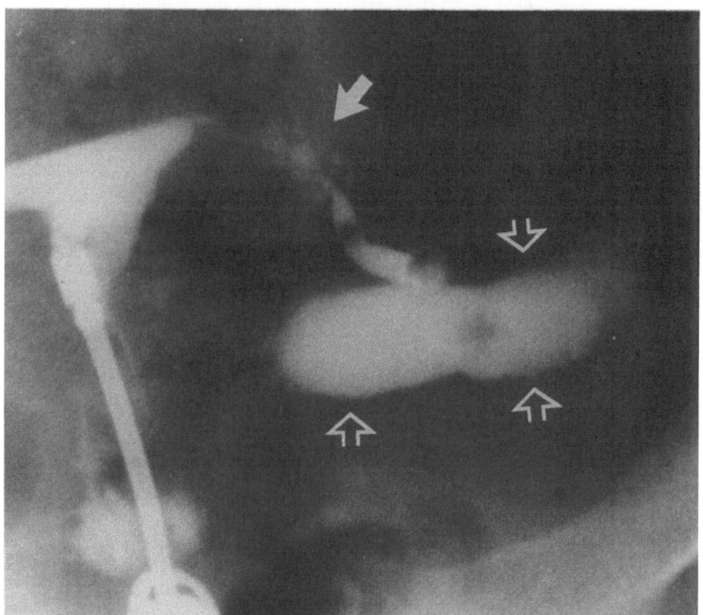

FIGURE 14-15. Salpingitis isthmica nodosa. Anteroposterior film reveals a cluster of fine grapelike diverticula filled with contrast material (arrow). These surround the isthmic portion of the fallopian tube. There is an associated distal tubal occlusion with hydrosalpinx (open arrows).

and Crohn's disease.[2] Antibiotics should be given to the patient after hysterosalpingography when tubal disease is demonstrated in order to avoid an exacerbation of pelvic inflammatory disease.

Postoperative Changes of the Uterus and Fallopian Tubes

Surgical correction of some congenital anomalies, such as bicornuate and septate uterus, can be evaluated by hysterosalpingography. When tubal reanastomosis, reimplantation, and tuboplasty are contemplated, preoperative hysterosalpingography is required to ascertain the current status of the tubes. After surgery the success or failure may best be evaluated by repeat hysterosalpingography.

References

1. Winfield AC, Wentz AC: Diagnostic imaging of infertility. Baltimore, Williams & Williams, 1987.
2. Siegler AM: Hysterosalpingography. New York, Harper & Row, 1967.
3. Yoder IC: Hysterosalpingography and Pelvic Ultrasound Imaging in Infertility and Gynecology. Boston, Little, Brown and Co, 1988.
4. Wolf DM, Spataro RF: The current state of hysterosalpingography. RadioGraphics 1988; 8:1041–1058.
5. Karasick S, Goldfarb AF: Peritubal adhesions in infertile women: Diagnosis with hysterosalpingography. AJR 1988;152:777–779.

15

Magnetic Resonance Imaging and Computed Tomography of the Uterus

CYNTHIA L. JANUS

The development of computed tomography (CT), and more recently magnetic resonance imaging (MRI), and the rapid technologic advances in these areas over the past decade have made it possible to image the pelvis in exquisite detail.[1–3] Although ultrasound is a well-established screening modality for evaluating the pelvis, it cannot provide specific tissue characterization and is not adequate for staging pelvic malignancies. Computed tomography allows better evaluation of the deep pelvis and pelvic sidewalls but is also limited by a relative lack of soft tissue contrast resolution. Further disadvantages include the fact that it involves ionizing radiation and the need for intravenous contrast with its associated small risk of contrast reaction.

Magnetic resonance imaging is an excellent and increasingly utilized modality with which to evaluate the female pelvis.[2,3] Not only can it depict the zonal anatomy of the uterus in exceptional detail, but it can demonstrate changes in uterine appearance in response to the hormonal environment during different phases of the menstrual cycle[4–6] and with age.[3] The ability to image in multiple planes and lack of ionizing radiation are other advantages of this modality.

Basic Principles

Both CT and MRI are computer-based modalities that image the body in thin tomographic slices. In the basic model of a CT scanner, a narrow beam of x-rays passes through the patient in a linear fashion. The x-rays that are not absorbed are detected by a radiation detector that scans synchronously with the x-ray beam. This scan sequence is repeated at multiple angles around the patient and results in a series of profiles that reflect the attenuation properties of the portion of the body being scanned. A series of transverse tomographic images of the patient is reconstructed using a complex mathematical algorithm.[7]

While CT requires ionizing radiation, MRI is based upon the interaction between radiowaves and hydrogen nuclei in the body in the presence of a strong magnetic field. When the body is placed in the magnetic field within the bore of a magnetic resonance imager, a small percentage (<1/1,000,000) of the protons tend to align in the same direction as the field. When subjected to radiowaves of a certain frequency, the protons absorb the energy and their alignment is changed. They subsequently realign with the field as they release the radiowave energy that constitutes the MR signal. Tissues absorb and then release this radiowave energy at different characteristic rates, and evaluation of an MR image requires knowledge of the normal relative signal intensities of the tissues being imaged. There are several factors that determine the signal intensity of any given tissue. Two of the factors, TR and TE, are machine settings selected by the radiologist. TR is the length of time allowed the protons to become aligned with the direction of the main magnetic field, or the time between the administered radiofrequency pulses. TE is the amount of time allowed for the absorbed radiowave energy to be released. Two other important factors are tissue qualities termed T_1 and T_2. The time constant T_1 indicates how rapidly a specific tissue will be ready to absorb the radiowave energy to which it is exposed, and the time constant T_2 indicates how quickly a specific tissue will lose the radiowave energy it has absorbed. Finally, the number of protons in a given tissue and the motion or flow of these protons are additional factors that influence the signal intensity of a tissue.[8,9]

As in CT, an MR image represents a composite of picture elements (pixels) generated by a computer. While in CT the numerical value of each pixel represents x-ray attenuation by a given volume element (voxel) of tissue, the numerical value of each pixel of an MR image represents the intensity of the MR signal emitted from that voxel of tissue.

Preparation and Protocol

Patients undergoing CT should fast 4 hours or more prior to the exam since in most cases they will be receiving intravenous contrast. Oral and rectal contrast will also be administered prior to scanning to opacify bowel loops.

There is no patient preparation for a MRI exam. It is helpful, however, for the patient to have a partly distended urinary bladder because this facilitates evaluation of the bladder wall and determination of possible invasion in staging of gynecologic malignancies. Although not entirely essential, motion artifact from peristalsis can be minimized if the patient refrains from eating just prior to the exam.

A potential risk of MRI is the potential projectile effect of forceful attraction of ferromagnetic objects to the magnet.[10,11] Contraindications to the exam include the presence of a ferromagnetic foreign body in the orbit, cerebral aneurysm clips, or a cardiac pacemaker. Caution must also be taken in certain patients with embedded shrapnel. Exam quality is severely degraded by motion artifacts secondary to the relatively slow acquisition time, and some patients who are acutely ill or cannot cooperate cannot be scanned. In addition, because of the small bore of the magnet, patients who are severely claustrophobic and some obese patients cannot be studied.

The usual imaging sequences for evaluation of the uterus include T_2-weighted images in the sagittal plane and T_1 and T_2-weighted images in the axial projection.[2] Zonal anatomy of the uterus and cervix is best demonstrated on the T_2 sequence, whereas T_1 images best delineate the uterus and parametria from the surrounding fat. The sagittal projection is useful in depicting extent of tumor in the cephalocaudal plane and for evaluating invasion of the bladder and rectum. The transverse projection is essential for assessment of parametrial extension and lymph node enlargement.[2,3] Ovarian detail is best visualized utilizing transaxial and coronal planes.

Normal Anatomy of the Female Pelvis

On CT scan the normal uterus is relatively homogeneous and similar to muscle in density (Fig. 15-1). It is possible to differentiate the endometrial region from the myometrium with contrast-enhanced scanning, but this is not reliably seen in all patients.

With MRI, the zonal anatomy of the uterus can be differentiated on T_2-weighted sequences as a central high-signal intensity area representing the endometrium and an area of moderate but lower signal intensity peripherally corresponding to the myometrium (Fig. 15-2). The junctional zone is visualized between these two areas as a low-signal-intensity band and is currently believed to correspond to the inner one third of the myometrium.[2,3] On T_1-weighted images the uterus is featureless and displays a homogeneous medium signal intensity.

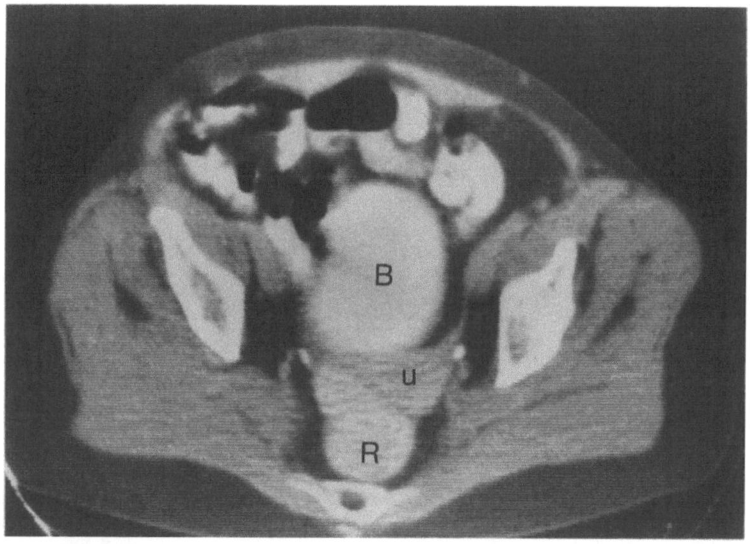

FIGURE 15-1. On CT scan the uterus is seen in axial projection and is of similar density to the body musculature. U = uterus, B = urinary bladder, R = rectum.

In women of reproductive age, the endometrial region and junctional zone are prominently seen (Fig. 15-3A) and both increase in width, with the most rapid change in diameter seen in the periovulatory period (days 8 through 16)[6] (Fig. 15-4A). Just prior to menstruation there is a general decrease in signal intensity from the endometrium (Fig. 15-5A), and during menstruation discrete areas of low signal intensity representing blood clots can be seen within the endometrium. Endometrial atrophy has been shown in patients taking oral contraceptives.[4] Magnetic resonance images of the uterus in premenarchal girls and postmenopausal women not taking exogenous hormones demonstrate an atrophic or absent endometrium measuring 2 mm or less in thickness without the cyclic changes.[3] The junctional zone appears thin or may not be visible as a distinct structure. On T_2-weighted images, the myometrium has an intermediate signal intensity, lower than that in women of reproductive age. In postmenopausal women taking exogenous estrogens, the uterus has an appearance similar to that of women of reproductive age.

Magnetic resonance images of the cervix demonstrate a central area of high signal intensity that represents the region of the endocervical glands surrounded by a low signal area representing the fibrous cervical stroma (Figs. 15-2 and 15-6). The pericervical tissue is of medium signal intensity on T_1 sequences (Fig. 15-6) and, in women of reproductive age, of increased signal intensity to varying degrees on T_2 images.

The vagina is best delineated on T_2-weighted images in the transverse plane.[12] The lateral fornices are the landmark for the upper one third (Fig. 15-7). The middle third of the vagina is found at the level of the bladder base and the lower third at the level of the urethra (Fig. 15-8). In premenarchal girls, the central high signal intensity of the mucus is thin and the vaginal walls is of low signal intensity. In women of reproductive age a central area of high signal intensity representing mucus and epithelium surrounded by the low-signal-intensity vaginal wall is seen on T_2 images in the early proliferative phase. The thickness of the mucous portion of the vagina increases in the early secretory phase and there is usually some loss of contrast differentiating the central mucus from the medium-high signal intensity of the vaginal wall. The vaginal wall is of high signal intensity in pregnant patients, particularly during the second and third trimester. In postmenopausal women

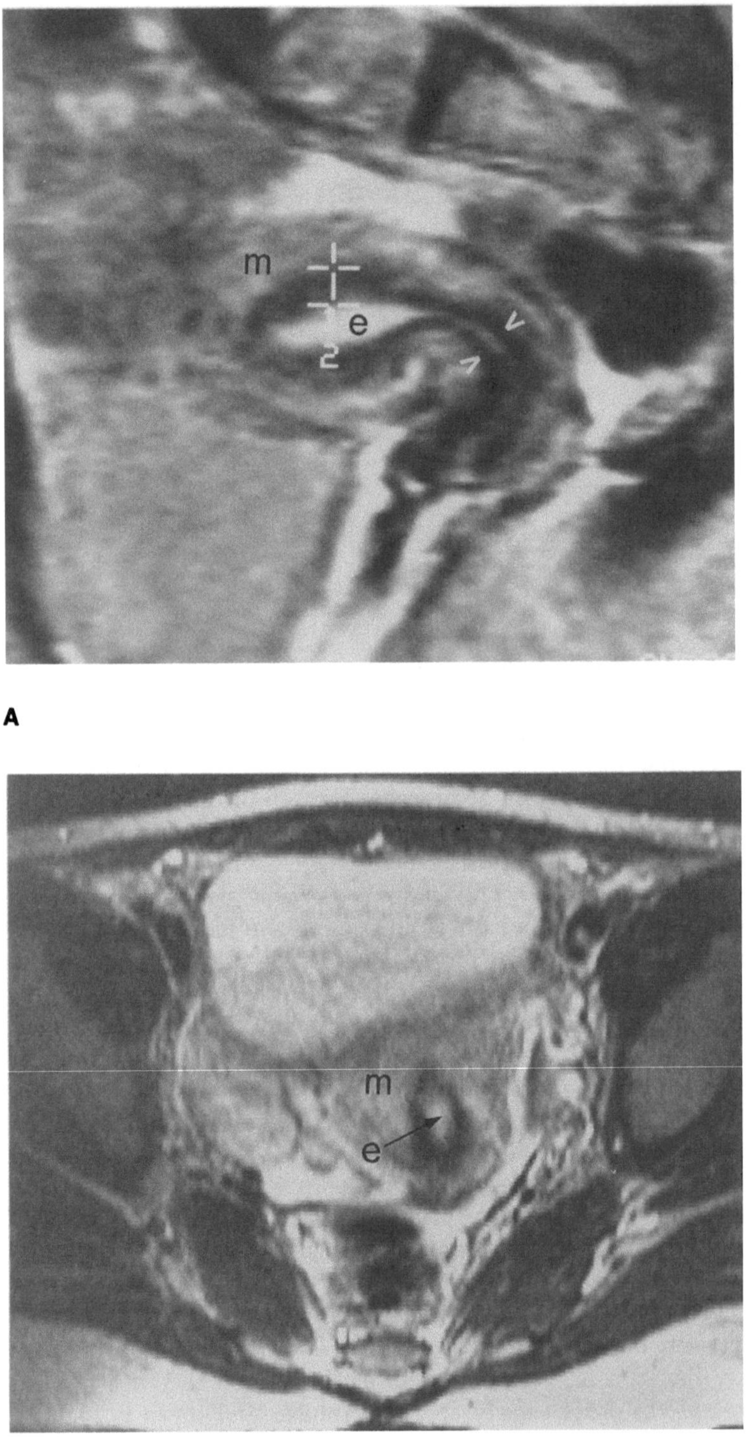

A

B

FIGURE 15-2. Zonal anatomy is well delineated on T_2-weighted sequences in the sagittal (A) and axial (B) projections. e = endometrium, m = myometrium, >< = cervix; junctional zone demarcated by cursors on sagittal view.

FIGURE 15-3. Day 4. (A) The endometrial region (e) is well seen but is relatively thin during the early part of the menstrual cycle. (B) Small developing follicles are seen in the ovaries. O = ovary (between arrows) with follicles. (*continued*)

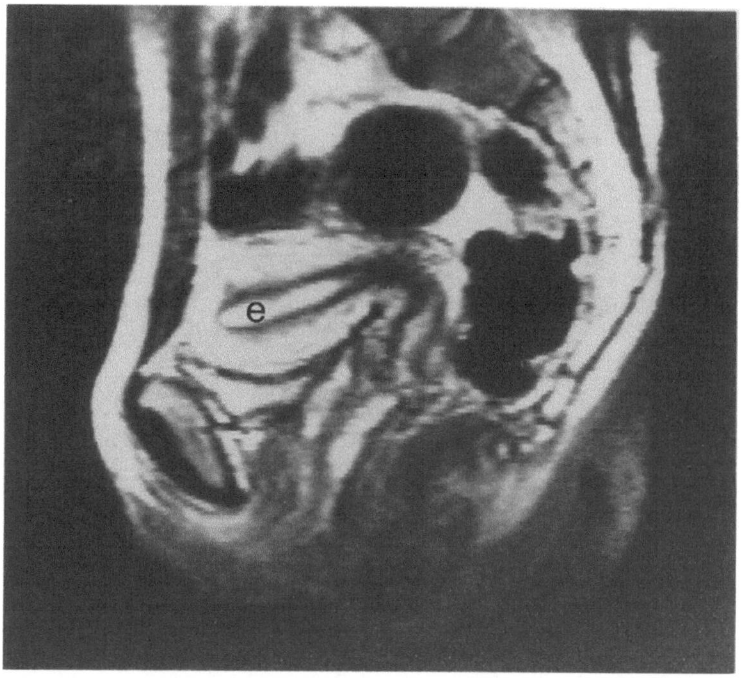

A

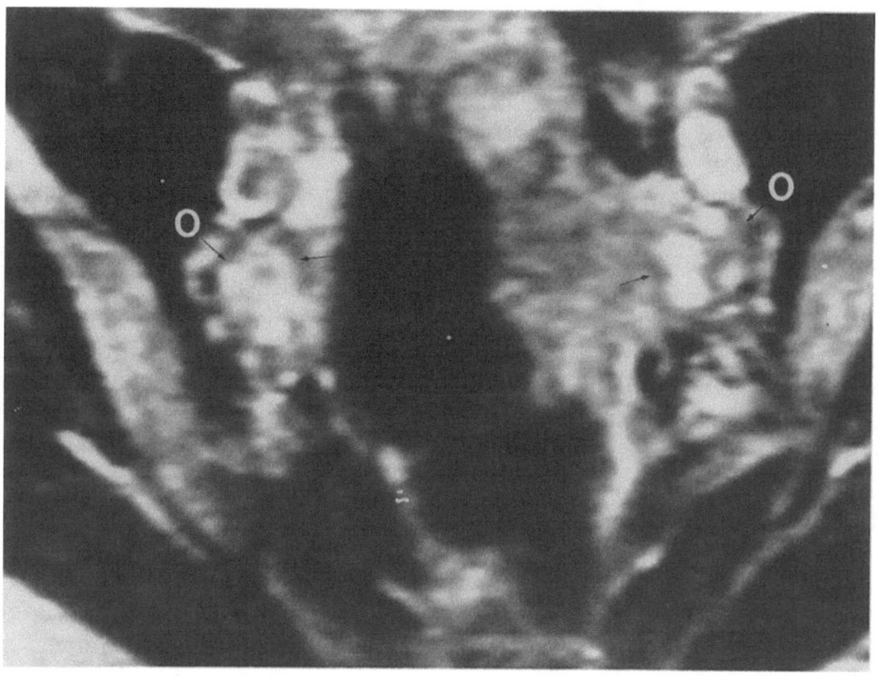

B

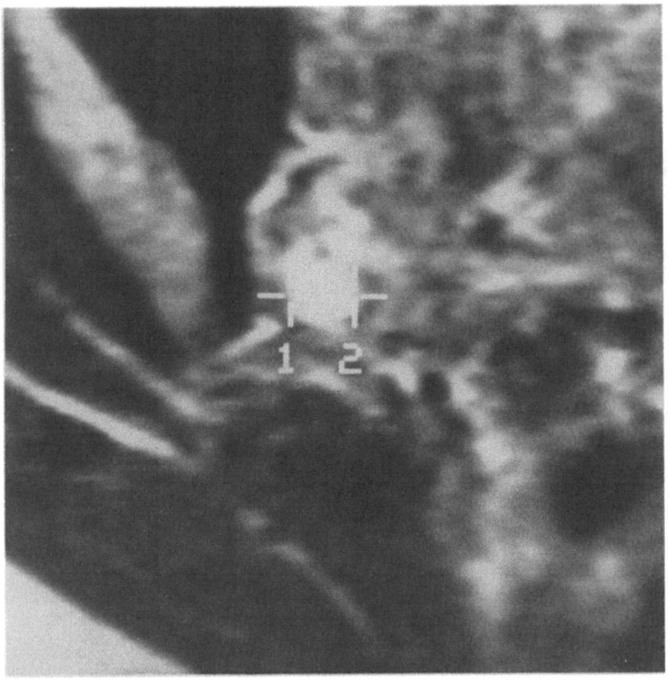

C

FIGURE 15-3. (*continued*) (C) A small developing dominant follicle (between cursors) is seen in the right ovary.

not taking estrogens, on the other hand, the central high-signal mucus portion is thin and the vaginal wall is of low signal intensity.

The ovaries can be recognized on MRI in most women of reproductive age. They are of low to medium signal intensity on T_1-weighted images. On T_2 sequences the ovaries are of very high signal intensity and individual follicles can be appreciated[6] (Figs. 15-3B, 15-3C, 15-4B, 15-4C, and 15-5B). The ovaries are less frequently seen on CT scan and their internal architecture cannot be delineated.

Congenital Abnormalities

Congenital anomalies of the uterus are rare but their clinical significance can be great. Until recently, a combination of physical exam and radiographic studies, particularly sonography and hysterosalpingography, were used to delineate these anomalies. These often gave inconclusive or incomplete information, however, and surgery was necessary for definitive evaluation.

The advent of MRI has made it possible to delineate many of these abnormalities in a noninvasive manner.[13,14] Absence of the vagina and/or uterus[12,13,15] is easily diagnosed (Fig. 15-9). Hematometria and hematocolpos[13,16] are characterized by uterine and/or vaginal distention with fluid that is usually of medium to high signal intensity on T_1-weighted images secondary to the presence of blood and of high signal intensity on T_2 sequences.

Uterine septations can be visualized as low-signal bands within high-signal endometrium. Diagnosis of bicornuate uterus can be made by evaluation of the configuration of the uterine fundus and by identifying the medium- to high-signal myometrium separating the endometrial cavities.

Uterine Abnormalities

Leiomyoma and Leiomyosarcoma

The most common uterine tumor is the leiomyoma, which occurs in approximately 20 to 30% of women during their reproductive years. In certain clinical settings, such as infertility[17] or recurrent abortion or prior to myomectomy, accurate assessment of the number, size, and location of leiomyomas is important. With the use of MRI, fibroids as small as 0.5 cm can be visualized and characterized as submucosal, myometrial, or subserosal in location. In general, leiomyomas are well circumscribed and sharply delineated from adjacent normal myometrium. Uncomplicated fibroids have low signal intensity on T_1- and T_2-weighted images (Figs. 15-10 and 15-11), whereas those with hyaline or myxomatous degeneration may display variable signal intensities.[3,17]

On CT scan, both leiomyoma and leiomyosarcoma may appear as uterine enlargement with inhomogeneous areas of decreased attenuation (Fig. 15-12A) and there may be foci of calcification. Computed tomography features are similar to those of endometrial cancer, and preoperative differentiation may not be possible. Although MRI may also not be able to differentiate a complicated leiomyoma from leiomyosarcoma or other malignant uterine tumor, by its ability to delineate zonal anatomy MRI can more precisely define the extent of the tumor[18] (Fig. 15-12B).

Adenomyosis

This condition is characterized by the presence of endometrial glands and stroma deep within the myometrium. Frequency of oc-

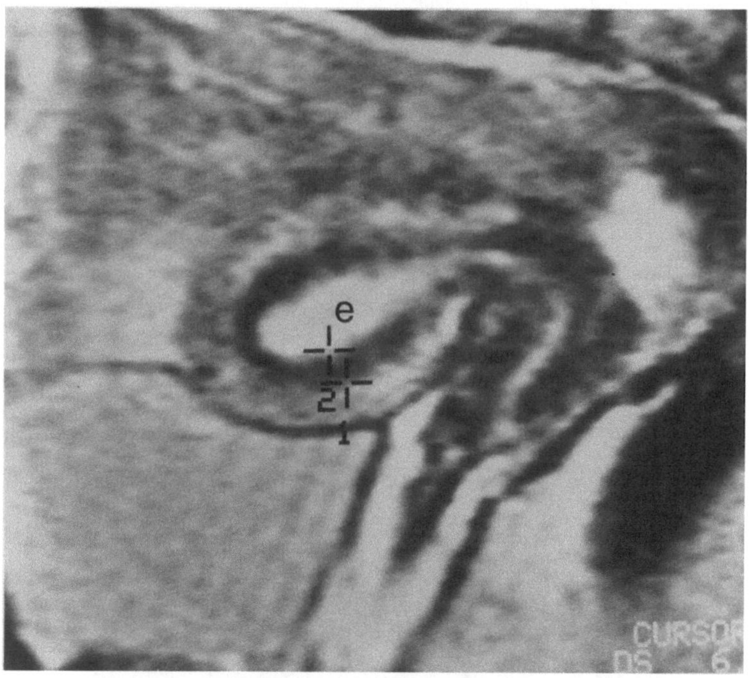

A

FIGURE 15-4. Day 12. (A) The endometrial region is very prominent during the periovulatory period. The most rapid increase in width of this zone occurs during days 8 to 16 of the cycle. e = endometrium; junctional zone between cursors. *(continued)*

FIGURE 15-4. (*continued*) (B) The dominant follicle (f) is seen in the right ovary. (C) After ovulation the borders of the follicle (f) appear crenated.

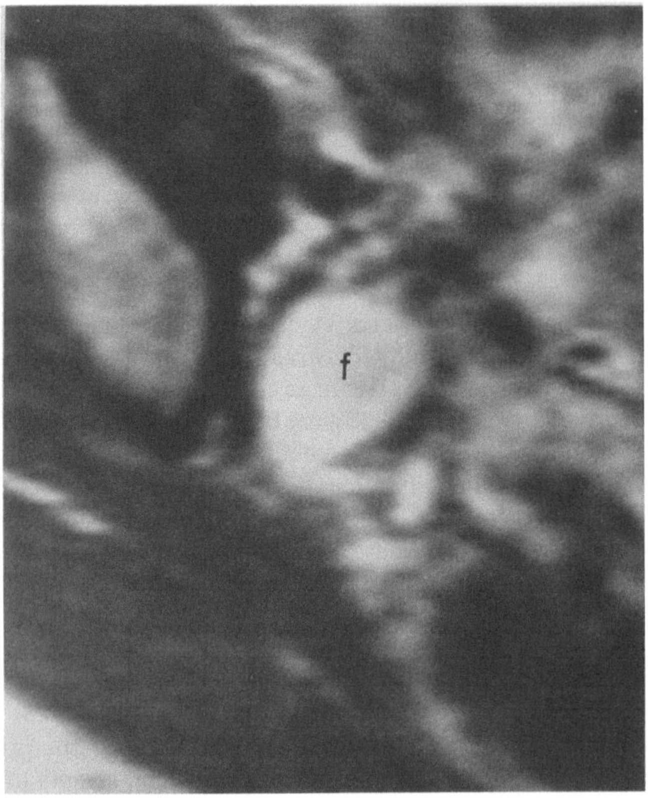

B

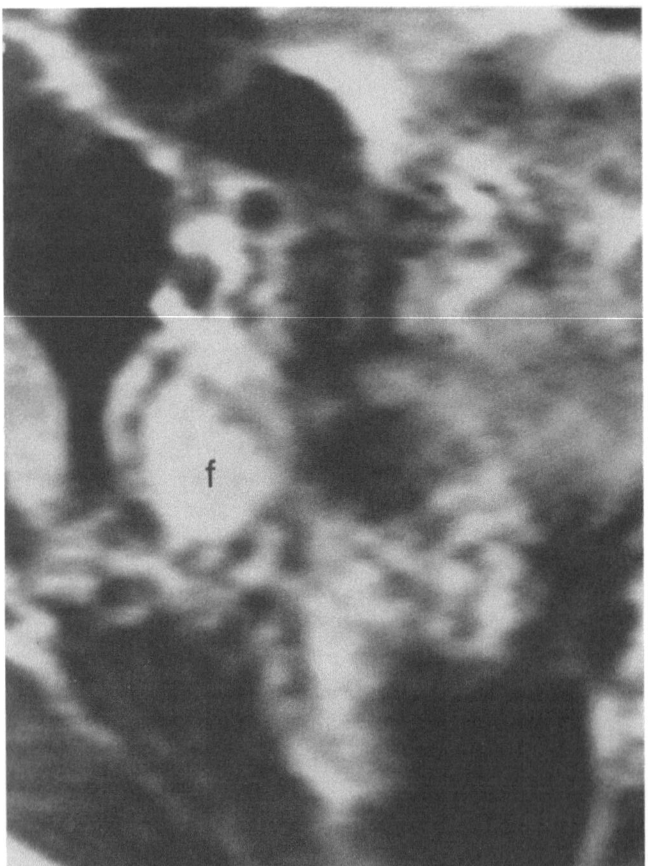

C

FIGURE 15-5. Day 24. (A) A decrease in signal intensity of the endometrial zone (e) prior to and during menstruation secondary to vasospasm and necrosis occurring during this period. Junctional zone between cursors. (B) A corpus luteum (c) is evident at this time.

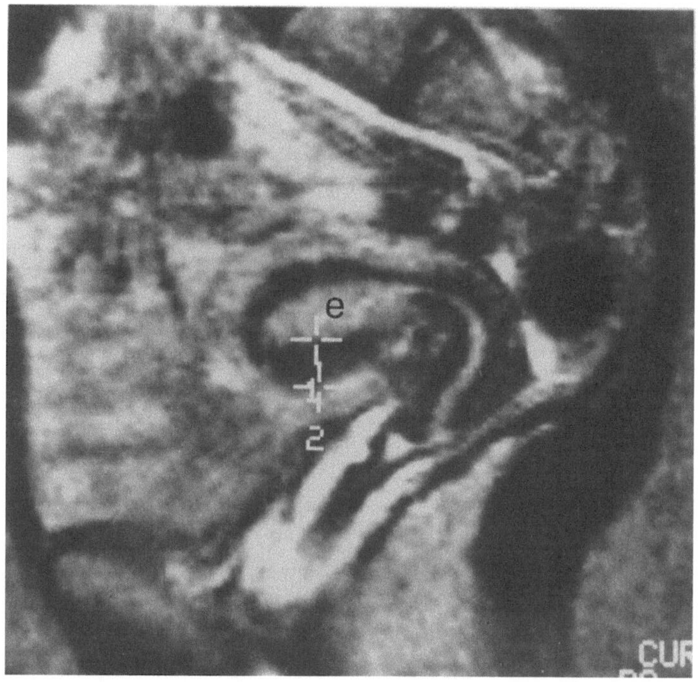

A

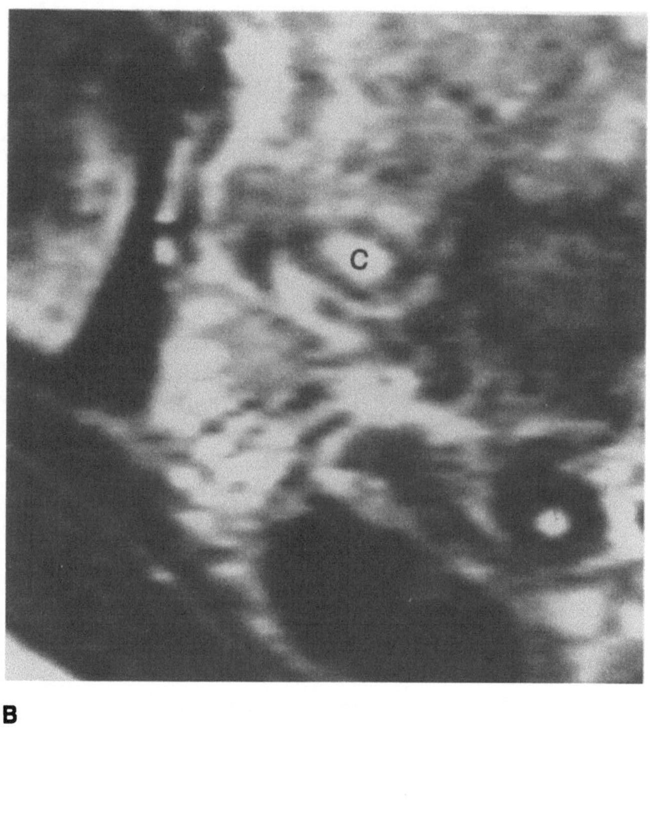

B

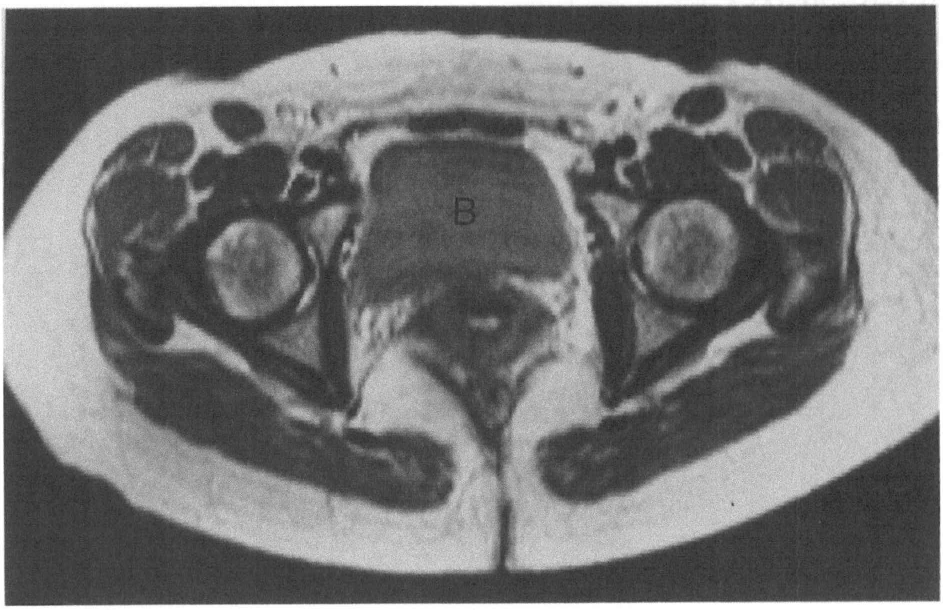

FIGURE 15-6. The pericervical tissue is of medium signal intensity on T_1-weighted sequences.

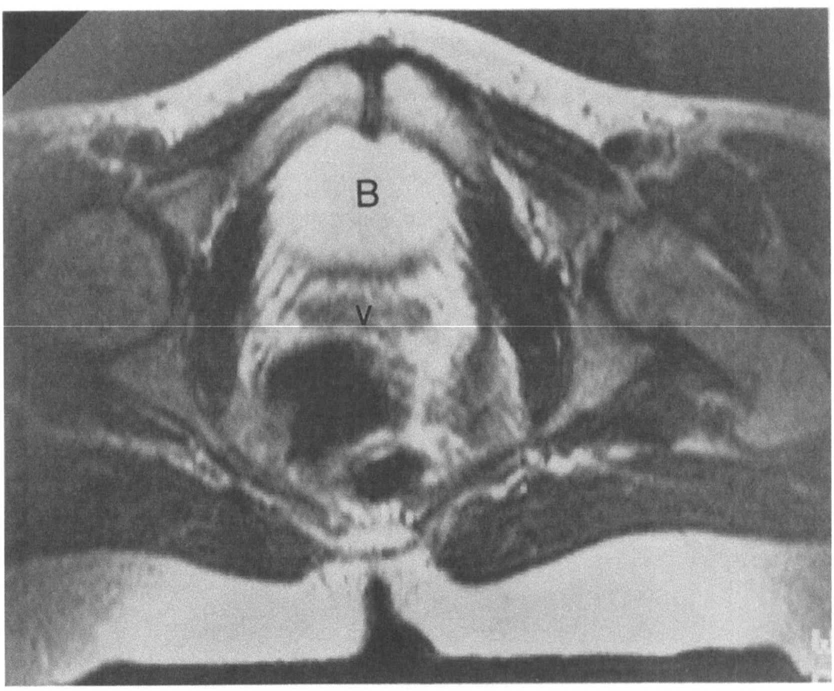

FIGURE 15-7. The vaginal fornices demarcate the upper one third of the vagina. V = vagina, B = urinary bladder.

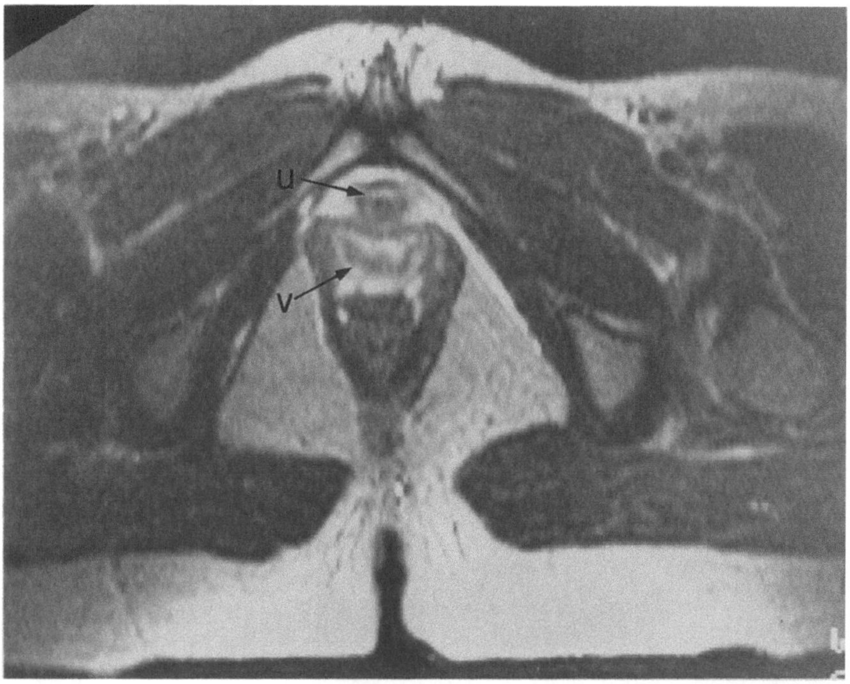

FIGURE 15-8. The urethra is seen anterior to the lower one third of the vagina. V = vagina, U = urethra.

currence increases in multiparous women. While endometriosis may follow the cyclic changes occurring in the normal endometrium, the endometrial glands in adenomyosis are of the basalis type and not affected by hormonal stimulation.

In diffuse adenomyosis the uterus is enlarged. On T_2-weighted images there is thickening of the junctional zone seen as a wide band of low signal adjacent to the endometrium, which maintains its normal high signal intensity. This corresponds to the site of abnormality seen at pathology around the uterine cavity and comprised of diffusely hypertrophied myometrium.

Focal adenomyosis may simulate a leiomyoma on gross inspection, and these tumors are sometimes referred to as adenomyomas.[19] However, adenomyotic foci interdigitate with the normal smooth muscle. On T_2-weighted images, this may appear as a localized area of low signal intensity poorly delineated from the surrounding myometrium.[19] The diagnosis of adenomyosis cannot be made with CT scan.

Gestational Trophoblastic Neoplasia

The category of gestational trophoblastic neoplasms (GTNs) represents a spectrum of pathology, including the benign hydatidiform mole, the invasive mole (chorioadenoma), and the highly malignant choriocarcinoma.[20] Elevated levels of human chorionic gonadotropin (hCG) help to diagnose trophoblastic disease, and serial measurements are a good indicator of tumor activity but give no information as to the site of the tumor. Sonography can give information as to uterine and tumor size and CT scan can demonstrate metastatic involvement, but both are limited in delineating the uterine

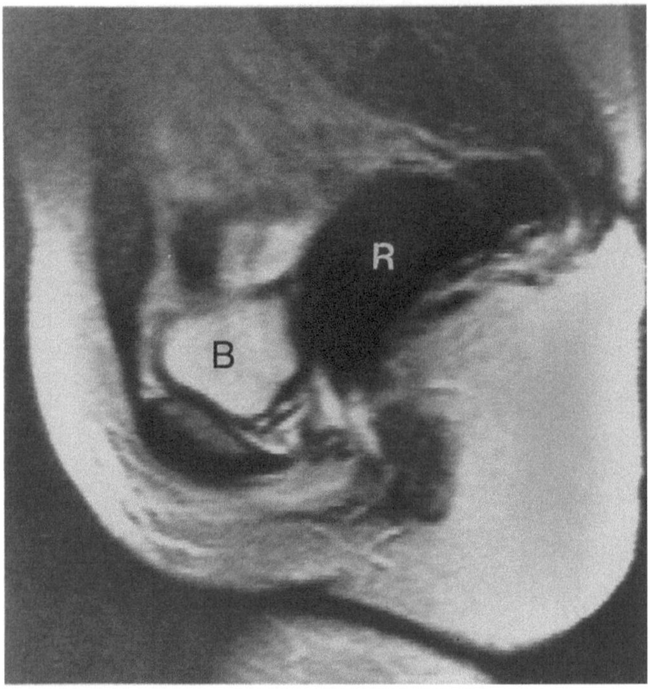

A

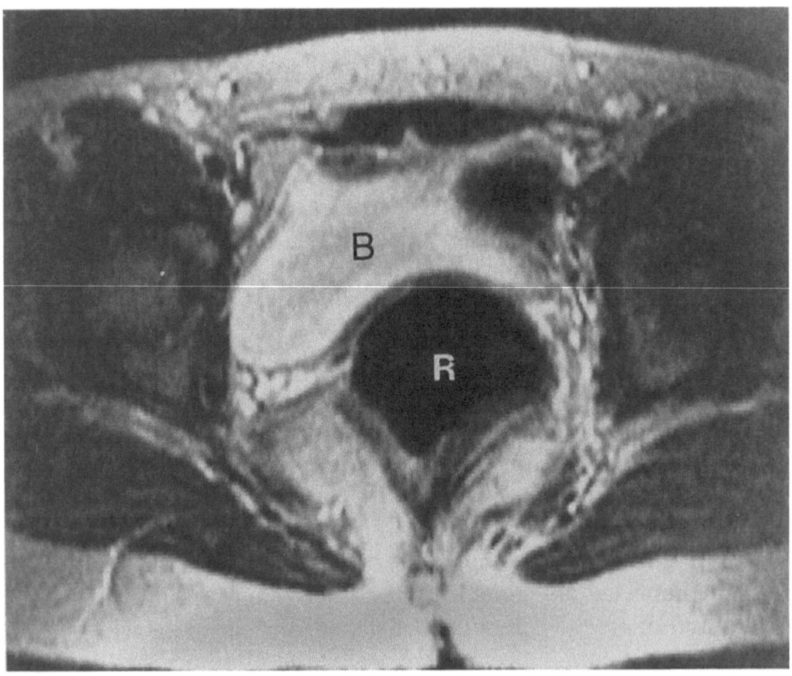

B

FIGURE 15-9. Absence of the uterus is easily appreciated on sagittal (A) and axial (B) views from a T_2-weighted sequence. B = urinary bladder, R = rectum.

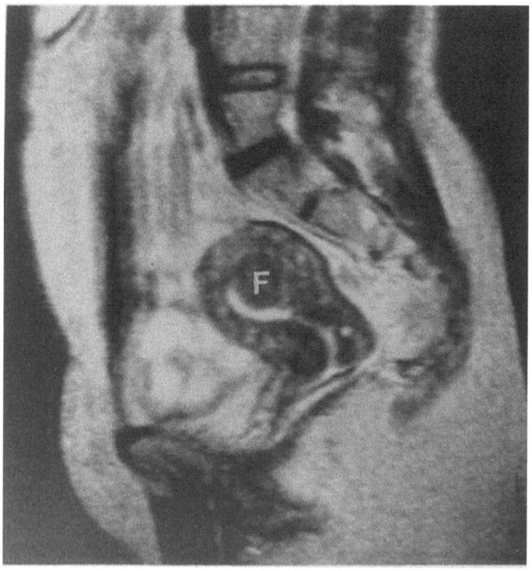

FIGURE 15-10. Uncomplicated fibroids (F) demonstrate low signal intensity on T_2-weighted sequences, as seen on sagittal view.

zonal involvement (Fig. 15-13A). Angiography has been used in the past but is an invasive procedure and is no longer widely used in this setting. On T_2-weighted MR images, a uterine GTN is usually seen as a hypervascular mass and the degree to which it obliterates the normal zonal anatomy is readily demonstrated (Fig. 15-13B). On T_1-weighted images, the signal intensity of the tumor is often similar to that of normal myometrium, but may contain areas of high signal intensity indicating the presence of blood (Fig. 15-13C). Tortuous vessels in the region of the tumor and adjacent myometrium and in the adnexae are usually seen, and associated theca-lutein cysts are easily demonstrated on MRI.[20,21] Magnetic resonance imaging is particularly valuable in detecting a uterine GTN when the tumor deeply invades the myometrium but does not extend to the endometrial surface, so that uterine curettage specimens are nondiagnostic. It can be used to rule out a uterine source for elevated hCG levels in patients with extrauterine germ cell tumors. This is especially valuable when surgery is planned. Finally, along with serial hCG levels, MRI is used to assess response to treatment.

In evaluating the patient with GTN of the uterus for metastatic disease, CT is considered by many to be the imaging procedure of choice because of its availability and scanning speed. Magnetic resonance imaging should be used, however, to evaluate the uterus as the site of the primary disease, to

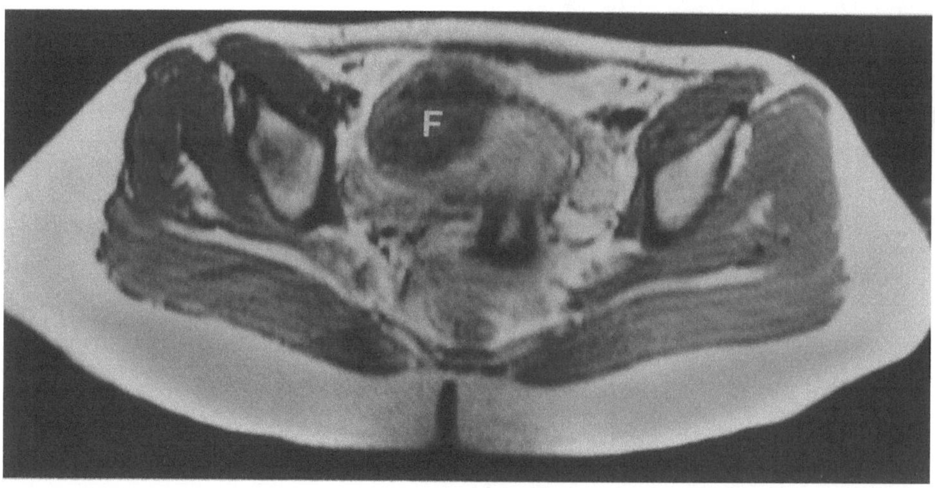

FIGURE 15-11. A fibroid (F) is seen at the right lateral aspect of the uterus and displays a low signal intensity on this axial T_2-weighted image.

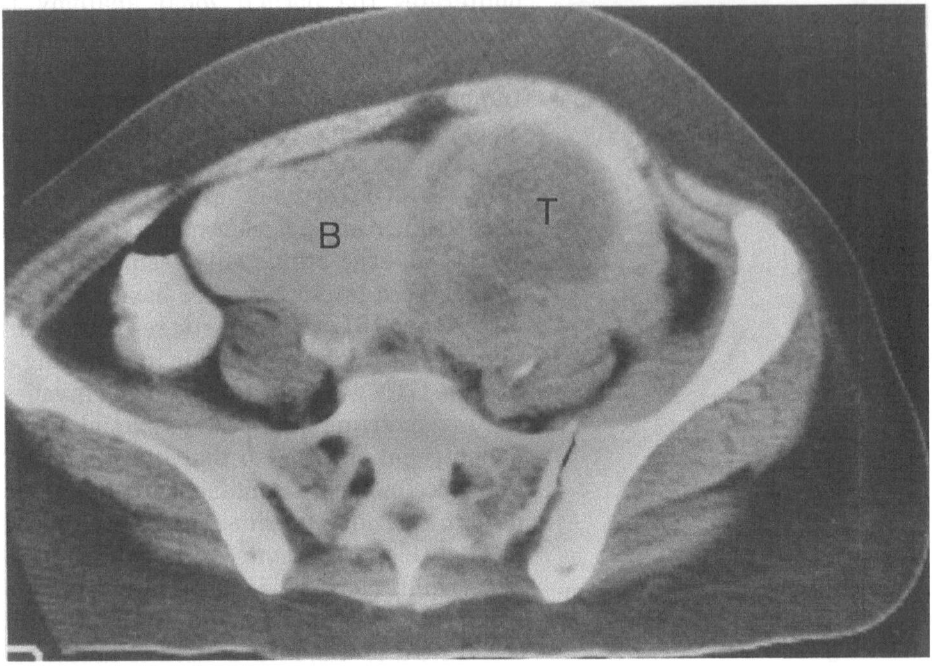

A

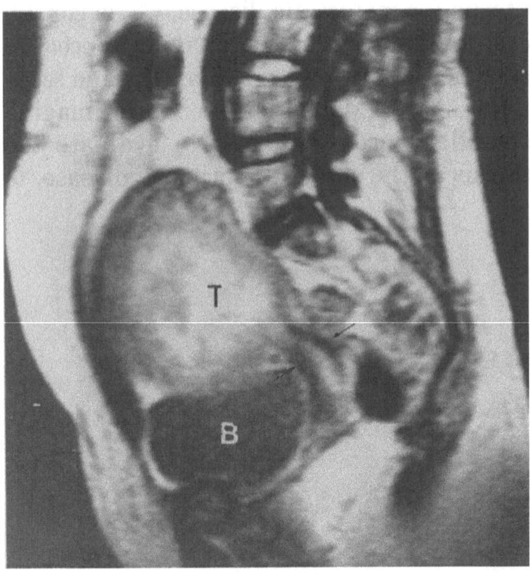

B

FIGURE 15-12. (A) Leiomyosarcoma on CT scan usually has a nonspecific appearance of decreased density. (B) T_2-weighted sagittal view from an MRI exam demarcates the large heterogeneous but mainly high-signal-intensity tumor occupying most of the body of the uterus. The cervix (between arrows) is normal. T = tumor, B = urinary bladder.

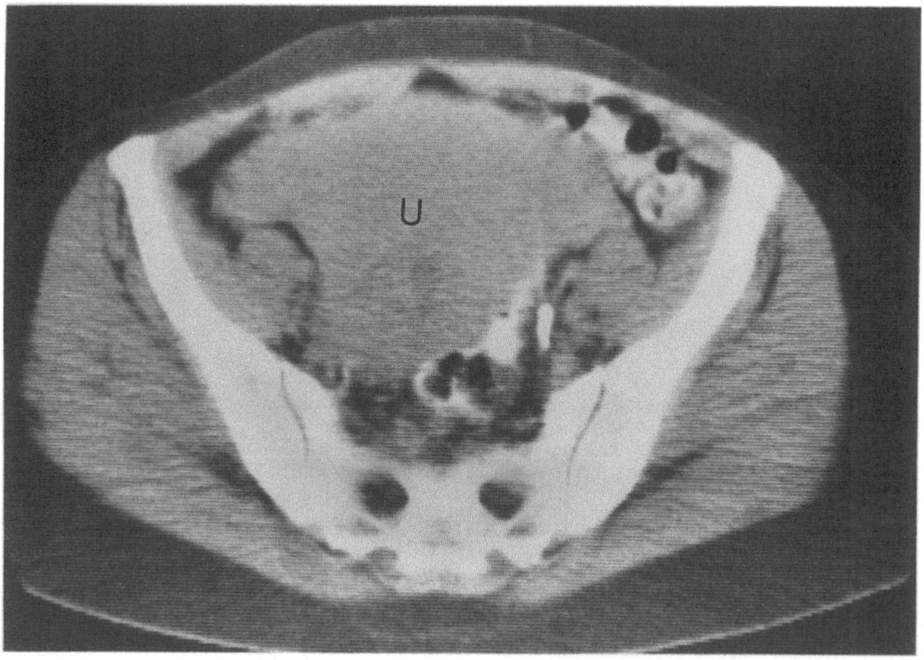

A

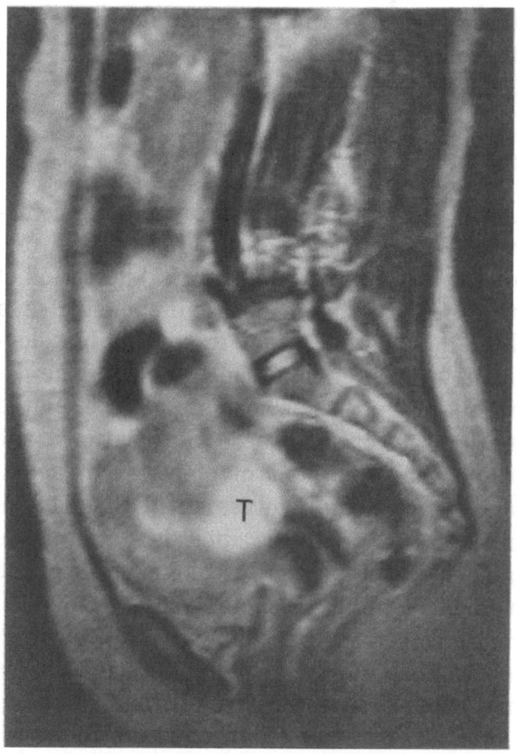

FIGURE 15-13. (A) CT scan shows nonspecific enlargement of the uterus in this case of choriocarcinoma. U = uterus. (B) The size and extent of the tumor mass is more precisely delineated on T_2-weighted sagittal views from an MR scan. T = tumor. (*continued*)

B

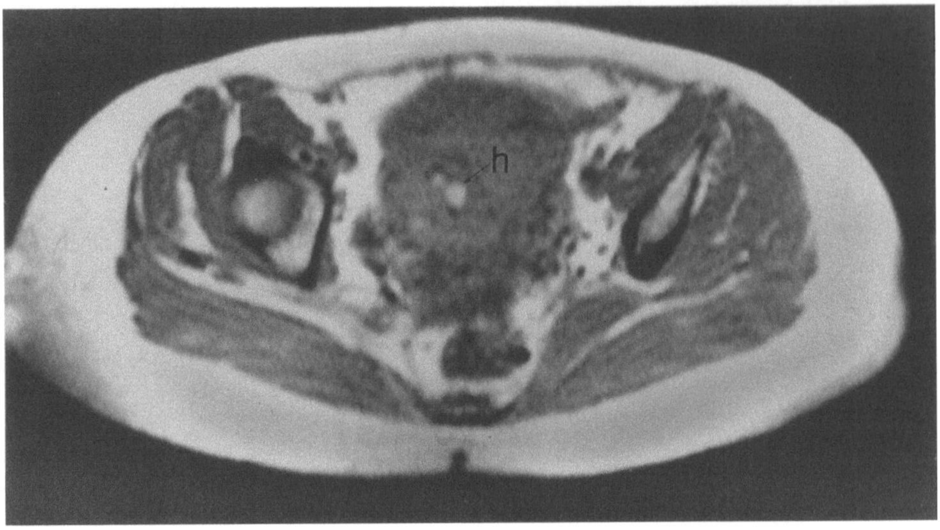

C

FIGURE 15-13. *(continued)* (C) The mass is not as well delineated on the T_1 sequence, but small high-signal-intensity areas representing hemorrhage are easily appreciated. h = hemorrhage.

follow response to treatment, and when there is a contraindication to the use of intravenous contrast.

Endometrial Carcinoma

Although the MRI appearance of endometrial carcinoma is nonspecific and the modality is not sensitive for initial detection of the disease, the modality demonstrates a high degree of accuracy in staging histologically proven endometrial cancer.[22] The patient's prognosis has been found to be related to the stage of the disease, the location of the tumor and depth of myometrial invasion, and lymph node involvement. Other factors influencing prognosis include patient age and the cell type and histologic grade of the tumor. Lymph node metastases and increasing myometrial invasion signal a poor prognosis. It is now known that uterine size is not necessarily indicative of stage. While in some cases uterine enlargement is due to advanced stage III or IV disease. it may also be related to the presence of fibroids or to estrogen therapy.[22]

On CT scan endometrial cancer may appear as diffuse or focal uterine enlargement and/or an abnormal area of decreased density in the endometrial region[23,24] (Fig. 15-14). It has been used to depict the actual tumor mass in patients with suspected advanced disease and in aiding in staging of patients with an equivocal pelvic exam. It has been particularly helpful in delineating unsuspected omental metastases and lymph node involvement. Computed tomography has been useful in judging patients' response to chemotherapy or radiation therapy and in evaluating patients for recurrent disease.

On MRI endometrial carcinoma most often manifests as expansion of the central area of high signal intensity on T_2-weighted images (Fig. 15-15A). It can also be seen as an area of decreased attenuation within the high-signal endometrial region[22] (Fig. 15-15B).

Preservation of a normal-appearing junctional zone indicates that the tumor is confined to the endometrium. Superficial

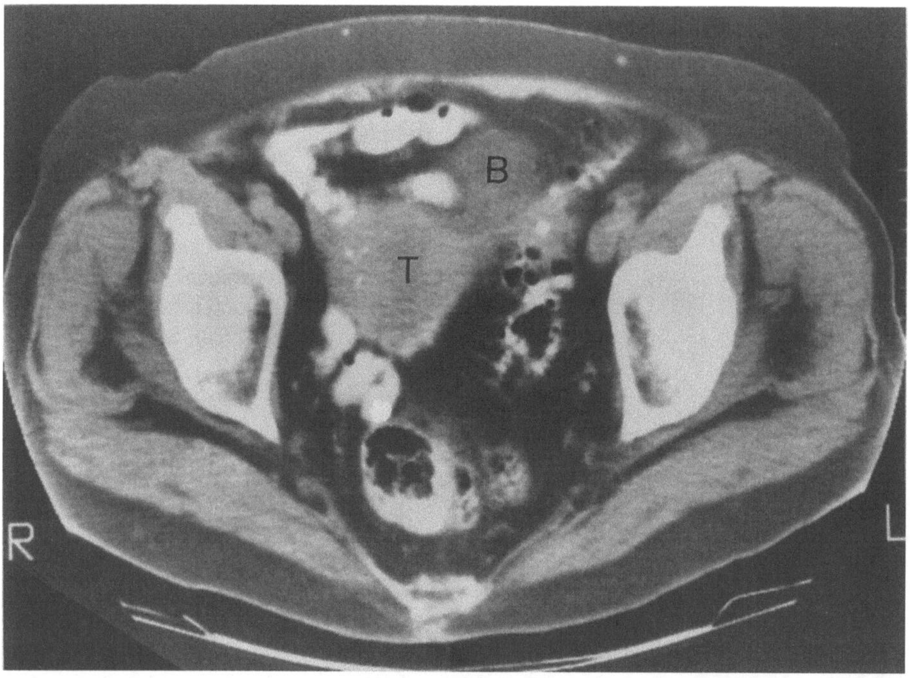

FIGURE 15-14. On CT scan, endometrial carcinoma may appear as an area of decreased density. T = area of tumor involvement in the uterus.

myometrial invasion is depicted by segmental interruption of the junctional zone. Since the junctional zone may not be visible in some postmenopausal women, an irregular interface between endometrium and myometrium may also be used to indicate superficial myometrial invasion (Fig. 15-15C). With deep myometrial penetration, the tumor extends across more than one half of the myometrial thickness.[22]

In conclusion, MRI can reliably reveal the depth of myometrial invasion and thus aid the gynecologist in determining whether surgical sampling of pelvic and aortic lymph nodes is indicated. It can be of value in helping to make the decision between radiation and surgical treatment. As the availability of MR scanners increases, the modality will undoubtedly supplant CT in the work-up of patients with endometrial cancer.

Carcinoma of the Cervix

Accuracy in staging carcinoma of the cervix is important not only in determining prognosis but also in choosing optimal treatment protocols. Treatment of stage I and limited stage II disease includes surgery, radiation therapy, or both.[25–27] There are many variations in tumor size and location included in the stage I category and, therefore, knowledge of exact tumor size and location can aid in making the decision between surgery alone or combined with radiation therapy. Radiation therapy alone is the treatment choice for most patients with more advanced disease (stage IIA or greater). It is of great importance, therefore, to determine the presence or absence of parametrial involvement.

International Federation of Gynecology

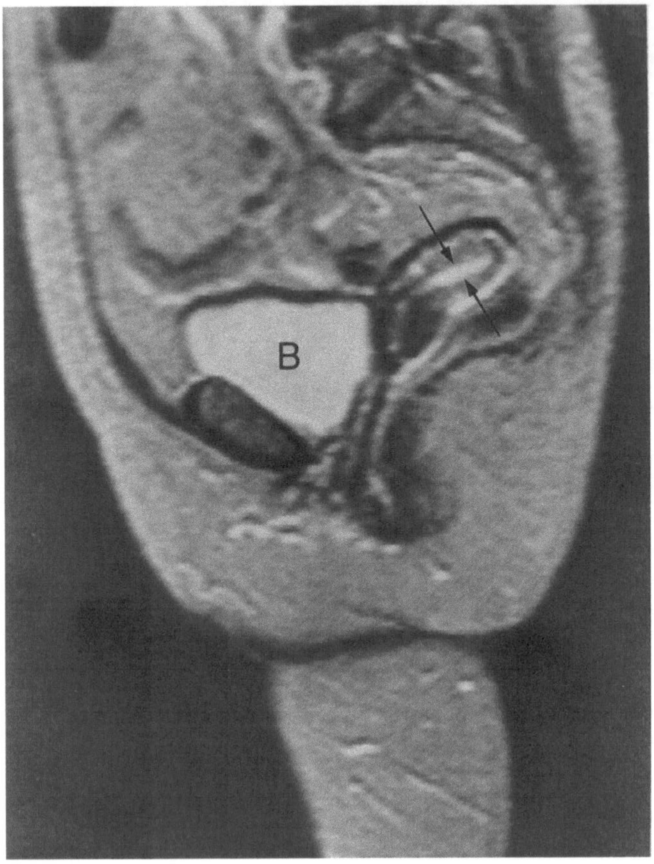

A

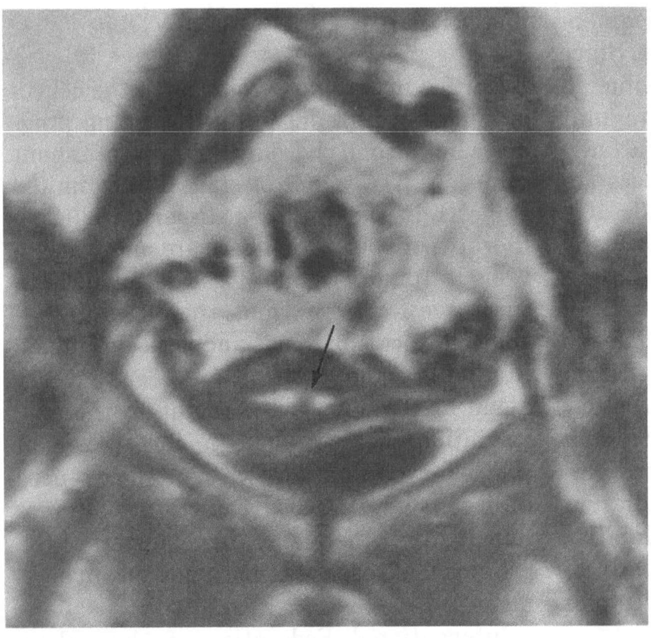

B

FIGURE 15-15. (A) Endometrial carcinoma is often seen on MRI as (nonspecific) expansion of the high-signal-intensity central zone (between arrows). B = urinary bladder. (B) An area of decreased attenuation in the endometrial region (arrow) on a coronal view of the uterus of another patient is nonspecific and may be seen with endometrial cancer, polyp, or blood clot. (*continued*)

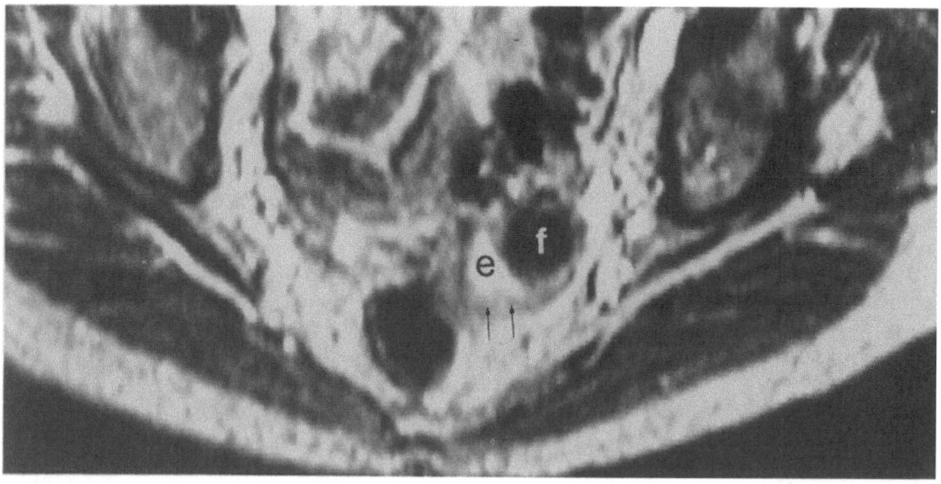

C

FIGURE 15-15. (*continued*) (C) An area of superficial myometrial invasion (small arrows) manifests as partial obliteration of the junctional zone on MRI. e = endometrial tumor, f = fibroid.

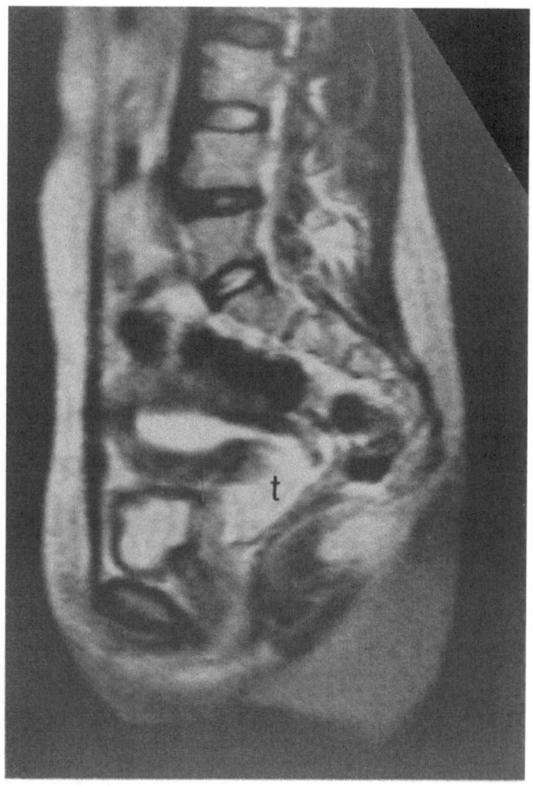

FIGURE 15-16. A bulky high-signal-intensity mass representing a large cervical cancer replaces the normal low-signal cervical fibrous stroma. The anterior and posterior cervical lips are obliterated. t = cervical tumor.

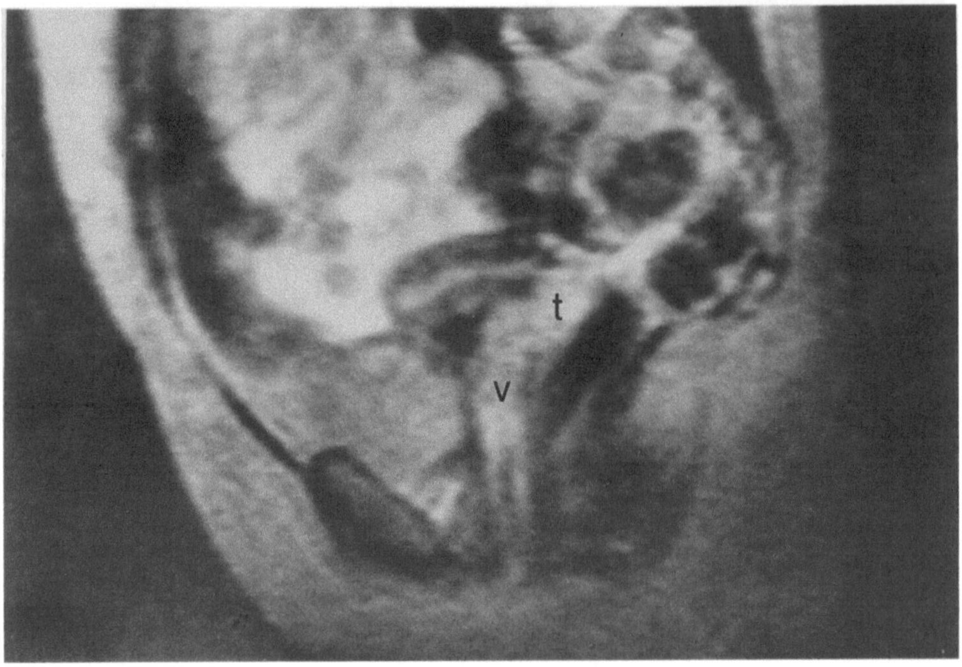

A

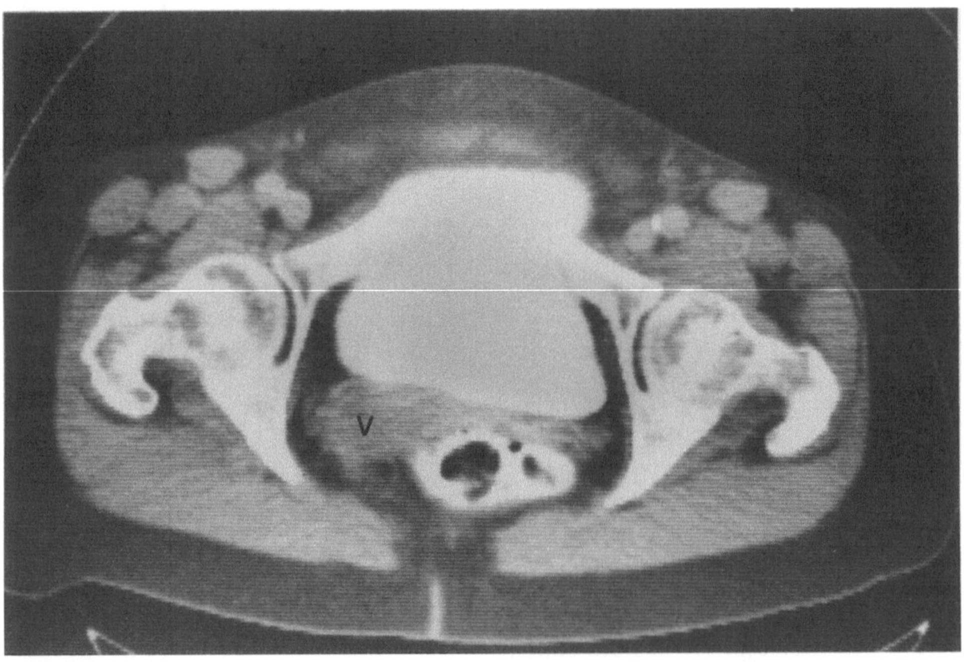

B

FIGURE 15-17. (A) A high-signal-intensity cervical mass is seen to extend to the vagina on this T_2 sagittal view. t = cervical tumor, v = vagina. (B) On CT scan the tumor manifests an asymmetric enlargement of the vagina. (V). The ability to image the uterus in the sagittal plane makes it easier to delineate the vagina from the cervix.

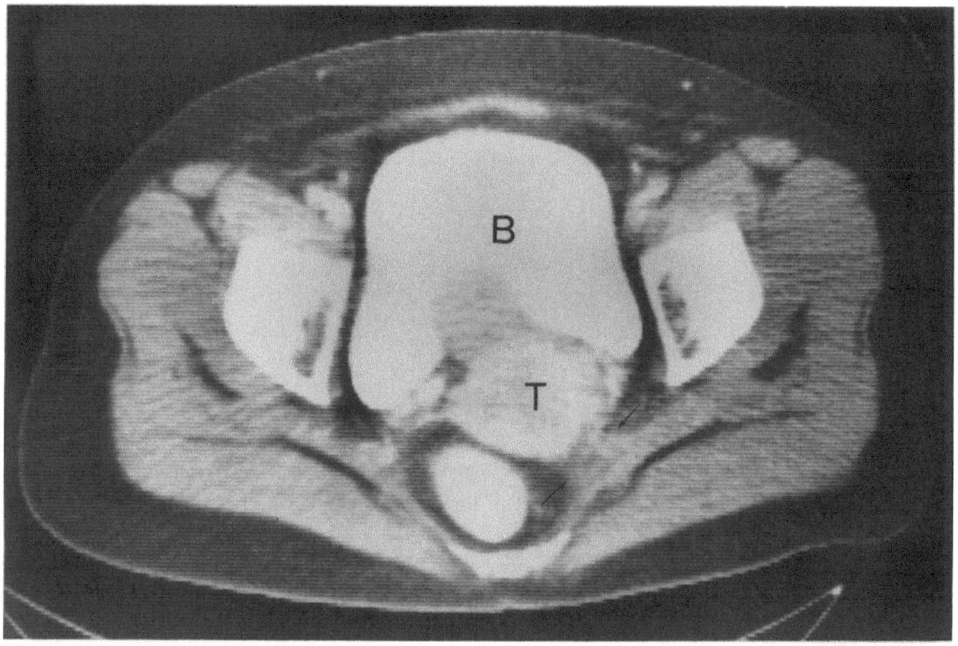

A

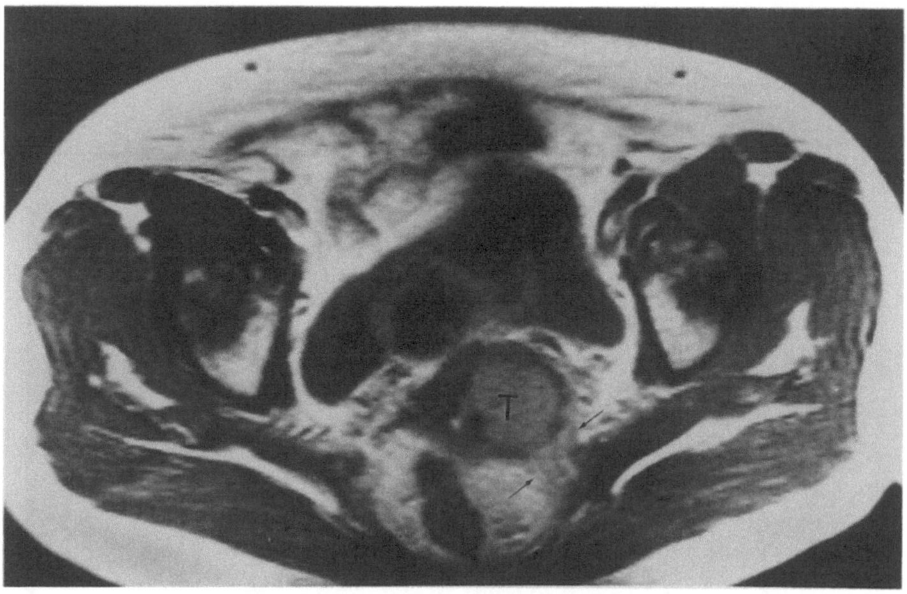

B

FIGURE 15-18. Extension of a cervical tumor to the left pelvic sidewall (T in A, arrows in B) is seen on CT scan (A) and MRI (B). B = urinary bladder.

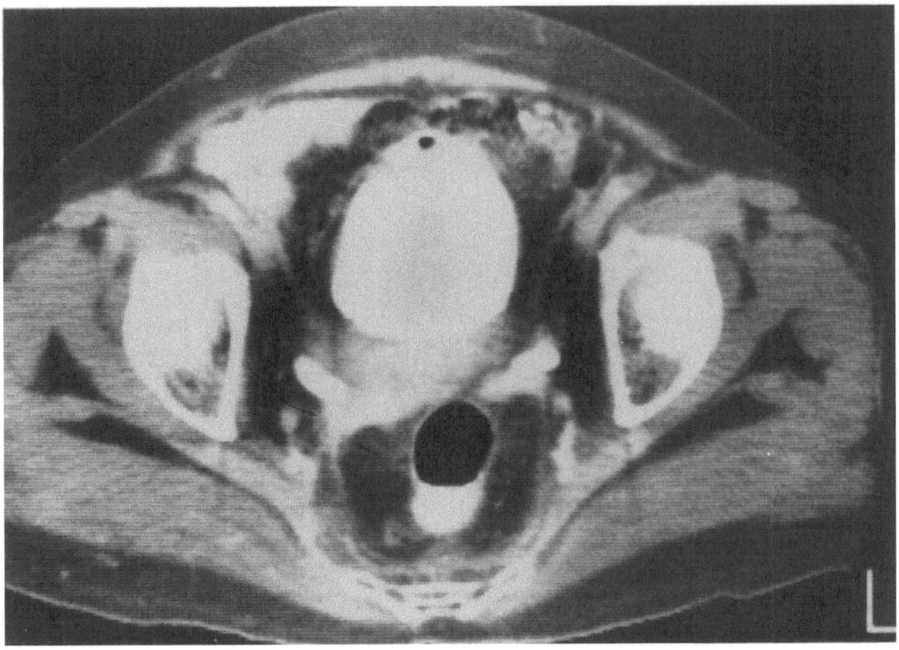

A

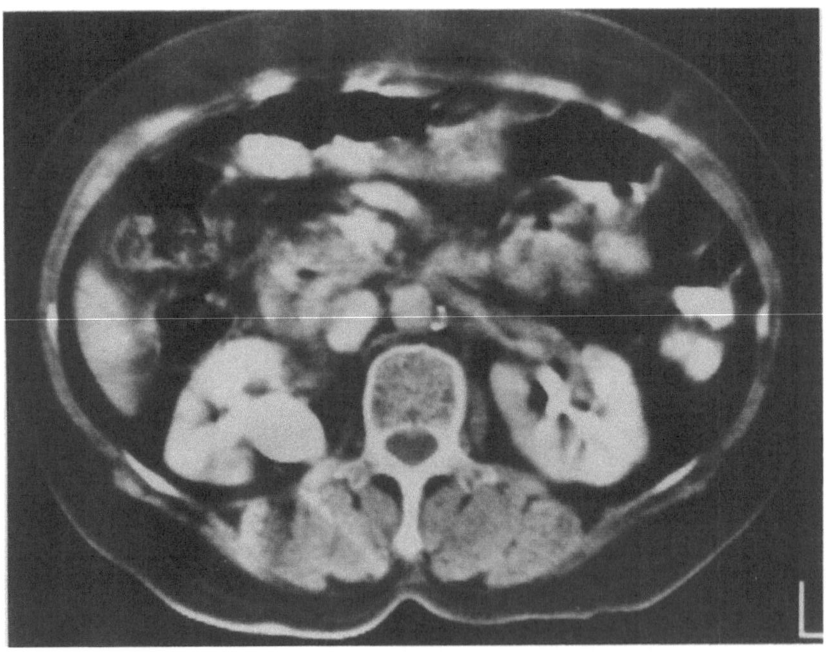

B

FIGURE 15-19. Right parametrial extension (arrows) on CT scan (A) causing right hydronephrosis (B). (C) MRI shows extension of cervical tumor into right parametrium (arrows). (*continued*)

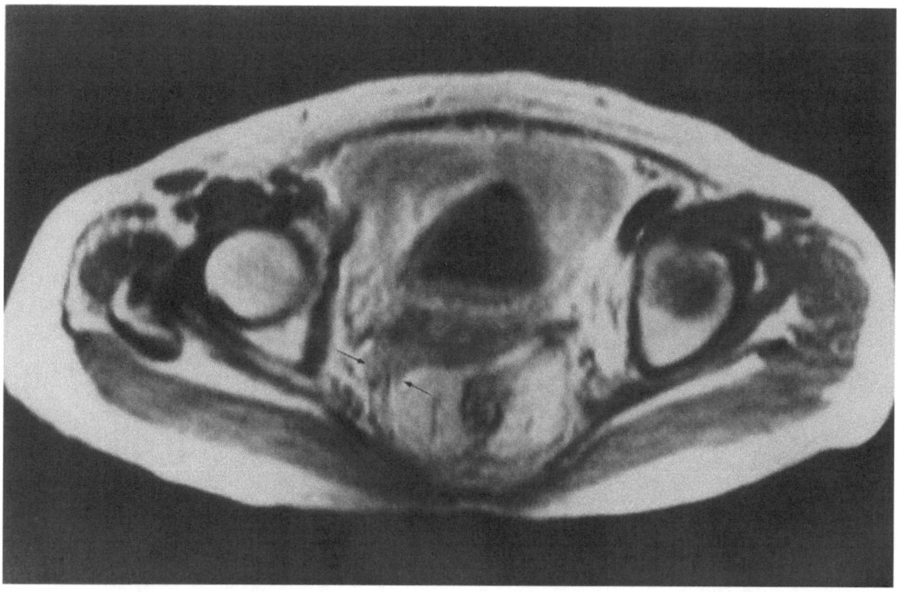

C

FIGURE 15-19. *(continued)* (C)

and Obstetrics (FIGO) clinical staging of cervical carcinoma is often inaccurate because of the limitations in assessing the parametria, pelvic sidewalls, and lymph node chains.[23,25]

A cervical mass may be seen on CT as enlargement of the cervix that may be homogeneous or heterogeneous, overlapping in appearance with the cervical fibroid.[23,25,27] It generally will not be recognizable unless it is large enough to deform the cervical contour, and CT is not generally indicated in patient with carcinoma in situ. Computed tomographic scanning has some limitations in staging, with difficulties in differentiating stages IB and IIB and in understaging pelvic sidewall involvement.[25,27] It does play a role in delineating tumor extent for radiation therapy planning and in diagnosing lymph node metastases or recurrent disease.[28] Accuracy for CT scanning in staging cervical carcinoma is generally reported to be between 60 and 70%.[25,27]

An advantage of MRI is its ability to demonstrate the cervical tumor in multiple planes. Tumor location, depth of cervical invasion, and extension to the uterine corpus can be assessed. However, MRI cannot be used as a screening modality because nabothian cysts, hemorrhage, and changes secondary to biopsy may all have an appearance similar to cervical cancer.

A cervical neoplasm can be recognized on T_2-weighted images as a high-signal-intensity mass that can be delineated from the normal low-signal-intensity cervical stroma[3] (Fig. 15-16). The tumor is classified as stage IB if it is confined to the cervix and/or uterine corpus. The outer low-signal-intensity cervical stroma is sharply demarcated from the pericervical tissue. With tumors extending into the upper two thirds of vagina (stage IIA), there is loss of the normally low signal intensity from the vaginal wall (Fig. 15-17). With pericervical-parametrial involvement (stage IIB), in addition to loss of the low-signal-intensity cervical stroma there is irregularity of the lateral cervical margin, and a parametrial mass or area of abnormal signal is identified on T_2-weighted images. The tumor is staged as IIIA if the high-signal-intensity mass ex-

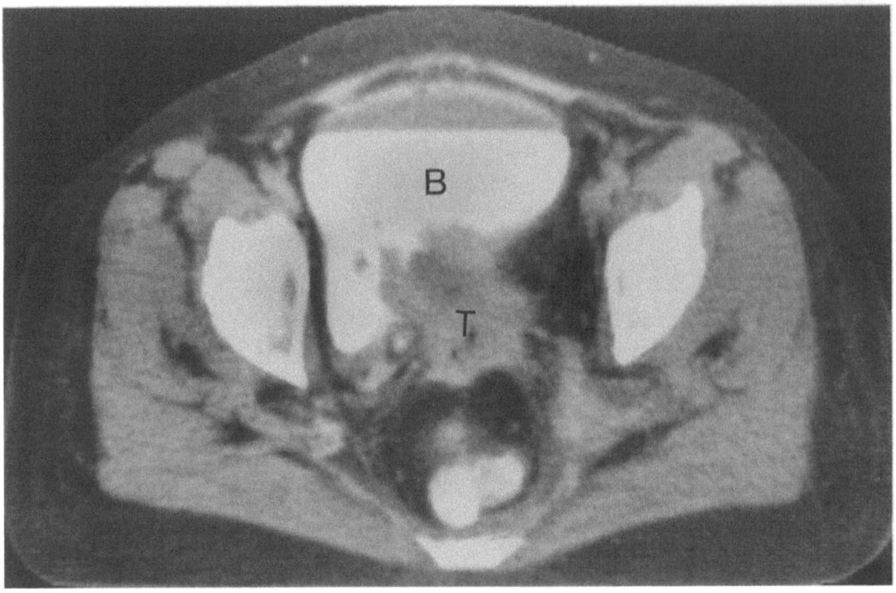

A

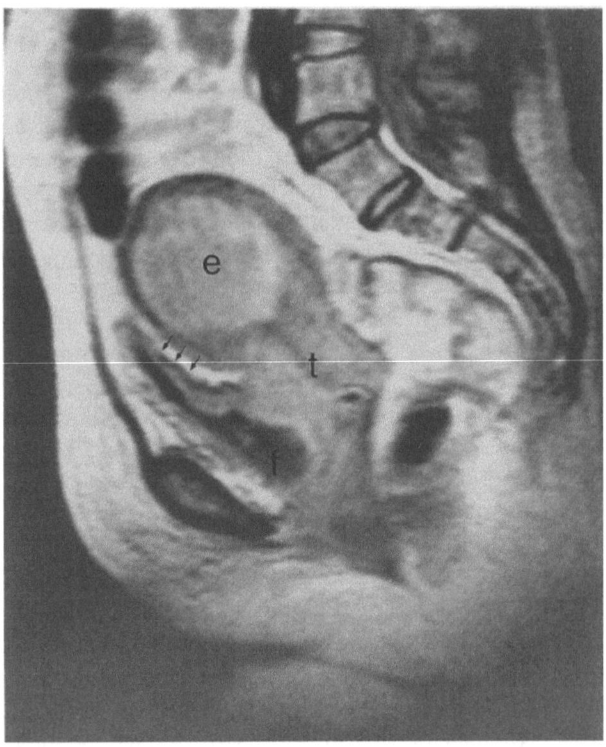

B

FIGURE 15-20. (A) Invasion of the wall of the urinary bladder (B) by cervical carcinoma is seen on CT scan as an irregular interface between the tumor (T) and the bladder. (B) On MRI, this is seen on T_2-weighted sagittal view as loss of the fat plane between the tumor and the bladder and obliteration of the low-signal-intensity line representing the bladder wall (small arrows). t = cervical tumor, e = obstructed endometrial cavity, f = foley catheter balloon.

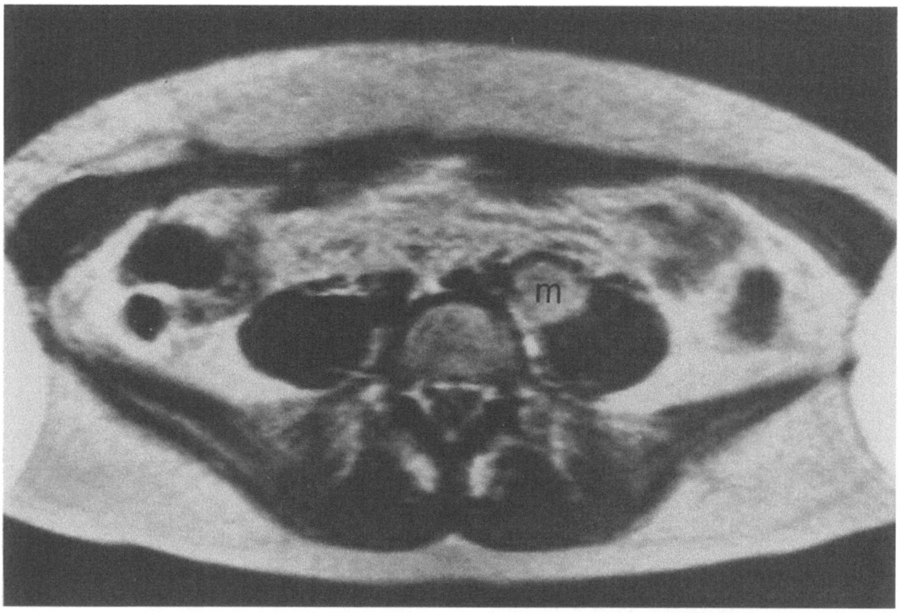

A

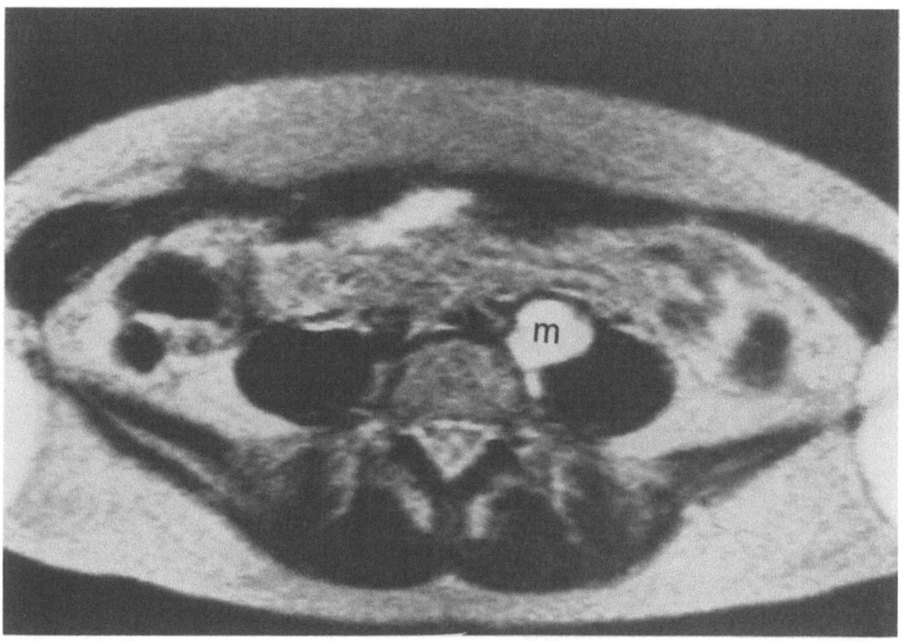

B

FIGURE 15-21. Axial MRI views show a moderate-signal-intensity mass (A) that increases in signal intensity with more T_2 weighting (B) and represents metastatic involvement of the left common iliac lymph nodes from a primary cervical carcinoma. m = malignant lymph nodes.

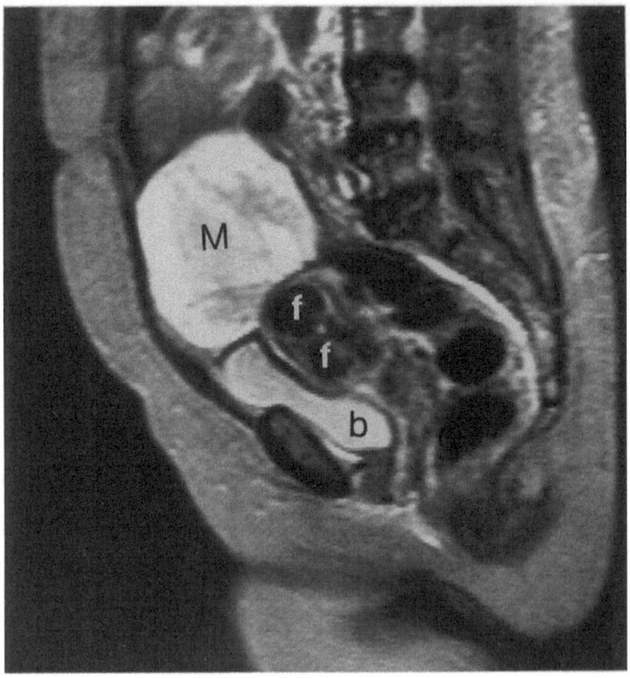

A

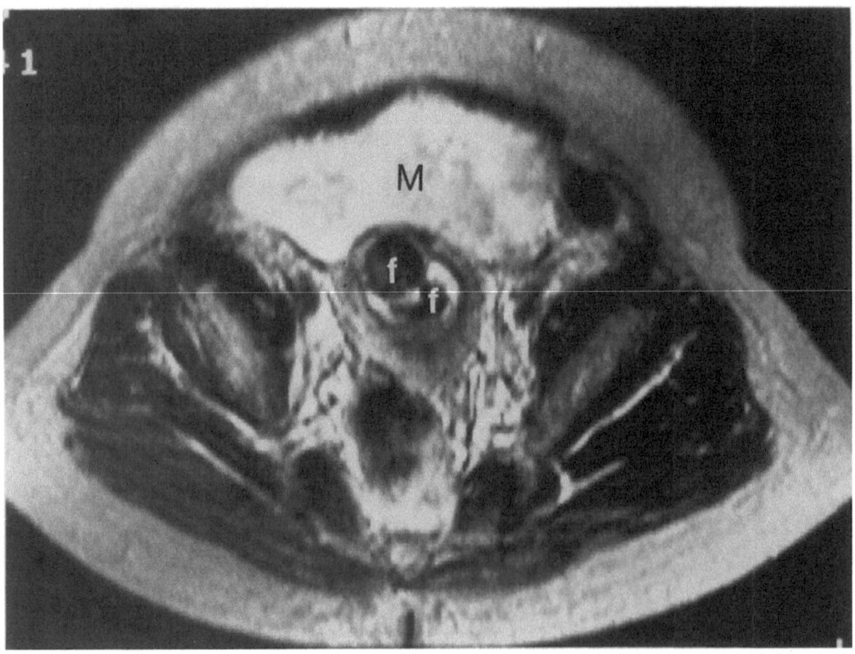

B

FIGURE 15-22. Sagittal (A) and axial (B) views demonstrate the presence of a large complex ovarian mass (M). The uterus is also abnormal, containing multiple fibroids (f). b = urinary bladder.

tends to the lower third of the vagina and as stage IIIB if it extends to the pelvic sidewall (Fig. 15-18) or is associated with ureteral obstruction at the level of the cervix (Fig. 15-19). Tumor involvement of bladder (Fig. 15-20) or bowel wall (stage IVA) is shown by loss of the normal low signal of the wall of these structures on T_2-weighted images and associated loss of fat planes between them and cervix. Lymph node metastases may be identified by nodal enlargement and variable increase in signal intensity on T_2 sequences[29,30] (Fig. 15-21).

Ovarian Tumors

At present MRI has a limited role in evaluating ovarian tumors. In cases in which ultrasound is equivocal, the ability of MRI to delineate uterine zonal anatomy and to image the pelvis in multiple planes can help determine whether a pelvic mass is of uterine or ovarian origin (Fig. 15-22). Axial and coronal views are most helpful in this regard. Magnetic resonance imaging can add to the characterization of an ovarian mass by its ability to distinguish the presence of fat as in a dermoid, or blood as in endometriomas.

Conclusion

Ultrasound is likely to remain the initial procedure for evaluating clinically suspected pelvis masses. Because of its availability and rapid scanning times, CT is useful in evaluating complex pelvic masses and in staging gynecologic malignancies. Magnetic resonance imaging has an important role in evaluating patients in whom ultrasound results are inadequate. It is extremely valuable in evaluating congenital anomalies of the genitourinary tract. Already, it is adding important information in the staging of gynecologic malignancies and may supplant CT in this regard in the future, particularly as the technology continues to advance and faster scanning techniques become available. Finally, through its excellent tissue characterization and possibly through the future use of spectroscopy, MRI may add valuable information to our understanding of the changes that occur in the normal uterus under the hormonal influences that occur in the menstrual cycle and with aging.

References

1. Blies JR, Elias JH, Kopecky KK, et al: Assessment of primary gynecologic malignancies: comparison of 0.15 T resistive MRI with CT. AJR 1984;143:1249–1257.
2. Heiken JP, Lee JKT: MR imaging of the pelvis. Radiology 1988;166:11–16.
3. Hricak H: MRI of the female pelvis: a review. AJR 1986;146:1115–1122.
4. McCarthy S, Tauber C, Gore J: Female pelvic anatomy: MR assessment of variations during the menstrual cycle with the use of oral contraceptives. Radiology 1986;160:119–123.
5. Haynor DR, Mack LA, Soules MR, et al: Changing appearance of the normal uterus during the menstrual cycle: MR studies. Radiology 1986;161:459–462.
6. Janus CL, Wiczyk HP, Laufer N: Magnetic resonance imaging of the menstrual cycle. Magn Reson Imaging 1988;6(6):669–674.
7. Ter-Pogossian MM: Physical principles and instrumentation, in Lee JKT, Sagel SS, Stanley RJ (eds): Computed Body Tomography. New York, Raven Press, 1986, pp 1–7.
8. Pykett IL, Newhouse JH, Buonanno FS, et al: Principles of nuclear magnetic resonance imaging. Radiology 1982;143:157–168.
9. Fullerton GD: Magnetic resonance imaging signal concepts. RadioGraphics 1987; 7(3):579–596.
10. Schwartz JL, Crooks LE: NMR imaging produces no observable mutations or cytotoxicity in mammalian cells. Radiology 1983; 139:583–585.
11. Wolff S, Crooks LE, Brown P, et al: Tests for DNA and chromosomal damage induced by nuclear magnetic resonance imaging. Radiology 1980;136:707–710.
12. Hricak H, Chang YCF, Thurnher S: Vagina: Evaluation with MR imaging. Part I: Normal anatomy and congenital anomalies. Radiology 1988;169:169–174.
13. Dietrich RB, Kangarloo H: Pelvic abnormalities in children: assessment with MR imaging. Radiology 1987;163:367–372.

14. Mintz MC, Thickman DI, Gussman D, et al: MR evaluation of uterine anomales. AJR 1987;148:287–290.
15. Togashi K, Nishimura K, Itoh H, et al: Vaginal agenesis: classification by MR imaging. Radiology 1987;162:675–677.
16. Vainright JR, Fulp CJ, Schiebler ML: MR imaging of vaginal agenesis with hematocolpos. J Comput Assist Tomogr 1988;12(5):891–893.
17. Dudiak CM, Turner DA, Patel SK, et al: Uterine leiomyomas in the infertile patient: preoperative localization with MR imaging versus US and hysterosalpingography. Radiology 1988;167:627–630.
18. Janus CL, White M, Dottino P, et al: Uterine leiomyosarcoma—magnetic resonance imaging. Gynecol Oncol 1989;32:79–81.
19. Togashi K, Nishimura K, Itoh K, et al: Adenomyosis: diagnosis with MR imaging. Radiology 1988;166:111–114.
20. Hricak H, Demas BE, Braga CA, et al: Gestational trophoblastic neoplasm of the uterus: MR assessment. Radiology 1986;161:11–16.
21. Mirich DR, Hall JT, Kraft WL, et al: Metastatic adnexal trophoblastic neoplasm: contribution of MR imaging. J Comput Assist Tomogr 1988;12(6):1061–1067.
22. Hircak H, Stern JL, Fisher MR, et al: Endometrial carcinoma staging by MR imaging. Radiology 1987;162:297–305.
23. Lee JKT, Balfe DM: Pelvis, in Lee JKT, Sagel SS, Stanley RJ (eds): Computed Body Tomography. New York, Raven Press, 1986, pp 393–413.
24. Dore R, Moro G, D'Andrea F, et al: CT evaluation of myometrium invasion in endometrial carcinoma. J Comput Assist Tomogr 1987;11(2):282–289.
25. Walsh JW, Vick CW: Staging of female genital tract cancer, in Walsh JW (ed): Computed Tomography of the Pelvis. New York, Churchill Livingstone, 1985, pp 163–184.
26. Rubens D, Thornbury JR, Angel C, et al: Stage IB cervical carcinoma: comparison of clinical, MR and pathologic staging. AJR 1988;150:135–138.
27. Whitley NO, Brenner DE, Francis A, et al: Computed tomographic evaluation of carcinoma of the cervix. Radiology 1982;142:439–446.
28. Walsh JW, Amendola MA, Hall DJ, et al: Recurrent carcinoma of the cervix: CT diagnosis. AJR 1981;136:117–122.
29. Togashi K, Nishimura K, Itoh K, et al: Uterine cervical cancer: assessment with high-field MR imaging. Radiology 1986;160:431–435.
30. Waggenspack GA, Amparo EG, Hannigan EV: MR imaging of uterine cervical carcinoma. J Comput Assist Tomogr 1988; 12(3):409–414.

16

Hysteroscopy and Hysteroscopic Surgery

Avner Hershlag and Alan H. DeCherney

Historical Remarks

Why has it taken 120 years for hysteroscopy to attain its present position as an acceptable mode of diagnosis and treatment of intrauterine pathology? While access to the uterine cavity was always desired, unlike other cavities the uterus contained a potential space only, measuring about 3 to 8 mL. The frustration caused by the lack of good illumination and no means for distention of the uterine cavity led Munde in 1880[1] to describe hyststeroscopy as a "fleeting glimpse of the endometrium [that] is of little clinical value," compared to the information derived from "using the tip of the index finger, the proverbial "edge of the gynecologist." Indeed, only a glimpse of the uterine cavity was available when Panteloni[2] performed the first hysteroscopy in 1869 using reflected candlelight from a concave mirror. Although the modern era of endoscopy started with Max Nitze's cystoscope in 1879[3] the breakthrough was not until 50 years later when Rubin, in 1925,[4] used for hysteroscopy a modified cystourethroscope with CO_2 as the distention medium. He also suggested using a cutting instrument through a second channel, and noted that the endometrial cavity is less vascular in the postmenstrual stage. Subsequent developments of the panoramic hysteroscope were all further modifications of the cystoscope. Illumination of the cavity improved rapidly following the introduction of cold light fiberoptics by Fourestier and colleagues.[5] Proper visualization and the ability to perform intrauterine surgery were enhanced with the introduction of a high-molecular-weight dextran (Hyskon) as a distention medium, by Edström and Fernström in 1970.[6] Besides the panoramic hysteroscope, two additional instruments that were developed in the modern era are the contact hysteroscope, by Marleschki[7] in 1968, later improved by Vulmiere and Barbot,[8] and the microhysteroscope by Hamou and Coupez in 1980.[9]

Indications and Contraindications

Hysteroscopy is gradually becoming a common diagnostic tool used not only in patients with infertility or pregnancy wastage or abnormal uterine bleeding, but also in a variety of other situations, including retained foreign bodies and more recently in obstetrics (Table 16-1). The direct view of the uterine cavity afforded by hysteroscopy offers some distinct advantages over other diagnostic methods such as dilation and curettage (D&C), endometrial biopsy, hysterosalpingography, and ultrasound. The main strength of hysteroscopy, is the option for intrauterine surgery, allowing concomitant treatment of diagnosed lesions. Possible hysteroscopic procedures are listed in Table 16-2.

Should hysteroscopy be a routine diagnos-

TABLE 16-1. Diagnostic indications for hysteroscopy

- Abnormal uterine bleeding
  - Premenopausal
  - Postmenopausal
    - directed biopsies
    - staging endometrial cancer
  - Planned myomectomy
- Infertility or pregnancy wastage
  - Abnormal hysterosalpingography (HSG)
  - Recurrent pregnancy loss
  - Second trimester pregnancy loss
  - Long-standing infertility and normal HSG
- Foreign bodies
  - Intrauterine device
  - Other foreign bodies (bone)
  - First trimester termination failure
- Obstetric indications (investigational)
  - Directed chorionic villous sampling
  - Amnioscopy
  - Fetoscopy, embryoscopy (intrapartum)

tic procedure? In over 300 fertile women requesting a transcervical sterilization procedure, the incidence of intrauterine lesions was 13%.[10] However, the rate of abnormal findings was 58 to 71% and 40 to 62% when hysteroscopy was done for abnormal uterine bleeding and infertility, respectively.[10]

TABLE 16-2. Hysteroscopic procedures

- Intrauterine
  - Abnormal uterine bleeding
    - excision of polyps
    - excision of myomas
    - directed biopsies (uterus, cervix)
    - endometrial ablation
  - Infertility and pregnancy wastage
    - lysis of adhesions
    - resection of uterine septum
    - excision of myomas and polyps
  - Foreign bodies
    - IUD retrieval
    - foreign body removal (bone)
    - removal of retained products of conception
- Intratubal
  - Infertility
    - cannulation of the tubes with proximal obstruction
    - gamete intrafallopian transfer (GIFT)
    - tubal preembryo transfer (TPET)
  - Sterilization

Hysteroscopy is contraindicated with intrauterine pregnancy (if obstetric intervention is not contemplated), acute pelvic inflammatory disease (PID), and invasive carcinoma of cervix, and when the surgeon is inexperienced. Relative contraindications include menstruation, cervical stenosis, dense intrauterine adhesions, and postpartum or postabortal uterus.[11]

Hysteroscopic Instruments

There are three types of rigid hysteroscopes in current use: the panoramic hysteroscopes, the contact hysteroscope, and the Hamou microhysteroscope.

The *panoramic hysteroscopes* are all modified cystoscopes. The hysteroscope consists of a telescope and an outer sheath of stainless steel. The telescopes range from 4 to 6 mm in diameter, the metallic sheaths from 5.2 to 8 mm. The optical systems are designed to give a direct view (0°) or a range of oblique views, of which the 30° lens is the most versatile (Fig. 16-1 and 16-2). Operating capabilities are afforded by flexible instruments such as scissors, biopsy forceps, and coagulating electrodes, inserted through a side channel (Fig. 16-3). A modified sheath

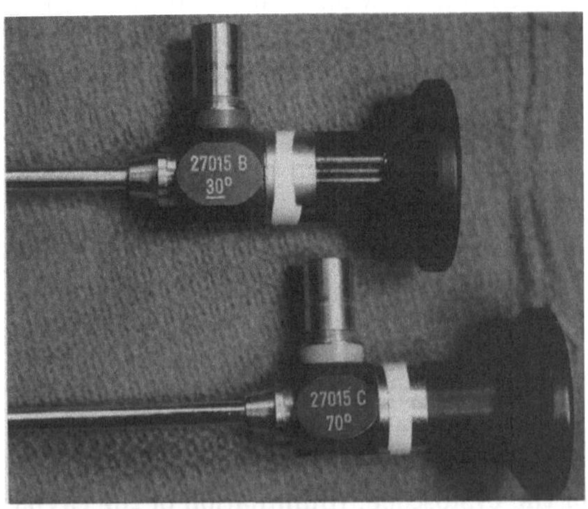

FIGURE 16-1. Thirty-degree and 70° panoramic hysteroscopes. (Courtesy of Karl Storz Endoscopy–America, Inc., Culver City, CA.)

FIGURE 16-2. Different deflection of the lenses of 30° and 70° panoramic hysteroscopes. (Courtesy of Karl Storz Endoscopy–America, Inc., Culver City, CA.)

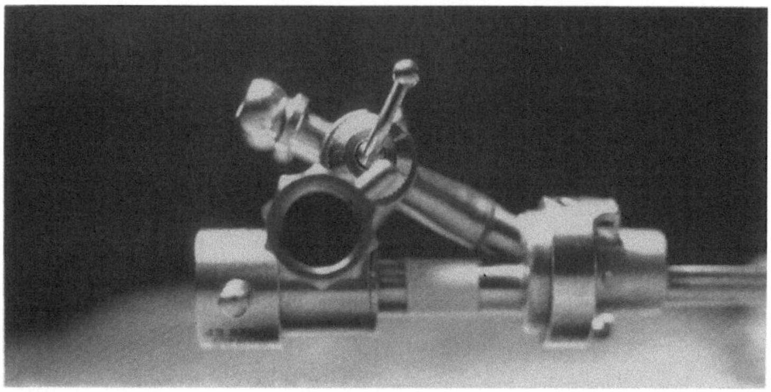

FIGURE 16-3. The operating side channel. (Courtesy of Karl Storz Endoscopy–America, Inc., Culver City, CA.)

with a deflecting mechanism allows control of laser fibers and also tubal cannulation (Fig. 16-4). The resectoscope requires introduction through a wider sheath (Figs. 16-5 through 16-8). Three companies market panoramic hysteroscopes: Olympus Corporation of America, Karl Storz Endoscopy–America Inc., and Richard Wolf Medical Instrument Corporation. A typical set of instruments for operative hysteroscopy is shown in Figure 16-9. As its name implies, the major advantage of the panoramic hysteroscope is its ability to provide an excellent view of the uterine cavity and a realistic appreciation of the size of lesions and their localization in relation to the tubal ostia, along with operative capabilities. The major drawbacks are the need for constant flow of distention medium and the need for cervical dilation—and therefore anesthesia.

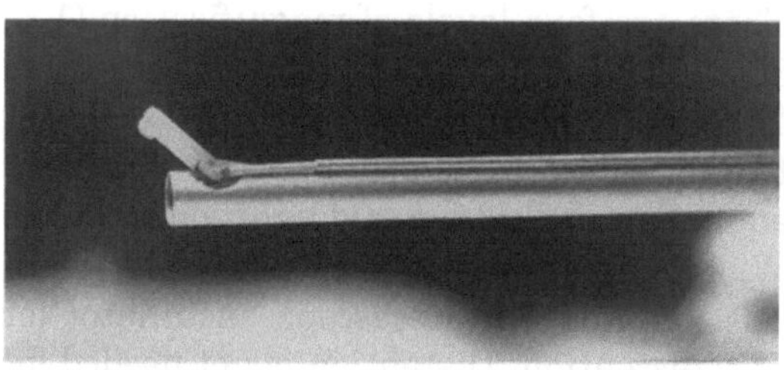

FIGURE 16-4. A sheath with a deflecting mechanism for laser fibers and tubal cannulation. (Courtesy of Karl Storz Endoscopy–America, Inc., Culver City, CA.)

The *contact hysteroscope* consists of three components: the lens system, a cylindrical chamber, and an eyepiece (magnification of 1.6×). An optical glass stem surrounded by metal serves as a light guide and a magnifying optical system. The coating on the end of the rod lens prevents blood from interfering with vision. These hysteroscopes are available in 4-, 6-, and 8-mm outer diameter, manufactured by MTO company in Paris and imported and distributed in the United States by Advanced Biomedical Instruments (Woburn, MA). Advantages of the contact hysteroscope include: a clear image, since blood and tissue fluids are squeezed away from the object in view; very little distortion of the image, because the angles of entry and

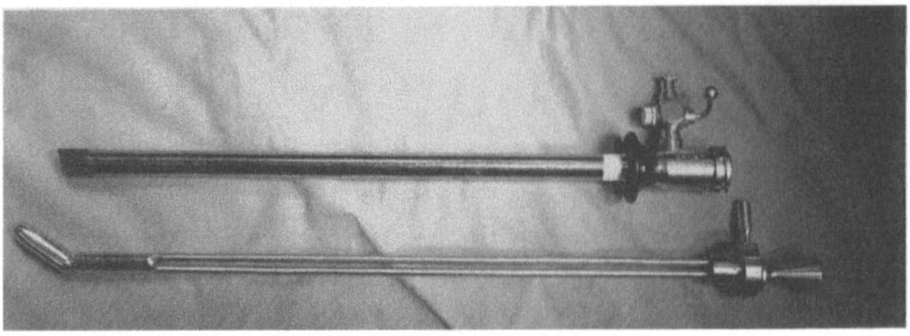

FIGURE 16-5. The wider introducer and scope necessary for using the resectoscope. (Courtesy of Karl Storz Endoscopy–America, Inc., Culver City, CA.)

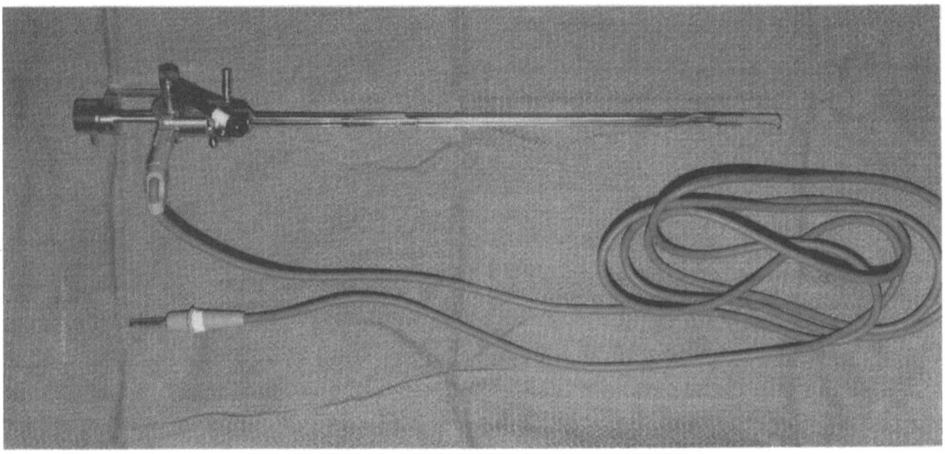

FIGURE 16-6. A modified urologic resectoscope. (Courtesy of Karl Storz, Endoscopy–America, Inc., Culver City, CA.)

exit of light rays are identical, and all rays are transmitted through the same channel[12]; no need for distention media; and no need for cervical dilation when the 4-mm scope is used in nulliparous women and the 6-mm scope in multiparous.[13] Disadvantages of contact hysteroscopes are the lack of a panoramic view and of operative capabilities. Photodocumentation is available from Baggish and Barbot[8,14,15] and Valle.[16]

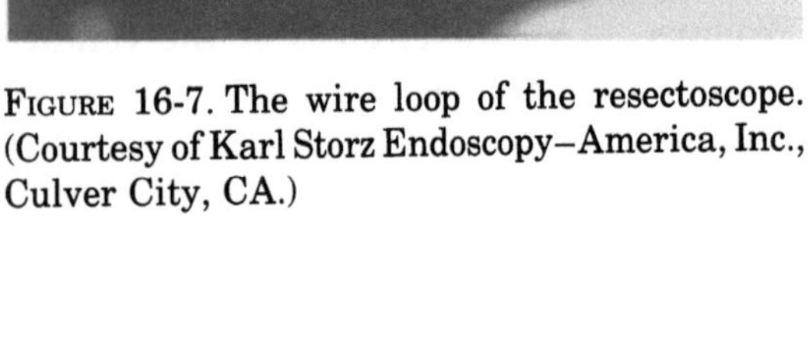

FIGURE 16-7. The wire loop of the resectoscope. (Courtesy of Karl Storz Endoscopy–America, Inc., Culver City, CA.)

The *Hamou microhysteroscope* utilizes a 4-mm telescope and a 5.2-mm outer sheath. There are four levels of magnification (1 ×, 30 ×, 60 ×, and 150 ×), the first two designed for panoramic hysteroscopy and the latter two for microscopic examination. Ancillary equipment includes a cold light source of 150 W, a flexible fiberoptic cable for light transmission, and a hysteroscopic insufflator for CO_2 distention. Distention is not needed for the microscopic settings. Cervical dilation is infrequent and anesthesia can be minimized. Hamou performed all hysteroscopics in the

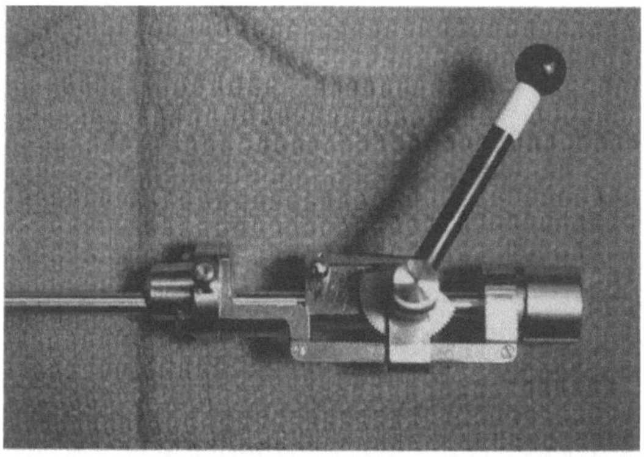

FIGURE 16-8. The handle of the resectoscope, allowing manipulation of the wire loop. (Courtesy of Karl Storz Endoscopy–America, Inc., Culver City, CA.)

office with no anesthesia.[17] The higher magnification allows for a detailed "cytologic" view, especially when using Lugol's solution to stain the squamocolumnar junction. However, microhysteroscopy has not replaced colposcopy. Photodocumentation is available from Taylor and Hamou.[18]

Technique of Hysteroscopic Examination

Hysteroscopy is best performed in the proliferative phase, since in the secretory phase endometrial thickness and folds make interpretation difficult and bleeding likely. The type of hysteroscope chosen depends on the diagnostic capabilities required, anticipated intrauterine surgery, and availability. The magnifications of the Hamou microhysteroscope make it the preferred instrument when visualization of the endocervix is desired in combination with a panoramic view of the uterus. The contact hysteroscope, which necessitates no distention media and no light source, is most useful in the office setup. Panoramic hysteroscopes are the most

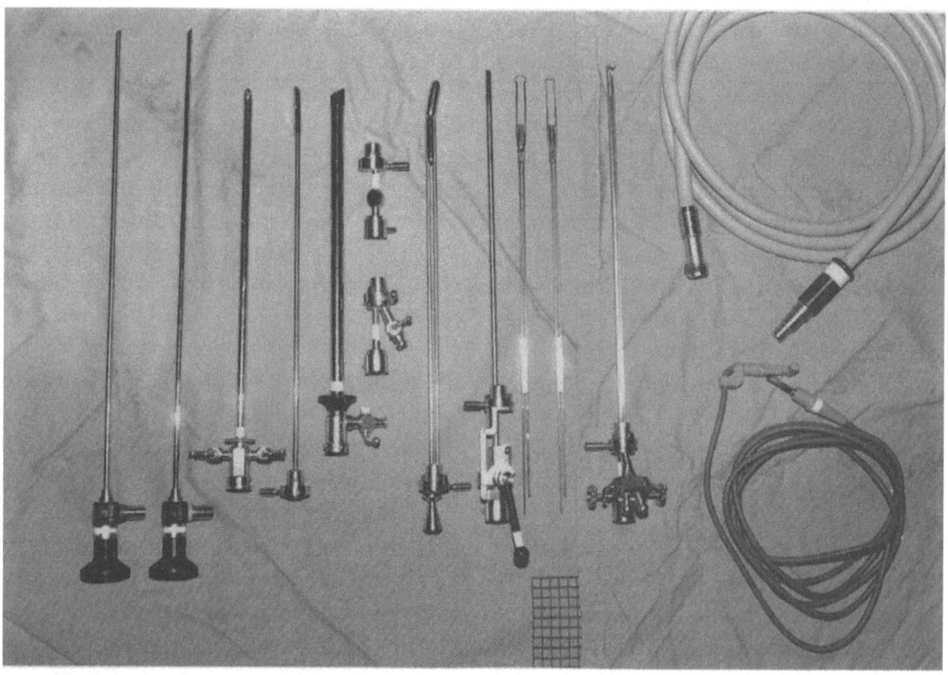

FIGURE 16-9. A full set of hysteroscopic equipment. (Courtesy of Karl Storz Endoscopy–America, Inc., Culver City, CA.)

widely used, especially for a combined diagnostic and therapeutic procedure.

For the panoramic hysteroscope some cervical dilation is usually required, and a paracervical block should be done if the patient is not under general anesthesia. A weighted speculum is inserted into the vagina, the cervix thoroughly cleansed with sterile solution, and the anterior lip of the cervix grasped with a tenaculum. The cervix should be dilated just enough to allow introduction of the hysteroscope with the cervix tightly surrounding it to prevent excessive leakage. Following insertion into the cervical canal, the instrument is guided into the uterine cavity under direct view. Distention medium is instilled continuously through a side channel. A panoramic view aids in orienting the surgeon and should be followed by a systematic inspection of all four walls. Special attention should be paid to the cornual areas surrounding the tubal ostia.

Concomitant laparoscopy is recommended whenever operative hysteroscopy is performed, especially using the resectoscope. It reduces the chances of uterine perforation, and keeps the bowel away to prevent thermal injury to the bowel and perforation. Ultrasonography has been used to help localize the tip of the hysteroscope, especially with 5% dextrose in water as the distention medium,[19] but has not gained wide usage.

Office hysteroscopy may be performed using the contact hysteroscope or a small panoramic hysteroscope with minimal inconvenience to the patient. According to Siegler,[20] in 75% of patients a hysteroscope with an outside sheath of 5 mm (No 15 French) can be introduced into the uterine cavity without the need for a paracervical block or anesthesia.

Distention Media

Low-viscosity fluids, such as 5% glucose in water or saline solution, are inexpensive and provide good uterine distention and visualization. Saline should not be used with electrosurgical instruments. The fluid may be instilled via a 1-L infusion bag, with the pressure controlled by a blood pressure cuff inflated to 80 to 120 mm Hg. A 10-minute procedure will use 150 cc. Pressures may have to be increased to visualize the tubal ostia.[21] Dextran and normal saline are miscible with blood. Low-viscosity fluids are especially good for short office procedures, and are not recommended for operative hysteroscopy. Hyperglycemia, accompanied by proportional hyponatremia, was reported recently as a complication where the procedure lasts 20 minutes or longer.[22]

Carbon dioxide provides excellent visibility of the uterine cavity and allows good photographic documentation of findings. Insufflation equipment allows delivery pressures of up to 200 mm Hg and a maximal flow rate of 100 mL/minute. For a routine hysteroscopy, pressures should be set at 100 mm Hg and flow rates at 40 to 60 mL/minute. Troublesome gas bubbles may obscure the view if there is fluid or mucus within the uterine cavity. In addition, surgical manipulations are more difficult when CO_2 is used, because bleeding rapidly becomes a hindrance. The use of CO_2 is considered safe, and no changes in electrocardiograms, P_{CO_2}, or pH have been reported.[23–25] Experimental evidence shows that in dogs increasing the flow rate to above 400 mL/minute results in the appearance of tachypnea and arrhythmias, and a flow of 1,000 mL/minute for 60 minutes is lethal.[26] Insufflators used for laparoscopy should never be used for hysteroscopy. Shoulder pain is common but can be minimized by placing the patient in a slight Trendelenburg position during the procedure.

Dextran 70, or Hyskon, is 32% dextran (molecular weight of 70,000) in 10% dextrose. Hyskon is optically clear, electrolyte free, nonconductive, biodegradable, and immiscible with blood, the latter quality making it particularly suitable for operative hysteroscopy. Hyskon is introduced into a side channel via intravenous tubing. It may be administered from a manually operated 50-cc syringe, or from an electric pump that the surgeon operates by a foot pedal, where the pressure may be increased to 600 mm Hg

(Fig. 16-10). Transtubal leakage may be a problem when intrauterine pressure exceeds 150 mm Hg. Concerns relating to the possibility of retrograde passage of endometrial tissue into the peritoneal cavity, leading to endometriosis, have not been substantiated. The high viscosity of Hyskon causes stickiness that upon drying may jam the instruments if not thoroughly washed with warm water after every use. The following complications have been described using Hyskon: anaphylaxis, at an estimated incidence of 1 in 1,000,[27] thought to be due to immunoglobulin G antibodies generated previously in response to beet sugars[28]; a noncardiogenic pulmonary edema, caused by toxicity to the pulmonary capillaries or by an immunologic mechanism[29,30]; and disseminated intravascular coagulation (DIC).[30] It is recommended to keep the volume of Hyskon used at less than 300 cc in order to avoid complications. The potential of Hyskon for infection is unknown. While rapid bacterial growth in vitro in the presence of Hyskon has been shown,[31] others have failed to duplicate these results.[32] Leakage of the medium through the dilated cervix continues to be a problem. Gimpelson[33] devised a four-tooth tenaculum with four knobs positioned proximal to each tooth to prevent reflux.

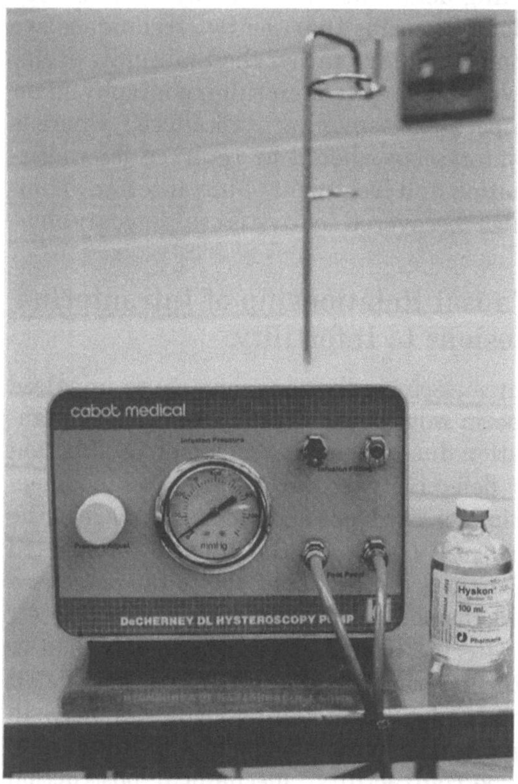

FIGURE 16-10. The DeCherney DL pump for automatic instillation of the Hyskon solution, operated by a foot pedal. (Courtesy of Cabot Medical Corporation, Longhorne, PA.)

Complications

Complications of hysteroscopy include uterine perforation, bleeding, and cervical laceration. If a resectoscope is used and bowel is in close proximity to the uterine wall, a bowel wall burn and possibly perforation may occur. An inadvertent bowel biopsy was reported during hysteroscopic lysis of adhesions using a small biopsy forceps.[34] Concomitant laparoscopy would allow direct observation of the uterine wall while keeping the bowel away. Inadvertent hysteroscopy during pregnancy resulted in premature rupture of membranes and preterm labor.[35] The complications of the procedure and the distention media are listed in Table 16-3.

TABLE 16-3. Complications of hysteroscopy

Procedure related
Cervical laceration
Uterine perforation
Bleeding
Bowel injury
Premature rupture of the membranes and preterm labor in undiagnosed pregnancy
Complications of distention media
5% dextrose in water
hyperglycemia (+ hyponatremia)
CO_2
shoulder pain
Hyskon
anaphylaxis
noncardiogenic pulmonary edema
DIC

Hysteroscopic Procedures

Infertility and Pregnancy Wastage

Hysteroscopy is most commonly performed in the infertile patient with an abnormal hysterosalpingogram. It should also be considered in patients with infertility in whom a thorough work-up failed to explain their infertility. Such a work-up would typically include a history and physical, recording of basal body temperatures, a midluteal progesterone measurement, an endometrial biopsy, a hysterosalpingogram, a semen analysis, and a postcoital test. A single pregnancy loss in the second trimester and habitual abortion, or three consecutive pregnancy losses in the first trimester, warrant evaluation of the uterine cavity.

Hysterosalpingography allows visualization of the tubes and their patency and provides an outline of the uterine cavity, and may have an added therapeutic benefit.[36] The main drawbacks are the exposure to ionizing radiation, the risk of anaphylaxis, and discomfort to the patient. Conversely, hysteroscopy allows direct visualization of the uterine cavity, identification of the specific type of lesion found and its exact location, and immediate therapeutic intervention.

The differential diagnosis of intrauterine filling defects on hysterosalpingography includes air bubbles, mucus, polyps, submucous fibroids, intrauterine synechiae, a septum, and (rarely) asynchronous uterine contractions.[37] The presence of air bubbles may lead to a suspicion of a myoma, polyp, or adhesion, and local uterine spasm induced by contrast material may resemble a müllerian fusion defect.[38] Conversely, excessive amounts of contrast material may obscure intrauterine lesions. Erroneous interpretation on hysteroscopy may be caused by artifacts due to dislodging of strips of endometrium during the cervical dilation preceding the procedure. Anchored at the two poles, these may be misdiagnosed as adhesions; anchored at one pole, as polyps.[39] Distinguishing between polyps, myomas, and thick shaggy endometrium may be difficult. Thick strips of endometrium have a shredded appearance and can be easily dislodged by gently manipulating the end of the scope. Polyps have a smooth free end and may be distinguished from fibroids by their gentle undulations when the pressure of the distending medium decreases.

While several studies comparing hysteroscopy to hysterosalpingography advocate the diagnostic superiority of hysteroscopy,[38,40–42] others do not support that conclusion.[6,43–45] Some of the differences in conclusions may be accounted for by variations in technique, and by experience in performing either procedure. The middle-of-the road stand is taken by Fayez and associates,[37] who concluded from investigating 400 patients with infertility that hysterosalpingography was as accurate as hysteroscopy in the diagnosis of normal or abnormal uterine cavities, whereas the nature of the intrauterine filling defect was revealed by hysteroscopy alone. It seems that the two techniques are complementary rather than mutually exclusive, and that hysterosalpingography is an important screening procedure, whereas hysteroscopy should be reserved for confirmation and treatment of intrauterine anomalies discovered by hysterosalpingography.

Causal Relationship of Intrauterine Lesions to Infertility

Intrauterine adhesions may cause impaired sperm migration or problems with implantation due to defective endometrial function or defective vascularization of the endometrium on the basis of occlusion damage of the arterial vessels of the uterus.[46] Repeated abortions are probably due to the decreased dimensions of the uterine cavity.

Uterine fibroids leiomyomas may lead to sterility through one of several postulated mechanisms, including: interference in sperm transport, myometrial irritation and hypercontractility, mechanical obstruction of the cornual or cervical regions, endometrial atrophy or ulceration over a submucous myoma, and vascular changes.[47] Similar theories have been suggested for the etiology of fetal wastage. Besides the vascular changes mentioned, myomas may cause increased uterine irritability and contractility, either because of rapid growth and degeneration, or because of alterations in oxytocinase activity.[47] Infertility in patients with uterine septa may be caused by implantation of an embryo onto an avascular septum. About 20% of patients with müllerian fusion defects have reproductive problems.[48]

Access to the Fallopian Tubes

Recent studies of proximal tubal obstruction failed to confirm luminal occlusion histologically in many cases. Instead, some patients have an amorphous material forming a cast of the tubal lumen.[49] Thus, it seems that at least in some, if not most, cases of proximal tubal obstruction the nature of the obstruction is different than in the more distal disease and that resection and reanastomosis could be replaced in the majority of cases by transcervical fallopian tube cannulation. Hysteroscopic cannulation of the proximal tube using a ureteral catheter or a No 19 epidural catheter and lavage, and lysis of peritubal adhesions using the biopsy forceps, have resulted in conceptions.[50–52] Tubal patency in selected cases of proximal tubal obstruction has also been reestablished fluoroscopically using guided wires,[53] polyethylene catheters,[53] or modified balloon angioplasty catheters.[54] Hysteroscopy may be utilized for performing transcervical gamete intrafallopian transfer (GIFT) and tubal preembryo transfer (TPET), thus avoiding laparoscopy and general anesthesia.

Intrauterine Adhesions (Asherman's Syndrome)

Pregnancy is the most common predisposing factor for the development of intrauterine adhesions, occurring in over 90% of cases. Denudation of the basal layer during a postpartum curettage seem to be the most common trigger. A particularly vulnerable period is between 1 and 3 weeks postpartum or postabortion, whereas curettage in the first 48 hours is less conducive to intrauterine scarring.[55,56] Other etiologic factors described include infection, vascular damage, tuberculosis, and constitutional reasons.[57]

The most common symptoms of intrauterine adhesions are infertility (43%), amenorrhea (37%), hypomenorrhea (31%), and habitual abortion (14%). Hysterosalpingography reveals filling defects that are characterized by their irregularity, angulated forms with sharp contours, and homogeneous opacity.

Before operative hysteroscopy was available, treatment of intrauterine adhesions consisted of vaginal lysis of adhesions by D&C, or an abdominal hysterotomy followed by insertion of an intrauterine device (IUD) for 4 to 8 weeks and treatment with conjugated estrogens for 1 month. The overall results were 55% pregnancy rate and 61% term pregnancy rate compared to 45 and 30%, respectively, with no treatment. Normal menses were achieved in over 90% of the patients.[57] Hysteroscopic lysis of adhesions results in a pregnancy rate of 60.3%, a term pregnancy rate of 73%, and restoration of normal menses in 87.2% of patients. The reproductive outcome correlates, to some degree, with the severity of adhesions.[58]

Hysteroscopy should be performed in most cases with a concomitant laparoscopy. Prophylactic antibiotics are given. A panoramic view starting at the internal os allows orientation, although adhesions may make this difficult by obscuring the tubal ostia. The latter may come into view in the process of dividing the adhesions and should be looked for during the procedure. We divide adhesions with the resectoscope, set at 30 W. Severe adhesions that do not allow a reasonable view of the uterine cavity may necessitate a hysterotomy. Others have used scissors,[58–60] the outer sleeve of the hysteroscope,[61] or the Hamou microhysteroscope.[62] Following the surgery, a No 8 or 10 Foley catheter is inserted for 1 week, in addition to treatment with conjugated estrogens (10 mg/day) for 40 days, to help the endometrium regenerate.

Obstetric complications in patients who have previously been treated for intrauterine adhesions include placenta accreta and increta, cervical pregnancy, uterine sacculation, and dehiscence. These may result in cesarean hysterectomies, preterm delivery, and neonatal death.[63,64] A thorough uterine exploration at delivery, a complete initial suction curettage for abortions, and avoidance of curettage during the vulnerable period, 1 to 3 weeks postpartum, may help prevent this syndrome.

Hysteroscopic Metroplasty

Of all müllerian anomalies, only the septate uterus is amenable to hysteroscopic metroplasty. Abdominal metroplasty has involved either a wedge resection of myometrial tissue, actually removing the septum (Jones), or incising the septum without removing it (Tompkins). Such procedures performed on patients with habitual abortion have resulted in a 95% conception rate, 73% term pregnancy rate, and 77% viability. However, this procedure involves a laparotomy and increased risk of pelvic adhesion formation and mandates cesarean delivery.[65]

Hysteroscopic metroplasty has practically replaced the abdominal techniques. The preoperative hysterosalpingogram should be available in the operating room. Concomitant laparoscopy is mandatory to rule out a bicornuate uterus (in the presence of which hysteroscopic surgery should not be employed). Laparoscopy should also help reduce the risk of uterine perforation and injury to bowel or bladder. When CO_2 is used for uterine distention, bleeding may obscure visualization. Thus, it is recommended to inject Pitressin (vasopressin) at the lateral fornices (20 units Pitressin in 50 cc normal saline; inject 6 to 8 mL). When Hyskon is used, the cervix should be dilated to allow egress of the solution.

Flexible, semirigid, and rigid scissors have all been successfully used by different surgeons to resect uterine septa.[66–68] We use an 8-mm No 24 French wire loop urologic resectoscope, applying a cutting current of 30 W.[69] Laser fiberoptics [argon, neodymium:yttrium-aluminum-garnet (Nd:YAG), or potassium–titanyl phosphate (KTP)] have been used as well.[70] Only bare fibers are used, since CO_2-conducting fibers may cause bubbling or gas embolization. Resection is carried out until the tissue is flush with the neighboring endometrium. In a survey of 300 patients from multiple series using scissors or the resectoscope, term pregnancy rate was 62%.[71]

With complete uterine septa (extending to the cervical canal), cervical incompetence is prevented by limiting resection to the uterine part of the septum above the internal os. The other cervical cavity may be identified and blocked by a Foley balloon inflated to 10 cc to prevent escape of the distending medium.[72]

March and Israel[68] resect septa (over 3 cm wide at the top of the uterus) by performing a series of incisions with the scissors along one lateral margin until up to within 0.5 cm of the normal myometrium. After the other lateral margin is gradually incised in the same manner, the remaining short notch is incised.

Intraoperative bleeding is best managed by identifying the bleeding point and directly coagulating it. If the bleeding site cannot be identified, tamponade by a 5- or 10-cc balloon Foley catheter for about 6 hours is very effective in stopping the bleeding.[73]

Broad-spectrum antibiotics should be administered perioperatively, and mild analgesics will control the pain. We encourage the patient to conceive on the second cycle after surgery. Estrogen treatment is not necessary, unless the septum has a very broad base.

Intrauterine Foreign Bodies

Foreign bodies located within the uterine cavity or lodged in the uterine wall have included IUDs, bones, a Heyman capsule, and a plastic suction curette.[74,75] A flat plate x-ray and ultrasound are unreliable in locating a lost IUD, as is D&C, which is not only a blind procedure but often fails to retrieve the IUD. Hysteroscopy is invaluable in removing embedded and fragmented devices.[75,76] Patients with intrauterine bones following spontaneous or induced abortion present with secondary infertility, pelvic pain, or passage of bone fragments. Successful removal of bones by hysteroscopy has been reported to be followed by pregnancies in previously infertile patients.[77]

Abnormal Uterine Bleeding

Abnormal uterine bleeding may be conveniently divided into three categories: premenopausal, postmenopausal, and postabortal. Irregular uterine bleeding in the reproductive age group is more commonly termed "dysfunctional uterine bleeding" (DUB), because anovulation is the primary etiology. Basic work-up should thus include a basal body temperature, a progesterone determination, and an endometrial biopsy. An intrauterine lesion may be suspected with persistent DUB despite several months (3 to 6) of hormonal therapy. Historically, D&C has been used for diagnosis and treatment. However, not only does D&C miss a lesion in up to 35% of cases,[78] but in 60% of patients less than half the cavity was found to be curetted in post-D&C hysterectomy specimen.[79] Several authors have shown the superiority of panoramic hysteroscopy to curettage in making an accurate diagnosis of the uterine cavity.[80,81] Lesions commonly found in this age group include uterine polyps, submucous leiomyomata, and endometrial hyperplasia.

Post- or perimenopausal bleeding has been traditionally managed by endometrial biopsies combined with cervical curettage or a "formal" D&C. Hysteroscopy may be useful either as an initial screening procedure or with recurrent bleeding when a previous D&C failed to reveal pathology.[78] Complete correlation was found between the findings on hysteroscopy and a subsequent curettage in 49 postmenopausal patients.[82] In a large series published by Mencaglia and associates,[83] the Hamou microhysteroscope was used on an outpatient basis in 618 patients 45 years old or older. Hysteroscopy had an accuracy rate of 100% for endometrial neoplasia, 87.5% for high-risk hyperplasia, and 65.2% for low-risk hyperplasia. In their series, no endometrial lesion was detected by biopsy in 54.1% of cases, which casts serious doubt on whether a D&C should be performed in every case of postmenopausal bleeding. Rather, office hysteroscopy with directed biopsies seems to be a better diagnostic tool, besides being safer and more cost effective.

Hysteroscopy may be used at least for initial diagnosis, if not as an alternative to D&C in postabortal bleeding. Successful utilization of hysteroscopy has been reported in cases of incomplete abortion,[84] failure of first trimester termination,[85] and after a second trimester abortion.[86]

Should hysteroscopy be a standard procedure in patients with abnormal uterine bleeding? An abnormal uterine cavity was found in up to 57% of such patients.[87] It is therefore reasonable to suggest that, after an endocrine etiology is excluded, hysteroscopy will give the highest diagnostic yield and will allow appropriate treatment.

Hysteroscopic Myomectomy

Uterine leiomyomas are the most common solid pelvic tumors, occurring in up to 25% of women in the reproductive age group. Symptoms occur in less than half of the patients and include pelvic pain (34%), menorrhagia (30%), infertility (27%), previous fetal wastage (3%), and a pelvic mass (9%).[47]

The advantages of hysteroscopic versus abdominal myomectomy are the same as for metroplasty, and include the avoidance of a laparotomy, shortening hospital stay and cost, and avoiding a uterine incision with possible adhesions and a mandatory cesarean section should the patient conceive. Hysteroscopic resection of leiomyomas is possible with sessile as well as pedunculated myomas.[88,89]

Hysteroscopic resection starts with adequate cervical dilation to allow insertion of the urologic resectoscope, which is set at a 30-W cutting current. Hyskon is used for uterine distention. The loop of the resectoscope is placed on the far side of the tumor and is drawn toward the objective lens until one slice is completed. This "slicing" should be repeated until the myoma is shaved to the level of the surrounding myometrium. A polyp is cut off at the level of the cavity wall as well. Continuing appreciation of the tubal ostia provides orientation and prevents tubal damage. We consider concomitant laparoscopy mandatory to reduce the risk of uterine performation and thermal bowel injury. However, hysteroscopic removal of submucous fibroids in 300 patients with no laparoscopic control has been described.[90] The use of the Nd:YAG or argon laser is possible for pedunculated lesions.

The perioperative management should also include oral prophylactic antibiotics 48 hours after surgery, 10 mg of conjugated estrogens for 30 days to ensure endometrial growth in the ablated area, and the use of a 5- or 10-cc Foley balloon in the uterus for up to 6 hours to control bleeding. The size of some myomas may be decreased significantly with gonadotropin-releasing hormone analogs.[91] This requires treatment for at least 3 months. Such reduction in size may allow the myomectomy to be safely performed hysteroscopically, rather than abdominally, with submucous myomas.

Hysteroscopic myomectomy cures abnormal uterine bleeding in 97% of patients.[47] Resolution of infertility after myomectomy has been more difficult to document.

Intractable Uterine Bleeding

In a small group of patients, commonly used treatment modalities such as estrogens or hysterectomy or even the less common radiation therapy are contraindicated because of a severe cardiac or respiratory disease, a bleeding diathesis, alcoholism, and drug addition. Ablation of the endometrium is a viable alternative in such circumstances, provided endometrial neoplasia has been ruled out by biopsies.

DeCherney and associates[92] use the modified urologic resectoscope to cauterize the entire endometrium with electric coagulation current at 30 W. Satisfactory cauterization is obtained by moving the resectoscope in addition to the wire loop and its handle mechanism. The procedure requires 30 minutes or less, and in case of bleeding postoperative tamponade is achieved by a 5-cc Foley balloon for up to 6 hours. Of 21 women treated this way, 3 women died from their primary disease and all except for 1 of the rest remained amenorrheic. The advantages of the resectoscope are its availability, low cost, short time needed to complete the procedure, and small amount of Hyskon used. Since the resectoscope is a unipolar instrument, potential risks include bowel injury and transection of major vessels.*

Laser ablation of the endometrium is discussed in Chapter 17. Droegemueller and associates[93] performed cryotherapy using a freon probe (freezing point 39.8°C). However, hysterectomy revealed endometrial regeneration in the cornual areas. Subsequent use of nitrous oxide (freezing point 88.5°C) led to more widespread damage.[94] Schenker and Palishuk[95] achieved endometrial ablation with 10% formalin in rabbits. Selective arterial embolization, as used by Schwartz and associates,[96] is an option especially useful in emergencies, requiring highly qualified personnel.

Hysteroscopic Sterilization

Hysteroscopic sterilization was developed in an attempt to devise alternatives to laparoscopic sterilization that would be safer and faster, would require little or no anesthesia, could be performed in the office and therefore be cheaper, and would be potentially reversible. To satisfy these goals, multiple

* Electrocoagulation of the endometrium using a "rollerball" has been recently described.[92a]

techniques have been tried. These are listed in Table 16-4.

Patients to be excluded are those with a fixed retroversion of the uterus, leiomyomata, preexisting tubal disease, or a short postpartum interval.[97]

Preparation of the patient includes a prostaglandin antagonist to control the cramping and the use of prophylactic tetracycline. Tubal spasm during the procedure may be relieved by 1 mL of glucagon or terbutaline, acting as a tocolytic.[98]

Thermal burns have been attempted using both hot and cold cauterization. Electrocautery should be performed in midproliferative phase, using a No 6 or 7 French Silastic-tipped cylindrical electrode and a cauterizing current of 25 W for 8 seconds. In a multicenter study on 587 patients, there was a 57% success rate in performing the occlusion on first attempt. However, there were 25 major complications, including one death.[99] Electrocautery has been generally abandoned because of the high complication rate and the high failure rate due to regeneration of damaged epithelium. Droegemuller et al.[100] reported limited success using a cryoprobe to cause coagulation necrosis and subsequent scarring, first in baboons and then in humans prior to hysterectomy.

A variety of chemical agents have been used experimentally and therapeutically in order to cause inflammation and epithelial destruction with eventual luminal fibrosis. The classes of these agents are listed in Table 16-4. The most effective and least toxic are the cyanoacrylate compounds (tissue adhesives). These are instilled as a liquid that develops into a solid cast, which will start degrading slowly, so that by 16 weeks the tube is totally replaced by connective tissue.[97] A methyl derivative in the form of a special device called Femcept was described by Neuwirth et al.,[101] with a bilateral closure rate of 87% after two instillations and a pregnancy rate of 2 to 9%. The main drawback of these techniques is their necrotizing nature, making them essentially irreversible.

TABLE 16-4. Hysteroscopic sterilization methods

Thermal burns
Electrocautery
Cryocautery
Chemical agents
Caustic agents
Acids and bases
Sclerosing agents
Granuloma-producing agents
Adhesives
Polymer plugs
Silicone
Mechanical devices
Surgical nylon loop
Hydrogel device
Blockage of the uterotubal junction

Transcervical instillation of silicone that polymerizes to form an occlusive cast in the tube, currently marketed under the name Ovabloc (RSP Laboratories, Stamford, CT), has been used. Successful placement has been reported to be 83% by a multicenter study[98] and 91.3% by Loffer.[102] The main reason for failure is tubal spasm. Removal of the plugs is facilitated by a loop at the proximal end. However, the complete cast is easily broken, so the proximal part may be removed hysteroscopically but laparoscopy may be needed to retrieve the distal part. The degree of reversibility of this procedure is still unknown.

Several mechanical devices have been suggested. Surgicala nylon (1 mm) that has a loop at its distal end may be opened to varying widths to fit the intratubal diameter. The loop and a small proximal barb anchor the device. The loop can be inserted in an office setting without anesthesia using the Hamou microhysteroscope at a magnification of 20 ×. The device is loaded into a catheter that is then passed through the uterotubal junction. In 144 of 166 patients the procedure was completed on first attempt. There were four expulsions.[103] Another proposed device is a hydrogel fixed on a nylon skeleton with two small wings to prevent expulsion during the 30 minutes it takes to form the plug.[104] The inability to completely block the tubal lumen has resulted in a high failure rate. Another device is a 4-mm plug with four small metal hooks on the base that anchor the device into the endometrium at the uterotubal junction. In human prehysterectomy trials there was a 90% rate of tubal occlusion and 90% patency rate after removal.[105]

Hysteroscopic sterilization technology continues to evolve. Currently available methods that have been proven safe, easy to perform in the office setting, and reversible still have a high failure rate, thus making them inferior to the more commonly used laparoscopic techniques.

Use of the Hysteroscope in Obstetrics

Endoscopic chorionic villi sampling was described in 65 patients who subsequently underwent termination of pregnancy.[106] When six different sampling techniques were compared, the combination of ultrasonic guidance and direct vision endoscopic biopsy gave the best results.[107] The chorionoscope consists of a forward oblique telescope (30° angle) with a diameter of 3 mm. It is inserted into a sheath 3.7 × 5 mm with an instrument channel. A No 5 French flexible scissors for splitting the decidua and biopsy forceps are used, with saline infusion to distend the uterine cavity.

Intrapartum hysteroscopy was described by Petrikovsky.[108] He used a choledochofibroscope (5.8 mm diameter, 75° angle, 65 cm length) to determine umbilical cord pathology, the presence of meconium, and the integrity of a uterine scar. Such information may be useful in the intrapartum decision-making process.

Conclusions

Hysteroscopy has become a major diagnostic as well as treatment modality in the female patient. In the patient suffering from infertility or pregnancy wastage, it allows visualization as well as surgical treatment of intrauterine adhesions, submucous myomas, and uterine septa. It should be regarded as complementary rather than an alternative to hysterosalpingography. Access to the tubes allows better understanding of proximal tubal obstruction and possible relief of such obstruction without resorting to tubal surgery. The use of a transcervical approach for GIFT is becoming a viable option.

In the patient with abnormal uterine bleeding, hysteroscopy offers distinct advantages over the traditional "blind" D&C, both in directed biopsies and in treatment. Hysteroscopic myomectomy is extremely successful in restoring normal menses. Endometrial ablation in patients who are poor surgical candidates for hysterectomy induces permanent amenorrhea in almost all patients.

As hysteroscopy becomes more available, the practicing gynecologist should be cautioned against an overzealous use of hysteroscopic surgery. Intrauterine surgery should not be utilized prior to a full evaluation of a couple with infertility. Similarly, oligo- or anovulation should be ruled out in abnormal uterine bleeding in the reproductive age group before hysteroscopy is contemplated.

While hysteroscopy has already replaced abdominal surgery in treating most cases of small submucous myomas, intrauterine adheions, and uterine septa, it is rapidly capturing new frontiers not only in the treatment of infertility but also in obstetrics and, to some extent, in gynecologic oncology.

References

1. Munde PF: Minor Surgical Gynecology, New York, W Wood, 1880, p 99.
2. Panteloni D: On endoscopic examination of the cavity of the womb. Med Press Cir 1869;8:26.
3. Nitze M: Über eine neue Behandlungsmethode der Holden des mischlichen Korpers. Med Press Wien 1879;20:851.
4. Rubin JC: Uterine endoscopy: endometroscopy with the aid of uterine insufflation. Am J Obstet Gynecol 1925;10:313.

5. Fourestier M, Gladu A, Vulmiére J: Perfectionments de l'endoscope médicole. Presse Med 1952;60:1292.
6. Edström K, Fernström I: The diagnostic possibilities of a modified hysteroscopic technique. Acta Obstet Gynecol Scand 1970;49:327.
7. Marleschki V: Hysteroscopische Festellung der spontanen Perfesionsschwankungen am menschlichen Endometrium. Zentralbl Gynäkol 1968;90:1094.
8. Barbot J, Parent B, Dubuisson JB: Contact hysteroscopy: another method of endoscopic examination of the uterine cavity. Am J Obstet Gynecol 1980;136:721–726.
9. Hamou J, Coupez F: Microcolpohysteroscopic orifficielle et endocervicale. Paper presented at the Journées de la Societe française de Colposcopie, Paris, 1980.
10. Cooper JM, Houck RM, Rigberg HS: The incidence of intrauterine abnormalities found at hysteroscopy in patients undergoing elective hysteroscopic sterilization. J Reprod Med 1983;28:659–661.
11. Gomel V, Taylor PJ, Yuzpe AA, et al: Laparoscopy and Hysteroscopy in Gynecologic Practice. Chicago, Year Book Medical Publishers, 1986, pp 58–62.
12. Wheeler JM, DeCherney AH: Office hysteroscopy. Obstet Gynecol Clin North Am 1988;15:29–39.
13. Shapiro BS: Instrumentation in hysteroscopy. Obstet Gynecol Clin North Am 1988;15:13–21.
14. Baggish MS: Contact hysteroscopy: a new technique to explore the uterine cavity. Obstet Gynecol 1979;54:350–354.
15. Baggish MS, Barbot J: Contact hysteroscopy. Clin Obstet Gynecol 1983;26:219–241.
16. Valle RF: Hysteroscopy for gynecologic diagnosis. Clin Obstet Gynecol 1983;26:253–276.
17. Hamou JF: Microhysteroscopy. Clin Obstet Gynecol 1983;26:285–301.
18. Taylor PJ, Hamou JE: Hysteroscopy. J Reprod Med 1983;28:359–389.
19. Shaler E, Zuckerman H: Operative hysteroscopy under real-time ultrasonography. Letter to the editor. Am J Obstet Gynecol 1986;155:1360.
20. Siegler AM: Practical office hysteroscopy. The Female Patient 1987;12:77–103.
21. Quinones GR, Alvardo DA, Aznar RR: Tubal catheterization: applications of a new technique. Am J Obstet Gynecol 1974;114:674.
22. Carson SA, Hubert GD, Schriock ED, et al: Hyperglycemia and hyponatremia during operative hysteroscopy with 5% dextrose in water distension. Fertil Steril 1989;51:341–343.
23. Hulf JA, Corall IM, Knights KM, et al: Blood carbon dioxide tension changes during hysteroscopy. Fertil Steril 1979;32:193.
24. Hulf JA, Corall I, Strunin L, et al: Possible hazards of nitrous oxide for hysteroscopy. Br Med J 1975;2:511.
25. Lindenmann HJ: A symposium on advances in fiberoptic hysteroscopy. Contemp Obstet Gynecol 1974;3:115.
26. Lindenmann HJ, Mohr J, Gallinet A, et al: Der einfuss von CO_2-gas während des Hysteroscopie. Geburtshilte Frauenheilkd 1976;36:153.
27. Borten M, Seibert CP, Taymor ML: Recurrent anaphylactic reaction to intraperitoneal dextran 70 used for prevention of postsurgical adhesions. Obstet Gynecol 1983;61: 755.
28. Maddi VI, Wyso EM, Zinner EN: Dextran anaphylaxis. Angiology 1969;20:243.
29. Leake JF, Murphy AA, Zacur HA: Noncardiogenic pulmonary edema: a complication of operative hysteroscopy. Fertil Steril 1987;48:497–499.
30. Zbella EA, Moise J, Carson SA: Noncardiogenic pulmonary edema secondary to intrauterine instillation of 32% dextran 70. Fertil Steril 1985;43:479–480.
31. Berstein J, Mattox JH, Ulrich JA, et al: The potential for bacterial growth with dextran. J Reprod Med 1982;27:77.
32. Elkins TE, Trenthem L, McNeely SG, et al: Potential for in vitro growth of common bacteria in solutions of 32% dextran 70 and 1% sodium carboxymethyl cellulose. Fertil Steril 1985;43:477.
33. Gimpelson RJ: Preventing cervical reflux of the distension medium during panoramic hysteroscopy. J Reprod Med 1986;31:592–594.
34. Gentile GP, Siegler AM: Inadvertent intestinal biopsy during laparoscopy and hysteroscopy: A report of two cases. Fertil Steril 1981;36:402–404.
35. Cumming DC, Taylor PJ: Combined laparoscopy and hysteroscopy in the investigation of the ovulatory infertile woman. Fertil Steril 1980;33:475–478.
36. DeCherney AH, Kort H, DeVore GR: Increased pregnancy rate with oil-soluble hys-

terosalpingography dye. Fertil Steril 1980;33:407–410.
37. Fayez JA, Ghazie M, Schneider PJ: The diagnostic value of hysterosalpingography and hysteroscopy in infertility investigation. Am J Obstet Gynecol 1987;156:558–560.
38. Siegler AM: Hysterography and hysteroscopy in the infertile patient. J Reprod Med 1977;18:143.
39. Taylor PJ, Lewinthal D, Leader A, et al: A comparison of dextran 70 with carbon dioxide as the distentioin medium for hysteroscopy in patients with infertility or requesting reversal of a prior tubal sterilization. Fertil Steril 1987;47:861.
40. Englund S, Ingleman-Sundberg A: Hysteroscopy in the diagnosis and treatment of uterine bleeding. Gynecologia 1957;143:217.
41. Taylor PJ, Cumming DC: Hysteroscopy in 100 patients. Fertil Steril 1979;31:301.
42. Kessler I, Lancet M: Hysterography and hysteroscopy: a comparison. Fertil Steril 1986;46:709–710.
43. Wiswedel K: Routine hysteroscopy in infertility? S Afr Med J 1987;72:257–258.
44. Snowden EU, Jarrett JC II, Dawood MY: Comparison of diagnostic accuracy of laparoscopy, hysteroscopy and hysterosalpingography in evaluation of female fertility. Fertil Steril 1984;41:709.
45. Lübke F, Hindenburg HJ: Hysteroscopy as an examination method in sterility, in Siegler AM, Lindenmann JH (eds): Hysteroscopy: Principles and Practice. Philadelphia, JB Lippincott, 1984, p 173.
46. Polishuk WZ, Siew FP, Gordon R, et al: Vascular changes in traumatic amenorrhea and hypomenorrhea. Int J Fertil 1977;22:189.
47. Buttram VC, Reiter RC: Surgical Treatment of the Infertile Patient. Baltimore, Williams & Wilkins, 1985, p 206.
48. Rock JA, Jones HW Jr: The clinical management of the double uterus. Fertil Steril 1977;28:798.
49. Sulak PJ, Letterie G, Coddington CC, et al: Histology of proximal tubal occlusion. Fertil Steril 1987;48:437.
50. Sulak PJ, Letterie GS, Hayslip CC: Hysteroscopic cannulation and lavage in the treatment of proximal tubal occlusion. Fertil Steril 1987;48:493–494.
51. Daly DC, Soto-Albors CE, Aversa MA: Hysteroscopic detection and treatment of adhesions at the tubal ostium/uterine junction in infertile patients. Fertil Steril 1986;46:138–140.
52. Daniell JF, Miller W: Hysteroscopic correction of cornual occlusion with resultant term pregnancy. Fertil Steril 1987;48:490–492.
53. Platia MP, Krudy AG: Transvaginal fluoroscopic recanalization of a proximally occluded oviduct. Fertil Steril 1985;44:704.
54. Confino E, Friberg J, Gleicher N: Transcervical balloon tuboplasty. Fertil Steril 1986;46:963–966.
55. Jensen PA, Stromme WB: Amenorrhea secondary to puerperal curettage (Asherman's syndrome). Am J Obstet Gynecol 1972; 113:150.
56. Eriksen J, Koestel C: The incidence of uterine atresia after postpartum curettage: a follow-up examination of 141 patients. Dan Med Bull 1960;7:50.
57. Schenker JG, Margolioth EJ: Intrauterine adhesions: an updated appraisal. Fertil Steril 1982;37:593–610.
58. Valle RF, Sciarra JJ: Intrauterine adhesions: Hysteroscopic diagnosis, classification, treatment and reproductive outcome. Am J Obstet Gynecol 1988;158:1459–1470.
59. March CM, Israel R, March AD: Hysteroscopic management of intrauterine adhesions. Am J Obstet Gynecol 1978;130:653.
60. March CM, Israel R: Gestational outcome following hysteroscopic lysis of adhesions. Fertil Steril 1981;36:445.
61. Sugimoto O: Diagnostic and therapeutic hysteroscopy for traumatic intrauterine adhesions. Am J Obstet Gynecol 1978;131:539.
62. Hamou J, Salat-Baroux J, Siegler AM: Diagnosis and treatment of uterine adhesions by microhysteroscopy. Fertil Steril 1983;39:321.
63. Jewelwicz R, Khalaf S, Neuwirth RS, et al: Obstetric complications after treatment of intrauterine synechiae (Asherman's syndrome). Obstet Gynecol 1976;47:701–705.
64. Friedman A, DeFazio J, DeCherney A: Severe obstetric complications after aggressive treatment of Asherman syndrome. Obstet Gynecol 1986;67:864–867.
65. Holtz G: Prevention and management of peritoneal adhesions (modern trends). Fertil Steril 1984;41:497–507.
66. Chervenak FA, Neuwirth RS: Hysteroscopic resection of the uterine septum. Am J Obstet Gynecol 1981;141:351–353.
67. Valle RF, Sciarra JJ: Hysteroscopic man-

agement of the septate uterus. Obstet Gynecol 1986;67:253–257.
68. March CM, Israel R: Hysteroscopic management of recurrent abortion caused by septate uterus. Am J Obstet Gynecol 1987;156:834–842.
69. DeCherney AH, Russell JB, Graebe RA: Resectoscopic management of müllerian fusion defects. Fertil Steril 1986;45:726–728.
70. Goldrath MH: Hysteroscopic laser surgery, in Baggish MS (ed): Basic and Advanced Laser Surgery in Gynecology. Norwalk, CT, Appleton-Century-Crofts, 1985, pp 357–372.
71. Siegler AM, Valle RF: Therapeutic hysteroscopic procedures. Fertil Steril 1988, 50:685–703.
72. Rock JA, Murphy AA, Cooper WH: Resectoscopic techniques for the lysis of a class V complete uterine septum. Fertil Steril 1987;48:495–496.
73. DeCherney AH, Polan ML: Hysteroscopic management of intrauterine lesions and intractable uterine bleeding. Obstet Gynecol 1983;61:392.
74. Zipkin B, Rosenfeld DL: Hysteroscopic removal of a Heyman radium capsule. J Reprod Med 1979;22:133.
75. Sciarra JJ, Valle RF: Hysteroscopy: a clinical experience with 320 patients. Am J Obstet Gynecol 1977;127:340.
76. Valle RF, Sciarra JJ, Freeman DW: Hysteroscopic removal of intrauterine devices with missing filaments. Obstet Gynecol 1977;49:55–60.
77. Taylor PJ, Hamou J, Mencaglia L: Hysteroscopic detection of heterotopic intrauterine bone formation. J Reprod Med 1988; 33:337–339.
78. Corfman RS: Indications for hysteroscopy. Obstet Gynecol Clin North Am 1988;15:41–49.
79. Stock RF, Kanbour A: Prehysterectomy curettage: an evaluation. Obstet Gynecol 1975;45:537.
80. Gimpelson RJ, Rappold HO: A comparative study between panoramic hysteroscopy with directed biopsied and dilation and curettage. Am J Obstet Gynecol 1988;158:489–492.
81. Goldrath MH, Sherman AI: Office hysteroscopy and suction curettage: can we eliminate the hospital diagnostic dilation and curettage? Am J Obstet Gynecol 1984;152:220.
82. Raju KS, Taylor RW: Routine hysteroscopy for patients with a high risk of uterine malignancy. Br J Obstet Gynecol 1986; 93:1259–1261.
83. Mencaglia L, Perino A, Hamou J: Hysteroscopy in perimenopausal and postmenopausal women with abnormal uterine bleeding. J Reprod Med 1987;32:577–582.
84. Tchabo J: Use of contact hysteroscopy in evaluating postpartum bleeding and incomplete abortion. J Reprod Med 1984;29: 749.
85. Valle RF, Sabbagha RE: Management of first trimester pregnancy termination failures. Obstet Gynecol 1980;55:625.
86. Van Lith DAF, Sciarra JJ: Current status of hysteroscopy in gynecologic practice. Fertil Steril 1983;21:125.
87. Buttram VC, Reiter RC: Uterine leiomyomata: etiology, symptomatology, and management. Fertil Steril 1981;36:433–445.
88. Neuwirth RS: A new technique for and additional experience with hysteroscopic resection of submucous fibroids. Am J Obstet Gynecol 1978;131:91.
89. DeCherney AH: Treatment of irregular menstrual bleeding by hysteroscopic resection of submucous myomas and polyps, in Siegler AM, Lindenmann HJ (eds): Hysteroscopy: Principles and Practice. Philadelphia, JB Lippincott, 1984, p 173.
90. Hallez JP, Perino A: Endoscopic intrauterine resection: principles and technique. Acta Eur Fertil 1988; 19:17.
91. Coddington CC, Collins RL, Shawker TA, et al: Long acting gonadotropin hormone-releasing hormone analog used to treat leiomyomata uteri. Fertil Steril 1986;46:624.
92. DeCherney AH, Cholst I, Naftolin F: The management of intractable uterine bleeding utilizing the cystoscopic resectoscope, in Siegler AM, Lindenmann HJ (eds): Hysteroscopy: Principles and Practice. Philadelphia, JB Lippincott, 1984, p 140.
92a. Townsend DE, Richart RM, Paskowitz RA, et al: "Rollerball" coagulation of the endometrium. Obstet Gynecol 1990;76:310.
93. Droegemueller W, Greer B, Makowski E: Cryosurgery in patients with dysfunctional uterine bleeding. Obstet Gynecol 1978; 38:256.
94. Droegemueller W, Makowski E, Macsalka R: Destruction of the endometrium by cryosurgery. Am J Obstet Gynecol 1971;110:467.
95. Schenker JG, Palishuk WZ: Regeneration of rabbit endometrium following intrauterine

instillation of chemical agents. Gynecol Invest 1973;4:1.
96. Schwartz PE, Goldstein HM, Wallace S, et al: Control of arterial hemorrhage using percutaneous arterial catheter techniques in patients with gynecologic malignancies. Gynecol Oncol 1975;3:276.
97. Thatcher SS: Hysteroscopic sterilization. Obstet Gynecol Clin North Am 1988;15:51–59.
98. Cooper JM, Rigberg HS, Houck R, et al: Incidence, significance and remission of tubal spasm during attempted hysteroscopic tubal sterilization. J Reprod Med 1985;30:39.
99. Darabi KF, Roy K, Richart RM: Collaborative study on hysteroscopic sterilization procedures: final report, in Sciarra JJ (ed): Gynecology and Obstetrics. Philadelphia, Harper & Row, 1978, vol 6, p 81.
100. Droegemueller W, Greer BE, Davis JR, et al: Cryocoagulation of the uterine cornua. Am J Obstet Gynecol 1978;131:1.
101. Neuwirth RS, Richart RM, Bolduc LR, et al: Trials with the Femcept method of female sterilization and experience with radiopaque methyl cyanoacrylate. Am J Obstet Gynecol 1983;145:948.
102. Loffer FD: Hysteroscopic sterilization with the use of formed-in-place silicone plugs. Am J Obstet Gynecol 1984;149:261.
103. Hamou J, Gasparri F, Scarselli GF, et al: Hysteroscopic reversible tubal sterilization. Acta Eur Fertil 1984;15:123.
104. Brundin J: Intratubal devices, in Hafez ESE, van Os WAA (eds): Biodegradables and delivery systems for contraception. Progress in Contraceptive Delivery Systems Series, vol 1. Lancaster, England, MTP Press. 1980, pp 123–137.
105. Hosseinian AH, Morales WA: Clinical application of hysteroscopic sterilization using uterotubal junction blocking devices, in Zatuchni GI, Shelton JD, Goldsmith A, et al (eds): Female Transcervical Sterilization. Philadelphia, Harper & Row, 1972, p 31.
106. Hahemann H, Mohr J: Antenatal foetal diagnosis in genetic disease. Bull Eur Soc Hum Genet 1969;3:47.
107. Rodeck CH, Morsman JM: First-trimester chorion biopsy. Br Med Bull 1983;39:338.
108. Petrikovsky B: Intrapartum hysteroscopy. Aust NZ J Obstet Gynecol 1986;26:249.

17

Hysteroscopic Laser Ablation of the Endometrium

MILTON H. GOLDRATH

Menorrhagia represents a widespread clinical problem, and is the leading cause of elective hysterectomy in women with a normal uterus.[1–4] Various methods are used to treat this condition, including dilation and curettage (D&C), hormone therapy, ergot derivatives, antifibrinolytic agents, and nonsteroidal antiinflammatory drugs.[5] While oral contraceptives appear to be the most effective approach to treatment, their usefulness is limited by side effects and contractions to therapy.

Until recently, women who did not respond to pharmacologic intervention were limited to one of two options: hysterectomy or continued cycles of heavy menstrual bleeding. With the advent of hysteroscopic laser surgery in 1979,[6] women with menorrhagia refractory to other methods of therapy were offered an alternative to hysterectomy. For the first time, it was possible to destroy the endometrial tissue without making any incision. Laser ablation has also been employed in recent years to treat patients with moderate-sized fibroids. Thus, the number of patients for whom this procedure is potentially beneficial is quite sizable.

Reproducing Asherman's Syndrome with Laser Treatment

The primary objective of laser surgery is to limit uterine bleeding and produce amenorrhea—in essence, replicating Asherman's syndrome. First described by Fritch in 1894, Asherman identified this syndrome in 1948 as a condition involving uterine synechiae, which usually result from postabortal or postpartum uterine curettage.[7] The syndrome is associated with amenorrhea and infertility.

It soon became apparent that any technique capable of reproducing Asherman's syndrome might be used to correct excessive uterine bleeding. In subsequent years, a variety of chemical and physical methods were used for this purpose, the majority of which proved unsuccessful. While radium and cryocoagulation were both associated with some limited success, radium is no longer used in the absence of malignancy, and the cryocoagulation procedure described by Droegemueller[8,9] was abandoned because of the potential for producing painful hematometra. More recently, DeCherney and Polan have successfully destroyed the endometrium and produced amenorrhea through the use of a high-frequency coagulating current.[10] This technique has since been modified and used by numerous authors.

With the development of optical fibers capable of transmitting laser energy, lasers have steadily gained acceptance over the past decade for use in performing endoscopic and hysteroscopic surgical procedures. One such laser, the neodymium:yttrium-aluminum-garnet (Nd:YAG), is particularly well suited to photocoagulation of the endometrium as a result of its high power and transmission through fiberoptics. Its deep-tissue penetration (up to 4 mm beneath the surface) and extensive heating distinguish it from the carbon dioxide and argon lasers.

Only a small percentage of the power of the Nd:YAG laser is absorbed by the tissue, the majority being diffused backward (30 to 40%) and forward (25 to 30%) as the laser energy scatters. Because of the danger posed to the surgeon from backscatter after tissue impact, use of special safety lenses or a protective filter over the eyepiece of the hysteroscope is essential.

Surgical Procedure

The human uterus is characterized by a relatively thick myometrium (usually 1.5 cm or thicker) and thin endometrium (usually less than 1 mm), thus providing an ideal environment for laser surgery. The fact that only about 5% of the uterine thickness must be ablated in order to destroy the endometrium provides a substantial safety factor.

Physiologic saline is currently used to distend the uterus and has proven an excellent visualizing medium. Previous experience with dextran (Hykson) demonstrated that its high viscosity was associated with large bubbles that obscured viewing. In addition, it discolored with the intense heat of the laser.[6] Gaseous distending media are useful for diagnostic hysteroscopy, but should *not* be used for therapeutic hysteroscopic surgery because of the risk of fatal gas emboli.[11]

The procedure is performed with the patient under general or spinal anesthesia. A continuous-flow hysteroscope is used. The fluid flows in by gravity only and is removed by low suction to provide a constant flow of fluid at low pressure. This is important not only for the removal of blood and gas bubbles from the operative field, but also for the prevention of problems with excess fluid absorption,[6] an earlier-noted complication.

Endometrial photovaporization and photocoagulation are performed under direct visualization using a power output of 50 W continuous wave. The entire endometrial lining is treated, commencing in the cornual area and extending across the fundus and down the anterior and posterior walls to the isthmus, terminating 4 cm from the external cervical os. To avoid treating the endocervix, the level of the internal os (4 cm) is marked on the sheath of the hysteroscope. Care is also taken to avoid perforation near the tubal ostia, the thinnest portion of the uterus. This procedure takes 30 to 40 minutes in a normal-sized uterus, longer in a larger uterus.

Endometrial ablation results in pitting and charring of the treated areas. After a long healing period, synechiae develop that are similar to those described by Asherman[7] (Figs. 17-1 and 17-2).

Pretreatment with Danazol

Danazol was initially used in an attempt to duplicate the hormonal state of a patient who develops Asherman's syndrome.[6] Since postabortal or postpartum endocrine activity is negligible, it was believed that by placing a patient on danazol for 3 weeks prior to surgery and for 2 weeks following surgery, healing might occur before the endometrium had an opportunity to regenerate.

It has subsequently been discovered that healing takes up to 6 months following the procedure. Thus, danazol is no longer used

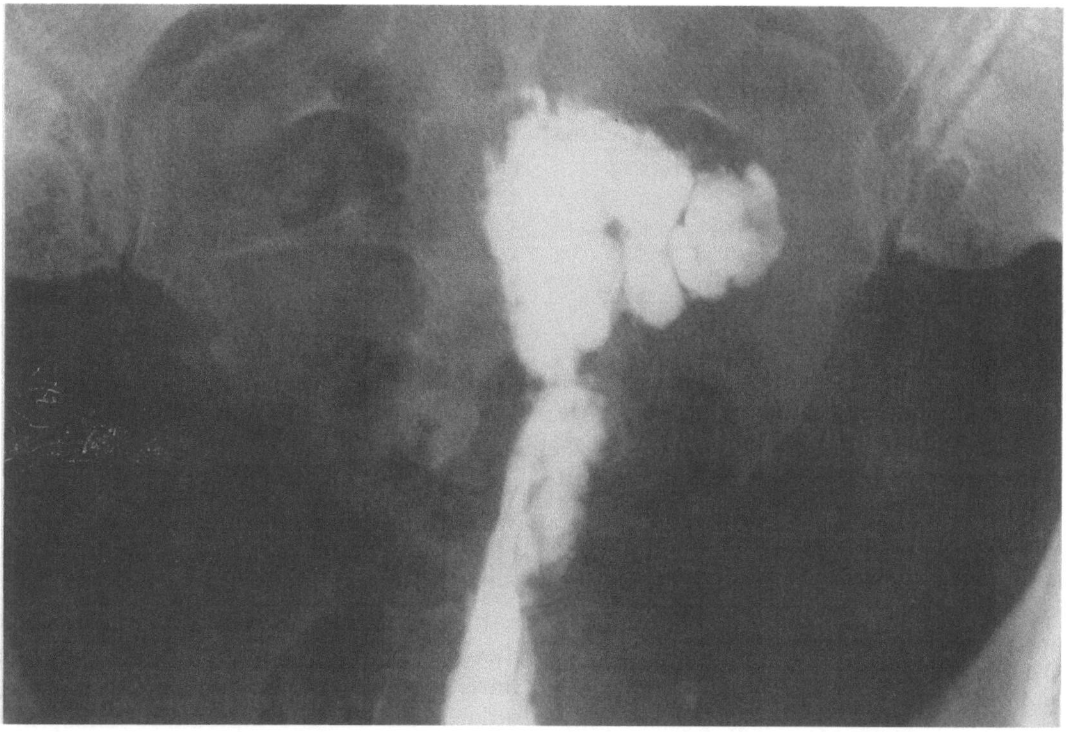

FIGURE 17-1. Hysterosalpingogram 4 months after laser ablation. Note beginning synechiae.

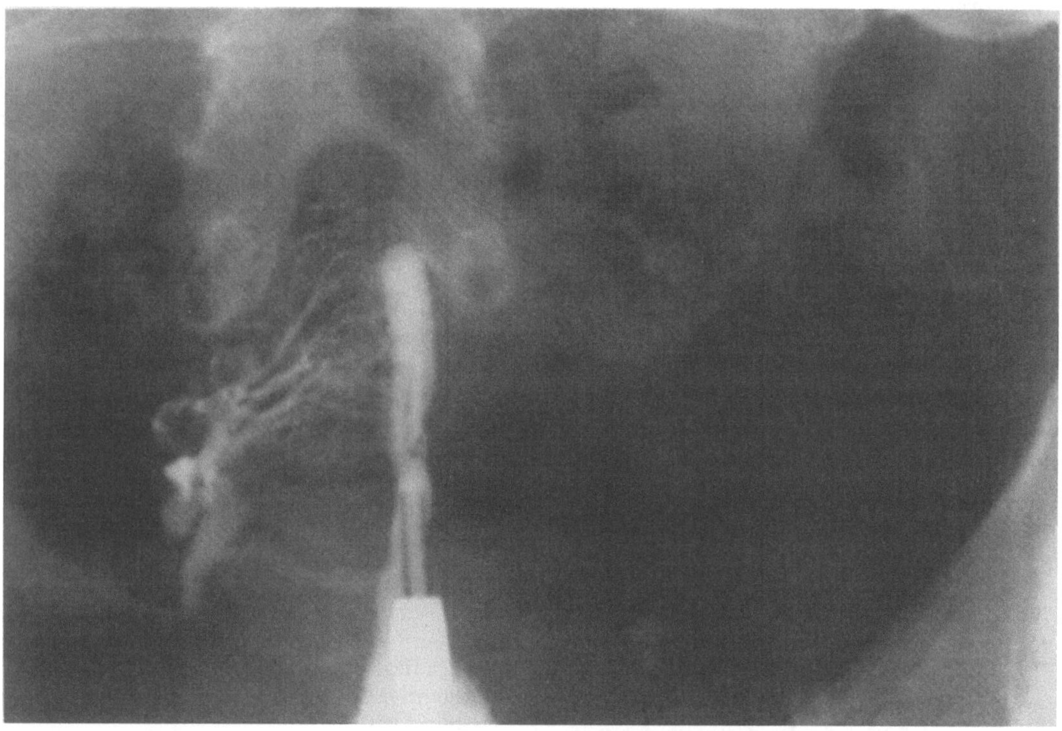

FIGURE 17-2. Hysterosalpingogram 18 months after laser ablation. Note contracted cavity and intravasation of contrast material.

postoperatively. However, the preoperative use of danazol greatly facilitates ablation of the endometrium, since the steroid decreases the thickness of the endometrial lining.

Clinical Experience with Hysteroscopic Laser Ablation

At our center, 324 women ages 12 to 53 years have undergone 335 hysteroscopic laser procedures for the treatment of menorrhagia from January 1979 thru March 1988. More than 50% of the patients (n = 176) had fibroids, most of which were submucous fibroids or adenomyosis. The criteria for patient selection include: (a) excessive and disabling bleeding, (b) inability or unwillingness to use other therapies, (c) candidacy for hysterectomy, (d) future childbearing not desired, and (e) D&C within 6 months of the surgical procedure to rule out adenocarcinoma of the endometrium or cancer precursors.

Most patients have fared extremely well postoperatively. The procedure is usually performed on an ambulatory basis, with the patients discharged 2 to 3 hours later. Although severe cramping typically occurs for the first 2 hours following surgery, it has recently been noted that the use of a paracervical block of bupivacaine hydrochloride (Marcaine) 0.25% with epinephrine 1:200,000 upon completion of the procedure greatly reduces the postoperative pain. Overall, the postoperative symptoms resemble those associated with a D&C. All patients have experienced varying degrees of serosaguineous discharge that lasts for approximately 4 weeks. None of the patients has become febrile.

Results

Overall, the results have been excellent in 292 patients, 50% of whom became amenorrheic and the remainder of whom experienced only a few days of spotting per cycle (Table 17-1). Seven patients reported good results, meaning they were satisfied with treatment and experienced what they considered to be normal menstrual bleeding. Of the 22 patients who had poor results, 10 of 11 who were retreated are currently amenorrheic and 1 had a hysterectomy. One patient elected to endure her symptoms. The remaining 10 patients underwent hysterectomy and were not retreated.

Forty-five of the patients had clotting disorders, all of whom did well following surgery. Eighteen of the patients were receiving warfarin (Coumadin) therapy, 10 had von Willebrand's disease, 5 were renal transplant patients, 4 had immunologic thrombocytopenic purpura, 3 were renal dialysis patients, an additional 3 had storage-pool disease (a platelet disorder), and 1 had Bernard-Soulier syndrome, also a platelet disorder. One was a symptomatic hemophilia carrier.

Out of a total of 19 hysterectomies, eight were in patients considered laser successes and eight in laser failures. In two, laser use was terminated because of the presence of large, submucous myomas. One patient had a hystrectomy 2 weeks later for cervical bleeding. The latter complication deserves special comment. At the time, it was our practice to continue administering danazol for 2 weeks after the procedure. The patient was taking warfarin. Ten days following surgery, excessive, uncontrollable bleeding occurred, necessitating hysterectomy. It is now known that the dose of warfarin must be re-

TABLE 17-1. Results of hysteroscopic laser surgery in danazol-pretreated women

Result	No. of patients
Excellent	292
Good	7
Poor	22
Retreated (laser)	11
Amenorrhea, 10	
Hysterectomy, 1	
Hysterectomy (not retreated)	10

Reprinted from ref. 14, with permission.

TABLE 17-2. Effects of single-dose medroxyprogesterone acetate

Administration of medroxyprogesterone acetate	No. of patients (%)				Total no. of patients
	Scant flow	None	Good	Poor	
No	115 (54.2)	79 (37.3)*	5 (2.4)	13 (6.1)	212
Yes	28 (27.7)	70 (69.3)*	2 (2.0)	1 (1.0)	101
Total	143	149	7	14	313

Reprinted from ref. 14, with permission.
*$p = 0.00001$ ($\chi^2 = 29.4$, $df = 3$).

duced in patients receiving danazol therapy. This underscores an important precaution in the prescribing of danazol: *coadminister with warfarin with great care.* Danazol potentiates the action of warfarin, greatly increasing the prothrombin time.[12] If the two drugs must be administered concomitantly, the warfarin dosage should be reduced by half. It is also advisable to monitor prothrombin times every 3 days through the course of combination therapy. The use of long-acting gonadotropin-releasing hormone (GnRH) agonists rather than danazol will eliminate this complication in patients taking warfarin.

Of the early failures, one had large submucous myomas, two had adenomyosis, and one failed for unknown reasons. Among the late failures, one involved a significantly enlarged fibroid and three had adenomyosis. Thus, five of the eight hysterectomies that failed had adenomyosis.

Because of the possibility that adenomyosis was associated with endometrial regeneration, it was decided to administer a single 150-mg dose of medroxyprogesterone acetate (Depo-Provera) to patients with suspected adenomyosis or fibroids on the day of surgery. Sixty-nine percent of the patients treated with medroxyprogesterone acetate became amenorrheic, as compared with only 37% of those who did not receive treatment (Table 17-2). This difference with highly significant ($p < 0.00001$). Quite possibly, the use of danazol or GnRH agonists for 4 to 5 months following surgery might also inhibit endometrial regeneration, but this requires additional study. Such treatment may not be warranted because of the undesirable side effects and increased cost of the drugs.

Complications

The major complication observed with this procedure has been fluid overload, which led to frank pulmonary edema in two cases. However, it should be noted that this problem occurred early when a "push" infusion method was used to distend the uterus. We now use a continuous irrigation technique that involves gravity feed of fluid into the uterus and active suction out of the uterus, with the rate of flow adjusted to distend the uterus and control any bleeding.

Following surgery, profuse bleeding developed in 13 patients after withdrawal of irrigating pressure. The bleeding was controlled by balloon tamponade using a Foley catheter inserted into the uterus.[13] Three patients had postoperative urinary tract infections related to the catheter drainage. In one patient endomyometritis developed 1 week postoperatively. Although asymptomatic, she was hospitalized for 2 days and treated with intravenous antibiotics. The purulent discharge stopped immediately.

All patients are followed 1 and 3 months postoperatively by suction curettage; if hematometra is found, suction curettage is performed weekly until the condition resolves. Experience suggests that if hematometra has not developed within 3 months following hysteroscopy, it will rarely develop subsequently. Routine suction curettage revealed 12 cases of postoperative hematometra. Only two patients had delayed hematometra.

Three patients had allergic reactions to danazol. The reaction was mild in two cases, and therapy was continued. Only one patient had to discontinue treatment because of a severe allergic reaction.

Laser ablation was not completed in one patient because of an instrument-related uterine perforation (not laser). The patient returned to the hospital 2 months later, at which time the laser procedure was completed.

Conclusion

The data accumulated over the past decade suggest that hysteroscopic laser surgery offers a highly viable alternative to hysterectomy in patients with menorrhagia who are not good candidates for other forms of treatment. Our current experience indicates that this procedure is remarkably safe and effective and involves minimal postoperative morbidity and discomfort. The preoperative use of danazol to enhance endometrial atrophy has greatly facilitated the success of this endoscopic intervention.[14]

References

1. Cole P, Berlin J: Elective hysterectomy. Am J Obstet Gynecol 1977;129:117.
2. D'Esopo DA: Hysterectomy when the uterus is grossly normal. Am J Obstet Gynecol 1962;83:113.
3. Parrot M: Elective hysterectomy. Am J Obstet Gynecol 1971;113:531.
4. Dyck FJ, Murphy FA, Murphy JK, et al: Effect of surveillance in the number of hysterectomies in the Province of Saskatchewan. N Engl J Med 1977;296:1326.
5. Nilsson L, Goran R: Treatment of menorrhagia. Am J Obstet Gynecol 1971;110:713.
6. Goldrath MH, Fuller TA, Segal S: Laser photovaporization of endometrium for the treatment of menorrhagia. Am J Obstet Gynecol 1981;140:14.
7. Asherman JC: Amenorrhea traumatica (atretica). J Obstet Gynaecol Br Emp 1948;55:23.
8. Droegemueller W, Makowski E, Macsalka R: Destruction of the endometrium by cryosurgery. Am J Obstet Gynecol 1971;110:467.
9. Droegemueller W, Greer B, Makowski E: Cryosurgery in patients with dysfunctional uterine bleeding. Obstet Gynecol 1971;38:256.
10. DeCherney A, Polan ML: Hysteroscopic management of intrauterine lesions and intractable uterine bleeding. Obstet Gynecol 1983;61:392.
11. Baggish MS, Daniell JF: Death caused by air embolism associated with neodymium:yttrium-aluminum-garnet laser surgery and artificial sapphire tips. Am J Obstet Gynecol 1989;161:877–878.
12. Goulbourne IA, Macleod DAD: Interaction between danazol and warfarin: case report. Br J Obstet Gynaecol 1981;88:950.
13. Goldrath MH: Uterine tamponade for the control of acute uterine bleeding. Am J Obstet Gynecol 1983;147:869.
14. Goldrath MH: Use of danazol in hysteroscopic surgery for menorrhagia. J Reprod Med 1990;35(suppl):91–96.

18

Laser Surgery of the Cervix for Intraepithelial Neoplasia

V. CECIL WRIGHT

Understanding Laser Light

In 1917, Einstein theorized that through an atomic process a unique kind of electromagnetic radiation could be produced.[1] When that radiation is in the visible, ultraviolet, or infrared spectrum, it is termed "laser" light (*l*ight *a*mplification by *s*timulated *e*mission of *r*adiation). Laser instruments produce such radiation by subjecting an active medium, for which the laser is named, to external stimulation and amplification. This creates a laser beam with a particular wavelength.

Laser light possesses four unique characteristics:

1. It is coherent. This means that all waves are exactly in step and phase with each other, in both space and time.
2. When unfocused it is highly collimated, indicating that all waves are virtually parallel to each other and therefore the beam diverges only slightly over long distances.
3. It is monochromatic, signifying that all waves have virtually the same wavelength and also the same energy.
4. It is the "brightest" existing light (i.e., has the highest luminosity).[2] It is slightly ironic that some "laser" light is invisible.

Tissue Response to Laser Light

The three most common lasers used in surgery are the carbon dioxide laser (CO_2; 10.6-μm wavelength), the neodymium:yttrium-aluminum-garnet laser (Nd:YAG; 1.064-μm wavelength) and the argon laser (0.48-μm wavelength). The wavelength of the photonic energy determines the proportion of a given laser beam that is scattered, reflected, absorbed, or transmitted by a particular type of tissue. The extinction length refers to the thickness of a substance that is required to absorb 90% of the incident laser energy. In water, the CO_2 laser beam has an extinction length of 0.03 mm. This indicates that 90% of the CO_2 laser beam's energy is absorbed within a depth of 0.03 mm of water.[3] The CO_2 laser energy is therefore noted for its surface absorption with negligible scattering, reflection, or transmission and thus is an excellent wavelength for excising or vaporizing tissues containing water. Such tissue characteristics are not exhibited by the argon or Nd:YAG laser beams, which penetrate much more deeply into tissue and are more useful for coagulation effects.

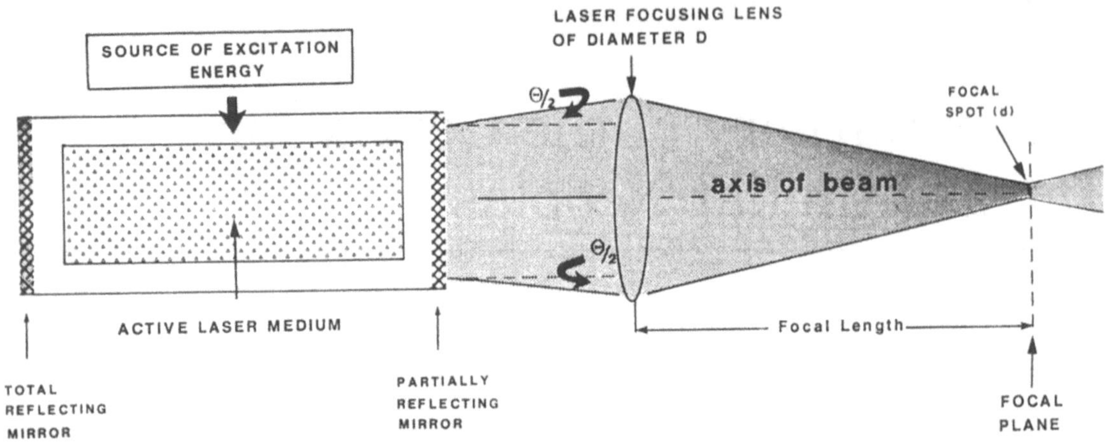

FIGURE 18-1. Components of the laser optical system. The laser light is captured by a laser focusing lens as it exits the optical cavity. The effective laser beam diameter (*d*) is a function of the beam divergence (θ) and the focal length of the laser light–capturing lens. (Reprinted from ref. 18, with permission.)

Surgical Delivery Systems

For surgery the laser energy produced by stimulated emission is captured by a lens as it exits the optical cavity (Fig. 18-1). This focuses the laser beam to a small effective beam diameter. The CO_2 laser beam is utilized in surgery via two basic delivery systems. One combines the laser beam with the optical system of an operating microscope (Fig. 18-2), enabling the surgeon to perform laser microsurgery. The other delivery system transmits the laser beam through the internal mirror sequence of a multiarticu-

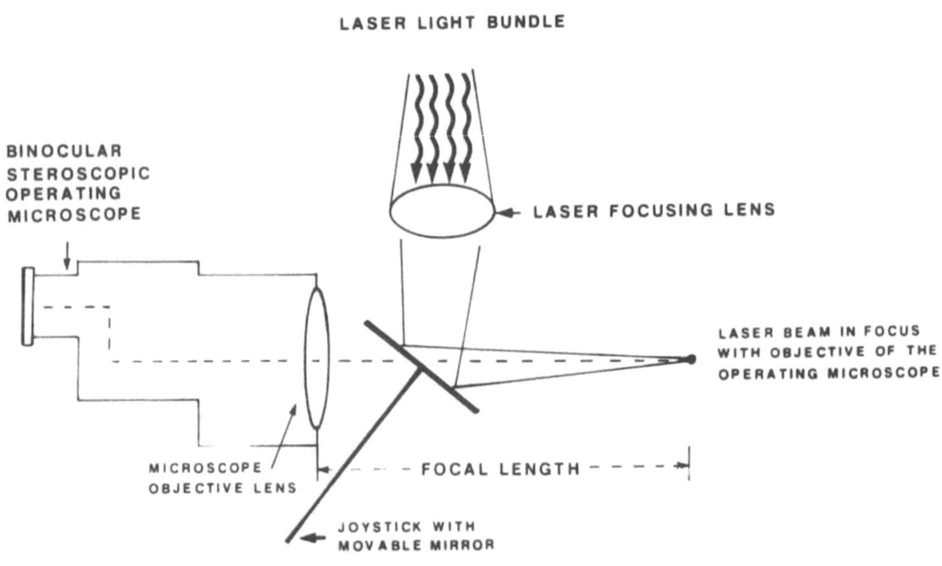

FIGURE 18-2. Basic instrumentation of the laser for microsurgery. With a matched lens system (i.e., a laser light–capturing lens of 300 mm and an objective microscopic lens of 300 mm) the smallest effective laser beam diameter is produced at the focal plane. In this situation the smallest beam diameter is 0.5 mm. (Reprinted from ref. 18, with permission.)

lated arm to various penlike handpieces that are used for freehand surgery. When the CO_2 laser is attached to the operating microscope or colposcope, an excellent working distance (from scope to tissue) is 300 mm, which equals the focal length of the main objective lens of the operating microscope. With a matched lens system (objective lens and laser light–capturing lens, both of 300-mm focal length), the smallest effective laser beam diameter with the microscope in focus is usually about 0.5 mm. This provides an appropriate spot size for rapid incision or excision. A larger beam diameter is obtained by defocusing the beam via a variable-spot-size mechanism, which effectively mismatches the scope and laser lenses.

Biologic Effects of CO_2 Laser Energy

The average power density, which determines the depth of vaporization by the laser beam per unit of time, is easily determined from the beam power and focal spot diameter. Surgical laser systems typically produce between 20 and 100 W of beam power. Power is chosen by selecting various settings on the individual instrument's control panel. A handy formula for calculating the approximate average power density (P) for a transverse electromagnetic mode (TEM_{00}) laser beam (one with gaussian intensity distribution) incorporates the total beam power in watts (W) multiplied by 100 divided by the square of the effective beam diameter (d) in millimeters[2]: that is, $P = 100\ W/d^2$. Beam diameter is measured by a test firing at 10 W for 0.1 second on a suitable target, such as a tongue depressor.

The power density of a CO_2 surgical laser determines whether the beam can be used to vaporize and thus destroy tissue or to incise or excise like a scalpel. Carbon dioxide laser energy applied to tissue containing water causes instantaneous boiling of intracellular and extracellular water, which forms steam. Water within the tissue evaporates at the boiling temperature corresponding to the ambient pressure, at least 100°C. This is called the vaporization threshold. Above the level for reliable boiling (100 W/cm^2), the power density determines the depth of tissue that will be vaporized per unit of exposure time. The difference between the vaporization and excisional-type procedures is only the end result: a two-dimensional or three-dimensional defect with or without an excised specimen. The basic tissue reaction to CO_2 laser energy is the same, thermovaporization. During this reaction, smoke and steam are produced that must be removed by custom suction and filtration systems.

When the laser is properly employed the amount of thermal damage surrounding the vaporized defect or crater should be less than 100 μm (Fig. 18-3). Minimizing exposure time is a crucial principle because heat conduction and tissue injury increase with time.[2] The best quality defect results by removing the largest volume of tissue per unit of exposure time (i.e., using as high a power density as can be controlled by the laser surgeon). However, high power density coupled with a short exposure time minimizes heat conduction and thus coagulates less along

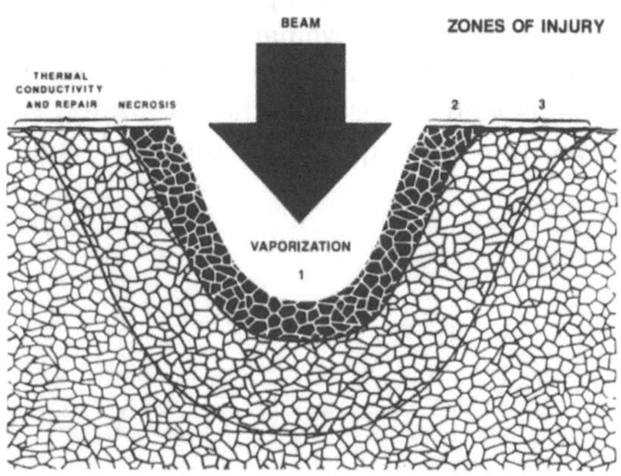

FIGURE 18-3. Zones of injury produced by the carbon dioxide laser beam. Zone 1 is the zone of vaporization, which produces a crater shaped like the TEM (transverse electromagnetic mode) of the laser. Zone 2 is the zone of thermal necrosis (50 to 100 μm) and Zone 3 is thermally injured tissue which is regenerated (100 to 500 μm). (Reprinted from ref. 18, with permission.)

the crater edge; therefore hemostasis is lessened. A balance between hemostasis and rapid vaporization must be achieved by the laser surgeon.

The most important surgical principle, which takes maximal advantage of the unique tissue properties and produce the best healing of the CO_2 laser, is: eliminate or destroy only the volume of tissue that is diseased or unwanted and do it as quickly as possible. It is the ability to excise and destroy tissue in a variety of configurations to match the distribution of disease with excellent hemostasis and little damage to healthy surrounding tissue that makes CO_2 laser microsurgery so valuable in treating the cervix.

Cervical Intraepithelial Neoplasia: The Distribution of Disease

The area susceptible to the development of cervical intraepithelial neoplasia (CIN) and invasive cancer lies between the original squamocolumnar junction and the histologic os. This area of susceptibility averages 2.5 cm in linear length. The transformation zone is only a variable-sized part of it. With increasing age of the woman, the area of susceptibility, including the transformation zone as well as disease when it occurs, recedes into and up the endocervical canal (Fig. 18-4). In the young and the premenopausal woman, the histologic os is located below the anatomic internal os. In adolescent years the original squamocolumnar junction is often located wide out on the ectocervix. In older women the histologic os can be located above the anatomic internal os, and the original squamocolumnar junction is usually located near the external os (Fig. 18-4).

Cervical intraepithelial neoplasia develops in the population of contaminated cells in the transformation zone that by metaplasia become atypical neoplastic epithelium. The linear extent of CIN lesions varies from 2 to 22 mm. Most lesions are less than 10 mm in linear length.[4] Rarely does CIN extend higher than 1.5 cm up the canal, and almost never more than 2 cm.[4,5] Therefore, when planning tissue destruction or removal for a particular patient, the surgeon must consider that the more exposed the columnar epithelium appears, the less likely is disease to be located high within the endocervical canal. The more endocervical the columnar epithelium, the more likely is involvement high in the canal. This is very important and helps determine the procedure necessary to remove CIN. Cervical intraepithelial neoplasia also extends into cervical crypts in many cases, with the percentage of cases and depth of involvement increasing with disease severity, regardless of location of the lesion. Studies have shown disease to extend

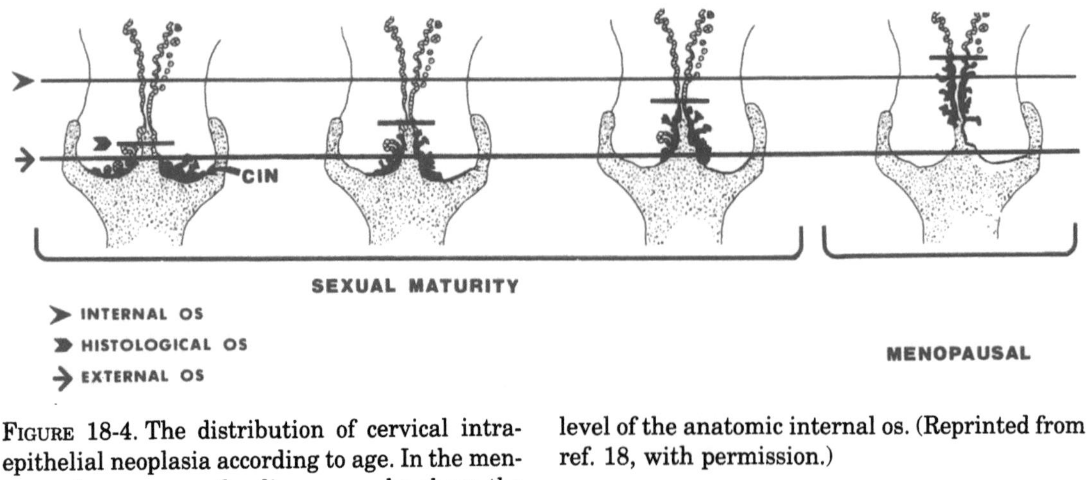

FIGURE 18-4. The distribution of cervical intraepithelial neoplasia according to age. In the menopausal age group, the disease can be above the level of the anatomic internal os. (Reprinted from ref. 18, with permission.)

TABLE 18-1. Maximum depth of cervical crypt involvement with CIN III

Crypt extension (mm)	No.	%	%	%
0	26	11.8		
0.1–0.9	140	63.6	86.8	95.9
1–1.9	25	11.4		
2–2.9	20	9.1		
3–3.9	5	2.3		
4–4.9	2	0.9		
5–5.9	2	0.9		
All cases	220	100.0		

Reprinted from ref. 11, with permission.

5.2 mm within crypts, with an average depth of involved crypts of 1.24 mm irrespective of whether CIN occupies the ectocervix or the endocervical canal.[5,6] However, in one series conducted by the author and pathologist Dr. Eleanor Davies, it was demonstrated that 95.9% of CIN III lesions do not have crypt involvement beyond 2.9 mm (Table 18-1). Therefore destruction of tissue or extirpation to a depth of 3.8 mm should eradicate all involved crypts in 99.7% of patients.[6] An extra few millimeters accounts for errors in measurement. With an understanding of linear length and crypt involvement, the surgeon can use the CO_2 laser to excise or destroy CIN by a planned approach with excellent healing and without interfering with the anatomic internal os in reproductive-age women. Therefore an incompetent cervix should not result.

Accepting the known dimensions of CIN, one can conceptualize the three-dimensional geometry of the disease-bearing tissue in a given patient, which is based on the cylinder, and apply appropriate laser operative techniques to remove or destroy it.[7–9]

Laser Vaporization of Ectocervical Intraepithelial Neoplasia

Laser vaporization is employed for well-defined ectocervical CIN regardless of the linear extent of the lesion or the histologic grade of the disease. The criteria to be fulfilled are: (a) the colposcopist must be certain from the qualitative assessment of the transformation zone that no invasive cancer is present; (b) the entire atypical transformation zone must be colposcopically defined; (c) correlation must exist between cytology, colposcopy, and histology findings, and all must indicate only squamous CIN; and (d) the CIN lesion must occupy the ectocervix at or below the level of the external os with no extension into the endocervical canal. With these criteria fulfilled the colposcopist can be assured that no invasive cancer is present and that it is safe to destroy the disease-bearing tissue without yielding a specimen.

Figure 18-5 shows in two dimensions the geometry of ectocervical disease. The 5-mm upright arrows indicate potential crypt involvement to that depth, and the distance between outer arrows reflects lesion diameter. Two defects are shown in section that would incorporate the entire lesion with a few millimeters of extra tissue: a domed cylinder and a cone. Although both configurations encompass obvious and potential disease, the volume (cone volume $= \frac{1}{3}\pi r^2 h$ and cylinder volume $= \pi r^2 h$, where r is the radius of the base and h is the height) of the domed cylinder is less and the cone is deeper and wider. Much unnecessary tissue would be excised with a cone-shaped defect when cure, not just diagnosis, is the goal. With the CO_2 laser beam the surgeon can create the defect necessary to eliminate disease and spare normal tissue. This defect is appropriate for those ectocervical lesions that have a lesion radius of 8 mm or less as measured from the innermost point of the disease (Fig. 18-6).

An office or clinic setting without anesthesia or analgesia is usually adequate. However, approximately 20% of all patients will require local infiltration of anesthesia into the ectocervix. Alternately, a paracervical block could be administered. A commonly used local anesthetic agent is commercial 1 or 2% lidocaine (Xylocaine) with epinephrine. This has a pH of 4.05. However, injection of this substance into the cervix causes a marked stinging sensation that can

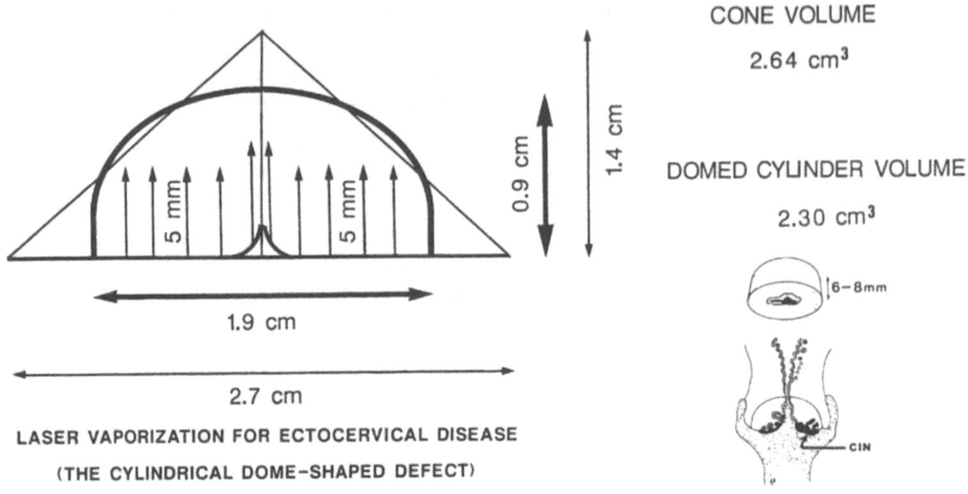

FIGURE 18-5. The geometry of ectocervical CIN. The dome-shaped cylindrical approach of vaporization accounts for the distribution of disease (linear length and crypt extension) and spares more normal tissue than does a conical approach. (Reprinted from ref. 18, with permission.)

be quite uncomfortable to the patient. The author has found that the addition of sodium bicarbonate USP to the lidocaine injection greatly reduces the discomfort. The commercial sodium bicarbonate is supplied in a 50-cc container (50 mEq/mL or 1 mEq/mL) containing an 8.4% solution, which equals 84 mg/mL of sodium bicarbonate. At a dilution of 1:10, (e.g., 0.1 cc of sodium bicarbonate USP injection is added to 10 cc of 1% lidocaine and epinephrine), a pH of 7.37 results. The mechanism of diminished pain upon cervical injection of the mixed solution is unknown; however, it is theorized to be related to the increase in the pH.[10]

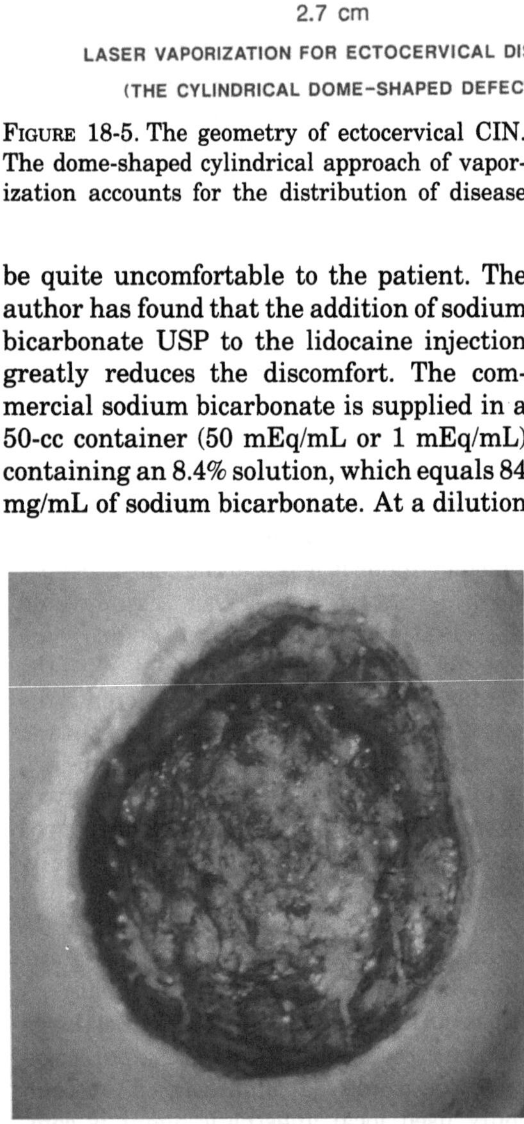

FIGURE 18-6. The complete vaporized dome-shaped defect with CIN for lesions with a radial linear length not exceeding 8 mm.

The cervix is exposed through a large bivalve speculum and the transformation zone is reidentified by washing it with a 3 to 4% solution of acetic acid. The CO_2 laser is attached to be operating microscope with a matched lens system of 300 mm. An objective lens in the operating microscope or colposcope of 300 mm equals the working distance (Fig. 18-2). This gives a comfortable working distance for the laser surgeon and allows adequate space between the microscope and the target should any manipulation be required. The operating microscope or colposcope should have variable magnification with a minimum setting of 3.0 to 7.5×, since higher magnifications will give an inadequate diameter of view and will therefore make the procedure technically difficult. A power density range of 750 to 1,000 W/cm² for a TEM_{00} mode beam is used.

First the lesion is encircled. Then the beam is rapidly moved in the horizontal, vertical, and diagonal directions by manipulat-

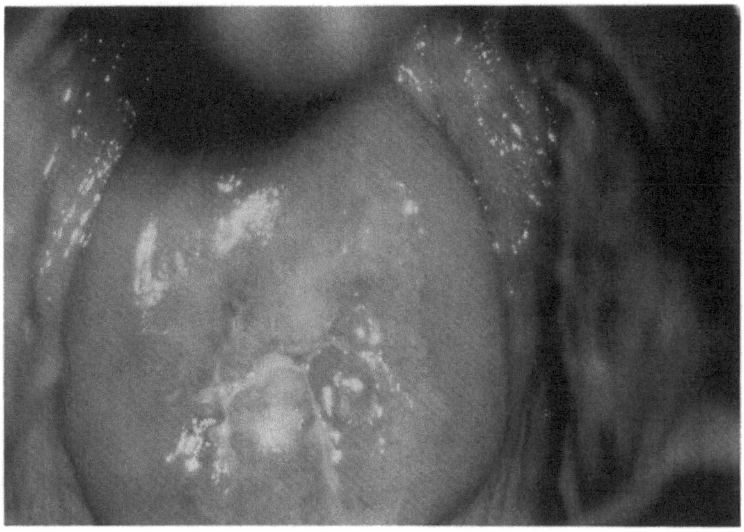

Figure 18-7. A large area of CIN extending widely onto the ectocervix, as noted by the aceto-white epithelium.

ing the joystick, overlapping each previous linear defect. This prevents furrowing. The laser vaporization is usually started at the 6 o'clock position, and an adequate depth of 6 to 8 mm is achieved before moving upward. A minimum effective laser beam focal spot diameter of 2 mm is required for laser vaporization. Smaller beams destroy less tissue at a time, produce furrowing, and increase bleeding, all of which lengthen the procedure. Destroying the cervical tissue to a measured minimum depth of at least 6 to 8 mm at every point along the lesion and transformation zone, particularly near the os, is essential to eliminate all disease. Frequently, the maximum depth of the defect from the most external plane of the ectocervix to the top of the dome above the cylinder extends 10 to 12 mm (Figs. 18-5 and 18-6). After completion the vaporized surface is coated with Monsel's solution (ferric subsulphate). When properly performed most vaporization procedures for ectocervical lesions with a linear length of 8 mm or less and producing the typical dome-shaped defect require 1 to 5 minutes only.[11]

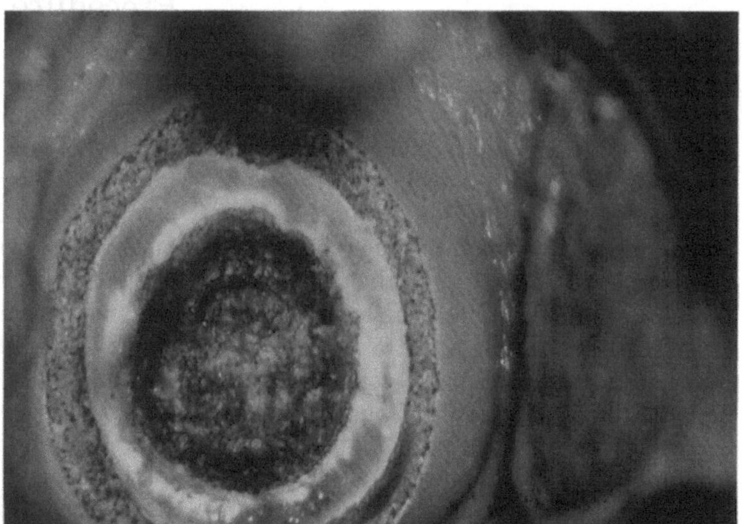

Figure 18-8. The central part of the lesion is vaporized as the traditional dome-shaped defect. The outer margin is outlined with the laser beam 2 to 3 mm beyond the disease.

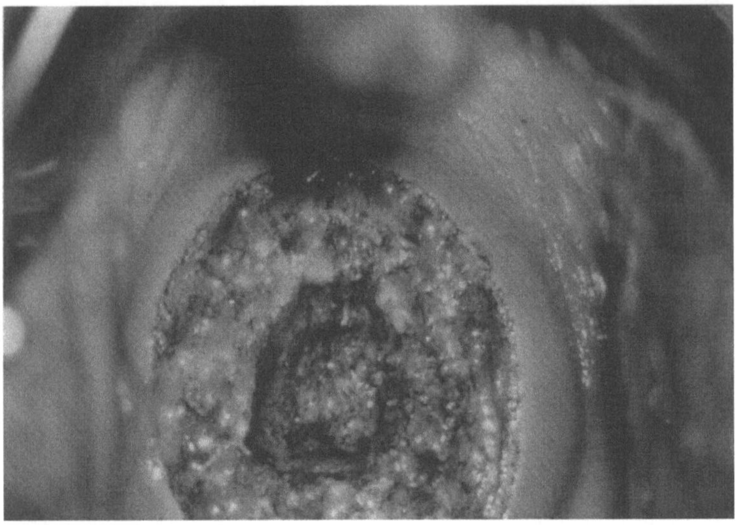

FIGURE 18-9. The final result of central vaporization performed as in Figures 18-7 and 18-8, with further peripheral vaporization to a lesser depth. The final appearance is a "cowboy hat" configuration.

When ectocervical lesions have a linear length greater than 8 mm the basic dome-shaped defect is vaporized centrally as previously described, followed by peripheral vaporization to a lesser depth (4 to 6 mm) (Figs. 18-7 through 18-9). If necessary the laser vaporization can be carried anteriorly, laterally, or posteriorly from the cervix onto the vagina to destroy the entire lesion, but with a vaporized depth not exceeding 1 mm into the vaginal tissue. Vaginal intraepithelial neoplasia is an epithelial disease only and there are no crypts to harbor disease. Only superficial vaporization is required. The central dome-shaped defect with shallower peripheral vaporization produces a "cowboy hat" or tiered configuration (Fig. 18-9). This defect is preferred to a very large domed defect, which may not fully regenerate and which produces a new squamocolumnar junction well out on the ectocervix, exposed to the vaginal environment. With the tiered defect normal tissue is spared, the cervix fully regenerates, and the new squamocolumnar junction should be located at the level of the external os in 3 weeks. This makes follow-up with colposcopy and cytology relatively easy. This vaporization procedure can also be accomplished within 3 to 6 minutes.

Should bleeding occur during the vaporization procedure and not be stopped by the thermal effect of the beam, the vapor plume suction attachment is simply detached from the speculum and a long plastic (Yankhauer) suction tip is substituted to suck the blood out of the field. The laser beam is applied to the base of the bleeder and the heat effect is used to seal the vessel. In order for the evacuation system to suck both blood and plume at the same time, a trap is placed in the suction line. The blood will go into the trap and the plume will continue on through the filtering system.

Laser Cylindrical Excision Procedure

The difference between conical and cylindrical volume is more dramatically illustrated in Figure 18-10, which shows the geometry of an endocervical CIN lesion extending 1.5 cm up the canal. Since disease can also extend into cervical crypts to 5 mm (shown by horizontal arrows), to include all obvious and potential disease an excised specimen would either by cylindrical in shape, with a diameter of 1.2 cm, a height of 1.6 cm, and a volume less than 2 cm^3, or conical in shape, with a base of 2.4 cm, a height of 2.7 cm, and a volume of more than 4 cm^3. With a cone-shaped defect twice as much tissue is included, most of it with little likelihood of bearing disease.[8,9,11]

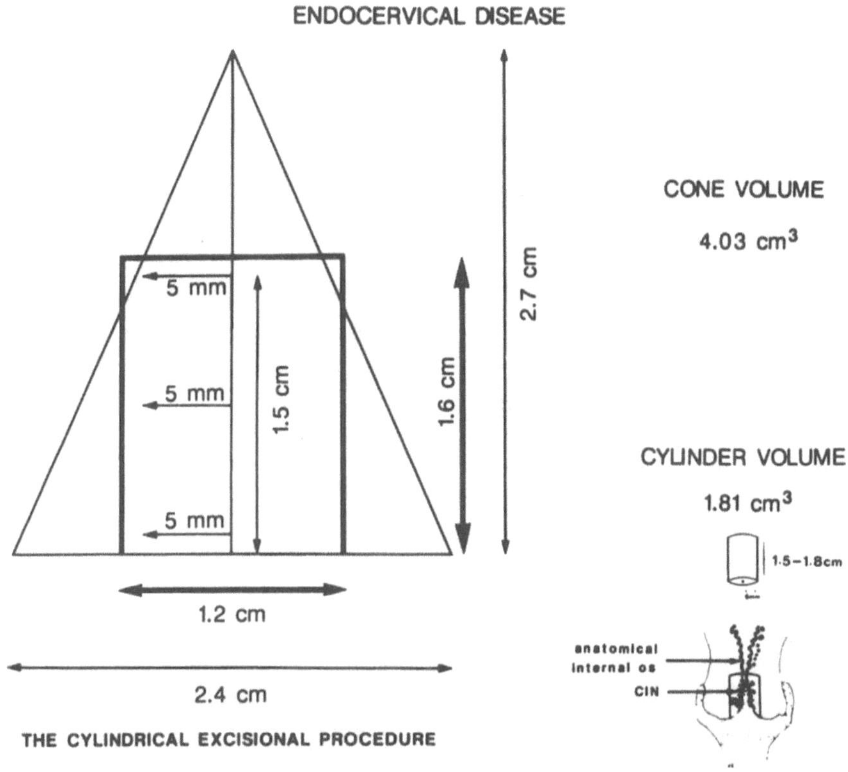

FIGURE 18-10. The geometry of CIN occupying the endocervical canal. Horizontal arrows indicate possible crypt involvement. The cylindrical approach removes half as much volume of tissue as would be required of a cone that encompassed all disease. It provides for both accurate diagnosis and a high cure rate. (Reprinted from ref. 18, with permission.)

The laser cylindrical excision procedure is performed: (a) when discrepancies between cytology, colposcopy, and histology exist; (b) for lesions located within the endocervical canal that require tissue for histologic investigation; (c) when cytology or colposcopy suggests possible invasive cancer that has not been confirmed by colposcopically directed biopsy; or (d) when adenocarcinoma in situ is diagnosed on colposcopic biopsy. This procedure replaces the usual scalpel conization of the cervix, in which a cone-shaped tissue specimen is obtained.

The author has found that 50% of all patients with severe dysplasia and/or carcinoma in situ have lesions that occupy in part or whole the endocervical canal. These data are based upon referrals to an abnormal Papanicolaou smear clinic. That is, 50% of the high-grade lesions will require a diagnostic excision because of endocervical involvement. Others have found that 2.5% of patients under the age of 30 required excision for adequate assessment of abnormal Papanicolaou smears.[12] It is well known that colposcopic examination is more often unsatisfactory in patients with advanced age. It is further noted that nearly one half of postmenopausal women will have unsatisfactory colposcopic examinations.

The cylindrical excision procedure is usually performed with the patient under general anesthesia in a properly equipped operating room. The patient is usually of an ambulatory day care status. The instrumentation includes a CO_2 laser with the articulating arm attached to an operating microscope with an objective lens of 300 mm (Fig. 18-2). A 0.5-mm beam diameter is used, which is usually the smallest spot diameter that can be produced with a 300-mm objective lens. A weighted speculum is placed in the vagina and a tenaculum is placed on the cervix at the 12 o'clock position. This gives adequate exposure. Hemostatic sutures are placed at the 3 and 9 o'clock positions and a 1% Pitressin solution (not exceeding 1 pressor unit) is injected into the cervical stroma. The cervix is washed off with a 3 to 4% solution of acetic acid.

The ectocervical os is circumscribed using the laser beam in pulses or in continuous mode with a 6- to 8-mm radius measured from the os. Then beginning at the 12 o'clock position, heavy single-tooth forceps or an Iris hook are placed in the specimen for downward traction. An incision depth or specimen height of 1.5 to 1.8 cm is achieved by moving the beam around the circle in a clockwise fashion at a power density of 10,000 to 20,000 W/cm.^{2} The incision is made parallel to the endocervical canal. The anticipated height is achieved in a particular arc using opposing traction before moving onward. It is important not to keep encircling the cervix in order to achieve adequate height. This will prolong the procedure, cause unnecessary tissue handling, and result in excessive thermal damage to the excised part of the specimen. Once around is enough. When adequate height is achieved a scalpel is used to cut the specimen transversely at the apex while the surgeon looks through the microscope. The specimen is marked (cut) at the 12 o'clock position for pathologic identification and placed in fixative for the pathology laboratory (Fig. 18-11). A 2-mm beam diameter is then used to quickly flash over the white-appearing apex, using the brief thermal effect to seal any blood vessels. Should there be any bleeding, the fluid is sucked out of the field, with the blood going into a trap and the plume onward through the suction filtering equipment. The defect sides and base are coated with Monsel's solution. When properly performed the entire procedure takes approximately 6 minutes, of which 3 minutes is spent excising the specimen with the CO_2 laser beam.[8,9,11]

This procedure can alternately be done through a bivalve speculum under paracervical block (see description for local anesthesia for laser vaporization). With this technique the Pitressin solution is injected into the cervical stroma. Once adequate height is achieved with the CO_2 laser beam, a tonsil snare can then be fed around the cylindrical specimen onward toward the apex; when the apex is reached the snare will cut off the specimen in a transverse manner. In the author's opinion, it is unacceptable to use the CO_2 laser beam to cone the apex and subsequently to vaporize the coned apex to produce a cylindrical defect, because the worst histology is located at the cephalad level and evidence of invasive cancer may be destroyed. After the snare severs the specimen a 2-mm diameter beam is then used to seal the blood vessels at the apex as after scalpel cut.

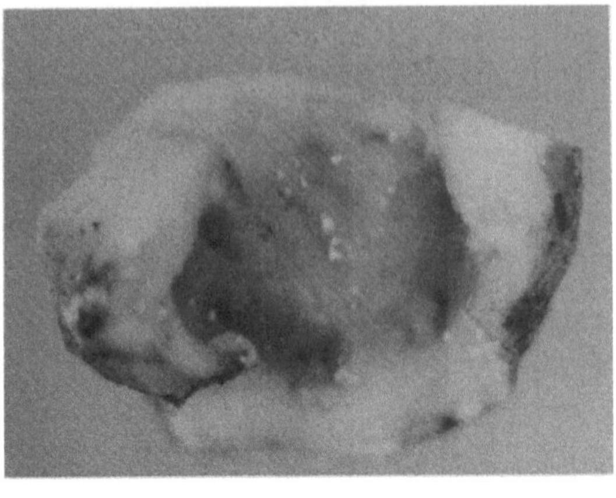

FIGURE 18-11. The cylindrically excised specimen is opened at the 12 o'clock position for pathologic orientation. The only thermal effect, of 50 to 100 μm, is on the outer surface and does not interfere with histologic interpretation. Note clean upper cut.

Laser Combination Procedure

The same magnitude of difference between cone and cylindrical specimen volumes is noted with cases of endo- plus ectocervical disease (Fig. 18-12). The principle of maximal linear extent suggests that with wide ectocervical expression the disease is unlikely to extend high into the canal. Thus a

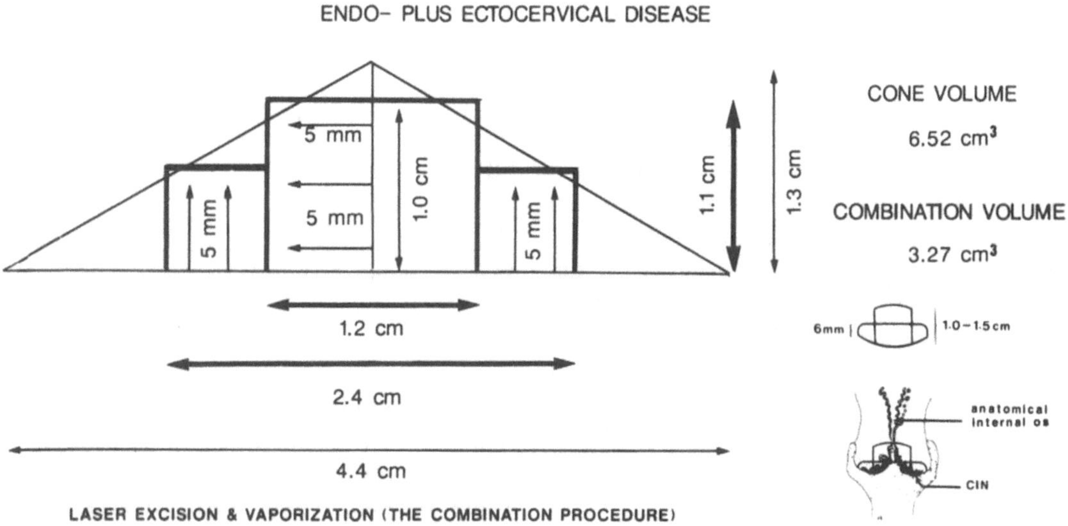

FIGURE 18-12. The geometry of CIN occupying the ectocervix and endocervical canal. To maintain the cervix and provide tissue for histology, this disease configuration requires central cylindrical excision followed by peripheral vaporization. This removes only one-half the volume of tissue that would be required by a conization to produce a cure. (Reprinted from ref. 18, with permission.)

complex combination of cylinders could be used (moderately tall and slender in the center but wide and shallow on the periphery), or an extremely wide cone could be used. The difference in the volume of tissue required to include all potential disease is considerable. The combination of vaporization and excision is intended to remove or destroy only the required tissue while still maintaining an outer rim of cervix so that regeneration will occur from all injured surfaces.[8,9,11]

This procedure involves both laser vaporization (i.e., destruction) and the excision of a central cylinder of tissue. It is appropriate when a large lesion on the ectocervix also extends beyond colposcopic vision into the endocervical canal, or when findings from cytology and colposcopy suggests CIN but a diagnostic-therapeutic specimen from the canal is necessary to completely evaluate this area and eliminate the possibility that an invasive malignancy is present. The height of the cylinder removed in cases with considerable ectocervical expression is usually less than in cases of pure endocervical disease, about 1.0 to 1.5 cm.

The procedure is performed as ambulatory day care surgery with patients under anes-

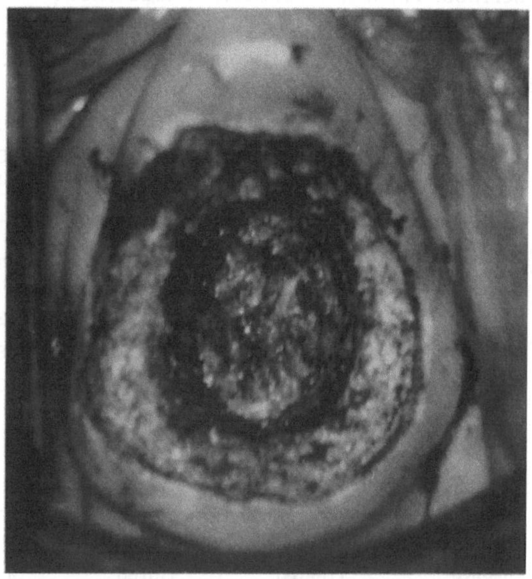

FIGURE 18-13. The cervix after central laser excision and peripheral vaporization. Again a "cowboy hat" appearance is noted. The flanges will vary in width depending upon the extensiveness of the ectocervical component, but on the cervix a depth of 5 mm is required.

thesia. Removal of the shallow central cylinder for pathologic interpretation is done first. Following excision, laser vaporization of the diseased ectocervical tissue and the remainder of the transformation zone, peripheral to the central cylinder, is performed to a depth of 5 to 6 mm. This procedure also produces a "cowboy hat"–shaped defect (Fig. 18-13). The relative configuration of the defect (top and flanges) will vary depending upon the disease distribution. Should the excised specimen verify only CIN disease, then this therapy usually will suffice. Should invasive cancer be identified in the canal by histology, the patient is managed accordingly.

TABLE 18-3. Results of one laser surgery (vaporization, excision, and combination)

Diagnosis	Cases	Failures	% Cured
CIN I	644	44	93.2
CIN II	699	34	95.1
CIN III	984	35	96.4
Total	2,327	113	95.1

Reprinted from ref. 11, with permission.

Results of Laser Surgery for CIN

Between 1978 and 1986, 2,327 patients with CIN who were evaluated in a prospective study in the Abnormal Pap Smear Clinic at St. Joseph's Hospital, London, Ontario, were treated by one of the three methods described. No other method of treatment was offered to these patients as a first intervention. The location of disease determined the laser procedure performed.[11]

Table 18-2 shows the results after one laser surgery. One vaporization procedure cured 93.4% of CIN I, 95.6% of CIN II, and 96.9% of CIN III cases. One excision or combination procedure cured 92.6% of CIN I, 93.8% of CIN II, and 96.0% of CIN III cases. Table 18-3 summarizes the success rates after one laser surgery with all procedures combined. Tables 18-4 and 18-5 illustrate the diagnosis of persistent disease after laser vaporization (Table 18-4) and laser excision (Table 18-5). In all procedures, condyloma accounted for increased failures in the lower grade lesions. Table 18-6 illustrates the management of persistent disease after the first laser surgery. Thirteen hysterectomies were done (0.56%). One case of persistent CIN III was managed by a third laser vaporization. The first follow-up examination, done at the 3-month anniversary, including colposcopy and cytology, was very accurate in predicting the persistent disease. That is, greater than 90% of all patients with persistent disease were identified at the first follow-up visit, with all remaining such patients identified at the second visit except in one case, the latter being identified at the third visit (9 months).

After laser surgery, in 90% of the cases the new squamocolumnar junction formed at the external os and in all cases the cervix appeared to regenerate to its original or near-original mass (Fig. 18-14). No case of cervical stenosis occurred in a menstruating woman. Table 18-7 illustrates posttreatment bleeding. Most patients with bleeding needing attention were managed in the clinic (7.8% of

TABLE 18-2. Cure rates after one treatment

	Vaporization			Cylinder and combination		
Disease	Treated	Cured	% Cured	Treated	Cured	% Cured
CIN I	469	438	93.4	175	162	92.6
CIN II	504	482	95.6	195	183	93.8
CIN III	481	466	96.9	503	483	96.0
Total	1,454	1,386	95.3	873	828	94.8

Reprinted from ref. 11, with permission.

TABLE 18-4. Histologic diagnosis of persistent disease following primary laser vaporization

	Total cases		Persistent disease		Histologic diagnosis of persistent disease			
Initial diagnosis	No.	%	No.	%	CIN I	CIN II	CIN III	Condyloma
CIN I	469	32.2	31	6.6	22	3	1	5
CIN II	504	34.7	22	4.4	12	4	2	4
CIN III	481	33.1	15	3.1	4	3	6	2
Total	1,454	100.0	68	4.7	38	10	9	11

Reprinted from ref. 11, with permission.

TABLE 18-5. Histologic diagnosis of persistent disease following primary laser excision

	Total cases		Persistent disease		Histologic diagnosis of persistent disease			
Initial diagnosis	No.	%	No.	%	CIN I	CIN II	CIN III	Condyloma
CIN I	175	20.1	13	7.4	4	2	1	6
CIN II	195	22.3	12	6.2	3	5	2	2
CIN III	503	57.6	20	4.0	2	7	11	0
Total	873	100.0	45	5.2	9	14	14	8

Reprinted from ref. 11, with permission.

the excision group and 1.9% of the vaporization group). Of the excisional group, 1.7% required hospital admittance for bleeding, as compared to 0.9% of the vaporization group. One patient in the excisional group required a blood transfusion. Five patients required cervical suturing under anesthesia. No patient required a hysterectomy to control bleeding either during the procedure or postoperatively.

The complication of bleeding associated with scalpel conization is well recognized. A review of the literature indicates that hemorrhage occurring at the time of the procedure or within the first 10 days varies between 2.1 and 17%.[13–17] Of those who had troublesome bleeding, the transfusion rate varied between 8.3 and 78.6%. In those patients who had troublesome bleeding, the hysterectomy rate varied between 2.0 and 18.8% for the control of bleeding.[13,15] The outcome with these laser approaches to surgery for CIN is much better.

Table 18-8 illustrates the pregnancy outcome in 195 patients within the group (142 vaporization and 53 laser excision patients) who became pregnant after their surgery. There was no increase in the premature delivery or cesarean section rates over standard patients.

TABLE 18-6. Management of persistent disease

Initial diagnosis	Cases	Failures	Second laser treatment	Second and third laser treatment	Focal cautery	Hysterectomy[a]
CIN I	644	44	40	0	2	2
CIN II	699	34	31	0	1	2
CIN III	984	35	25	1	0	9
Total	2,327	113	96	1	3	13[b] (0.56%)

Reprinted from ref. 11, with permission.
[a] Hysterectomy for benign reasons: 20 (no CIN in specimens).
[b] All extirpated specimens contained CIN III histologically.

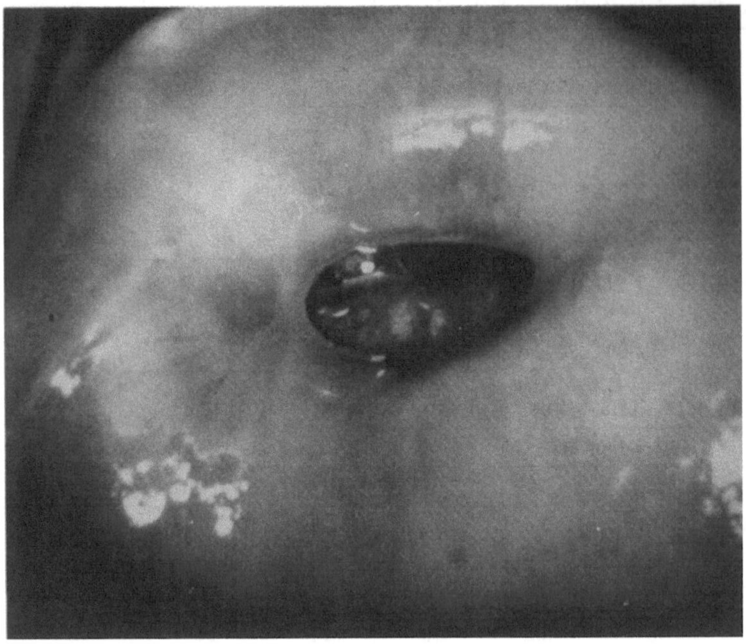

FIGURE 18-14. The healed cervix after laser surgery. This is the typical appearance regardless of the laser surgical procedure. The volume removed is restored because the surgical defects leave a base and a rim of cervix from which reepithelialization can occur. The new squamocolumnar junction becomes relocated at the external os in 90% of the cases.

TABLE 18-7. Posttreatment bleeding management location

Procedure	Total cases	Clinic or hospital outpatient		Hospital inpatient	
		No.	%	No.	%
Excision	873	68	7.8	15	1.7
Vaporization	1,454	28	1.9	13	0.9
Total	2,327	96	4.1	28	1.3

Reprinted from ref. 11, with permission.

Conclusion

These three procedures—vaporization of a dome-shaped cylinder, excision of a cylindrical specimen of endocervix, and a combination of both—appear to be effective for eliminating CIN regardless of histologic grade, lesion size, or disease location.[8,9,11] This is probably because the surgical approaches are designed to remove appropriate tissue volumes based on the solid (i.e., three-

TABLE 18-8. Pregnancy after laser surgery

	Vaporization	%	Laser excision	%	Total	%
Total pregnancies	142		53		195	
Term deliveries (37+ wks)	116	93.5	42	89.4	158	92.4
Premature deliveries (20–36 wks)	8	6.5	5	10.6	13	7.6
Spontaneous abortion	12	8.4	1	1.8	13	6.7
Therapeutic abortion	3	2.1	3	5.6	6	3.1
Extrauterine pregnancy	3	2.1	2	3.7	5	2.7
Vaginal deliveries	106	85.5	44	93.6	150	87.7
Total cesarean sections	18	14.5	3	6.4	21	2.3
Fetal death (23 wks)	1	0.8	0	0.0	1	0.6

Reprinted from ref. 11, with permission.

dimensional) geometry of CIN. The defects created always have a base and sides to promote healing, and the laser creates little damage to surrounding healthy tissue as well. The geometry of disease and the surgical defect are cylindrical in shape and differ substantially from the traditional cone in theory, practice, and results. These data suggest that removing with the laser a planned volume of diseased tissue based on the geometry of CIN does not appear to increase the premature delivery rate or produce an incompetent cervix. This is likely related to the fact that the three laser procedures are designed to remove disease but not encroach upon the anatomic internal os in the reproductive female.[8,9,11]

Restoration of near-normal volume of tissue and an optimal location of the new squamocolumnar junction at the level of the external os allow effective follow-up assessments by colposcopy and cytology (Fig. 18-14). Therefore, any persistent or new CIN disease is easily diagnosed and retreatment easily facilitated by further laser vaporization or excision.

References

1. Einstein A: Zur quantum theorie der strahlung. Phys Zeit 1917;18:121.
2. Wright VC, Riopelle MA: Surgical CO_2 Laser Fundamentals. Houston, Biomedical Communications, 1988.
3. Polanyi TG: Physics of surgery with lasers. Clin Chest Med 1985;6:179.
4. Przybora LA, Plutowa A: Histological topography of carcinoma in situ of the uterine cervix. Cancer 1959;12:268.
5. Scott RB, Reagan JW: Diagnostic cervical biopsy technique for the study of early cancer: Value of the cold-knife conization procedure. JAMA 1956;160:343.
6. Anderson MC, Hartley RB: Cervical crypt involvement by intraepithelial neoplasia. Am J Obstet Gynecol 1980;55:546.
7. Wright VC, Riopelle MA: The geometry of cervical intraepithelial neoplasia as a guide to its eradication. The Cervix 1986;4:21.
8. Wright VC, Riopelle MA: Laser cylindrical excision to replace conization. Am J Obstet Gynecol 1984;150:704.
9. Wright VC, Riopelle MA: Laser surgery for cervical intraepithelial neoplasia: Principles and results. Am J Obstet Gynecol 1983;145:181.
10. McKay W, Morris R, Mushlin R: Sodium bicarbonate attenuates pain on skin infiltration with lidocaine, with or without epinephrine. Anesth Analg 1987;66:572.
11. Wright VC: Carbon dioxide laser surgery for the cervix and vagina: Indications, complications and results. Comp Ther 1988;14:54.
12. Shingleton H, et al: Outpatient evaluation by colposcopy, biopsies and endocervical curettage. Am J Obstet Gynecol 1970;108:429.
13. Van Negell JR Jr, Parker JC: Diagnostic and therapeutic efficiency of cervical conization. Am J Obstet Gynecol 1976;124:134.
14. Bostofte E, et al: Conization by carbon dioxide laser or cold knife in the treatment of cervical intraepithelial neoplasia. Acta Obstet Gynecol Scand 1986;65:199.
15. Jones HV III, Buller RD: The treatment of cervical intraepithelial neoplasia by cone biopsy. Am J Obstet Gynecol 1980;137:882.
16. McCann SW, Michael A, Crapanzano JT: Sharp conization of the cervix: Observations of 501 consecutive patients. Obstet Gynecol 1969;33:470.
17. Claman AD, Lee NL: Factors that relate to complications of cone biopsy. Am J Obstet Gynecol 1974;120:124.
18. Wright VC, Riopelle MA: Gynecologic Laser Surgery: A Practical Handbook. Houston, Biomedical Communications, 1982.

19

Congenital Absence of the Uterus and Vagina

Albert Altchek

Definition and Incidence

Congenital absence of the vagina is referred to as Rokitansky syndrome, Mayer-Rokitansky-Kuster-Hauser syndrome, vaginal agenesis, and vaginal aplasia. There are several recent reviews of the condition.[1–4] The classical reference is the Jones and Rock text.[5] Absence of the uterus in genetic females occurs as part of the Rokitansky syndrome. About 9–10% of cases of this syndrome may have a functioning uterus but without a vagina. The syndrome is focused on a vagina because its absence is more readily noticed than the absence of the uterus.

Absence of the uterus occurs in genetic males with a female phenotype (androgen insensitivity). There is a short vagina present.

The incidence is uncertain and could only be determined by careful examination of each newborn female, probably requiring passing a urethral catheter into the vagina. Incidences based on hospital admissions reflect factors of patient selection. It has been estimated to occur in anywhere from 1:1,000 to 1:10,000 infants, with the usual estimate as about 1:5,000. In the average gynecologic practice, there may be one case every 10 years. With primary amenorrhea (pregnancy being ruled out) included in patient selection, the incidence is about 15% and is second to gonadal dysgenesis (Turner's syndrome).

Pathology

The usual pathology is complete absence of the vagina and of the uterus. Actually, there is a bilateral swelling of the müllerian (paramesonephric) ducts at the pelvic brim of about 1 to 2 cm width and 2 to 3 cm length that is the nonfused area that should have been the uterus (uterine anlagen, uterine bulb, vestigial uterus). It has a reddish color on laparoscopy and biopsy shows normal myometrial cells. There is no endometrium or, in the unusual case with endometrium, it is usually a nonfunctioning basalis. This vestigial uterus may be flat against the lateral pelvic wall retroperitoneally or have a short mesentery. If the vestigial uterus is prominent, there may be a round ligament of the uterus present. Even if small, there is always a round ligament of the ovary leading to a normal ovary. Although it is thought that normal fallopian tubes are always present, in fact I have observed by laparoscopy that there is variation and gradation to almost complete absence of tubes. The vestigial uterus and adnexae are high at the pelvic brim and may be overlooked if the laparoscopist (or sonographer) inspects only the central true pelvis (Fig. 19-1). A well-known artist's sketch incorrectly depicts these structures as being central and deep in the true pelvis. If there is an absent kidney, these structures are less well developed on that side.

FIGURE 19-1. Laparoscopic view of the pelvis in congenital absence of the vagina. The vestigial unfused uteri are high at the lateral pelvic brim with the adnexae. They are not in the deep true pelvis.

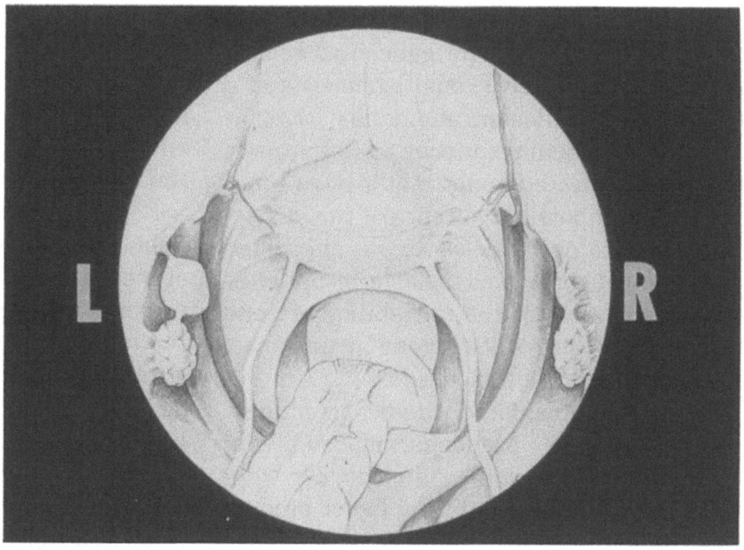

The vulva has a normal female appearance, including the clitoris, labia majora and minora, urethra, vestibule, and perineum. Where the hymen should be is an imperforate wall corresponding to the outer wall of the hymen. There is a normal female phenotype, psyche, ovulation, hormonal pattern, and karyotype. General growth and development are normal. Extragenital anomalies are frequent.

About one third of cases have abnormal kidneys, usually agenesis of one kidney or ectopic location. There may also be solitary, fused, or horseshoe kidneys; abnormalities of the collecting system; malrotation and solitary pelvic ectopic kidneys; and malfunctioning kidneys.[1]

About 12% have skeletal anomalies, usually of the vertebrae. As part of an evaluation, an intravenous pyelogram is often done, which would reveal lumbar anomalies; however, cervical and thoracic anomalies may be more frequent. Less often, there may be obvious extremity anomalies. Routine x-rays of the hands show that more than half have abnormalities, including brachymesophalangy of digits 2 through 5, small distal phalanx of digit 1, long proximal phalanx of digits 3 and 4, and long metacarpals of digits 1 through 4. There may also be carpal abnormalities.[6]

The practical problem is that the patients are young and therefore radiation is avoided if possible. Routine complete skeletal x-rays would probably indicate greater than 12% skeletal anomalies.

Other less common anomalies include syndactyly, absence of a digit, thenar hypoplasia, inguinal and femoral hernias with ovaries, cardiac anomalies, deafness, facioauriculovertebral spectrum, cleft palate, bowel malrotation, and situs inversus.

Etiology

There are many conflicting and overlapping theories regarding etiology, including variable manifestations of a single underlying genetic defect that can be expressed in genital, renal, and vertebral anomalies, including absence of both kidneys.[1] Rare familial occurrences have suggested both autosomal recessive and dominant inheritance patterns. It may be that there are many causes, including "polygenic/multifactorial" inheritance as well as "etiologic heterogeneity."[4] Some have a pragmatic approach of considering isolated Rokitansky syndrome cases as "sporadic," whereas those cases with additional other anomalies that are familial are considered as "familial."

In two families, there seemed to be an autosomal dominant single gene that caused unilateral or bilateral renal agenesis and a spectrum of müllerian anomalies from bicornuate or didelphys uterus to Rokitansky syndrome. Affected males might have renal agenesis and possibly absence of the vas deferens or seminal vesicles. It was suggested that this congenital adysplasia is underdiagnosed because of the variability of gene expression, the reduced gene penetrance, and the lethal renal anomalies.[7]

Some have considered it to be the mildest form of female pseudohermaphroditism, with inhibition of müllerian duct development by müllerian-inhibiting factor production.[8]

Rokitansky syndrome has a close association with absent kidney. Unilateral and or bilateral renal agenesis or severe dysplasia is called hereditary renal adysplasia (HRA). It is an autosomal dominant trait with incomplete penetrance and variable expression. Renal sonography has been recommended for parents, siblings, and other close relatives when there is an individual with unilateral or bilateral renal agenesis. The risk for recurrence of bilateral renal agenesis in siblings is about 3 to 5%, but in the offspring of affected or obligate heterozygotes for HRA, the risk of bilateral severe renal adysplasia is 15 to 20%.[9] Renal agenesis is usually related to mesonephric development and therefore to internal genitalia anomalies. Unilateral renal agenesis may be associated with uterine defects, hypoplasia, and premature deliveries. Bilateral and unilateral renal agenesis can be expressions of hereditary renal dysplasia and since the penetrance of HRA is not 100%, for unaffected parents who have had a child with bilateral renal agenesis or HRA the risk of having another child with severe bilateral disease ranges between 3.5 and 15 to 20%. It is suggested that in girls with unilateral renal agenesis the Rokitansky syndrome should be suspected, and it may be "prudent to investigate HRA in the families of" women with Rokitansky syndrome.[10]

There has also been a suggestion of an association with errors of galactose metabolism and premature menopause.[11] Some have adopted an all-inclusive concept involving several genes in combination with the environment.

Regardless of theories of etiology, it would seem that there is a final common pathway of mechanism. There is a disturbance of the mesonephric (wolffian) duct. Although it is the male internal genital duct, it forms the stimulus and framework for the progressive development of the paramesonephric (müllerian) female duct. The state of development of the müllerian duct in the adult with Rokitansky syndrome corresponds to the fetus at 9 weeks after the last menstrual period or 7 weeks after fertilization. Since the mesonephric duct also is involved with the metanephric (adult) kidney, its malfunction leads to absence of the kidney. At this time of embryologic life the mesenchyme of different parts of the body and vertebrae is being organized and correspondingly may be disturbed.

Diagnosis

The diagnosis is easily overlooked. The presenting complaint is primary amenorrhea in an otherwise normal female. Patients and mothers are often shy about discussing it, hope that menstruation will develop, and hope the doctor will ask about it. When the patient or the mother mentions it, even in an offhand manner, an appointment should be made for a careful history and physical examination. This does not necessarily imply that an expensive and lengthy series of laboratory tests will be done. Furthermore, definitions of what constitutes primary amenorrhea should not hinder the physician. When the question is raised the patient should be examined promptly regardless of age, development, or standard recommendations regarding time of evaluation.

We as physicians tend to be "upbeat" and

encouraging. When a casual reference is made regarding absence of menstruation we sometimes simply note the development of secondary sex characteristics and we reassure without a complete examination. We don't think of the diagnosis.

The diagnosis is not easy to make. The patients have a normal female phenotype, sexual orientation, karyotype, and ovulatory endocrine evaluation. Unfortunately, extensive and expensive hormone tests are often done before a complete physical examination. Even a casual inspection of the vulva can be misleading since there is a normal clitoris and labia. What would be the outer wall of the hymen is present, but without an opening. It may have a ruffled or relatively smooth surface, a false opening suggesting an outer layer opening and inner layer intact wall, and/or a dimple (see Figs. 19-2 and 19-3).

The experienced clinician can make a relatively reliable diagnosis by rectal examination. In the usual case, there is no uterus. Sometimes, what feels like a bilateral sacrouterine ligament can be felt meeting in the midline. These are in fact the vestigial remnants of the underdeveloped müllerian ducts.

For confirmation, a pelvic and abdominal sonogram may be done. An experienced sonographer is important because there have been incorrect reports of a uterus (because of stool in the rectum) and of absence of ovaries (because the adnexae are high and lateral at the pelvic brim). The sonographer should also check for the presence and location of kidneys. If it is desired to avoid radiation or dye allergy, an intravenous pyelogram may be omitted. It is not necessary to do a karyotype or an endocrine work-up in the usual case. (In addition, genetics departments may refuse to use their laboratories to verify an XX pattern.)

As a routine, a serum testosterone is often done to rule out androgen insensitivity male pseudohermaphrodism (testicular feminization). There may be a shallow, blind vagina and no uterus present on rectal examination. There may be a family history of amenor-

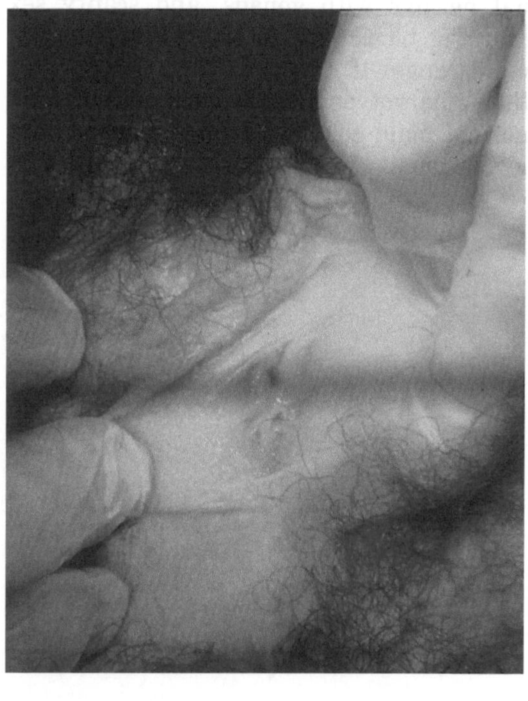

FIGURE 19-2. The vulva in congenital absence of the vagina is deceptively normal in appearance, with a normal clitoris and labia. What would be the outer wall of the hymen may have a ruffled surface.

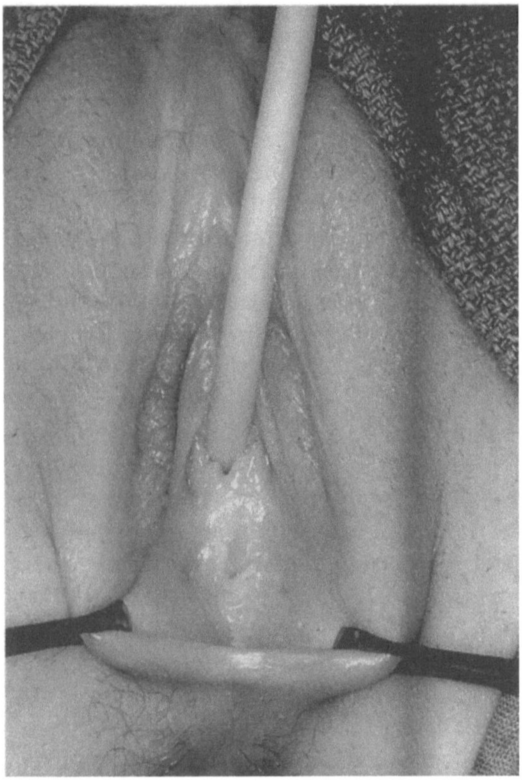

FIGURE 19-3. Congenital absence of the vagina showing a false opening where the hymen should be.

rheic females, unilateral or bilateral inguinal hernias with gonads, and scanty sex hair. The blood testosterone is in the normal male range. If this condition is suspected then a karyotype must be done since the abnormal male gonads have a propensity for malignant change and must eventually be removed. It is less common than Rokitansky syndrome. Pediatric surgeons tend to recommend gonadectomy before puberty with puberty developed by exogenous estrogens. If there is partial androgen insensitivity,

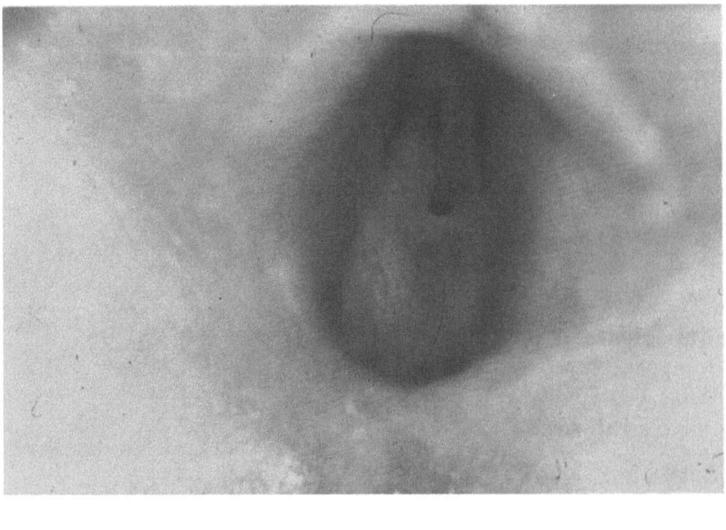

FIGURE 19-4. Microperforate hymen. The hymen is smooth and may bulge. With microperforate hymen, there is a minute opening just below the urethral meatus.

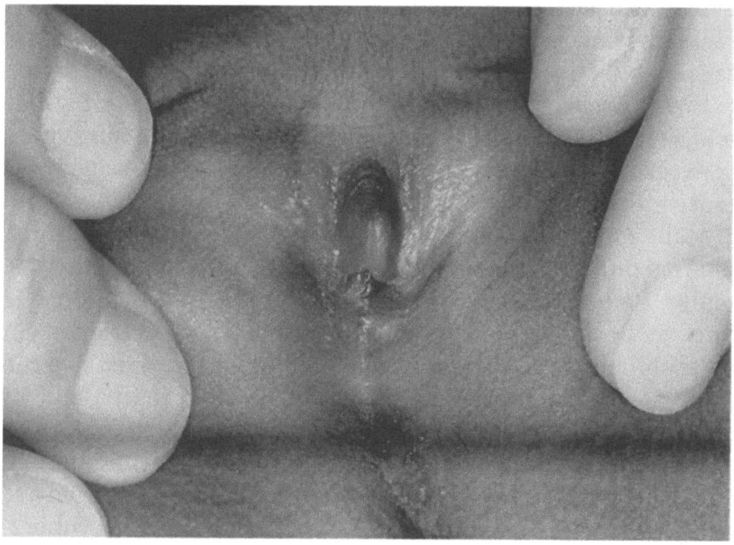

FIGURE 19-5. Agglutination of the labia minora may occur in the prepubertal child and distort the vulva.

there is clitoral enlargement and there may be masculinization at puberty. This is another reason to perform the gonadectomy before puberty.

The differential diagnosis includes imperforate or microperforate hymen (Fig. 19-4) and transverse vaginal septum at various levels. The imperforate hymen may bulge from collected mucus or later from menstrual blood, which may be felt on rectal examination. The microperforate hymen has a tiny opening just under the urethra that may be seen when the hymen is retracted posteriorly by an applicator stick. The distal transverse vaginal septum may be thick and not bulge.

Agglutination of the labia minora may occur in the prepubertal child and is identified by the distortion of the vulva (Fig. 19-5).

The diagnosis may be made more difficult by the Klippel-Feil syndrome (Rokitansky syndrome with fused cervical vertebrae). The patient has a short, fused neck that superficially resembles the webbed neck of gonadal dysgenesis (Turner's syndrome), which includes streak ovaries, XO karyotype, a small uterus, sexual infantilism, and short stature.

Other diagnostic problems occur with atypical variations of Rokitansky syndrome. About 8 to 10% of cases have a functioning uterus, and with this usually a normal cervix. Although it has not been emphasized in the literature despite its extreme significance, with a normal cervix I have always found a surrounding cap of vaginal mucosa. Initially, a cervix and uterus will be felt on rectal examination. With the onset of menstruating, the vaginal cap will distend and can be felt on rectal examination.

One of my patients with an absent vagina and a functioning uterus reported menstrual periods with small volume but extending many days. I had her return when she was menstruating and found the blood emerging drop by drop through a pinpoint opening too small to probe. When I did the McIndoe procedure I could not find any trace of the narrow fistula.

In two cases with a functioning uterus and an absent cervix, the fundus could be felt on rectal examination.

Magnetic resonance imaging (MRI) is not necessary for the usual case. However, it is even more reliable than sonography since it gives better tissue differentiation. It seems to be the only noninvasive method of detection of cervical stenosis when a uterus is present.[12] Historically, exploratory laparotomy was used and later routine laparoscopy. Even the latter is not necessary today. However, it is used if there is pain or the presence of a pelvic mass or if the diagnosis is uncertain.

Final decisions regarding management and surgery still must be made on the basis of experience, judgment, and clinical and operative findings.

Male pseudohermaphrodites with microphallus and flat perineum are reared as females. Before the age of gender identity (before age 2) the vulva is feminized surgically by mobilizing redundant scrotal folds to a more posterior and lateral position, by incising the phallic urethra, and by recession of the phallus and resection of redundant tissue. The gonads are also removed. At puberty, estrogen is administered to develop secondary sexual characteristics. In late adolescence or early adulthood, a McIndoe procedure is used to create a vagina. Because of the flat perineum, vaginal dilation is not possible.[13]

Management

Background: The Year 1938

The year 1938 was momentous for the origination of both the standard operation and the standard nonoperative treatment. Archibold H. McIndoe (a plastic surgeon) and J. B. Bannister (a gynecologist) published their landmark paper from London on what is now the standard operation for absent vagina and is referred to as the McIndoe procedure.[14] They dissected a space, lined it with one large sheet of split-thickness skin graft, and held it in place with an oval-shaped, hard plastic mold that was sewn in for 6 months. Their work was based on reconstruction of World War I injuries. They acknowledged the previous forgotten work of a New York City surgeon, Robert Abbe, who in 1898 used many small pieces of Thiersch (split-thickness) skin grafts over a stuffed "thin French rubber pouch" to line the dissected vaginal space, and recommended dilatation with a wax candle bougie for "many months" to avoid postoperative contraction.[15] McIndoe took his patients back to surgery 6 months later to open the perineum and remove the mold. His long-term series results were good except that he was always plagued by an approximately 2% chance of rectal fistulas. This was probably due to pressure necrosis. Although the McIndoe procedure is often referred to, in fact, at present, usually a hard plastic mold is not used initially and it is not embedded for many months.

In 1938, Virgil S. Counseller, Chief of the Division of Surgery of The Mayo Clinic, reported five cases of congenital absence of the vagina and two cases of traumatic obliteration of the vagina treated with a similar technique.[16] The differences were that Counseller used a flexible rubber tube rather than a hard rubber mold and that he removed the tube in from 10 to 14 days and replaced it with a hard Bakelite stent that was worn for 6 months. McIndoe completely closed the vulva over the mold and left it in place for 6 months. Although Counsellor acknowledged advice from McIndoe and referred to it as the McIndoe procedure, there has been speculation about who originated the procedure.

A 1989 report from the Mayo Clinic noted that "we continue to consider the McIndoe operation as the procedure of choice for most patients with vaginal agenesis."[17] This was despite the problems encountered. At the Mayo Clinic from 1976 to 1985, there were 50 patients with congenital absence of the vagina who were treated by a modified McIndoe procedure. There were two rectovaginal fistulas and one graft failure. Five patients required additional reconstructive vaginal operations. Of 47 who responded to a survey, 40 considered the operation to be functionally successful. A few noted inadequate vaginal lubrication. Short-term complications included a distal urethrotomy, the need to take a second graft, three hematomas, reoperation to replace a stent, cough with expulsion of the graft, and a rectovaginal fistula due to pressure. Late complications (1 month after surgery) included 1 rectovaginal fistula from trauma of the stent, 5 cases with neovaginal stricture, 1 with a symptomatic rectocele, 1 with keloid formation at the donor site, 6 with urgency or stress urinary incontinence, and 26 with granulation tissue.[17] The Johns Hopkins group and others have had even better results with the McIndoe operation.[4,5] The McIndoe procedure is the most frequently used operative procedure for congenital absence of the vagina.

Also in 1938, Robert T. Frank, Chief of Gynecology at the Mount Sinai Hospital in New York City, devised the technique of placing pressure against the dimple using graduated hard tubes for 30 minutes three times daily to eventually invaginate the area.[18] Perhaps this technique was a reaction to the complications and deformity sometimes resulting from his previous 1927 Frank-Geist ("satchel handle") operation in multiple stages involving the creation of full-thickness tubed pedicles from the inner thigh that were inverted into a surgically created vaginal space.

In 1981, J. M. Ingram, the Chairman of Obstetrics and Gynecology at the University of South Florida in Tampa, Florida, devised a bicycle seat with graduated dilators to enable the patient to sit on the dilator rather than push it in with her hands.[19] These dilator techniques usually require a nontender, pliable dimple, intense determination of the patient, and extraordinary emotional support, and usually require a daily 2-hour pressure application to be successful. It is safe, although the patient has to avoid urethral dilatation. Cystitis sometimes occurs, and occasionally the invaginated vagina may tend to prolapse.[17]

Other Operative Procedures

Amnion Membrane

Amnion membrane has been used instead of split-thickness skin graft to line the new vagina. It seems to be bacteriostatic and to encourage host epithelial growth, and apparently does not cause an immunologic reaction. The amnion cells eventually die. Hughes and Spence reported cases of human amniotic membrane as a lining for the newly dissected vagina and cited 37 previously reported cases.[20] One was a 12-year-old girl with a functioning uterus and upper hematocolpos with a distal 5-cm aplasia. She did not use the vaginal dilator at first and developed an upper pyocolpos (presumably from stricture). I had a similar case of a 15-year-old who had a functioning uterine fundus without a cervix and had undergone previous unsuccessful vaginal surgery. In order to avoid a donor site scar and the need for postoperative bed rest, I used amnion membrane (four times instead of the usual two applications). There was an initial good result, but after discharge from the hospital she did not keep the mold in and the new vagina closed down.

I also used amnion membrane in an ambulatory patient with a previous partial take of a McIndoe procedure after curretting the granulations. The amnion did not hasten reepithelialization of the surrounding surviving split-thickness skin graft. It finally healed without scarring with continous used of a vaginal mold.

Amnion has lost favor because of the fear of transmission of acquired immunodeficiency syndrome and hepatitis. Edmonds wrote "and so this very successful technique must now be abandoned."[3]

Vaginal Pouch

Also in 1938, L. R. Wharton of Johns Hopkins, in four cases with a small vaginal pouch, dissected a vaginal space and without mentioning it pushed in the surface dimple disc of epithelium to cover part of the distal neovagina. The space was left open by a condom-covered balsa wood mold until epithelialization occurred, presumably from the disc.[21] He later abandoned the method in 1950 in favor of split-thickness skin graft, believing that the latter had less scar formation.

B. H. Sheares in 1960, at the University of Malaya in Singapore, found a patient with two very small openings on each side of the midsagittal plane that could be probed to 6 cm. He interpreted this to be the distal vestigial müllerian ducts. He found buried islands of embryonic squamous epithelium and gland acini in the vesicorectal septum. Using Wharton's principles, he described and reflected a racket-shaped flap from the vaginal dimple or pouch. Based on his observation, he used a Hegar cervical dilator on each side of the midline to form a double-barreled tunnel, resected the vesicorectal septum, pushed in the dimple flap, and used a silver or stainless steel mold with a urethral groove. Even after 6 months, he advised twice-daily obturator dilation indefinitely despite sexual activity. He assumed that epithelialization resulted from upward growth of vestibule and dimple epithelium and from buried islands of epithelium that were now at the neovaginal surface. One of his 18 cases had a uterus, hematometra, and a cervix that was exposed. She later became pregnant and had an unplanned term vaginal delivery followed by another vaginal delivery. In a series of 10 cases with the McIndoe procedure one patient (with uterus) had three subsequent term cesarean sections.[22]

Surgical Vulvar Pouch

Another option is the surgical vulvar pouch, devised by Professor Williams of Oxford.[23] The advantage is that it is a relatively short, easy, and safe operation that usually does not need vaginal dilators postoperatively. This is probably why the English national health service recommends it. Its proponents claim that their patients are happy with it; however, patients are not given a choice and patient satisfaction has not been studied. The disadvantage is that the neovaginal angle is downward and different, the vulva looks unusual (suggesting a urogenital sinus), the urinary stream has a distorted dribble, and the labia minora are compressed anteriorly (sometimes requiring excision). To avoid urinary distortion others have built the perineum less high and have increased the depth with dilators for the dimple. This method has not been popular outside of England. The principle of the Williams vulvovaginal pouch is used when there are congenital cloacal anomalies, or with previous pelvic exenteration and/or radiation. Edmonds, also of London, observed that the Williams vulvovaginoplasty "is used much less frequently today" because of its unnatural angle and distorted external genitalia.[3]

I have used the Williams operation in cases in which the patient has had previous unsuccessful vaginal surgery for repair of vesical and ureteral fistulas, and dense scar tissue that would make a vaginal space dissection hazzardous. When I do the Williams operation, I do not create a bridge as high as Professor Williams does in order not to distort the urinary stream. Although the vulva pouch is short, there may also be a small vaginal pouch or one may be created by dilators.

Alternative Approaches

Continuously throughout the world, physicians have been fascinated by this unusual anomaly and have attempted many ingenious approaches. It was generally recognized that a space could be dissected but the problem was to keep it open. In the late 19th century, various European surgeons used segments of rectum and intestine to line the new vagina.

In 1904, James Baldwin of Columbus, Ohio, proposed lining the dissected space with a double loop of ileum or sigmoid colon that would later be converted to one structure.[24] Generally, bowel is no longer used since it requires a laparotomy with potential hazzards. Nevertheless, a few enthusiasts in Yugoslavia still use sigmoid colon to construct a surgical vagina, except when there is a solitary pelvic kidney (because of the proximity of blood vessels).[25]

Recently, an operation was devised in Italy by Vecchietti in which a laparatomy is done to extend a wire through the abdomen, down through the potential vaginal space, and out through the dimple, where an olive is attached. A device on the abdomen keeps the wire under constant tension so that the external olive invaginates the dimple. It is analogous to continous external dilator pressure on the dimple. This technique has been used in Europe but not in the United States.

Practical Clinical Management

My practice is a referral practice. The patient complains of primary amenorrhea, and has usually been seen by one or more physicians. They are uncertain of the diagnosis or suspect it but are reluctant to give the bad news to the patient. Very often a battery of expensive hormone tests have been done. The patient and her mother are very anxious and want to know the diagnosis that moment. I have to tell them that the patient's body was incompletely developed, that there is no functioning uterus and vagina, that she will never menstruate or be pregnant, and that unless something is done she will not have a sex life. The moment she hears this, she developes a fixed stare and glaze in her eyes, her jaw drops, and she goes into a state of depression. It is as if she were struck by a thunderbolt. The mother usually cries. One realizes that from that point on it is difficult to "get through" to the patient or communicate with her.

In about 6 weeks the reaction of depression usually abates and one can discuss the various options.[26] I speak to the patient with one or both parents present so that they can reinforce my suggestions at home. I point out that:

1. It is almost impossible to find any person who is perfect physically and mentally.
2. She is fortunate not to have a life-and-death anomaly such as a cardiac or lung problem.
3. She is intelligent, attractive, and young and has the whole world in front of her.
4. One day, she will find a mutually attractive male who will marry her for herself, and, if they wish, they can adopt children.
5. Many of my patients have actually had successful careers, marriages, and adoptions.
6. About 15% of all married couples (in which the wife has a uterus) do not achieve pregnancy.
7. She is fortunate to have a supportive family.

I also advise that she not discuss this with friends or relatives who will not understand the problem and will spread distorted gossip. It should be discussed only with professional persons, and I suggest a psychiatrist. Usually the patient recoils at the suggestion. I explain that she is not "crazy," but she needs someone to talk to. In practical reality, I find that I personally have to give continuous emotional support. I am always concerned about the possibility of suicide, which I relate privately to the parents. This has not occurred in any of my patients nor have I found any references in the literature despite the usual severe initial depression.

The mother often suffers as much as the patient with a sense of guilt because of the delivery of an incompletely formed daughter. The mother usually denies any illness or medications in pregnancy. The mother also fears the wrath of the patient.

After creation of a vagina, I have to be a practical counselor regarding social activities. To begin with the patient often beams with delight because she feels she has been returned to the human race. At that point the depth of her previous depression can be finally appreciated. She is no longer a "freak of nature." Most patients do not rush to be sexually active. They are happy simply knowing that they have the potential for sex. I caution that they should not tell their boyfriends since it might drive them off. One patient could not resist telling her consort, who, despite being a physician, left her. The patient should not be deprived of what would have been otherwise her usual behavior. I do suggest, however, that when there is serious talk of marriage she find out whether her potential mate desires children and that she inform him that if they wish to have children it would be by adoption. Depending on the individual patient, I sometimes add that with in vitro fertilization using her egg and her husband's sperm and a surrogate uterus, she could have a biologic child. Unfortunately, there are emotional and legal hazards. Theoretically a uterine transplant would enable her to menstruate and become pregnant, however, this has not yet been reported.

In general, after the initial reaction of depression the patients tend to mature emotionally rapidly and apply themselves to their studies and/or work. Most mothers tell me that their emotionally immature daughters seem to "grow up overnight." I encourage them to seek satisfaction in accomplishment, which is a form of sublimation and builds self-esteem.

The patient is presented with various options, which are usually: (a) do nothing (except observation and superficial psychotherapy), (b) use vaginal dilators, (c) perform Williams' operation, or (d) perform McIndoe operation.[27,28]

In the vast majority of cases, there is no urgency of treatment because there is no functioning uterus. A functioning uterus requires urgent surgery in order to create an exit for menstrual blood. In the usual case, the first option is not to do anything since there is no urgency and the condition is asymptomatic. Furthermore, it gives the patient time to emotionally digest the situation.

Another option is the vaginal dilator technique described by Frank.[18] With the use of progressively enlarging dilators first being pressed posteriorly and later cranially, sometimes a space can be invaginated. It requires the presence of a dimple or small pouch, nontender, supple tissue, and compulsive determination. Shortly after being introduced the technique fell out of favor, although Ingram repopularized it using a bicycle seat to support the vaginal dilator.[19] It has a good chance of success if the patient sits on the dilator for 2 hours daily. Extraordinary determination and emotional support as well as appropriate tissue are essential. Many authorities recommend the dilator approach initially because it is not dangerous.

Most patients are not successful with dilators and dislike them. The Mayo Clinic group has been the only one to express dissatisfaction with it because in two cases it was unsatisfactory "to the patient because of its propensity to evert or prolapse to the introitus."[17] Nevertheless, I suggest that the patient try it for whatever it might accomplish (even if only to make the surgery easier by creating a dimple), to test the patient's emotional ability in advance and to acquaint her with the use of a vaginal mold after surgery. Very rarely one sees a patient in whom vigorous coital attempts have developed a 5-cm vagina (when the tissues are flexible) or have dilated the distal urethra.

The McIndoe Operation

"Only those with considerable experience should attempt to surgically reconstruct a vagina because there is very little margin for error."[4] However skillful surgical construction of the vagina is not for the usual specialist in gynecology. In the usual gynecologic office, one case might be seen every 10 years. Most specialists have not had adequate training in learning this surgery. Even many reproductive endocrinology and infertility fellowships do not give adequate training. After learning the procedure, maintenance of one's ability depends on continuing activity. Aside from the surgery, the unusual long-term preoperative and postoperative management together with the psychological factors play a crucial role. The McIndoe operation should be done only by those with adequate experience since the first operation gives the best chance of success, since there is the potential for serious complication such as rectal or bladder fistula, and since this is a rare condition. Nevertheless, the usual gynecological specialist must understand the patient's options and the problems involved and might wish to participate in the pre- and postoperative management.

Before doing the McIndoe procedure, the patient should know the various options, should understand what the procedure entails, including the long hospitalization and the prolonged need to use a vaginal stent or mold, should be motivated and emotionally mature, and should be very desirous of having it done. Sometimes mothers are anxious to have the patient operated upon as soon as possible to make the patient as normal as possible. It must be understood that the decision is up to the patient.

With my usual patient, a year or longer goes by from the time of diagnosis to the actual procedure. This gives the patient time to adjust emotionally and to be sure of her commitment. The diagnosis is usually made when the patient is investigated for amenorrhea, which might be at any age from 13 to 17, with an average of about 15. I try to wait until the patient is at least 17, although

I have had good results with emotionally mature 16-year-olds. I suggest that the surgery be done during summer vacation to avoid missing school and making up excuses. Most patients request surgery after finishing high school and before starting college so as to have later summers free. In past years gynecologists would wait until the patient was about to be married; however, the patient's reduced self-esteem and depression tended to reduce her ability to meet men. In addition, with changes in social mores young women may wish the option of sexual activity prior to or without formal marriage. The oldest patient I operated upon was a 42-year-old woman who had the diagnosis made 20 years previously. She had been told erroneously that nothing could be done. This poor soul suffered continue torment and rejected any interest that any man ever showed in her. She was referred after she sought care again when an ardent suitor insisted on marriage.

Surgical Details

Preoperative Considerations

Preoperatively the surgeon should:

1. Review the local anatomy—including the size of the dimple, the flexibility of the tissue, and the upper edge of the transverse perineal fascia, which is determined by palpation and rectal examination.
2. Review the location of the kidneys.
3. Examine the patient in her swimsuit to plan the site of the donor skin graft for the least conspicuous area.
4. Make certain that the patient understands the procedure and her responsibilities.

A low-residue diet is started the day before surgery and a routine enema is given. Elastic stockings are placed before surgery. An alternating electric air mattress or a sponge rubber "egg crate" mattress is made ready. The room should be air conditioned.

The surgeon should be unhurried and have good assistance. The operating room should have a head light, long narrow retractors and instruments, several vaginal molds, a comfortable surgeon's seat, and skin grafting instruments.

A single pelvic kidney increases the hazard of the surgery. I have operated and found that the rectal finger was the best guide in avoiding the kidney and also that the kidney often is not rigidly fixed. I did not find it necessary to do a laparotomy, as some have suggested.

The Klippel-Feil syndrome of congenital fusion of the cervical vertebrae, short neck, and low posterior hairline has been described, usually until renal malformation. I had such a patient. At first glance, she resembled a typical Turner's syndrome patient. The anesthetic problem was the question of stability of the neck while turning the patient under general anesthesia from her side for taking the graft to supine for the vaginal dissection. The surgical problem was that she had a single pelvic kidney. The psychological problem was an obvious shoulder and neck deformity of which the patient was self-conscious. Most cases do not have an obvious deformity. Despite all the problems she has had a successful McIndoe procedure. The point is that thoughtful individualization is necessary. General endotracheal anesthesia is used.

Laparoscopy is considered for a pelvic kidney, for any uncertainty about the diagnosis and the pelvic anatomy, and for pelvic pain. If there is hematometra in either vestigial uterus then laparotomy with resection may be required.

Obtaining and Preparing the Graft

Some surgeons do the dissection of the vaginal space before taking the skin graft to determine how large a cavity is made and therefore how large a graft to take. This is especially helpful if experience is limited. With adequate experience and therefore confidence of being able to dissect an ample space, the skin graft is taken first. This helps keep patient movement to a minimum; otherwise the patient would be in lithotomy position, then on the side, and finally in lithotomy position again. In addition, time is saved as one surgeon sews the skin graft over the mold while another performs the vaginal dissection.

The patient is placed securely in the Sims position with the upper knee flexed to expose the buttock and with the skin under mild tension. Some surgeons place the patient in prone position in case grafts are to be taken from each buttock. Either the traditional drum dermatome or the electric dermatome may be used (Fig. 19-6). The former gives a more uniform thickness graft, is more tedious to use, and requires a dermatome cement glue. The later uses a mineral oil lubrication for the skin. The usual recommended width is 10 cm (4 inches), and the length is twice the length of the neovagina (20 cm or 8 inches). Jones and Rock noted "the graft should be at least 22 or 24 cm long and 10 cm wide."[5] My usual practice is to make a capacious vagina to allow for possible postoperative stricture; therefore, more than one graft strip may be required. A 10-cm wide graft gives a tight two-finger-width tube. The split-thickness skin graft is 18/1,000th to 20/1,000th of an inch in thickness.

An alternative donor site is the posterior inner thigh. In former years, a standard textbook recommended an anterior thigh graft because of technical ease; however, the permanent mark is disastrous from a psychological viewpoint. The patient will notice it when she undresses and cannot hide it at the beach. Other surgeons still take the graft from the abdomen.

The raw donor site is covered with one layer of Xeroform gauze with a light outer dressing. The Xeroform is never pulled off; it will gradually peel off spontaneously in about 12 days, after which the separated edges are trimmed. An alternative is a transparent Tegaderm covering.

The graft is sewn over a vaginal mold, raw surface outermost, using interrupted vertical mattress sutures of 5-0 absorbable synthetic material. Jones and Rock use a soft foam rubber mold carved from a 10 × 10 × 20-cm block to twice the size of the neovaginal cavity. It is covered with two condoms. An alternative is to roll up a sheet of foam rubber to the desired width. This method is the simplest.

In addition to their method, I also use two other methods. One is to drape the sutured graft over a wide Ferguson speculum (hollow tube), insert it in the neovaginal space, stuff it with one or several pieces of foam rubber, and withdraw the speculum (Fig. 19-7).[27] The other method is to use an inflatable (adjustable) vaginal stent made of a silicon elastomer shell filled with polyurethane foam with a fill tube and a "through-tube" for drainage (Mentor Corp.). I prefer a made-to-

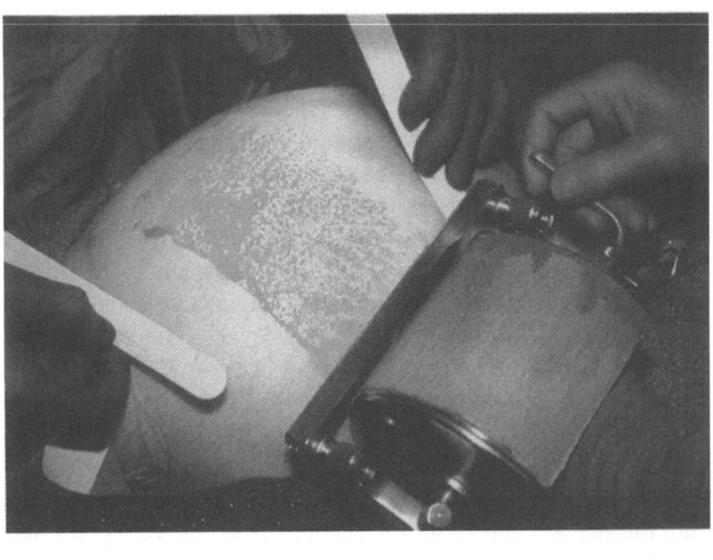

FIGURE 19-6. Taking the skin graft from the buttock.

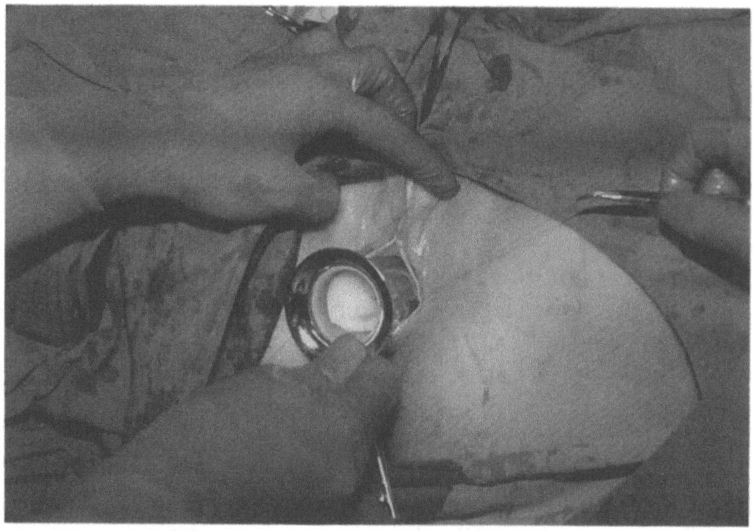

FIGURE 19-7. The graft is placed over the speculum and inserted into the neovaginal space. The speculum contains the sponge rubber.

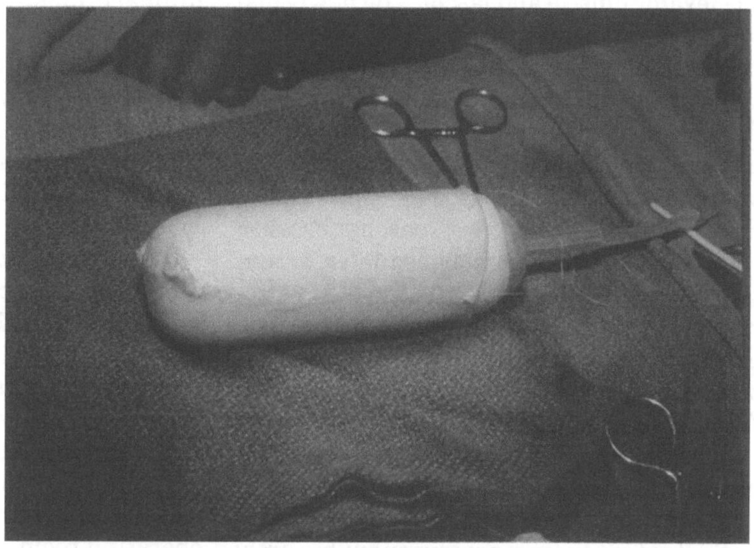

FIGURE 19-8. The inflatable stent covered with a graft.

order stent of 12 or 14 cm length and 4.5 cm width. It saves carving time but requires care to avoid puncturing (Fig. 19-8).[27]

Once the surgeon has taken the graft and dressed the donor site, the patient is carefully turned and placed in lithotomy position.

Preparing the Neovaginal Space

A Foley catheter is made indwelling in the bladder, and the first of several rectal examinations is made. A transverse incision is made through the posterior part of the dimple and on the anterior edge of the transverse perineal muscle and fascia. This avoids injury to the urethra, and the dimple epithelium is later pushed in to cover the urethral floor. In the Wharton procedure, the incision is more anterior so that the dimple surface is used to cover the distal posterior neovaginal wall. Initially there is often brisk bleeding, which I assume is due to a fusion of the bulbocavernous plexus.[27] This has not been mentioned by others. Great care is taken to avoid urethral, bladder, or rectal injury, because the local anatomy varies. There may be a midline sagittal septum running anterior to posterior, and dissection may be easier on either side. Some have advised using Hegar dilators on either side; however, improper placement could cause injury.

After dissecting to about a depth of 3 cm there is a dramatic change as one reaches a plane of cleavage. Anteriorly, there is a smooth, gray surface with small bright vessels on its surface with the bladder above it. Posteriorly, there is a red, rough, "raw hamburger" surface with the rectum underneath. These landmarks have also not been described adequately. Bleeding may occur from large lateral and posterior vessels near the levator muscle and fascia. The newly dissected distal neovagina tends to have persistent small vessel bleeding circumferentially at the introitus. The dissection is carefully continued to high in the neovaginal apex, where the peritoneum is reached. The surgeon should avoid exposing more than 2 cm of peritoneum for two reasons: because the tight vulvar sutures placed later to contain the mold may cause the mold to perforate into the peritoneal cavity, and because excessive exposure can result in a later enterocele. Meticulous hemostasis is essential. Any hematoma will prevent the graft from taking, will cause graft necrosis, and will increase the chance for infection.

There is an increase in operative difficulty with short stature; a narrow subpubic arch; strong levator muscles; a short, rigid perineum; a pelvic kidney; a deep cul-de-sac; or previous unsuccessful surgery.[17] I have found that cutting the levator muscles causes brisk bleeding.

Graft Placement and Final Steps

After the neovaginal space has been dissected, the skin graft is placed by one of the above techniques. The edges of the graft are sewn by interrupted sutures to the skin of the introitus.

In former years a hard mold was utilized, but most now use a soft mold to avoid pressure necrosis of the urethra. The graft-bearing mold is kept in the neovaginal space by two or three heavy braided silk sutures closing the labia majora. Although these sutures are painful, they prevent postoperative cough expulsion of the mold and graft and death of the graft.

Subprapubic bladder drainage is used and the Foley catheter removed. Be certain that the suprapubic tube is patent and do not perforate the inflatable mold when it is introduced.

Early Postoperative Care

Postoperatively, the patient is kept flat in bed for 48 hours. The house staff and nurses must be reminded because most patients get out of bed the next day. The suprapubic tube drainage must be measured periodically to be certain there is no obstruction. If there is any question of obstruction the Foley catheter is kept in. A urinalysis and culture is done, especially if there is only one kidney. The intravenous drip is usually continued for 48 hours with a broad-spectrum antibiotic.

The low-residue diet is continued. Some also prescribe a binding agent. Elastic stockings have already been placed on the legs. The donor site is reinforced with a second dressing if necessary; however, the initial dressing directly on the raw surface is left in place. Vaginal bleeding is immediately reported. Small amounts are observed. Active or arterial bleeding requires a return to the operating room. The temperature is checked every 4 hours.

After 48 hours, the head of the bed is tilted up 30° and the patient is "log rolled" by the nurse to either side.

From 7 to 9 days later the patient is returned to the operating room, where the labial sutures are removed, the mold is removed, the neovagina is inspected for graft take, a new soft mold is inserted, and the suprapubic tube is removed.

One or 2 days later, the patient is instructed on removal, washing, and reinsertion of the mold. She is told never to leave the mold out for more than a few minutes. At this time or 6 weeks later, a firm stent of polythylene or silicone is used. The patient is often terrified and needs emotional support. She must learn how to do this by herself and be responsible. I will not let the patient leave the hospital until she can do this. After 1 or 2 days of practice, she may be discharged home. It is interesting that some patients insist on wearing gloves for a long time as part of copying precisely what I show them. When the patient gets out of bed, after the second trip to the operating room, there may be severe orthostatic hypotension, which also has not been reported in the literature but is present in all cases to some degree.

Long-Term Postoperative Care

I usually have the patient telephone daily for the first few days after discharge and have her return in 2 weeks. She always has an extra mold, usually the hard, plastic Counsellor mold. Most dislike the inconvenience of the soft molds.

I continue to see the patient at 2- to 4-week intervals for several months. The mold is kept in constantly except for daily washing. Leaving the mold out for a few hours can result in stricture of the new vagina. Lack of motivation and discomfort on later attempts to reinsert the mold may cause failure. Swimming is limited by the condition of the donor site, where sunburn, hot water and soap are avoided. The patient needs continuous emotional support.

After about 3 to 6 months, the stent is removed for progressively longer daily intervals. Any difficulty in reinsertion means that the removal intervals have to be reduced. Progressively, the stent is left out on alternate nights.

There is extraordinary variation in the tendency to postoperative stricture. Some patients require that the stent be left in for 1 to 3 nights weekly for more than a year. Continous indefinite observation is important. Coital activity is a significant factor.

The main long-term complication is stricture of the neovagina. Suggestions in the literature vary from an optimistic discontinuation of the vaginal mold after 8 weeks, when sexual activity may begin, to McIndoe's belief that the tendency for stricture passes after 6 months,[14] to Sheares' recommendation of twice-daily vaginal dilation even after 6 months and marriage ("This is the most important detail in the follow-up and should be continued indefinitely").[22] The Mayo Clinic report of 50 cases found 5 cases with a tendency toward progressive contracture of the neovagina if not coitally active or using a vaginal stent dilator, even several years after the operative procedure.[17]

The other long-term problem is emotional support, although with time patients generally do very well. Orgasm seems to be achieved as normal. Other problems are relatively rare, including condylomata acuminata, enterocele, squamous cell carcinoma, and leiomyoma of vestigial uterus.

Although some keep repeating the statement that the transplanted skin becomes vaginal mucosa, I find that my inspection, palpation, and biopsies still show skin. (One of my biopsies showed persistent sweat glands.) Reports of the graft turning into vaginal mucosa may be due to undermining of the skin by upward growth of vaginal-type cells from the introitus, especially in areas where the skin graft did not take. I have also done biopsies of the wall where the vagina should be at the start of surgery, and these reveal vaginal mucosal type cells.

Congenital Absence of the Vagina with a Functioning Uterus and Normal Cervix

Most cases (about 90%) of congenital absence of the vagina also have an absence of a functioning uterus. About 9% have a functioning uterus with a normal cervix. I have always found a cap of vaginal mucosa around the cervix. This is a critical finding in management since such cases have a reasonably good prognosis for menstruation and pregnancy if vaginal drainage is accomplished and if there is no retrograde menstrual flow with secondary pelvic endometriosis. Some previous authors did not describe or report this finding, and therefore their results are difficult to interpret. Even some contemporary authorities have not adequately emphasized this finding in management. The finding of a cervix and a vaginal cap as a unit also must have significance in embryologic development, which as yet has not been explained.

About 1% of cases have a functioning uterine body but without a cervix and without a vagina. A recent series of 90 cases of vaginal agenesis found hematometra in 7 cases (7.7%), which agreed with a literature review of 10%, implying the presence of a functioning uterus.[29]

The English tend to be conservative regarding a functioning uterus with a cervix and an absent vagina. Professor Edmonds (who succeeded Professor Sir Christopher J. Dewhurst in London) recommended a laparatomy to drain the hematometra and a coincident noncommunicating vaginoplasty. The patient is placed on continous oral contraceptives to prevent menstruation, and after 6 months a uteroneovaginal anastomosis is performed.[3] Other English authors have dissected a space, kept it open for a few days to evacuate all the old menstrual blood, and then lined it with a split-thickness skin graft.

Jones and Rock believe that congenital absence of the vagina with the presence of a uterus (which is the usual description) may in fact be an extreme form of transverse vaginal septum, since urologic anomalies are rare with the latter. Their illustration shows a normal uterus and cervix that is surrounded by a cap of vaginal mucosa. This finding of the cervix and cap is critical and was understated regardless of the definition. They recommended a vaginal space dissection and the placement of a hollow, hard plastic tube with a bullous upper end that is left in place for 4 to 6 months to permit epithelialization.[5] Their procedure does not require a skin graft, postoperative bed rest, or the voluntary use of a mold while the stent remains in place, although later daily vaginal dilatation for 2 to 4 months is advised. Thus their procedure would be advantageous for a nonsexually active, poorly motivated, young adolescent.

It has been suggested that with a high (usually thick) transverse vaginal septum surgical resection should be delayed until the upper vagina is distended by mucus or blood.[30] After resection of the septum, if there is an extended raw area, a hard plastic hollow tube with a bullous upper end is made indwelling in the vagina for 4 to 6 months to permit epithelialization,[30] similar to the Jones and Rock approach for the presence of a uterus and cervix with a congenital absence of the vagina.[5]

My experience is that with a normal cervix there has always been a cap of vaginal mucosa. Usually, this cap will progressively distend to contain menstrual flow without retrograde flow into the peritoneal cavity. The uterus and cervix and distended cap can be felt by rectal examination and visualized by sonography or MRI. The distended cap will progressively extend downward toward the introitus, and rectal examination will demonstrate a wide, tense, cystic mass that conceals the uterus.

I recommend surgery when the condition is discovered in adolescence even if the cap is not yet distended, to avoid the small chance of retrograde menstruation and secondary endometriosis. The choice would be the McIndoe operation or the use of the Jones hard plastic tube. If the condition is discovered later, with a distended cap, I perform the McIndoe operation. If the sac is large enough then it may be mobilized and flaps brought down to the introitus, thus avoiding a skin graft as well as the need for postoperative bed rest. Such a decision requires experience and judgment.

Functioning Uterine Fundus without a Cervix—with or without a Vagina

Congenital absence of the vagina in association with the presence of a functioning uterine fundus but absent cervix is very rare, and occurs in about 1% of cases of Rokitansky syndrome. The differential diagnosis of a functioning uterine fundus and absent vagina with a cervix (and vaginal cap) versus without a cervix is difficult on clinical examination before the onset of menses. On rectal examination in a patient without a cervix, the uterine fundus is felt high up. After the uterine fundus begins to menstruate in cases of absence of a cervix, there is immediate retrograde menstrual flow if the fallopian tubes are patent because of the relatively thick, unyielding uterine wall, and pain occurs early. On rectal examination, no cystic mass is felt below the uterus. Markham et al., in 1987, were the first to report MRI of cervical agenesis combined with vaginal agenesis.[12] Both rectal examination and sonography had suggested erroneously that a cervix was present. The MRI picture correctly showed a distal cervical agenesis, and this offered "a significant new approach in the care of patients presenting with cervical agenesis." A hysterectomy was done.[12]

With great insight, Rock[2] described two types of cervical agenesis. In type 1 there are vestiges of the cervix with small inclusions of endocervical-type tissue within a fibrous cervical stroma. Type 2 has a lower uterine segment that terminates in a peritoneal sleeve well above the normal vaginal communication. For type 2 cervical agenesis Rock advised primary hysterectomy.[2]

The differential diagnosis of cervical atresia versus a high transverse vaginal septum in an otherwise patent vagina is difficult. If the latter is present and there is an accumulation of menstrual blood, then clinical examination may disclose a cystic mass below the uterus, and this may be confirmed by sonography or MRI. Even then there may be a hematotrachelos with an imperforate external cervical os that may cause confusion.

There are two schools of thought regarding management of congenital absence of the cervix with a functioning uterine fundus. The Johns Hopkins and general consensus views are that an abdominal hysterectomy should be done as a primary procedure. The reasons are the very poor chance of preserving fertility, the risk of severe morbidity from infection, the tendency for surgical fistulas (fundus to vagina) to close, and the frequent secondary endometriosis.[2,30–32]

Jacob and Griffin reported two personal cases of congenital atretic cervix with noncanalized cervical stroma with a normal uterine cavity and the presence of a vagina. After creating an endometrial-vaginal fistula twice in the first case, there was an eventual hysterectomy for endometriosis. The second case has had cyclic menses and mild dysmenorrhea for 43 months.[33]

Jacob and Griffin's literature review of 41 cases of congenital cervical atresia indicated that 13 had incomplete vaginas, including 5 with congenital absence of the vagina, 2 with atretic vaginas, 1 with upper vaginal atresia, 1 with an annular constriction of the upper vagina, 1 with a 5-cm vagina, 1 with a 2-cm deep vagina, and 2 with 1-cm vaginal dimples. Of the 5 with congenital absence of the vagina and surgical fistula creation, two died. One, age 13, died with multiple abdominal abscesses and evisceration 7 weeks after the initial creation of a fistula with a rubber catheter stent, followed by two abdominal reexplorations. The other fatality was due to septic shock at age 16 following hysterectomy done for sepsis 8 months after a McIndoe procedure with creation of a fistula. Of the other three with congenital absence of the vagina, two had later hysterectomies and one (with a hypoplastic cervix) was well at 21 months postoperatively. Of the eight other cases of incomplete vaginas with creation of fistulas, four had primary hysterectomies, and had a later hysterectomy. Three have not had adequate length of follow-up, including one case of 17 months with cyclic menses and two with repeat surgery.

Of the 28 patients with normal vaginas, 7 had primary hysterectomies, 7 had later hysterectomies, 3 had inadequate follow-up, 8 had follow-up of 3 years or less, and one had a 10-year follow-up. There were only two women who became pregnant. One had a solid, cordlike cervix and a normal uterus and vagina with suspicious areas of endometriosis on one ovary. A fistula was formed and polyethylene tubes placed. Six years later, there was a successful cesarean section (case of Zarou). The other had a duplex vagina with a visible cervix in the left and none on the right. There was a uterus didelphys with a right hematometra and atresia of the right cervix and upper vagina. A transvaginal drainage of the right hematometra was done. She had a term pregnancy twice in the left uterus with a cesarean section each time (case of Monks). This second case had a normal left cervix and therefore should have not been included in this review. Therefore, of 40 cases of cervical atresia with surgical creation of a fistula, there was only one pregnancy.[33]

The editorial note appended to the above review discussed fistula formation with insertion of an indwelling stent with the hope that epithelialization will occur: "This procedure never works." Most cases eventually have a hysterectomy because of the problem of keeping the fistula open. An immediate hysterectomy is in the best interest of the patient. "Perhaps an imaginative surgeon can devise a source of exogenous epithelium which will prevent stenosis while providing egress for menstrual blood, ingress for sperm, and a barrier to infection."[33]

The second case of successful pregnancy with cervical atresia was reported by Fraser in 1989 and was not included in the Jacob and Griffin review.[34] The patient had a normal vagina and a functioning uterine fundus and only partial cervical atresia. There was a 1.5-cm long vaginal cervix but the 1.5 cm of upper cervix was replaced by loose areolar tissue. Hysterescopy showed complete occlusion at the midpoint of the endocervical canal. Equalized portions of the lower uterus and upper cervix were excised and sutured together and a tube left in place for 3 weeks. Despite recurrent, severe pelvic endometriosis, pregnancy occurred 3 years later with a complete placenta previa and a cesarean section. The author suggested that the presence of endocervical mucus in this case prevented ascending infection. He recommended hysterectomy for complete cervical atresia or rudimentary fibrous cervix.[34]

In his 1986 review of congenital atresia of the cervix, Farber found 35 cases all presenting with primary amenorrhea and cyclically recurrent or chronic pelvic pain.[35] In 25 cases, there was hematometra, hematosalpinx, and pelvic endometriosis or adenomyosis due to menstrual blockage. The vagina was absent in 14 cases. There was a tendency toward multiple anomalies of the vagina, uterus, and adnexae. Of the 35 cases, 10 had primary hysterectomies. In the other 25 cases there was a surgical attempt to create a fistula. There were two postoperative deaths from sepsis. Nine had a later hysterectomy. Only nine of the 25 had a follow-up of more than 3 years. These cases are difficult to manage because they are rare, there may be multiple gynecologic anomalies, there is often retrograde menstruation with pelvic endometriosis, and long-term followup may be lacking. Farber had two personal cases of congenital absence of the vagina with cervical atresia and a functioning uterus. A rubber "T" tube was made indwelling for 3 months to keep the surgical uterovaginal fistula patent. The vagina was developed with a Wharton technique. Asymptomatic menstruation has continued for 8½ years.[35]

Cukier et al. reported a 15-year-old with congenital absence of the vagina, a blind-ending hypoplastic cervix, and a normal uterine cavity by ultrasound. Using a McIndoe procedure with a series of stents covered with split-thickness skin grafts, there has been menstruation for 21 months.[36]

In cases of a functional uterus with cervical atresia, Edmonds advised hysterectomy if there is no cervical tissue in the presence of widespread endometriosis and a rudimentary uterus.[3] If there is a cervical band of tissue and a uterus present then the creation of a passage and an indwelling stent to allow fistula formation "seems reasonable," although there is the risk of death from sepsis. In cases of absent vagina, a functional uterus, and cervical atresia, a neovagina is created and 6 months later (using oral contraseptive pseudopregnancy) a uterine anastomosis is made. In four or five cases, initial menstruation was established. Edmonds acknowledged that the Johns Hopkins group advises primary hysterectomy.[3]

I have had two (unreported) cases of congenital absence of the vagina and of the cervix (type 2) with a functioning uterine fundus. Both patients and their parents wished to have everything possible done to preserve the uterus. The first case was an emotionally mature 17-year-old. I did a McIndoe vaginal construction and a simultaneous laparotomy to create a uterovaginal fistula, which was also lined with skin graft and temporarily kept open by a plastic tube. Several prophylactic fistula dilatations were done. She continued to menstruate and was placed on oral contraception, but was lost to follow up after about 5 years. The second case was an emotionally immature 15-year-old who had had a previous unsuccessful vaginal operation elsewhere and continued to have cyclic pain. I dissected a vaginal space and lined it with several layers of amnion to avoid a skin graft. A uterovaginal fistula was made at laparotomy when endometriosis was found. During the hospitalization, the patient did well. After going home she stopped using the vaginal stent, with resulting stenosis of the upper neovagina. Despite several attempts to reopen the fistula from below, she eventually required an abdominal hysterectomy.

Functioning Uterine Anlagen

Cyclic or chronic pelvic pain with or without a pelvic mass suggests hematometra in a uterine anlage. Murphy et al.,[37] in their review of the literature of Rokitansky syndrome, found three cases with a unilateral functioning anlage. They reported the first case of bilateral functioning anlagen. Rectal examination showed bilateral adnexal mass. Laparoscopy revealed two separate anlagen connected by a bridge of tissue spanning the base of the pelvis. Laparotomy was done to remove both structures. The larger left anlage also had adenomyosis.[37] The general consensus is that functioning uterine anlagen (müllerian bulbs, vestigial uteri) that derived from thickening of the müllerian (paramesonephric) ducts should be excised.

Singh and Devi of India, in 1980, extending previous work of Chakravarty, performed a surgical fusion of müllerian bulbs and vaginoplasty with a uterovaginal fistula in three cases of Rokitansky syndrome. In two cases scanty menstrual flow was achieved. Presumably one of the bulbs (anlagen) was functional, and a Wharton technique was used for the neovagina. "Proper and adequate hormone therapy" was given.[38] In 1983, they reported on eight cases who had simultaneous reconstruction of the uterus and vagina.[39] All had menses; one of these also had a successful pregnancy. The patient, age 15, had constant pelvic pain and on rectal examination had a tender, left globular mass. The latter was a large, cystic müllerian bulb (6 × 8 cm) with hematometra. It was surgically united with the small (2 × 3-cm) solid right bulb. There was no cervix or vagina. A "T" tube was placed in the new uterine cavity and connected to the surgical neovagina. Although most of the vagina stenosed, a fine uterovaginal fistula persisted and menstruation appeared. She became pregnant and was delivered by cesarean section. The authors first considered the anomaly to be hematometra in a double uterus without cervix and vagina. They later called it a nonfused müllerian system with absence of the vagina. The first diagnosis is more descriptive and probably correct, since their illustrations show the uterine anlagen deep in the true pelvis, close to each other and connected by a "short bent fibrous band."[39] The usual uterine anlagen are high on the lateral pelvic walls and far apart. This would be the third pregnancy with an absent or incompletely developed cervix and the first after fusion of a double uterus without cervix or vagina.

References

1. Griffin JE, Edwards C, Madden JD, et al: Congenital absence of the vagina. The Mayer-Rokitansky-Kuster-Hauser syndrome. Ann Intern Med 1976;85:224–236.
2. Rock JA: Anomalous development of the vagina. Semin Reprod Endocrinol 1986;4:13–32.
3. Edmonds DK: Congenital malformations of the vagina and their management. Semin Reprod Endocrinol 1988;6:91–98.
4. Berg PA, Breen JL, Gregori CA: Congenital absence of the vagina—the Mayer-Rokitansky-Kuster-Hauser syndrome. Adolesc Pediatr Gynecol 1989;2:73–85.
5. Jones HW Jr, Rock JA: Anomalies of the mullerian ducts, in: Reparative and Constructive Surgery of the Female Generative Tract. Baltimore, Williams & Wilkins, 1983, pp 146–186.
6. Strubbe EH, Thijn CJ, Willemsen WN, et al: Evaluation of radiographic abnormalities of the hand in patients with the Mayer-Rokitansky-Kuster-Hauser syndrome. Skeletal Radiol 1987;16:227–231.
7. Biedel CW, Pagon RA, Zapata JO: Mullerian anomalies and renal agenesis: autosomal dominant urogenital adysplasia. J Pediatr 1984;104:861–864.
8. Ghirardini G, Segre A: Vaginal agenesis (Mayer-Rokitansky-Kuster-Hauser syndrome): recent etiopathogenetical and anatomical views. Clin Exp Obstet Gynecol 1982;2:98–102.
9. McPherson E, Carey J, Kramer A, et al: Dominantly inherited renal adysplasia. Am J Med Genet 1987;26:863–872.
10. Opitz JM: Editorial comment: Vaginal atresia. (von Mayer-Rokitansky-Kuster or MRK anomaly) in hereditary renal adysplasia (HRA). Am J Med Genet 1987;26:873–876.
11. Cramer, DW, Ravnikar VA, Craighill M, et al: Mullerian aplasia associated with maternal deficiency of galactose-1-phosphate uridyl transferase. Fertil Steril 1987;47:930–934.
12. Markham SM, Huggins GR, Parmely TH, et al: Cervical agenesis combined with vaginal agenesis diagnosed by magnetic resonance imaging. Fertil Steril 1987;48:143–145.
13. Rock JA, Jones HW: Construction of a neovagina for patients with a flat perineum. Am J Obstet Gynecol 1989;160:845–853.
14. McIndoe AH, Bannister JB: An operation for the cure of congenital absence of the vagina. J Obstet Gynaecol Br Emp 1938;45:490–494.
15. Abbe R: New method of creating a vagina in a case of congenital absence. Med Rec 1898;64:836–838.
16. Counseller VA: Congenital absence and traumatic obliteration of the vagina and its treatment with inlaying Thiersch grafts. Am J Obstet Gynecol 1938;36:632–638.
17. Buss JG, Lee RA: McIndoe procedure for vaginal agenesis: results and complications. Mayo Clin Proc 1989;64:758–761.
18. Frank RT: The formation of an artificial vagina without operation. Am J Obstet Gynecol 1938;35:1053–1055.
19. Ingram JM: The bicycle seat stool in the treatment of vaginal agenesis and stenosis: a preliminary report. Am J Obstet Gynecol 1981;140:867–873.
20. Hughes EG, Spence JEH: Human amniotic membrane: its use as an allograft in vaginal construction. Adolesc Pediatr Gynecol 1988;1:39–41.
21. Wharton LR: A simple method of constructing a vagina: report of four cases. Ann Surg 1938;107:842–854.
22. Sheares BH: Congenital atresia of the vagina: a new technique for tunnelling the space between bladder and rectum and construction of the new vagina by a modified Wharton technique. J Obstet Gynaecol Br Emp 1960;67:24–31.
23. Williams EA: Congenital absence of the vagina. A simple operation for its relief. J Obstet Gynaecol Br Commow 1964;71:511–512.
24. Baldwin JF: The formation of an artificial vagina by intestinal transplantation. Ann Surg 1904;40:398–403.
25. Novak F, Kos L, Plesko F: The advantages of the artificial vagina derived from sigmoid colon. Acta Obstet Gynecol Scand 1978;57:95–96.
26. Altchek A: Psychologic aspects of adolescent obstetric and gynecologic conditions, in Heacock DR (ed): A Psychodynamic Approach to Adolescent Psychiatry. The Mount Sinai Experience. New York, Marcel Dekker, 1980; pp 239–278.
27. Altchek A: What to do when there is no vagina. Contemp OB/GYN 1988;32:50–51, 54, 57–59, 63–64, 66, 68.
28. Altchek A: Pediatric and adolescent gynecology, in Nichols DH, Evrard JR (eds): Ambulatory Gynecology. Philadelphia, Harper & Row 1985, pp 20–62.
29. Salvatore CA, Lodovicci O: Vaginal agenesis:

an analysis of ninety cases. Acta Obstet Gynecol Scand 1978;57:89–94.
30. Rock JA, Azziz R: Genital anomalies in childhood. Clin Obstet Gynecol 1987;30:682–696.
31. Niver DH, Barrett G, Jewelewicz R: Congenital atresia of the uterine cervix and vagina: three cases. Fertil Steril 1980;33:25–29.
32. Maciulla GJ, Heine MW, Christian CD: Functional endometrial tissue with vaginal agenesis. J Reprod Med 1978;21:373–376.
33. Jacob JH, Griffin WT: Surgical reconstruction of the congenitally atretic cervix: two cases. Obstet Gynecol Surv 1989;44:556–569.
34. Fraser IS: Successful pregnancy in a patient with congenital partial cervical atresia. Obstet Gynecol 1989;74:443–445.
35. Farber M: Congenital atresia of the uterine cervix. Semin Reprod Endocrinol 1986;4:33–38.
36. Cukier J, Batzofin JH, Conner JS, et al: Genital tract reconstruction in a patient with congenital absence of the vagina and hypoplasia of the cervix. Obstet Gynecol 1986;68:32s–36s.
37. Murphy AA, Krall A, Rock JA: Bilateral functioning uterine analgen with the Rokitansky-Mayer-Kuster-Hauser syndrome. Int J Fertil 1987;32:316–319.
38. Sing KJ, Devi L: Hysteroplasty and vaginoplasty for reconstruction of the uterus. Int J Gynecol Obstet 1980;17:457–459.
39. Singh J, Devi YL: Pregnancy following surgical correction of nonfused mullerian bulbs and absent vagina. Obstet Gynecol 1983; 61:267–269.

20

The Uterus Without Ovaries

DANIEL NAVOT AND MARYANNE C. WILLIAMS

The role of the uterus in reproduction may be divided into four major functions:

1. Throughout a series of well-defined maturational changes the endometrium prepares for embryo implantation.
2. Further endometrial and vascular changes accommodate the invading trophoblast and sustain the implanting embryo.
3. The uterus is adjusting to the growing fetoplacental unit, providing both nourishment and protection from physical injury.
4. There is a complex transition from a role of containment to expulsion in the later part of gestation and parturition, respectively.

All of the above functions are hormone dependent, and without the appropriate hormonal stimulation the uterus is a quiescent and, to a large degree, useless organ. The ovaries provide the appropriate hormonal environment for endometrial maturation and were considered obligatory for both implantation and pregnancy sustenance. Ovarian secretory products may play a more subtle role in later gestation and parturition.

With the advent of human in vitro fertilization (IVF)[1] and ovum donation[2–5] a new era has begun, an era in which women lacking ovaries or ovarian function may reproduce. With the wide application of these novel reproductive technologies a need has emerged to critically review reproductive treatment strategies in a wide range of pathologic conditions.

Patient Selection and Indications for Ovum Donation

Women in their reproductive life span who lack ovaries or have ovaries that are not functional make up an important group of sterile patients who may benefit from ovum donation (Table 20-1). By far the most common complaint that brings infertile couples to request ovum donation is the idiopathic form of premature ovarian failure, or "premature menopause." The latter is defined as permanent ovarian failure, characterized by elevated gonadotropin levels and occurring after menarche but before the age of 35 or 40 years.[6,7] Premature ovarian failure is relatively rare; however, approximately 4% of women reach menopause before the age of 30 years.[7] Undoubtedly a sizable proportion of these women may not have completed their families. Women with gonadal failure are relying on donated oocytes in order to conceive. There are several potential sources for ovum donors; most frequently, they are IVF patients who have supernumerary oocytes and are willing to give them away. Other ovum donors are family members, volunteers, or patients contemplating tubal

TABLE 20-1. Indications for ovum donations

Women with amenorrhea	Women with spontaneous cycles
Ovarian dysgenesis	Genetic indications
Turner's syndrome (45,XO)	Autosomal recessive
Pure or mixed ovarian dysgenesis (46,XX; 46,XY)	Autosomal dominant X-linked
Other chromosomal disorders	
Ovarian failure premature	Failed conventional profertility therapy
Idiopathic and immunologic	Ovulation induction, IVF, gamete intrafallopian transfer
Iatrogenic (surgical, x-ray, chemotherapy)	Oocyte abnormality
Menopause	Perimenopause

sterilization. A very important subgroup of patients who need donated ooyctes in order to conceive are women with ovarian malignancies who have been conservatively treated.

Preservation of the Uterus in Malignant Ovarian Tumors

There are several situations in which conservative surgery with preservation of the uterus may be justified. Treatment of ovarian malignancies has always included total abdominal hysterectomy and bilateral salpingo-oophorectomy. The aggressive nature of ovarian neoplasms and the pattern of their spread justifies this approach. Although there are conditions for which conservative surgery is indicated (Table 20-2), conservation was always related to presumably healthy ovarian tissue—mostly the contralateral ovary. *The concept of childbearing without ovaries is completely new and should be incorporated into clinical decision making*. Specifically, whenever preservation of ovarian tissue may compromise oncologic treatment but the uterus may be preserved, the latter option (bilateral salpingo-oophorectomy only) should be pursued. Very strict rules, however, should be observed and the patient should have thorough understanding of risks involved and consent to ovum donation if bilateral oophorectomy is contemplated.

There are some general rules that should always be adhered to.[8] The patient must be of childbearing age, of low parity, of general good health, and have no contraindications to pregnancy, and close follow-up should be available. Once these general rules have been observed, attention should be given to specific disease processes. Thus, conservative management in ovarian epithelial malignancies (stages Ia and Ib disease) may be contemplated if the tumor is well differentiated, there is no invasion of the capsule, and peritoneal washings are negative[8] (Table 20-2). Furthermore, subclinical tumor extension should be excluded by omental biopsy and assessment of pelvic and paraaortic lymph nodes. Bilateral oophorectomy should eliminate the 12% risk of microscopic metastases present in the contralateral ovary.[9] Dysgerminoma is another ovarian malignancy that, even if bilateral, could be conservatively treated. Radiation therapy may also be liberally applied (Table 20-2). Other germ cell and stromal tumors may also be treated with uterine conservation. The relatively young age of afflicted patients, the pattern of spread, and favorable response to aggressive chemotherapy [i.e., vincristine-dactinomycin-cyclophosphamide (VAC)] justify this approach (Table 20-2).

TABLE 20-2. Possible indications for uterine preservation

Ovarian malignancies	Nonmalignant conditions
Epithelial stage Ia, Ib[a]	Endometriosis
Dysgerminoma	Bilateral fibromas, teratomas, adenomas
Endodermal sinus tumor	
Embryonal carcinoma	
Immature teratoma	
Gonadoblastoma	
Granulosa cell tumor	

[a] See text.

A notable benign ovarian process in which the uterus should be preserved even if bilateral oophorectomy has been performed is severe debilitating endometriosis in a young individual (Table 20-2). However, any other benign process that does not leave viable ovarian tissue should not be considered as an automatic indication for hysterectomy.

Artifically Induced Endometrial Cycles in Ovarian Failure

Although various replacement protocols have been reported, they all endeavor to mimic the steroidal milieu of the normal menstrual cycle.[2,3,5] Essentially, estrogen and progesterone are administered sequentially. Lutjen et al., who reported the first pregnancy in a patient with premature ovarian failure, used oral estradiol valerate (E_2V) (Progynova; Schering) and vaginal progesterone suppositories for steroid replacement.[2] Following embryo transfer they utilized intramuscular progesterone injections and oral E_2V for the maintenance of pregnancy. Similar regimens utilizing oral micronized estradiol (Estrace; Mead Johnson)[10] or transdermal patches (Estraderm; CIBA Pharmaceutical Co.) are equally successful in inducing both normal endometrial morphology and physiologic adequacy, as evidenced by high rate of pregnancies.[11] Normal endometrial maturation, although obligatory, is not enough for successful outcome; the endometrium appears to have a temporally restricted capacity to sustain embryo implantation. The temporal window of endometrial receptivity in the human is restricted to days 16 to 19 of a 28-day cycle.[3,5] Asynchronous cycles in which embryo transfers are performed outside of the window of endometrial receptivity have so far failed to produce pregnancies.[3,5] Thus, it is crucial to accurately synchronize the developmental stage of donated embryos with the endometrial maturation of the recipient.[5]

There are several clinical settings in which adequate manipulation of the recipient's endometrial cycle may result in synchronization of otherwise asynchronous cycles. Three different protocols were devised for the induction of endometrial maturation. For the situation in which donor oocytes become available on very short notice, a short follicular phase protocol and an accelerated secretory transformation protocol were devised. For the circumstance in which oocyte retrieval is delayed, a long follicular phase protocol was devised. Figure 20-1 illustrates the three novel endometrial manipulation protocols, compared to a standard 28-day cycle (lowermost panel).

The short follicular phase protocol consisted of 6 days of estrogen administration (2 mg, three times daily; Estrace, Mead Johnson) before the addition of progesterone (25 to 50 mg/day; Carter-Glogan Laboratories). The accelerated secretory transformation cycle consisted of a standard 14-day follicular phase and, starting on day 15, 150 mg progesterone/day (intramuscularly) added to the estrogen regimen (Fig. 20-1). The long follicular phase protocol consisted of 3 to 6 weeks of unopposed estrogen stimulation before the addition of progesterone. A typical long follicular phase cycle (Fig. 20-1) had, in addition to the 14-day incremental estrogen dosage, 14 days of high-dose (6 mg/day) Estrace administration.

Figure 20-2 illustrates the serum hormonal patterns achieved throughout the various protocols. Estradiol levels were in the supraphysiologic range during the long follicular phase protocol, whereas endometrial exposure to estradiol was significantly shorter in the short follicular phase protocol (Fig. 20-2). Similarly, supraphysiologic serum progesterone levels were achieved throughout the accelerated secretory transformation protocol, whereas the other protocols have simulated a natural cycle in serum progesterone levels attained (Fig. 20-2). All three protocols have induced secretory

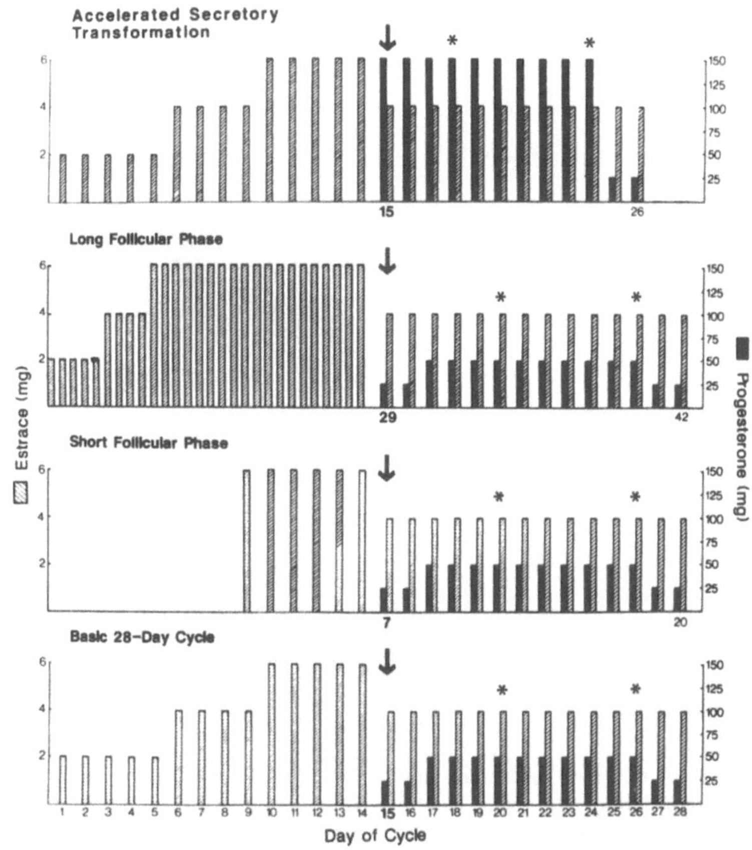

FIGURE 20-1. Sequential estrogen and progesterone replacement in the three study protocols and the control (basic) 28-day cycle. Note that the various cycles are normalized to day 15, the first day of progesterone administration (arrows). Progesterone administration actually commenced after only 6 days of estrogen in the short follicular phase, as opposed to 28 days of estrogen stimulation in the long follicular phase. (Reprinted from ref. 10, with permission.)

maturational changes similar to those described by Noyes et al.[12] Morphology of endometrial biopsies corresponded to chronologic criteria described above; that is, adequate endometrial response was uniformly achieved.

Embryo Implantation

To sustain embryo implantation, the endometrium must undergo specific maturational proliferation and differentiation events, including secretory transformation of the glandular elements, edema formation, vascular proliferation, and decidualization of the stroma. Implantation typically occurs on days 20 to 21. At that time the blastocyst comes into apposition with the epithelial lining, then attaches to it and burrows into the edematous, highly vascularized stroma. Thus, there are three components of the endometrium that act in concert throughout the process of implantation: (a) the secretory component, which provides the nourishing milieu to maintain the rapidly dividing embryo that entered the uterine cavity 2 to 3 days earlier; (b) the luminal epithelium,

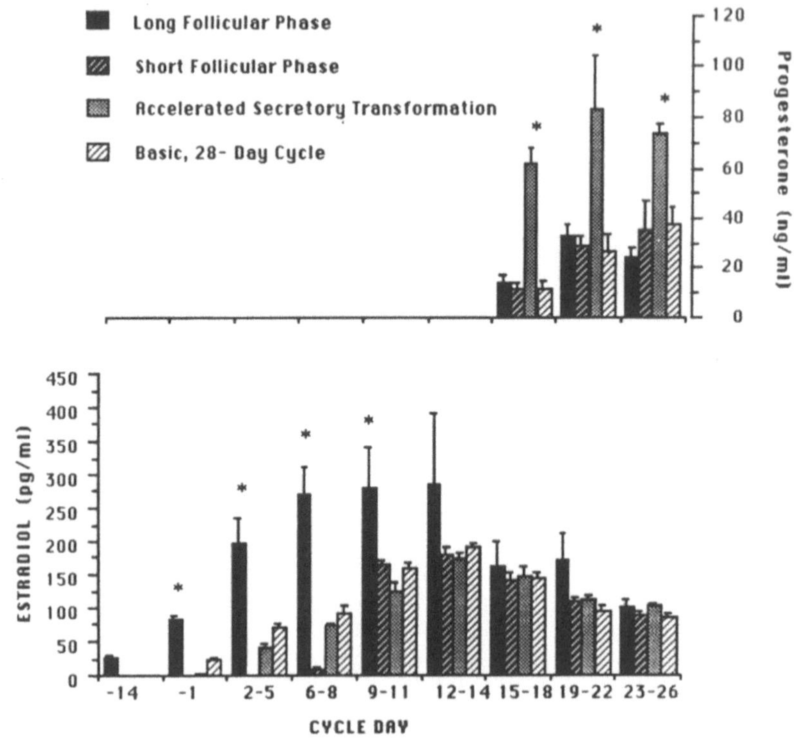

FIGURE 20-2. Mean (±SD) serum estradiol and progesterone levels throughout the cycle in the three treatment protocols and the control cycle. The asterisks in the lower panel denote significantly higher ($p < 0.05$) estradiol levels in the long follicular phase protocol than in the other protocols. The asterisks in the upper panel denote significantly higher progesterone levels in the accelerated secretory transformation protocol than in the other protocols. (Reprinted from ref. 10, with permission.)

which appears to be important in the development of uterine receptivity; and (c) the stroma, whose adequacy is the sine qua non for sustained implantation. All three components are dependent upon the prevailing steroidal milieu throughout the menstrual cycle. The ability to manipulate the endometrium according to clinical need, or availability of donated oocytes, allows for accurate timing of embryo transfer.

The need for proper synchronization between the developmental stage of the embryo and the endometrial differentiation of the recipient cannot be overemphasized.[3,5] For the 2- to 12-cell IVF embryo, the temporal window of endometrial receptivity probably extends from days 16 through 19. Within that window, day 17 or 18 may prove to be superior.[5] Simplified estrogen and progesterone protocols for endometrial manipulation, such as the short follicular phase protocol, allow for precise synchronization. Regular 28-day protocols are very difficult to synchronize, since hyperstimulated IVF cycles have short (typically 9-day) follicular phases. Thus, the artificial cycle of the recipient has to be initiated 4 to 5 days before the predicted menstrual period of the donor in order to synchronize the two cycles by days 16 to 19.[3] The short follicular phase should commence in the recipient 3 days after the establishment (not the prediction) of menstrual flow in the donor. Conversely, the long follicular phase allows for continuous estrogen stimulation before the onset of menstruation in the donor. Similarly, if unexpected donor eggs become available, progesterone administration (regular or high dose) to the recipient at the time of egg harvest may allow the endometrium to mature to day 16 or 17 by the time of embryo transfer. The three novel hormonal replacement protocols became, in 1989, the standard of care in Mount Sinai's ovum donation program in New York. All three protocols have proven that the morphologic integrity they create is also accompanied by functional capacity: the consistent production of an environment conducive of embryo implantation and sustenance of pregnancy. During the first 6 months of 1989 twenty pregnancies have been achieved throughout 50 ovum donation cycles. This exceptionally high success rate was made possible by the elimination of the need for elaborate synchronization. Utilization of these protocols allows for judicious use of supernumerary oocytes available for donation.

References

1. Steptoe PC, Edwards RG: Birth after the reimplantation of a human embryo. Lancet 1978;2:335.
2. Lutjen P, Trounson A, Leeton J, et al: The establishment and maintenance of pregnancy using in vitro fertilization and embryo donation in a patient with primary ovarian failure. Nature 1984;307:174.
3. Navot D, Laufer N, Kopolovic J, et al: Artifically induced endometrial cycles and establishment of pregnancies in the absence of ovaries. N Engl J Med 1986;314:806.
4. Buster JE, Bustillo M, Thorneycroft IH, et al: Nonsurgical transfer of in vivo fertilized donated ova to five infertile women: report of two pregnancies. Lancet 1983;2:223.
5. Rosenwaks Z: Donor eggs: Their application in modern reproductive technologies. Fertil Steril 1987;47:895.
6. Friedman C, Barrows H, Kim MH: Hypergonadotropic hypogonadism. Am J Obstet Gynecol 1983;145:360.
7. Tulandi T, Kinch RAH: Premature ovarian failure. Obstet Gynecol Surv 1981;36:521.
8. Di Saia PJ, Townsend DE, Morrow CP: Rationale for less than radical treatment for gynecologic malignancy in early reproductive years. Obstet Gynecol Surv 1974;29:581.
9. Munnell EW: Is conservative therapy ever justified in stage I cancer of the ovary? Am J Obstet Gynecol 1969;103:641.
10. Navot D, Anderson TL, Droesch K, et al: Hormonal manipulation of endometrial maturation. J Clin Endocrinol Metab 1989;68:801.
11. Droesch K, Navot D, Scott R, et al: Transdermal estrogen replacement in ovarian failure for ovum donation. Fertil Steril 1988; 50(6):931.
12. Noyes RW, Hertig AT, Rock J: Dating the endometrial biopsy. Fertil Steril 1950;1:3.

21

Management of Anovulatory Dysfunctional Uterine Bleeding in the Adolescent

ALBERT ALTCHEK

Until recently anovulatory dysfunctional uterine bleeding (DUB) was the most common urgent gynecologic problem of the adolescent. In recent years, with changing sexual mores and more frequent and earlier sexual activity, sexually transmitted disease and pregnancy complications may have become more common. Nevertheless DUB is still a frequent urgent problem.

A common scenario is when a hysterical mother and adolescent daughter burst into the physician's office screaming that she is "hemorrhaging" and demanding that the physician "do something!" Unless the physician has an overall plan in mind in advance, such an episode can be a harrowing experience for all concerned.

Dysfunctional uterine bleeding is abnormal bleeding due to lack of ovulation without organic structural gynecologic disease, without systemic disease, and without coagulation defects. It is a diagnosis of exclusion. It may be defined as excessive uterine bleeding due to anovulation in a woman of reproductive age who has ovaries that can produce estrogen.[1] Those are the only positive features.

The adolescent, compared to the older perimenopausal and usual gynecologic patient, has a different cause of anovulation, a different evaluation, a different prognosis, and a different management plan. The anovulation of adolescence usually is the failure of hypothalamic-pituitary maturation. The anovulation of the perimenopausal woman is often partial ovarian failure. With the adolescent, it is rare that a dilation and curettage is necessary because endometrial adenocarcinoma is extremely rare and because the bleeding is usually controlled by hormonal medication. In addition, curettage of itself does not help prevent recurrence of adolescent abnormal bleeding. Furthermore, it is desirable to avoid the physical and emotional trauma of a curettage in the adolescent. "In the adolescent patient, curettage should be performed only if underlying systemic disease has been ruled out and cyclic hormonal therapy has been unsuccessful in controlling the bleeding."[2] In the case of the perimenopausal woman, a curettage is usually the initial procedure to rule out endometrial adenocarcinoma, adenomatous hyperplasia, or polyp. In addition, the curettage not infrequently has a beneficial effect in reducing subsequent abnormal bleeding. In adolescence the long-term outlook is usually spontaneous ovulation and regular menses, whereas the perimenopausal woman hopes for a quiet menopause.

In our formal compartmentalized teaching, gynecology is completely separate from pediatrics and its subdivision of adolescent medicine. Residents in training do not get experience in examination, spectrum of pathology, and philosophy of management in other departments. In addition, many important concepts are relatively hidden in departmental divisions of gynecologic reproductive endocrinology and infertility, internal medicine, and pediatric endocrinology. Furthermore, standard textbooks of gynecology have only little material on adolescent DUB and often overlap it with perimenopausal DUB.

Background

Puberty

All females develop a negative estrogen (estradiol) feedback mechanism in midfetal life and it persists throughout extrauterine life. Estradiol inhibits follicle-stimulating hormone (FSH) secretion by an effect on the hypothalamus [decreasing gonadotropin-releasing hormone (GnRH) secretion] and also by directly inhibiting the anterior pituitary.[3] The "gonadostat" setting of the hypothalamic-pituitary system is very low (very sensitive to estrogen) in the prepubertal child. Aside from the negative estrogen feedback in prepuberty, there is also an unknown central intrinsic inhibitor that may be more important, perhaps located in the posterior hypothalamus, that suppresses GnRH secretion.

In early puberty, GnRH pulses increase, at first during sleep. This is the result of an increased estradiol tolerance of the hypothalamus–anterior pituitary and a decrease of the central intrinsic inhibitor. Gonadotropin-releasing hormone pulses later extend to the daytime. There are corresponding increases in gonadotropin, at first FSH and later luteinizing hormone (LH). Because of technical factors making LH pulses more precise than FSH pulses, investigators have usually measured the first. This may cause confusion because of the implication that LH increases before FSH.

Clinically, the median age for breast buds is 9;8 (years;months), pubic hair 10;5, and maximal growth 11;4. Menarche is at 12;8, on the down slope of the height spurt. The range is 9;1 to 17;7.[4] Other authorities consider that the mean age for the appearance of breast buds is 10;5, of pubic hair is 11;0, of onset of growth spurt is 11;5, and of the onset of menarche is 12;8 (the same) with a range of 10 to 16.[5] Throughout the world in the past century, there has been a progressively earlier menarche, thought to be related to better living conditions and nutrition.[4] Mixture of population may also play a role. In the United States, the mean age of menarche has remained the same for the past 20 years. There are family differences. There may also be differences in the sequence of events of early puberty development of growth acceleration, thelarche, pubarche, and menarche. The sequence of puberty development usually takes 2 to 4 years. Adrenarche, the adrenal androgen contribution that causes pubic and axillary hair, overlaps gonadarche in time but is apparently independent and may have its own variations.

Previously, it had been assumed that girls ovulate when menstruation first begins (menarche). In fact, half or more of girls have menarche without ovulation.[6] Ovulation depends on positive estradiol feedback, which may not develop until mid- or late puberty. Thus, all girls having menarche without ovulation are at risk for anovulatory DUB since they do not ovulate, they have ovaries that produce estrogen, and they are in an appropriate age group.

The positive estrogen feedback effect of persistent high levels of estradiol causing an LH surge and then ovulation is mediated through the anterior pituitary. Previous studies on rodents showed that the hypothalamus was the site of action. In primates, hypothalamic GnRH is necessary but the reflex center is the anterior pituitary.[3] It may involve an increase in GnRH receptor concentration.

The concept of anovulation is critical and paramount. It is true that there may be overlap between anovulation and ovulation, as, for example, an inadequate LH surge, luteinization without ovulation, and a short or inadequate luteal phase (often with a short proliferative phase). Such conditions may be difficult to diagnose with usual clinical testing.

It is generally believed that about 50% of

all menses are anovulatory in the first year, although some estimate it is as high as 80%.[7,8] It may take up to 5 to 7 years of menstrual life for ovulation to develop, and about 6% never ovulate.[6]

The Ovulatory Menstrual Cycle

Speroff et al. have eloquently described the normal ovulatory menstrual cycle.[9,10] The basic elements for ovulatory menstrual cycles revolve around estradiol functioning (a) as a negative and positive hormone feedback from the ovarian follicle to the brain (hypothalamus–anterior pituitary) and (b) as a local ovarian follicle mediator to increase gonadotropin sensitivity and to create an estrogenic local microenvironment. The follicle that will ovulate is selected in the first few days of the menstrual cycle, with early growth of granulosa cells, to become estrogen dominant. Other follicles grow but soon become atretic. In the early proliferative phase, the rise in FSH (which followed the decline with the onset of menstruation) causes preantral development, granulosa cell aromatization of androgen to estrogens, and an increase in FSH receptors. The dominant antral follicle is established about day 6 and it produces a rise in estradiol by day 7. The estradiol by negative feedback reduces FSH secretion and inhibits other follicles. The dominant follicle survives the low FSH because of an increased number of FSH receptors and high intrafollicular estrogen, and in addition LH receptors increase. Thus, the dominant follicle "assumes control of its own destiny" by its own estrogen and local polypeptide growth factors and by changing gonadotropin secretion by feedback to help itself and stop other follicles. At low levels estradiol has a negative feedback on LH, whereas at high levels it increases LH as it further reduces FSH. An estradiol concentration of 200 pg/mL for 50 hours can cause an LH surge, which sets off luteinization and progesterone formation. Ovulation occurs 24 to 36 hours after the peak estradiol level, 10 to 12 hours after the LH peak, and 34 to 36 hours after the onset of the LH surge.[9]

Luteinizing hormone causes the granulosa layer to luteinize and produce progesterone. The corpus luteum progesterone production reaches its maximum at about day 22, although there may be fluctuations. The luteal phase lasts from 12 to 17 days and requires LH. By an unknown mechanism, the corpus luteum starts a rapid decline by active luteolysis about 10 days after ovulation. Estradiol rises in the luteal phase and this may be the cause of the luteolysis. The rapid fall of progesterone and estrogen causes the menstrual period. With the start of a new cycle, there is a compensatory rise in FSH.

The endometrium during the proliferative phase under the influence of increasing estrogen regrows from 0.5 mm to 4 to 5 mm in thickness. Glandular epithelium and surface epithelium lining rapidly increase. The dense stroma has an initial edema and then a loose cellular composition as fluids are imbibed. The spiral vessels, aside from their branched, short, straight arterioles to the endometrial basalis, extend to the surface and form the glandular capillary plexus.[10]

In the secretory phase, under the influence of progesterone (which is antigrowth) added to estrogen, the endometrium remains the same height, glands secrete, spiral arterioles become more coiled, and the stroma becomes edematous. The basalis forms the inner one fourth, the edematous stratum spongiosum the middle half, and the superficial endometrial layer (stratum compactum) the outer one fourth.

Menstrual endometrial breakdown occurs because of the fixed life span of the corpus luteum and the consequent sudden drop of progesterone and estrogen. Endometrial thickness diminishes, there is a reduced blood flow, the spiral arterioles undergo rhythmic vasoconstriction, and there is interstitial hemorrhage as vessels break. The endometrial surface ruptures with bleeding. There is ischemic necrosis with a separation and desquamation between the basalis and spongiosum, leaving a thin, dense, raw endometrium. Bleeding stops because of transient thrombin-platelet plugs in vessels, collapse of matrix, tissue loss, and rising estrogen secretion in addition to arteriolar vasoconstriction.

Ovulatory menstrual bleeding is self-limited because events occur throughout the endometrium at the same time, because estrogen and progesterone result in a structurally stable endometrium, because progressive rythmic waves of vasoconstriction cause ischemic disintegration of premenstrual endometrium, and because bleeding is stopped by continued vasoconstriction, endometrial collapse, stasis, coagulation mechanisms, and increasing estrogen. There is no fibrous scarring of the healing endometrial surface because of local fibrinolysis and lysosome and other enzyme tissue dissolution.

Anovulatory Uterine Bleeding

The positive estrogen feedback that causes the LH surge and ovulation usually does not develop until mid- or late puberty. When menses begin in early puberty about half (some believe as many as 80%) of adolescents do not ovulate and do not produce progesterone and therefore are at risk for anovulatory DUB. The question is not why is there so much DUB but rather why do we not see more DUB? While mild irregular bleeding might not be reported to the physician, certainly severe bleeding would cause complaints.

It is assumed that the lack of ovulation and the late development of the positive estrogen feedback is due to the late maturation or lack of maturity of the hypothalamus–anterior pituitary. Mature GnRH pulsatile secretion has not yet developed and the hypothalamus-pituitary is unable to respond. This is an extension of the concept that puberty is maturation of the central nervous system.

With persistent anovulatory DUB the LH surge induced by exogenous estrogen was significantly less than in ovulating women. This occurred even though the anovulators had normal rises in LH and FSH after the injection of synthetic GnRH and even though the anovulators had apparently normal follicular development, as suggested by urinary estrogen excretion. Thus the failure to ovulate was attributed to less than normal sensitivity of the hypothalamus-pituitary to positive estrogen feedback (or the failure to completely develop a positive mechanism), as evidenced by an inadequate LH surge in response to estrogen.[11]

Nevertheless, although the usual explanation for the usual case of DUB is failure of the LH surge and therefore of ovulation, it may be too simplistic in that the LH surge is part of a complex interrelated series of events and not an isolated delayed development. In addition, peripheral factors may affect the mechanism. Theoretically cycle dysfunction may be due to an inability to respond to estradiol changes or an estradiol disturbance itself. Absence of decreased estrogen (as may occur in obesity) may result in inadequate FSH secretion early in the cycle, and lack of sufficient rise in estrogen may result in failure to set off the midcycle LH surge. In the ovary, the follicle may not develop and ovulate if the follicle estradiol secretion is insufficient since this increases FSH receptors and increases the FSH induction of LH receptors. Is it possible that the defect in some cases might be the inability of the formation of a dominant ovarian follicle that could produce a sustained high level of estrogen and some other signal hormone? This is the mechanism that is hypothesized to regulate the LH surge. The dominant preovulatory follicle tells the pituitary when it is ready. Perhaps the immaturity might be of the follicle as well as of the hypothalamus-pituitary.

Emans and Goldstein have hypothesized that abnormal DUB does not occur with most cases of anovulation because of a mature developed negative estrogen feedback and that the abnormal DUB cases do not have a mature negative feedback.[12] This is not the generally held view, which is that a negative feedback is present throughout life and begins in fetal life. Nevertheless, it could cause relatively regular menses based on estrogen withdrawal, but this would require a cyclic increase in circulating estrogen levels to trigger the negative effect of FSH reduction.

This concept could explain why most anovulators do not complain of abnormal bleeding. In early puberty and even before menarche, there may be crudely cyclic changes in FSH secretion.[13]

The direct mechanism of anovulatory DUB is the lack of ovulation, the lack of progesterone, and the unopposed estrogen effect on the endometrium. Abnormal bleeding can occur because of transient reduction of estrogen (estrogen withdrawal) or because of estrogen breakthrough bleeding. With the former, there is bleeding as large pieces partially and irregularly slough. With the latter, the endometrium has become too thick to be supported by estrogen and different superficial areas slough and heal over sporadically. There is excessive bleeding because there is an absence of a generalized action, of a structural compact stroma, and of vasoconstrictive waves together with fragile bulky tissue. There is also an absence of regrowth of surface epithelium and binding membrane from the endometrial basalis and cornual angles.[10]

If there has been heavy severe bleeding for several weeks, the endometrium eventually sloughs, leaving a thin, raw surface. Although dilation and curettage of the uterus is unusual today, older reports indicated that the endometrium is proliferative or may show various gradations of hyperplasia stimulation, such as polypoid hyperplasia, polyps, cystic-glandular hyperplasia, or occasionally adenomatous hyperplasia without atypia. The infrequent report of secretory endometrium may have been due to an incomplete ovulation of a luteinized follicle or to another cause of bleeding, such as coagulopathy.

There are two general clinical presentations of abnormal anovulatory DUB. One is complete irregularity in length of days of bleeding and in intervals between bleeding. This might occur with high estrogen levels. The second pattern is more characteristic. There is amenorrhea for 3 to 4 months followed by bleeding that persists for 3 to 4 weeks.[1] This might occur with persistent low estrogen levels. Anovulatory DUB is irregular, unannounced, without premenstrual molimina (breast swelling and tenderness, bloating, weight gain, irritability, premenstrual syndrome, etc.), and without painful periods. Ovulatory cycles tend to be regular, have premenstrual molimina, and have uncomfortable or painful periods. The latter is due to prostaglandin production in the secretory endometrium, which causes vascular and uterine spasm. Occasionally with ovulatory cycles, there may be staining for a few hours at ovulation.

The usual normal ovulatory menstrual period lasts from 4 to 6 days, with a variation of 2 to 8 days.[10] Some consider the mean to be 4 days, with bleeding over 7 days considered as prolonged.[14] The mean blood loss is 35 mL for the entire period, and 95% of women loose less than 60 mL. The normal total blood loss equates to about 10 to 15 soaked tampons or pads. Blood loss greater than 80 mL/cycle frequently causes anemia and is considered pathologic.[14] On a daily basis, this would be more than six pads. Some consider blood loss of over 150 mL to be abnormal. Unfortunately, it is difficult to measure blood loss and individual patients may grossly underestimate blood loss. The physician has to rely on the number of pads or tampons, the extent of soaking, the frequency of changing, and the amount of and size of clots. An indirect method is to determine the hemoglobin (or hematocrit) as well as the reticulocyte count and serum iron. Ovulatory menses are very regular and usually range from about 26- to 38-day intervals from the first day of one menses to the first day of the next menses. The variations are usually due to the proliferative phase, since menses occur 14 days after ovulation. Cycles less than 21 days are abnormal. Clinically, about 43% of women have irregular menses in the first year of menarche and this may persist for 5 years in 20%.[15] Obviously, the definition of irregularity will determine the incidence of irregular menses. In addition, 43% of adolescents are not considered to have abnormal DUB.

It would be difficult to get reliable data because of population selection, lack of follow-up, use of oral contraceptives, and the expense of testing.

Management

Evaluation

Who Should Be Evaluated?

It is sometimes difficult to draw the line between what is an acceptable variation of normal with adolescent vaginal bleeding versus what would be sufficiently abnormal to warrant investigation. Pediatricians and adolescent medicine physicians have to make this difficult choice and have to decide what is abnormal but requiring only observation versus what requires referral (and a vaginal examination). The pediatrician may refer the patient for evaluation if:

1. There is prepuberty vaginal bleeding. This may be due to a local anatomic cause, such as trauma, foreign body, severe vaginitis, or rhabdomyosarcoma
2. There is severe vaginal bleeding (more than six well-soaked pads daily), anemia, or intermenstrual irregular bleeding
3. There is severe pain
4. There are pregnancy symptoms or the adolescent is sexually active
5. Her mother took hormone pills during pregnancy
6. There is fever and pelvic discomfort
7. The physician and/or the patient has anxiety

The patient should be evaluated by a gynecologist or adolescent medicine physician with appropriate experience. The patient and her mother should be reassured that the patient will still be a virgin and that she will not have pain. She will feel a sense of pressure during vaginal examination.

History

The mother should be encouraged not to be in the examination room in order to obtain a first-hand account of the history from the patient and to ask delicate questions about such topics as sexual activities and substance abuse. A nurse should be present for assistance, reassurance for the patient, and legal protection of the physician.

The history might include (Table 21-1):

1. The present illness—when the present bleeding episode started, the severity of the bleeding, the number of pads and extent of soaking, and the presence of clots, pain, or fever
2. The past menstrual history—onset of menses, usual pattern
3. Secondary sex characteristic development
4. Systemic bleeding tendencies—spontaneous ecchymoses, hemorrhage after tooth extraction, and the like
5. Recent other illness or medications (which might affect coagulation)
6. Symptoms of anemia—weakness, faintness, shortness of breath, and the like
7. Possibility and symptoms of pregnancy
8. Past history—operations, illnesses, allergies, sexually transmitted diseases, pregnancies
9. Life-style—stress, weight change, eating disorders, intense athletic activity, substance abuse, sexual activity, contraception
10. Review of systems, especially neurologic headache, visual disturbances, anosmia,

TABLE 21-1. History outline

- I. Present illness
  - A. Present bleeding episode
  - B. Past menstrual history
  - C. Symptoms of blood loss
  - D. Possibility of pregnancy
  - E. Possibility of pelvic inflammatory disease
- II. Past history
  - A. General bleeding tendency
  - B. Secondary sex characteristic development
  - C. Illnesses
    1. Sexually transmitted diseases
    2. Operations
    3. Pregnancies
    4. Allergies
  - D. Life-style
- III. Review of systems
- IV. Family history

gastrointestinal chronic diarrhea, abdominal pain
11. Family history—polycystic ovarian disease (PCOD), premature menopause, diabetes

Examination

The examination begins with inspection of the soaked pad if the patient is bleeding. Then comes the general examination:

1. Blood pressure, pulse, height, weight, body type, fat distribution
2. Skin
   a. Androgen excess—acne, hirsutism, virilism, frontal balding
   b. Hypothyroidism—coarse, dry skin
   c. Hyperthyroidism—warm, moist skin
   d. Ecchymosis, petechiae
3. Thyroid enlargement
4. Breasts—size, tenderness, galactorrhea
5. Heart, lungs
6. Abdomen—hepatosplenomegaly
7. Axillary and pelvic hair
8. Vulva and pelvic region

The vulva is inspected for maturation, laceration, and amount of bleeding. The hymen orifice is visualized and gently palpated for diameter and flexibility to distention. In most cases, a virginal Pederson vaginal bivalve speculum will be used. Although narrow, it is as long as the standard adult speculum. The narrow but short speculum for children is inadequate for the adolescent. Infrequently, for the very narrow and rigid hymen a tubular vaginoscope (or hysteroscope) is necessary to visualize the vagina. Clots and blood have to be cleansed away. The cervix and vaginal walls are inspected for the source of bleeding. For the usual case of anovulatory DUB, there will be active, dark, menstrual-type bleeding coming from a closed, clean cervix. Sometimes, with severe DUB, the cervix may have a blue tint and be soft and the uterine fundus may have a boggy feel similar to that of pregnancy. Visualization is necessary to rule out vaginal laceration, neoplasm, and foreign body to detect cervical dilation and cervical polyp. Evaluation of abnormal bleeding requires a vaginal examination. Papanicolaou cytology may not be done because of active bleeding. If the patient is sexually active cultures for gonorrhea and chlamydia are considered. If feasible, a one-finger bimanual vaginal examination is done followed by a rectovaginal examination.

Laboratory Tests

The usual laboratory studies are a complete blood count and coagulation tests (platelet count, prothrombin time, activated partial thromboplastin time, and bleeding time) in order to detect anemia and coagulopathy. Significant blood loss despite a normal hemoglobin may be suggested by an increased reticulocyte count and low serum iron. If there is family, clinical, or laboratory suggestion of coagulation defect, then a hematologist may be consulted to detect rare and difficult-to-determine defects by specific protein assays of individual coagulation factors, lupus-like disorders, and platelet function tests. Other blood tests to consider (Table 21-2) are:

1. Blood (or sensitive urine) pregnancy human chorionic gonadotropin (hCG) quantitative test, to rule out ectopic or other pregnancy complication. Adolescents may consciously or unconsciously

TABLE 21-2. Laboratory tests to consider

Extent of blood loss: hemoglobin/hematocrit
Coagulation defect: complete blood count, platelets, prothrombin time, activated partial thromboplastin time, bleeding time
Rule out pregnancy: human chorionic gonadotropin
Rule out salpingitis: erythrocyte sedimentation rate
Additional possibilities
Thyroid function, thyroid-stimulating hormone
Prolactin
FSH, LH
SMA chemistries
Androgens

deny the possibility of pregnancy for fear of parental discovery or because they think they are too young or have had only infrequent coitus. In addition, adolescents often have irregular menses.

2. Thyroid function (triiodothyronine, thyroxine, and thyroid-stimulating hormone levels). Both hypo- and hyperthyroidism may result in anovulation. The thyroxine test is the most sensitive for hypothyroidism.
3. Prolactin. An elevated prolactin may result in anovulation, especially with galactorrhea.
4. Serum androgens [total and free testosterone, dehydroepiandrosterone sulfate (DHEAS), androstenedione, and 17-hydroxyprogesterone], if there are signs of androgen excess such as hirsutism, acne, or enlarged clitoris. Androgens may be elevated with PCOD.
5. FSH and LH. With partial ovarian failure, there may be unopposed estrogen without ovulation and an elevated FSH as well as LH. With PCOD, there is a tendency for LH to be twice as high as FSH. With early puberty anovulatory DUB, both are low or normal and the LH may be lower than the FSH.
6. Erythrocyte sedimentation rate (ESR). It is elevated with salpingitis, which may be a factor in abnormal bleeding.
7. SMA blood chemistries to rule out diabetes, renal disease, hepatitis, or other chronic systemic disease that may inhibit ovulation.

Additional tests for consideration include ultrasonography and computed tomography (CT) or magnetic resonance imaging (MRI). A sonogram of the pelvis and abdomen should be obtained if there is a question of an ovarian tumor, polycystic ovaries, or a congenital gynecologic anomaly. With the latter, there may be renal anomalies. Mononucleosis may cause an enlarged liver and spleen. A CT or MRI scan of the skull should be obtained if there is an elevated prolactin, significant headache, anosmia, visual disturbances, brain dysfunction, or galactorrhea.

Clinical judgment and individualization are required in ordering tests that may be expensive and frightening. The important factors are a good history, physical examination, blood count, and blood coagulation tests; clinical severity; duration of symptoms; and careful follow-up. Measurement of hCG and ESR at the same time would be prudent and relatively inexpensive.

Differential Diagnosis (Table 21-3)

Anovulatory DUB is a diagnosis of exclusion. Most cases of abnormal adolescent bleeding are due to DUB; about 5% are due to other causes. Depending on factors of case selection, pregnancy complications may be the main condition to rule out. In selected series of severe cases referred for hospitalization, about 20% had coagulation defects and about 5% had other pathology.[16]

"Von Willebrand's disease is the most common coagulopathy the physician treating an adolescent female is likely to encounter."[17] The clues are a family history of bleeding or bruising and a past history of childhood nosebleeds and easy brusing and menorrhagia at menarche. There is a prolonged bleeding time resulting from abnormal platelet function and decreased levels of factor VIIIC and factor VIIIAG. There are six variants. It is inherited as an autosomal dominant or recessive. Half of all cases may have menorrhagia. The bleeding time is the best indicator of clinical severity. Oral contraceptives may affect factor VIII testing. Collagen vascular disorders (lupus erythematosus, autoimmune disease, etc.) may resemble von Willebrand's disease.[17]

Thrombocytopenic purpura is another leading contender, and both can be discovered by routine coagulation tests (prothrombin time, activated partial thromboplastin time, platelet count, bleeding time). Infrequently, there may be a leukemia or bleeding due to oncology chemotherapy.

Chronic idiopathic thrombocytopenic pur-

TABLE 21-3. Differential diagnosis of abnormal adolescent vaginal bleeding

- Anovulatory DUB
- Pregnancy complications
  - Abortion
  - Ectopic pregnancy
  - Hydatidiform mole
- Blood dyscrasias
  - von Willebrand's disease variants
  - Thrombocytopenia
  - Leukemia
  - Aplastic anemia
  - Many rare varieties of disturbances of platelets, coagulopathies, and blood vessels
  - Possibly iron-deficiency anemia
- Gynecologic anatomic factors
  - Vulva—condylomata, trauma
  - Vagina
    - foreign body (diaphragm, tampon, sponge)
    - severe streptococcal vaginitis
    - trauma
    - hemangioma
    - congenital anomalies (double uterus and vagina with partial obstruction)
    - adenosis
    - rhabdomyosarcoma
    - clear cell adenocarcinoma
  - Cervix
    - cervicitis
    - polyp
    - adenosis
    - hemangioma
  - Uterus
    - intrauterine device
    - congenital anomalies
    - oral contraceptive abuse
- Gynecologic anatomic factors (*continued*)
    - rare: neoplasms, endometrial polyp, myoma, abortion
  - Fallopian tube—salpingitis, ectopic pregnancy
  - Ovary
    - salpingo-oophoritis
    - neoplasm, granulosa cell tumor
    - polycystic ovarian disease
  - Pelvic endometriosis
- Endocrinopathies
  - Hyper- or hypothyroidism
  - Hyper- or hypoadrenalism
  - Hyperprolactinemia
  - Diabetes mellitus
  - Polycystic ovarian disease
  - Premature ovarian failure
- Chronic systemic illness
  - Hepatic
  - Renal
  - Regional ileitis
  - Acquired immunodeficiency syndrome
- Medications
  - Anticoagulant
  - Chemotherapy
  - Excessive aspirin
- Life-style
  - Substance abuse
  - Eating disturbance
  - Stress
  - Excessive athletics
  - Rapid weight change, overweight, underweight
- Any cause of anovulation, with an ovary that produces estrogen and an intact uterus

pura, aside from causing severe uterine bleeding, may cause intracranial hemorrhage.[18] There are some rare coagulation defects that are not identified by conventional tests and require comprehensive hematologic evaluation, such as platelet storage pool disease.[19]

Functional ovarian neoplasms producing estrogen or hCG may cause abnormal bleeding.

Congenital anomalies may simulate DUB. With a double uterus, double vagina, and partial obstruction of one vagina, there will be external metrorrhagia as the partially obstructed side slowly discharges.

Iron-deficiency anemia as rare cause for abnormal bleeding has been described but is not understood, and cause and effect may overlap.

The Mild Case—Expectant Observation

In a mild case of DUB there is no anemia or active bleeding. The patient may be managed by expectant observation after evaluation. She is advised to keep a record of her menses (menstrual calendar), take iron tablets, have a nutritious diet, be of appropriate weight, reduce excessive stress, and maintain general health.

Recent prior life changes with personal social and health life stresses may be a factor in setting off DUB.[20] The patient is alerted to notice the onset of ovulatory cycles—regularity and premenstrual molimina. Recording of basal body temperature or blood progesterone is usually not feasible. The patient should be reevaluated periodically. If the original pelvic examination was inadequate or difficult or if the DUB persists, the pelvic exam should be repeated. If the bleeding is disturbing to the patient and interferes with her activities and especially schoolwork, then low-dose oral contraceptives may be prescribed for 3 or 4 months. An alternative is medroxyprogesterone acetate (MPA) tablets 10 mg daily for 10 days a month on days 16 through 25.

The Severe Case

Hormonal Hemostasis

Most cases of DUB are urgent problems seen during a time of heavy bleeding. After evaluation and making the presumptive diagnosis, the appropriate treatment is what I call "hormonal hemostasis,"[1] because the term is descriptive and concise. Patients with severe DUB may have heavy bleeding or light bleeding at the moment but are anemic.

In former years, some would prescribed separate tablets of ethinyl estradiol 0.02 mg and MPA 10 mg to avoid the stigma of oral contraceptive tablets. However, this is expensive, inconvenient, and confusing.[21] At present, there is general agreement that it is easier to use oral contraceptives. Although the theoretical deficiency is in progestin, progestin therapy is more effective when adequate estrogen is present. Furthermore, estrogen of itself is hemostatic. In addition, if there has been persistent heavy bleeding with a thin raw endometrium, progestin of itself is not effective.

To stop the bleeding, the patient should take one oral contraceptive tablet four times daily for 4 days. This will stop the bleeding usually within a day. If the bleeding does not stop, the diagnosis must be reconsidered, the patient reexamined, and consideration given to examination under anesthesia and a gentle uterine curettage.

Almost any combined estrogen-progestin oral contraceptive may be used, except for the new triphasics because of differing contents of each pill. The 21 rather than the 28 (with 1 week of placebo pills) package is used. The estrogen dose may be in the 30 to 50 μg (0.03 to 0.50 mg) range. For chronic use generally a less than 50-μg estrogen pill is preferrable.

Some of the oral contraceptives that may be used for "hormonal hemostasis" as one pill four times a day for 2 days include:

Lo/Ovral (21 tablets): 0.3 mg norgestrel; 0.03 mg ethinyl estradiol
Ovral (21 tablets): 0.5 mg norgestrel; 0.05 mg ethinyl estradiol
Demulen 1/35-21:1 mg ethynodiol diacetate; 0.035 mg ethinyl estradiol
Demulen 1/50-21:1 mg ethynodiol diacetate; 0.050 mg ethinyl estradiol
Levlen 21: 0.15 mg levonorgestrel; 0.03 mg ethinyl estradiol
Loestrin 1.5/30: 1.5 mg norethindrone acetate; 0.03 mg ethinyl estradiol
Norinyl 1 + 35 (21 tablets): 1 mg norethindrone; 0.035 mg ethinyl estradiol
Norinyl 1 + 50 21:1 mg norethindrone; 0.05 mg mestranol
Ortho-Novum 1/35 21: 1 mg norethindrone; 0.035 mg ethinyl estradiol
Ortho-Novum 1/50 21: 1 mg norethindrone; 0.05 mg mestranol
Ovcon 35 (21 tablets): 0.4 mg norethindrone; 0.035 mg ethinyl estradiol
Ovcon 50 (21 tablets): 1.0 mg norethindrone; 0.050 mg ethinyl estradiol
Enovid 5 mg: 5 mg norethynodrel; 0.075 mg mestranol
Enovid 10 mg: 9.85 mg norethynodrel; 0.15 mg mestranol

Another option is conjugated estrogen tablets (Premarin) 2.5 mg daily on days 1 through 25 and MPA 10 mg daily on days 14 through 25.

Relative containdications to oral contraceptives are infrequent in the adolescent but include active sickle cell disease, history of pulmonary embolism, active liver disease, active thrombophlebitis, and thromboembolic disorders. If the patient cannot tolerate oral estrogens or contraceptives because of nausea or if there is a contraindication, then the progestin norethindrone acetate (Norlutate) 5 mg three to four times daily will stop bleeding. Whereas MPA has a pure progesterone-type activity and is excellent for routine cycling, norethindrone acetate is much more effective in stopping heavy DUB.

Although oral contraceptives are approved by the Food and Drug Administration (FDA) for contraception and marketed for contraception, they may also be prescribed by the physician for other purposes that are generally recognized in the medical literature. Use of large doses, such as four tablets daily, for a short time to stop DUB is a standard recognized practice in adolescents and is considered effective and safe. Previous studies of mortality and morbidity with oral contraceptives were unfavorably biased by chronic use of older high-dose pills, by an older age group, and by the effect of smoking.

Unlike other oral contraceptives, Enovid 5 mg and 10 mg are FDA-approved for the treatment of hypermenorrhea, for the production of cyclic withdrawal bleeding, and for the treatment of endometriosis. Enovid was the original standard effective pill for DUB "hormonal hemostasis." The disadvantages are that it may cause nausea and that lower dose oral contraceptives are used for later cycling.

If the bleeding is very severe, the patient is very anemic, and/or heavy bleeding had been going on for more than a week previously, then in addition to the oral contraceptive, intravenous estrogen is given. The usual preparation is Premarin conjugated estrogen given as a 25-mg dose slowly every 4 hours until the bleeding stops, for a total of three doses. The Physicians' Desk Reference schedule is a repeat dose in 6 to 12 hours. It is approved for functional bleeding. Since the bleeding stops within hours, it is obviously not sufficient for cells to have grown. There must be some physical-chemical change with imbibing of fluid.

After 2 days of oral contraceptives as four pills a day (with or without intravenous Premarin), the bleeding will have completely stopped. If not, the diagnosis is incorrect. Following this I reduce the dose to one or two pills daily for 3 weeks. There is no bleeding during this time. The patient is placed on oral iron and recovers physically and emotionally. The pills are then stopped and there is withdrawal bleeding 2 or 3 days later. It is usually like a normal period. On day 5 of the cycle (first day of the flow is day 1), the patient is started on three or four cycles of low-dose combination oral contraceptive pills. If this is not done the patient usually has a recurrence of the DUB. If the withdrawal bleeding is very heavy, the low-dose oral contraceptive pills are started sooner (second or third day) and a more potent pill might be used (such as a 50-μg ethinyl estradiol). Table 21-4 reviews the emergency procedures for severe DUB.

TABLE 21-4. Emergency procedures for abnormal anovulatory DUB of adolescence

1. DUB is a diagnosis of exclusion
   - A. History
   - B. Physical examination, including vaginal examination
   - C. Laboratory
     1. complete blood count
     2. blood coagulation tests
     3. hCG, ESR
2. For severe bleeding start "hormonal hemostasis" immediately:
   - A. One oral contraceptive pill four times a day for 2 days.
   - B. For very severe bleeding, add Premarin 25 mg intravenous slowly.
   - C. The bleeding will stop within a day; if not the diagnosis is incorrect and the patient must be reevaluated and a dilation and curettage considered.
3. After the bleeding has stopped, and after 2 days of four oral contraceptive pills a day, reduce the dose to 1 or 2 pills daily and continue for 3 weeks. Following this, there will be withdrawal bleeding 2 or 3 days later.
4. Cycle for 3 to 4 months with low-dose oral contraceptive pills in standard fashion.

Other Treatment Approaches

Some authorities have a traditional "medical curettage" philosophy. The patient is placed on one contraceptive combined progestin-estrogen pill four times daily continously for 5 to 7 days. The bleeding subsides within 12 to 24 hours but the four pills daily are continued. (If the bleeding persists, the patient is examined under anesthesia and a dilation and curettage of the uterus is done. During this time, the patient is evaluated.) After 5 to 7 days of oral contraceptive therapy the pills are discontinued, and 2 to 4 days later there is withdrawal bleeding ("medical curettage") with severe cramps, heavy bleeding, and often discharge of chunks of tissue. On the fifth day of flow, the patient is started on standard low-dose cyclic combination oral contraceptive pills in a regimen of one/day for 3 weeks, stop 1 week, and repeat. Three such cycles are completed.[10] The rationale of the "medical curettage" is to remove the presumably extremely thickened and previously fragile endometrium. The problems are twofold: (a) the distress of the patient despite the reassurance of the physician, and (b) additional blood loss in a patient who may already be severely anemic. In addition, the rationale may not be correct for the adolescent who has had persistent, continous heavy bleeding for several weeks and has only a thin, raw endometrial surface.

Other authorities have another approach to stop acute, heavy bleeding. Conjugated estrogen is given in 20-mg doses intravenously at 2- to 4-hour intervals, which will stop the bleeding in 12 to 24 hours. The patient is then placed on oral conjugated estrogen in a dose of 2.5 to 3.75 mg daily for 3 weeks. Then, for an additional 7 to 10 days, the estrogen is continued with the addition of 10 to 20 mg oral MPA daily. When both medications are stopped, withdrawal bleeding occurs.[2]

Huffman divided adolescent DUB into three types:

1. Mild (no anemia)
2. Moderate (moderate anemia, hemoglobin over 9 g)
3. Severe (marked anemia, hemoglobin below 9 g); more than six well-soaked pads daily was considered as excessive bleeding.

For those patients with mild DUB, he advised reassurance, explanation, and reevaluation if the irregularity continued. For moderate DUB, he recommended hormonal therapy of separate pills of estrogen and medroxyprogesterone and if unsuccessful then testosterone. For severe DUB, he advised transfusion, intravenous Premarin (4/18 patients did not stop bleeding and required curettage), and oral estrogen-progestin. One case in his memory required hysterectomy.[21]

Goldstein performs a dilation and curettage if bleeding cannot be controlled within 24 to 36 hours.[22] About 20% of his patients required this, and usually endometrial hyperplasia or polyp was present. Less often, scanty curettings indicated a complete endometrial slough. Although most find that dilation and curettage is only rarely necessary, his series represents a selected referral group. Goldstein, over many years of experience, encountered two cases of DUB that did not respond to dilation and curettage. Both were controlled by uterine packing with iodoform gauze left in place for 24 hours. Only one patient required a hysterectomy—a 20-year-old who was on chronic anticoagulation, and of course this was not DUB.[22]

Megestrol acetate, a potent progestin, in a dose of 20 mg/day can reduce menstrual blood loss even in ovulatory menorrhagia, and in a dose of 40 mg/day in about 2 months can cause amenorrhea that can be continued for 3 to 4 months. It can also be helpful for those with blood dyscrasias.[14] Megesterol is usually used for palliation of endometrial carcinoma.

Nonsteroidal antiinflammatory drugs (NSAIDs) are usually prescribed to treat dysmenorrhea but have also been found to reduce menstrual bleeding. Nevertheless, they are not often used for adolescent DUB because they are not as effective as oral contraceptives and because of side effects. Mephenamic acid (500 mg three times a day for 3 to 5 days with menses) reduced mean menstrual blood loss from 160 ml to 127 ml (20%) in menorrhagia of unknown cause in parous women.[23] Excessive menstrual bleeding was defined as over 80 ml. Among 20 women, however, 3 had an increase in blood loss. Dysmenorrhea was improved in 10 of 13 women. Six women had side effects including nausea, vomiting, diarrhea, and vertigo.[23] Naproxen sodium can also reduce blood loss significantly. The mechanism of action of NSAIDs is unknown. It is possible that excessive blood loss may be due to an increased ratio of prostaglandin E_2 to prostaglandin $F_{2\alpha}$, although both increase.[24] Although there were insufficient cases for statistical analysis, mephenamic acid reduced blood loss with fibroids, anovulatory DUB, and von Willebrand's disease when treated[25] at menstrual periods continously for over 1 year. It can also reduce blood loss with an intrauterine device.

For patients who require cycling, who cannot or do not wish to take pills, and who do not have severe bleeding, intramuscular progestin may be used. A 250-mg dose of 17α-hydroxyprogesterone caproate has an approximately 12-day-long action. It will stop bleeding if present in a few days and cause withdrawal bleeding 10 to 14 days later. Another preparation is progesterone in oil, 100 mg intramuscularly, also given at 4-week intervals, which will cause withdrawal bleeding in a few days.[2] These intramuscular progestins are not contraceptive.

Continous Follow-up

Following 3 months of cyclic low-dose oral contraceptive therapy, there are some options:

1. If the patient is sexually active, then low-dose oral contraceptives may be continued if desired.
2. If the patient is not sexually active then:
   a. She may be observed for spontaneous onset of menses. The disadvantage is that after 3 months or longer, there may be a recurrence of DUB. Oral MPA 10 mg daily for 10 days may be given after 6 to 8 weeks to cause withdrawal bleeding.
   b. Medroxyprogesterone acetate may be given at 6-week intervals automatically. It will not prevent ovulation and spontaneous menstruation could still occur.
   c. Medroxyprogesterone acetate may be given at 4-week intervals automatically.

The patient should be aware of the options, participate in the decision, and understand that long term follow-up is important.

The Long Term

Persistent Anovulation

The natural history of 291 adolescents with severe DUB showed that change to normal was greatest in the first 2 years, but even by 4 years 50% still had DUB, and at 13 years 40% still continued with DUB.[26] The eventual results were slightly better if there were normal menses before the onset of DUB. "Although the duration of follow-up extended to 20 years, no patient developed normal periods after 10 years of persistent abnormality." There were 43 cases who had transfusions, 177 who had curettages, and 85 who had laparotomies. Four had hysterectomies for continued bleeding (ages 20 through 33), and four had hysterectomies for endometrial carcinoma (ages 23 through 33). After marriage, more than half presented with infertility and three fourths of those with continuing DUB were infertile. There was a poor obstetric history in those with or without continuing DUB.[26] It should be realized that the patients in this series (1941–1964) were in fact a select group, having been referred from other institutions in New York because of severe DUB, many having already had one curettage and transfusions. In addition, synthetic progestin was not available for management, which could have controlled the bleeding and prevented endometrial adenocarcinoma. Ovulation-inducing agents were not available for the anovulatory infertility.

It is generally agreed that chronic persistent anovulation may result in dysfunctional bleeding, amenorrhea, hirsutism, and infertility, and later in life endometrial hyperplasia and cancer, and possibly breast cancer.

Polycystic Ovarian Disease

Because of frequent lack of follow-up and different physician involvement, it is not generally appreciated that a significant number of DUB cases with abnormal bleeding go on into adult life with persistent anovulation, and many of these are later found to have "polycystic ovarian disease."[1] Previously it had been assumed that PCOD was a condition of the adult and not of the adolescent. The situation is confusing because:

1. Abnormal adolescent DUB is usually associated with excessive bleeding, whereas the PCOD of the adult is usually associated with oligomenorrhea.
2. Polycystic ovarian disease lacks a standard biochemical and endocrinologic definition and lacks a pathognomonic pathology, and there is no complete agreement on etiology.
3. Polycystic ovarian disease may wax and wave.
4. Not all PCOD cases have characteristic enlarged ovaries, and there is a lack of correlation of size and hormonal factors.
5. Based on sonography, 23% of women who considered themselves normal were found to have polycystic ovaries. On investigation, some of these women tended to have clinical and biochemical markers of polycystic ovarian syndrome.[27]

Nevertheless, it is logical to expect that anovulatory menses starting in adolescence may sometimes continue into adulthood with persistent anovulation, and that such cases may show a continuum or spectrum of change regarding progression and severity of anatomic and functional changes.

The Stein-Leventhal syndrome of PCOD was first described in 1935 and included anovulation, oligomenorrhea, hirsutism, obesity, and enlarged polycystic ovaries.[28] Since then there have been many arguments about what constitutes PCOD, how to define it, and what causes it. Although the ovarian pathology was characteristic, it was not pathognomonic and there were many variations on the theme. Furthermore, some thought it to be a primary ovarian disease, others considered it to be a hypothalamic disease, and still others thought it to be an adrenal disease. Many considered PCOD to be a specific disease.

The current concept is that PCOD is a symptom (like fever), that it results from chronic anovulation, and that there are many causes of chronic anovulation. Such persistent anovulation is associated with a "steady state" of gonadotropins and sex steroid, in contrast to the normal monthly cyclic variations.[29]

Is it possible to determine which cases of adolescent anovulatory DUB will continue into adulthood PCOD? There is a familial clustering of polycystic ovarian syndrome symptoms, and some families may show autosomal dominant inheritance.[30] Plasma androgens (especially testosterone) are an important determinant of body fat topography causing upper body fat predominance.[31]

After a single stimulation dose of gonadotropin-releasing hormone agonist (GnRHa), women with polycystic ovarian syndrome had a pattern similar to that of men, with greater early LH responses and lower peak FSH responses than normal women, as well as increased levels of androstenedione, increased 17α-hydroxyprogesterone, and normal to increased estrogen. These abnormal responses were unaffected by pretreatment with dexamethasone for adrenal suppression. Thus in PCOD there was a masculinized pituitary and ovarian response and a suggestion of abnormal ovarian 17-hydroxylase and $C_{17,20}$ progestin activities.[32]

Diagnosis

Some have diagnosed "polycystic ovary syndrome" in adolescence based on (a) menstrual disorders; (b) association with hirsutism, acne, or obesity; and (c) elevated free testosterone that did not decrease after adrenal suppression by dexamethasone.[33] It has been found that whereas normal pubertal girls have a daily nocturnal sleep surge of plasma LH, those with polycystic ovarian syndrome had desynchronized LH surges during the daytime, while the cortisol and prolactin profiles were normal. This was considered as evidence of central nervous system dysfunction as the initiating factor.[34]

The diagnosis of PCOD is not precise, and even the name varies since many use the terms "polycystic ovarian syndrome," "polycystic ovaries," "ovarian hyperandrogenism," and "Stein-Leventhal syndrome." The condition is usually characterized by a combination of:

1. Anovulation, menstrual disturbances, infertility
2. An elevated LH/FSH ratio of 2:1 or higher
3. Hyperandrogenism clinically and by hormone levels
4. Bilateral enlarged polycystic ovaries

It may be, however, that up to 20% of patients do not have a high LH/FSH ratio. This could be due to a high testosterone effect or perhaps insufficient testing. About 70% are disturbed by hirsutism and 30% have dysfunctional bleeding.

With persistent anovulation and persistent elevated estrogen, there is a continous elevation of LH. This causes an increased secretion of testosterone, androstenedione, dehydroepiandrosterone, DHEAS, 17-hydroxyprogesterone, and estrone. The androgens come from the ovary generally; however, the DHEAS comes from the adrenal. The ovarian secretion of estradiol is normal; however, the increased secretion of androstenedione, which is converted peripherally to estrone, increases the total estrogen level. Thus, there is an increase in the production of both androgens and estrogens. The testosterone reduces liver formation of sex hormone–binding globulin, which increases free testosterone and free estradiol and also inhibits hypothalamus and pituitary function.

In addition to the level of LH being high, its bioactivity is also increased. Aside from the elevated LH, there is a characteristic low FSH. This gonadotropin pattern may be due to the persistent high estrogen level of increased estrone and free estradiol. The negative estrogen feedback mechanism inhibits FSH production, and there may also be a contributing effect from inhibin from the polycystic ovary.[29]

In the ovary the persistent, although low, FSH causes the formation of multiple follicular cysts that last several months and are unable to ovulate. The surrounding granulosa cells atrophy, while the theca and stromal cells, in response to LH stimulation, luteinize and increase their secretion of testosterone and androstenedione. These androgens prevent estradiol from stimulating the granulosa cells to cause normal follicular development. This androgenic microenvironment results in numerous persistent arrested follicles 3 to 6 mm in size and atretic follicles.

The polycystic ovaries characteristically are moderately enlarged (often larger than the never-pregnant uterus) and have a white "oyster shell" appearance as a result of a thick capsule. Previously, it was thought that the capsule (tunica) prevented ovulation. The surface is smooth and bulges from the enclosed 20 to 100 cystic follicles. There is an increase in stroma, which may contain areas of luteinized islands of theca cells (hyperthecosis). The ovary may be of normal size and still have the characteristic histology.

Etiology/Pathogenesis

There are many theories regarding the initial event in the production of PCOD, including lack of a cycling center of the hypothalamus, adrenal hyperactivity, and primary ovarian disease. Others have indicated that PCOD is the final result of any chronic cause of lack of ovulation causing a steady state rather than the normal monthly hormonal cycling. Some divide it into an initial induction "generator" state and a chronic persistent "effector" state.[35]

New investigations have revealed that polycystic ovarian syndrome is often associated with a unique disorder of insulin action. There is insulin resistance and hyperinsulinism without fasting hyperglycemia, and insulin-stimulated glucose utilization is decreased. The mechanism is not understood. Dunaif and associates showed that the hyperinsulinism associated with polycystic ovarian syndrome may be due to the syndrome itself; is independent of obesity or impairment of glucose tolerance; and is not the result of decreased insulin clearance.[36] Insulin infusion in insulin-resistant women with polycystic ovarian syndrome can change peripheral sex hormone levels independent of gonadotropin release. It decreased testosterone, but did not increase it in normal women. Thus, there was no simple relationship between hyperinsulinemia and hyperandrogenism in PCOD.[37]

Recently a new two-part concept evolved about the origin of polycystic ovarian syndrome that also has implications for its definition and identification. The concept is, first, that anything that causes an insulin-resistant state will go on to develop an hyperinsulemic state and, second, in genetically susceptible individuals the hyperinsulinemia causes hyperandrogenism. Obesity is the most common cause of insulin resistance and in addition may increase preexisting insulin resistance of other causes. Therefore, it is understandable why obesity is found in 50 to 80% of polycystic ovarian syndrome cases, why "insulin resistance with hyperinsulinemia appears to be an almost universal finding,"[38] and why polycystic ovarian syndrome is often familial. Not all obese patients develop polycystic ovarian syndrome because they lack the second factor. Nonobese women with polycystic ovarian syndrome have other causes of hyperinsulinemia. "It might also be possible to identify women predisposed to developing polycystic ovarian syndrome, before clinical expression of the disease, by techniques designed to either raise or lower circulating insulin levels while monitoring serum androgens."[38] In cases of polycystic ovarian syndrome weight loss can reduce androgens. A therapeutic approach to predisposed women might be weight loss or dietary modification to reduce serum insulin.[38]

Conclusion

Almost half or more of adolescents do not ovulate when menarche begins. In mid- or late puberty the positive estrogen feedback mechanism develops in which a persistent high estrogen output from a dominant ovarian follicle sets off an anterior pituitary LH surge that causes ovulation and progesterone production. Cyclic estrogen-progesterone causes a stable endometrium with regular, predictable menstrual periods. The lack of ovulation in early puberty and menarche places half or more of adolescents at risk for abnormal bleeding due to DUB. The strange thing is not that abnormal DUB occurs, but rather why it does not occur more often.

Since our knowledge of the mechanism of ovulation is still limited, our approach is empiric. In cases of heavy bleeding, blood loss is stopped by "hormonal hemostasis" using large doses of oral contraceptives (or estrogen-progestin compounds), following which cycles are regulated for the next 3 or 4 months. It is expected that, after a time, the mechanism that controls ovulation (presumably maturation of the hypothalamus–anterior pituitary) will be capable of an LH surge. The adolescent is different than the perimenopausal woman, in whom a dilation and curettage or endometrial sampling is done at the start. Other possible causes of abnormal bleeding in the adolescent must be investigated.

Some adolescents never ovulate, perhaps because of failure in hypothalamic–anterior pituitary maturation or because of a peripheral modifying factor. This may happen in about 5% of all DUB patients or about 25 to 40% of those with severe abnormal DUB. Chronic anovulators, if untreated, may go on to have continous problems of DUB, oligomenorrhea, infertility, PCOD, and possibly endometrial hyperplasia and even endometrial adenocarcinoma in their 30s.

References

1. Altchek A: Dysfunctional uterine bleeding in adolescence. Clin Obstet Gynecol 1970; 20:633–650.
2. Dysfunctional Uterine Bleeding, ACOG Technical Bulletin No 134. Washington, DC, The American College of Obstetricians and Gynecologists, October 1989.
3. Speroff L, Glass RH, Kase NG: Neuroendocrinology, in Clinical Gynecologic Endocrinology and Infertility. Baltimore, Williams & Wilkins, 1989, p 51.
4. Speroff L, Glass RH, Kase NG: Abnormal puberty and growth problems, in Clinical Gynecologic Endocrinology and Infertility. Baltimore, Williams & Wilkins, 1989, p 409.
5. The adolescent obstetric-gynecologic patient, ACOG Technical Bulletin No 94. Washington, DC, The American College of Obstetricians and Gynecologists, July 1986.
6. Vollman RF: The Menstrual Cycle. Philadelphia, WB Saunders, 1977.
7. Apter D, Viinikka L, Vihko R: Hormonal patterns of adolescent menstrual cycles. J Clin Endocrinol Metab 1978;47:944–954.
8. McDonough PG, Gantt, P: Dysfunctional bleeding in the adolescent, in Barwin BN, Belisle S (eds): Adolescent Gynecology and Sexuality. New York, Masson Publishing, 1982, pp 59–78.
9. Speroff L, Glass RH, Kase NG: Regulation of the menstrual cycle, in Clinical Gynecologic Endocrinology and Infertility. Baltimore, Williams & Wilkins, 1989, p 91.
10. Speroff L, Glass RH, Kase NG: Dysfunctional uterine bleeding, in Clinical Gynecologic Endocrinology and Infertility. Baltimore, Williams & Wilkins, 1989, p 265.
11. Van Look PF, Hunter WM, Fraser IS, Baird DT: Impaired estrogen-induced luteinizing hormone release in young women with anovulatory dysfunctional uterine bleeding. J Clin Endocrinol Metab 1978;46:816–823.
12. Emans SJH, Goldstein DP: Dysfunctional uterine bleeding: polymenorrhea, Hypermenorrhea, in Pediatric and Adolescent Gynecology, ed 3. Boston, Little, Brown and Company, 1990, pp 221–242.
13. Faiman C, Winter JSD, Reyes FI: Patterns of gonadotropins and gonadal steroids throughout life. Clin Obstet Gynaecol 1976;3:467–483.
14. Mishell DR Jr, Fisher HW, Haynes PJ, et al: Menorrhagia. A symposium. J Reprod Med 1984;29:763–782.
15. Widholm O, Kantero RL: Menstrual patterns of adolescent girls according to chronological ages. Acta Obstet Gynecol Scand [Suppl] 1971;14:19–29.
16. Claessens EA, Cowell CA: Dysfunctional uterine bleeding in the adolescent. Pediatr Clin North Am 1981;28:369–378.
17. Bertolone SJ: Anemia and coagulation defects in adolescents. Semin Reprod Endocrinol 1988;6:35–43.
18. Hoots WK, Huntington D, Devine D, et al: Aggressive combination therapy in the successful management of life threatening intracranial hemorrhage in a patient with idiopathic thrombocytopenic purpura. Am J Pediatr Hematol Oncol 1986;8:225–230.
19. Walker RW Jr, Gustavson LP: Platelet storage pool disease in women. J Adolesc Health Care 1983;3:264–270.
20. Tudiver F: Dysfunctional uterine bleeding and prior life stress. J Fam Pract 1983; 17:999–1003.
21. Huffman JW, Dewhurst CJ, Capraro VJ: Abnormal genital bleeding in childhood and adolescence, in The Gynecology of Childhood and Adolescence, ed 2. Philadelphia, WB Saunders Company, 1981, pp 453–468.
22. Goldstein DP: Pediatric and adolescent gynecology, in Ryan KJ, Berkowitz R, Barbieri RL (eds): Kistner's Gynecology, Principles and Practice. Chicago, Year Book Medical Publishers, 1990, pp 610–670.
23. Dockeray CJ, Sheppard BL, Bonnar J: Comparison between mefenamic acid and danazol in the treatment of established menorrhagia. Br J Obstet Gynaecol 1989;96:840–844.
24. Jakubowicz DL, Wood C: The use of the prostaglandin synthetase inhibitor mefenamic acid in the treatment of menorrhagia. Aust NZ J Obstet Gynaecol 1978;18:135–138.
25. Fraser IS, McCarron G, Markham R, et al: Long-term treatment of menorrhagia with mefenamic acid. Obstet Gynecol 1983; 61:109–112.
26. Southam AL, Richart RM: The prognosis for adolescents with menstrual abnormalities. Am J Obstet Gynecol 1966;94:637–645.
27. Polson DW, Adams J, Wadsworth J, Franks S: Polycystic ovaries—a common finding in normal women. Lancet 1988;1:870–872.

28. Stein IF, Leventhal ML: Amenorrhea associated with bilateral polycystic ovaries. Am J Obstet Gynecol 1935;29:181–191.
29. Speroff L, Glass RH, Kase NG: Anovulation, in Clinical Gynecologic Endocrinology and Infertility. Baltimore, Williams & Wilkins, 1989, pp 213–232.
30. Lunde O, Magnus P, Sandvik L, Hglo S: Familial clustering in the polycystic ovarian syndrome. Gynecol Obstet Invest 1989;28:23–30.
31. Evans DJ, Barth JH, Burke CW: Body fat topography in women with androgen excess. Int J Obes 1988;12:157–162.
32. Barnes RB, Rosenfield RL, Burstein S, Ehrmann DA: Pituitary-ovarian responses to Nafarelin testing in the polycystic ovary syndrome. N Engl J Med 1989;320:559–565.
33. Moll GW Jr, Rosenfield RL: Plasma free testosterone in the diagnosis of adolescent polycystic ovary syndrome. J Pediatr 1983; 102:461–464.
34. Zumoff B, Freeman R, Coupney S, et al: A chronobiologic abnormality in luteinizing hormone secretion in teenage girls with the polycystic-ovary syndrome. N Engl J Med 1983;309:1206–1209.
35. Futterweit W: Hypothesis and summary, in Polycystic Ovarian Disease. New York, Springer-Verlag, 1984, p 163.
36. Dunaif A, Segal KR, Futterweit W, Dobrjansky A: Profound peripheral insulin resistance, independent of obesity, in polycystic ovary syndrome. Diabetes 1989;38:1165–1174.
37. Dunaif A, Graf M: Insulin administration alters gonadal steroid metabolism independent of changes in gonadotropin secretion in insulin-resistant women with the polycystic ovary syndrome. J Clin Invest 1989;83:23–29.
38. Nestler JE, Clore JN, Blackard WG: The central role of obesity (hyperinsulinemia) in the pathogenesis of the polycystic ovary syndrome. Am J Obstet Gynecol 1989;161:1095–1097.

22

Management of Adult and Perimenopausal Dysfunctional Uterine Bleeding and Current Use of Hormonal Replacement Therapy

HARRY H. HATASAKA AND LEON SPEROFF

In this chapter some of the controversies encompassing the clinical management of dysfunctional uterine bleeding and hormonal replacement therapy are confronted. An aggressive hormonal approach is promoted for both circumstances in general, yet the continued need for individualized assessment is emphasized.

Adult Dysfunctional Uterine Bleeding

Diagnosis

Not all abnormal uterine bleeding is "dysfunctional" uterine bleeding. Dysfunctional uterine bleeding (DUB) is defined as abnormal uterine bleeding with no identifiable organic cause. The presumptive diagnosis is made only after excluding pregnancy complications, pelvic infections and neoplasms, coagulopathies, trauma, medication-related bleeding, and systemic illness as sources of the abnormal bleeding. It is a relatively common condition comprising up to 50% of women with excessive menstrual blood loss.[1] The bleeding is often painless and unpredictable, ranging from long periods of amenorrhea to prolonged heavy flow. Dysfunctional uterine bleeding is classically attributed to anovulation and is more common at the extreme ages of menstrual function. It is generally estimated that 20% of patients with DUB are adolescents and about 50% are perimenopausal. The bleeding complications of anovulatory cycles can be rationally managed medically without resorting to the operating room. This time-tested approach relies on the known histologic and vascular principles of the endometrium.

Pathophysiology

Precise timing and quantities of estrogen and progesterone release conduct the recycling endometrium through its five phases: menstrual endometrium, proliferative phase, secretory phase, preparation for implantation, and the phase of endometrial breakdown.[2,3] The mitogenic effects of estrogen in the proliferative phase serve to progressively restore the structure and height of the thin menstrual endometrium. If left unchecked in this state of unopposed estrogen, asynchronous overgrowth of the endometrium leads to periods of amenorrhea followed by the excessive bleeding characteristic of estrogen breakthrough bleeding. It is only if ovulation occurs with its consequent corpus luteum formation and progesterone release that the endometrium can be stabilized. Despite the continued availability of estrogen, progesterone attenuates the effect of the estrogen by decreasing estrogen receptor replenishment[4] and converting potent estradiol to the less active estrone.[5] This results in a constant endometrial height during the secretory phase. If the appropriate ratio of estrogen to progesterone is maintained, the secretory phase culminates in an implantation phase that features a compact, structurally sturdy endometrium that is ideally constructed to accept an implantation. Ultimately, in the absence of pregnancy, the fixed life span of the corpus luteum lapses and the falling estrogen and progesterone levels initiate vasospasm of the nourishing spiral arterioles. Endometrial ischemia results with subsequent necrosis and desquamation.

Estrogen and progestin manipulations have predictable effects on endometrial physiology. An understanding of these effects can be used therapeutically to control DUB. The solitary withdrawal of either estrogen or progesterone is enough to initiate the bleeding sequence. An example of estrogen withdrawal bleeding is the flow that occurs after bilateral oophorectomy. Progesterone withdrawal bleeding is typified by the well-known progestin challenge test on an endometrium that has been primed by estrogen. The relative overabundance of either hormone over the other results in estrogen or progestin breakthrough bleeding. A classic example of estrogen breakthrough bleeding is the irregular bleeding of a woman with anovulatory polycystic ovaries who is subjected to unopposed endogenous estrogen stimulation of the endometrium. Progestin breakthrough bleeding is seen occasionally in oral contraceptive users after prolonged usage. The progestin dominance of the pill may lead to a thin, atrophic endometrium that is readily denuded, resulting in spotty breakthrough bleeding. It can be seen how estrogen alone can stop all categories of DUB temporarily by fostering thickening, repair, and synchronization of the endometrium. However, neglecting to stabilize the endometrium with a progestational agent will soon lead to estrogen breakthrough bleeding.

The average menstrual blood loss is about 30 mL; 80 mL is considered abnormal since the probability of iron-deficiency anemia becomes excessive at that level.[6] There are multiple mechanisms leading to the cessation of bleeding. Prolonged vasoconstriction, with vascular stasis and tissue collapse, limits blood loss. Meanwhile, estrogen-induced "healing" begins even before completion of the menses has occurred. Myometrial contraction surrounding uterine vessels does not play a large role in menstrual hemostasis as it does in postpartum hemorrhage.[7]

During the first day of menstruation, fibrin and platelet plug formation is the main hemostatic mechanism. Later, vasocontriction is the most important contributor to limiting blood loss. Vasoconstriction is in large part controlled by the balance of prostaglandins. Prostaglandin (PG) $F_{2\alpha}$ and thromboxane A_2 (TXA_2) cause vasoconstriction and platelet aggregation, whereas these effects are counterbalanced by PGE_2 and prostacyclin (PGI_2). Estrogen controls the synthesis of prostaglandins in the endometrium by inducing prostaglandin synthetase as levels of estrogen drop. The balance of the opposing prostaglandins is under the control of estrogen as well but the mechanism is poorly understood.[8] However, it has been postulated that defects in the normal vascular clotting mechanisms, such as excessive fibrinolysis of the plugs or abnormal ratios of the prostaglandins favoring vasodilatory and antiplatelet aggregation effects, may contribute to DUB.[9] This has led to the rational and successful use of antifibrinolytic agents and prostaglandin synthesis inhibitors to decrease blood loss from DUB.

Such imbalance of the endometrial hemostatic process may help explain in part the sizeable percentage of women diagnosed as having DUB who are ovulatory.[10]

Management

Successful management of adult DUB depends on systematic elimination of extrinsic sources of abnormal uterine bleeding. A thorough history and physical, including a pelvic exam with a Papanicolaou smear, are fundamental. A modicum of laboratory analyses serves to initiate the investigation. Pregnancy testing and hemoglobin assessment are required, but clinical appraisal determines the necessity of other radiologic and serum tests, such as a coagulation profile.

Once the probable diagnosis of DUB is made, a practical approach is employed. The immediate treatment of DUB cannot often await confirmatory tests of the etiology of anovulation. Punctual institution of hormonal therapy most efficiently leads to the eventual resolution of the DUB. In fact, failure of appropriate medical therapy in itself serves as a trial to exclude the diagnosis of DUB. Once the bleeding is controlled, a diligent search for the source of anovulation is in order; treating the source is more satisfactory than prolonged treatment of the symptoms. Prior to instituting hormonal treatment, however, a preliminary endometrial biopsy may be advisable. This serves to help exclude neoplasia, confirm anovulatory endometrium, and specifically elucidate the endocrine state of the endometrium in order to guide rational treatment. One rule of thumb is to biopsy all women with DUB beyond adolescence who have bleeding that is prolonged, frequent, or heavy.

The two main hormonal tools used in the management of DUB are estrogens and progestins. Estrogens come in oral, intramuscular, subcutaneous, intravenous, suppository, and transdermal forms commercially. Progestins come in oral, intramuscular, and suppository preparations. The oral forms are the simplest unless unusual clinical circumstances dictate otherwise. These tools are used with the goal of restoring synchronous endometrial events, vasomotor rhythmicity, and structural stability to the disordered anovulatory endometrium.

The program used must be individualized. Assessment should be made of the volume status and hematocrit along with the rapidity of flow and the age and reliability of the patient. Usually the category of the DUB (estrogen or progestin breakthrough or withdrawal) can be inferred from the patient's history, which helps in the selection of treatment. The options for initial therapy are progestin therapy, combined therapy, or estrogen therapy.

Progestin Therapy

This option takes advantage of the antiestrogen effects of progestins. It is best used in anovulatory women who continue to produce their own endogenous estrogen yet do not have an unstable overgrown endometrium. This situation typically occurs in the 40- to 50-year age range in most women at some time during their climacteric. Younger women who cannot or do not desire to undergo exogenous estrogen therapy for DUB may also be controlled with cyclic progestin therapy. It must be remembered, however, that cyclic progestins are not reliable contraceptives.

There are two clinical presentations that are amenable to progestin therapy. Oligomenorrhea can be managed using 10 mg of daily medroxyprogesterone acetate (MPA) for 10 consecutive days of each month to induce an orderly progestin withdrawal bleed. "Medical curettage" can be incited in a similar manner if dysfunctional menometrorrhagia or polymenorrhea is the presenting scenario. In this case endometrial stabilization is promoted by 10 to 14 days of MPA prior to withdrawal. Cyclicity is then achieved using MPA for 10 days each month.

Combined Therapy

Both estrogen and progestin together are the better choice if heavy endometrial build-up results in excess blood loss. The rapid proliferation of endometrial tissues and stabilization of lysosomal membranes due to the added estrogen promotes quicker control of the bleeding. Furthermore, when the combination is stopped, a more physiologic estrogen and progestin withdrawal bleed ensues. This results in a progressive reduction of endometrial height with each succeeding bleed. Any low-dose oral contraceptive can be employed, starting with a dose of one pill four times a day for 5 to 7 days. Usually the 21 pills of one pack are adequate and convenient. Flow is expected to significantly abate within 24 hours if DUB is the diagnosis. If it does not, dilation and curettage (D&C) is required.

When the regimen successfully arrests flow, a large amount of now stabilized endometrium awaits withdrawal. The patient needs to be prepared to expect a heavy flow 2 to 4 days after taking the last pill. If anemia is present, lesser doses of birth control pills can often be continued in immediate succession to the high stabilizing doses in order to defer the withdrawal bleed. The additional time can be used to correct the anemia and to search for the source of anovulation. After the withdrawal bleed is permitted, three cycles of combined oral contraceptives are used in the usual fashion to further decrease the endometrial height. Therapy can then be stopped and the pattern of bleeding observed. If dysfunctional bleeding resumes, progestin cycling can be implemented after again ruling out pregnancy. If contraception is desired, oral contraceptives can be used.

Estrogen Therapy

Because estrogen will temporarily stop all categories of DUB, it is helpful in situations of heavy, threatening blood loss. Estrogen is a necessary ingredient when the endometrium is insufficient in quantity to respond to the actions of progestins. Such an endometrium can result from prolonged heavy bleeding that denudes the lining down to the basalis. Also, chronic progestin therapy can engender a thin, fragile endometrium by overwhelming the mitogenic actions of endogenous estrogen. Indications for the use of estrogen therapy are:

Bleeding has been prolonged
Biopsy yields minimal tissue
The patient has been on progestin medication
Follow-up is uncertain

The acute bleeding in these circumstances can be slowed by the use of oral conjugated estrogens, from 1.25 mg to as much as 2.5 mg four times daily for 7 to 10 days. Use of 25 mg of intravenous conjugated estrogens every 4 hours up to three doses has also been found to be effective.[11] Estrogens induce mitoses within hours. If bleeding does not stop within 24 hours of intravenous conjugated estrogen administration or after 48 hours of the oral form, then D&C is indicated.

Other Medical Therapies for Dysfunctional Uterine Bleeding

Antiprostaglandins

Prostaglandin synthetase inhibitors can be quite effective in reducing blood loss. A decrease of approximately 30 to 50% is seen in women with menorrhagia.[12,13] Although their precise mechanism of action is unresolved, antiprostaglandins exert their effect at the level of the endometrial vasculature instead of on the hormonally responsive endometrial architecture. Therefore their use is especially beneficial in those DUB who have normal ovulatory endometrium.[14] Some popular regimens are listed in Table 22-1.

In general the prostaglandin synthetase inhibitors should be taken with food to minimize gastrointestinal irritation. The drugs are best started with the onset of menstruation because when taken prior to the onset of menses, a delay in the onset of bleeding with exacerbation of DUB may be seen.[15]

Antifibrinolytic Agents

Also effective in reducing excessive blood loss are the potent inhibitors of fibrinolysis, such as tranexamic acid and ε-aminocaproic acid. They inhibit plasminogen activators, which are essential to the fibrinolytic process and are found in the endometrium in high concentration at the initiation of menses. The antifibrinolytic agents offer a reduction of approximately 50% of menstrual blood loss and are more effective in those women with the greatest pretreatment menorrhagia.[16]

Tranexamic acid is given 1 g every 8 hours during the first 3 days of the menses, when

TABLE 22-1. Prostaglandin synthetase inhibitors in the treatment of menorrhagia

Medication	Dosing	Possible side effects
Ibuprofen (Motrin)	400 mg qid	Gastrointestinal distress, diarrhea, nausea, visual disturbances
Naproxen (Naprosyn)	500 mg stat then 250 mg q6–8 hr	Nausea
Mefenamic acid (Ponstel)	250 mg qid or 500 mg q8 hr	Gastrointestinal distress, headache, rare hematologic toxicity, dizziness
Indomethacin (Indocin)	25 mg tid	Gastrointestinal distress, nausea, fatigue, headache, corneal deposits, dizziness

Adapted from Jones G. S., Jones H. W. Jr.: Novak's Textbook of Gynecology, ed 10. Baltimore, Williams & Wilkins Co, 1981.

90% of blood loss normally occurs. Aminocaproic acid has a regimen that starts with 18 g/day for 3 days and finishes with 12, 9, 6, and 3 g on successive days. Side effects include headaches, dizziness, and gastrointestinal upset, but in general the drugs are well tolerated. The agents have been seldom used after early reports of clotting mishaps tarnished their status. These episodes, however, occurred when used in a few severely compromised patients and their safety has since been substantiated.[16,17] Their expense relegates them to second-line status, but when other modalities fail and surgery must be avoided, the antifibrinolytic agents may be cost effective.[7]

Other Agents

Danazol has been successfully used to treat DUB by exploiting its "pseudomenopause" rationale. It is effective if short-term control of DUB is a priority, but its cost, adverse metabolic effects, and side effects limit its long-term use.[18] A similar role is anticipated for the gonadotropin-releasing hormone (GnRH) agonists.

Ergot alkaloids have been found to be ineffective in mitigating blood loss in the nonpregnant uterus.[16]

Surgical Management

Dilation and curettage has an important role in ruling out organic disease when medical therapy fails to control DUB, but D&C is not first-line therapy and is not often curative. After curettage, blood loss is often temporarily reduced, but DUB is usually reestablished within four cycles depending on the etiology.[16] The time has come for more aggressive use of the hysteroscope. It adds little surgical time, yet provides far more information than blind curettage alone since intrauterine lesions are often missed without direct visualization.[19,20] Better still is the approach of screening those who may require surgery by a slow-instillation hysterosalpingogram or an office hysteroscopy combined with office endometrial sampling in order to more accurately identify those who can benefit from surgery. This will minimize nonessential surgery.[21,22] There are occasional indications for the treatment of DUB by endometrial ablation in women refractory to or unable to undergo more conservative therapy. Nonpathologic endometrium must be documented prior to its ablation using either the resectoscope or the neodymium:yttrium-aluminum-garnet (Nd:YAG) laser.[23–26] Objection has been raised to the nonselective use of endometrial ablation due to the risk of obscuring future endometrial carcinoma.[7] Furthermore, candidates should be aware of the significant incidence of persistent uterine bleeding of approximately 20% after endometrial ablation.[24]

The hysterectomy is reserved for those cases in which no organic pathology can be detected or corrected and in which medical control of the DUB cannot be achieved. Pathologic examination of such uteri most often reveals occult pathology such as intramural leiomyomata or adenomyosis. These findings are especially likely if excessive bleeding results in anemia or pain and when sounding reveals a uterine cavity greater than 8 cm.[27]

Perimenopausal Dysfunctional Uterine Bleeding

Dysfunctional uterine bleeding in the perimenopausal period (ages 40 to 55) of a woman's life presents special predicaments for both the woman and her physician. Endometrial hyperplasia and DUB can actually be considered physiologic since most women experience it several years prior to menopause. Therefore a hysterectomy for every such irregularly bleeding woman would be unimaginable.[28] The difficulty is in establishing when perimenopausal DUB is abnormal enough to merit specific attention.

As the menopause approaches, mean durations of menstrual cycles decrease and the variability of the intermenstrual time intervals increases. The shorter cycles are due in large part to an abbreviated proliferative phase.[29] Follicles produce less estrogen and the corpus luteum produces less progesterone. Finally, the luteal phase also decreases in length. Dysfunctional uterine bleeding can result from the altered, erratic hormone ratios as well as the subsequent anovulation. The bleeding pattern is not a reliable indicator of endometrial status. Some climacteric women may actually have endogenous estrogen excess as a source of their DUB. The four mechanisms that can produce this situation are:

1. Increased precursor androgen (tumors, stress, liver disease)
2. Increased aromatization (obesity, hyperthyroidism, liver disease)
3. Increased direct secretion of estrogen (ovarian tumors)
4. Decreased levels of sex hormone–binding globulin, allowing increased levels of free estrogen (hyperandrogenism, cortisol excess)

Because of the risk for endometrial neoplasia in this circumstance, some physicians advocate yearly endometrial sampling while any irregular bleeding persists, whereas others are satisfied with a wait-and-see approach. In between these extremes, various more selective approaches have been devised. In an attempt to introduce some objectivity into the evaluation, it is possible to follow gonadotropin, progesterone, and estradiol values periodically to assess the relative risk of continued unopposed endogenous estrogen production. However, these measures are not precise enough to gauge the true state of the endometrium while perimenopausal DUB continues.

We prefer a vigilant yet utilitarian approach. After comprehensive evaluation as for adult DUB, all perimenopausal women with dysfunctional uterine bleeding undergo an initial aspiration endometrial biopsy. Endometrial cancer can be expected in approximately 2% of women undergoing biopsies for abnormal bleeding. Endometrial hyperplasia is encountered about 15% of the time, endometrial polyps in 1% of cases, and normal histology half the time.[30] The disposable plastic aspiration curettage devices (Pipelle from Unimar, Wilton, CT; or the Z-Sampler from Zinnanti, Chatsworth, CA) have found utility in this situation. They are as accurate as the Vabra aspirator.[31] This method is used instead of the more risky and costly in-hospital D&C only if the uterus is normal on bimanual exam.[32] The devices are convenient and well tolerated. The uterine cavity can be sampled successfully approximately 90% of the time, often without the need for a stabilizing and sometimes painful cervical tenaculum. A nonsteroidal antiinflammatory agent is sometimes helpful when given at least 20 minutes before the procedure. After-cramps can be expected to diminish within about 10 minutes. When unable to enter the uterine cavity, when the uterus is abnormal to exam, and when medical therapy fails after office aspiration yields normal tissue, a formal hospital D&C procedure is essential.

When proliferative or hyperplastic endometrium that is free of atypia is encountered, cyclic progestin therapy is essential. Ten milligrams of MPA for the first 10 days of each months is well tolerated. Histologic regression of any hyperplasia must be documented by subsequent aspiration biopsy. For maximal diagnostic accuracy, formal D&C must be performed if hyperplasia remains (which it does in 2 to 5% of cases).

When hyperplasia reverses and bleeding becomes cyclic, periodic MPA is used until withdrawal bleeding stops. This event indicates the need for added estrogen. There is no need to defer an estrogen-progestin regimen if the woman begins to experience vasomotor symptoms before the cessation of menses. The symptoms are likely due to a relative decrease in estrogen.

This agenda fosters a smooth transition through the menopause and onto a long-term program of hormone replacement. No time is wasted at a critical period when the effects of estrogen deprivation are most serious. A woman should no longer fear the "change of life" as a time of suffering with an uncertain end point. The timely resolution of her climacteric symptoms allows the perimenopausal woman to retain her sense of self-control and to gain confidence in maintaining long-term interactions with her physician.

Hormonal Replacement Therapy—Current Concepts

It was not long ago that hormone replacement meant that a woman waited until she could no longer stand the distress of her hot flashes and dysphoria before presenting to her doctor, where she was temporarily given estrogen tablets alone until the barrage subsided. To our good fortune, investigation has since revealed previously unknown benefits of estrogen that have engendered a new approach with a brighter outlook. Instead of stopgap measures, sweeping preventive health care is now possible. Early and aggressive hormone replacement therapy (HRT) is now warranted.

In the United States there are over 40 million menopausal women. It is estimated that only 10 to 15% of these women are currently using HRT. This leaves the great majority unprotected from a number of long-term effects of estrogen deprivation.

Physiology of Estrogen Deprivation

The climacteric is a time of declining ovarian function sometimes lasting 20 years. It generally begins at about 40 years of age when ovarian estrogen begins to diminish. Bleeding ceases at the point of time labeled the menopause (range 48 to 55 years, with an average age of 50). Natural menopause is diagnosed retrospectively after 6 to 12 months of amenorrhea in women over the age of 45. The menopause is a time of relief for many women yet a time of depression for others, who regard the event as another reminder of our irrepressible biologic clocks. For all it is a certain indicator of the depletion of a most valuable commodity, estrogen.

Levels of estradiol in postmenopausal women fall to approximately 10 to 20 pg/mL, most of which is derived from peripheral conversion of estrone.[29,33] Estrone in turn is mainly derived from the conversion of androstenedione in the periphery. The more adipose tissue present, the higher will be the level of estrone that is derived from androstenedione. Androstenedione levels decrease to approximately half their premenopausal values as a result of a significant abatement of the ovarian contribution.[34] Declines are also seen in postmenopausal dehydroepiandrosterone (DHEA) and its sulfate (DHEAS), but circulating levels of testosterone tend to stay constant. Although there is less androstenedione being converted to testosterone in the periphery, the ovarian stroma compensates under the influence of higher levels of gonadotropins. Since estrogen shows a more pronounced decline than androgens, the ratio switches toward androgen dominance and facial hirsutism in elderly women often results.

Concurrent with estrogen decline, follicle-stimulating hormone (FSH) levels rise reliably most likely as a result of the diminution of inhibin. Inhibin is the nonsteroidal peptide produced in granulosa cells that normally effects negative feedback on FSH. It is not until the luteinizing hormone (LH) concentration rises to postmenopausal levels (>25 mIU/mL), joining the already high FSH levels (>40 mIU/mL), that a woman can be considered safe from an inadvertent pregnancy.

The decrease in estrogen has far-reaching effects, both acute and long term. Receptors for estradiol have been found in all genitourinary organs as well as in cardiovascular and gastrointestinal tissues. Other target organs include the skin, adrenal, and nerve tissues. Most recently, sensitive receptor assays have identified estradiol receptors on osteoblast-like cells in culture.[35] This adds support to the long-held suspicion of a direct estrogen effect on bone.

Corresponding to these receptor findings at the molecular level, estradiol deficiency produces a host of altered functions in the target tissues. Among these are hot flashes, memory changes, sleep difficulties, urinary incontinence, atrophic conditions, palpitations, atherosclerosis, and osteoporosis. Accordingly, estrogen replacement has potential benefits to the function of many of these end organs.

Benefits of Replacing Hormones

Vasomotor Symptoms

Most gratifying is the rapid relief of distressing hot flashes that usually accompanies hormonal replacement.[36] Some 85% of postmenopausal women will experience vasomotor flushing, which sometimes begins with a prodrome followed by reddening of the skin of the upper body. This coincides with a feeling of intense body heat that is not only measurable but can often be sensed by nearby friends and family. The flush concludes by perspiration and often a feeling of being chilled. Flushes usually last from seconds to several minutes but, rarely, can go on for an hour. A surge of LH but not FSH correlates with flushes,[37] but LH is not the cause of flushes since hypophysectomized women can still experience vasomotor flushes. The mechanism is poorly understood but appears to result from instability between the hypothalamus and the autonomic nervous system. Hot flashes may be as frequent as every 10 minutes or they may rarely occur. Within individuals they occur more frequently at night, when they can awaken some women and perhaps disrupt the quality of sleep of others. The addition of estrogen increases the amount of quality sleep stages.[38]

For most women, vasomotor flushing lasts for 1 to 2 years during the climacteric, but as many as 25 to 50% experience them longer than 5 years.

Psychological Well-Being

During the early menopause, women often report symptoms such as depression, fatigue, irritability, insomnia, forgetfulness, dizzy spells, headaches, joint and muscle pain, palpitations, and decreased libido. Reminiscent of "premenstrual tension syndrome," this symptom complex has been termed the "menopausal syndrome."

Study of the therapeutic effects of estrogen on these symptoms is confounded by subjectivity and a high rate of placebo responses. Campbell and Whitehead approached this problem by conducting a careful prospective placebo-controlled and doubly-blinded evaluation of the effect of estrogen on these somatic and psychological symptoms.[39] Most notable was a "domino" effect whereby relief of vasomotor symptoms initiates amelioration of a number of other menopausal symptoms. Long-term findings in this study revealed marked placebo effects of estrogen on sexuality, irritability, worry, hot flashes, and skin appearance, but also statistically significant improvement over placebo was noted in the relief of hot flushes, insomnia, short-term memory decrease, urinary frequency, and vaginal dryness. Jensen and Christiansen also noted significant dose-related reductions in menopausal symptoms by estrogen over placebo.[40]

Genitourinary Atrophic Conditions

The lower genitourinary tissues are replete with estrogen receptors and are quite estrogen dependent.[41] In the absence of estrogen, troublesome problems may surface as a result of tissue atrophy, including symptomatic vulvovaginal inflammation, shrinkage, and dryness with consequent dyspareunia. Also, urinary conditions may develop, including urgency and incontinence along with abacterial urethritis and cystitis. An increase in vaginal pH contributes to vaginitis.

Equally important, the supporting tissues and ligaments of the lower genital tract lose tensile strength and may add to pelvic relaxation. Lower systemic levels of estrogen than those required to prevent hot flashes usually suffice in preventing and reversing atrophic changes when given by any route.[42,43] It must be realized that if vaginal estrogen cream is chosen to treat atrophy and to assist in lubrication, significant systemic levels must be assumed because the transvaginal absorption of estrogen is substantial.[44]

Whereas genuine stress incontinence is not cured by estrogen replacement,[45] the hormone can improve urge incontinence and contributes to urethral epithelial thickening and to the maintenance of periurethral striated muscle tone.[46] Estrogen therapy can help greater than 50% of postmenopausal women with urinary incontinence.

With treatment, immediate resolution of most symptoms of atrophy should not be expected. Maximal restoration of genitourinary atrophy may take 6 to 12 months.

Nervous System

It is not generally appreciated that estrogen plays an important role in the nervous system. Estradiol receptors are located in portions of both the peripheral and central nervous systems. Estrogen plays a role in regulating both electrical effects directly and neurotransmitter release.[47] Besides vasomotor manifestations, other neurologic sequelae include balance and memory loss, numbness, and altered sensory perception. Improvement is noted in each of these with estrogen therapy.

Sexual Function

A sensitive topic that many (especially older) women are hesitant to broach with their physicians is that of altered sexual function. Yet most all climacteric women at some time will experience some difficulty with problems including dyspareunia, decreased libido, decreased ability to respond to sexual stimulation and dysfunctional interactions with their partners.[48] Many complex components, both psychological and physical, establish healthy sexual function, and estrogen has profound effects on many of these. Mood, sensory perception, and vaginal lubrication and distensibility among other factors can all be influenced by estrogen. It is likely that here too there is a "domino effect" of general heightened enjoyment of sexual function once dyspareunia is corrected by appropriate doses of estrogen.[49,50] Sexual activity itself enhances the beneficial support of estrogen by maintaining the circulation to vaginal tissues.

Musculoskeletal System—Osteoporosis

By now a common theme can be appreciated: The influences of estrogen are ubiquitous. The musculoskeletal system is no exception; all of its components are affected by estrogens.

Joints

When employed prior to the onset of joint disease, data demonstrate some protection against rheumatoid arthritis by estrogen replacement therapy.[51]

Muscle

A steady reduction in muscular strength with aging has been observed, but women on estrogen do not show this reduction.[52] This benefit contributes to a positive relationship with the other major components of the musculoskeletal system, the bones. Increased physical activity allowed by more muscular strength helps maintain bone quantity.

Bone

The loss of bone mass quantity, rather than a change in composition, together with evidence of fracture, is termed "osteoporosis." Because it results in significant morbidity and mortality, it has achieved due notoriety as a major health risk in mature women. Roughly 15 to 20 million people are afflicted with osteoporosis in the United States, accounting for an impressive 1.3 million new fractures every year.[53] As the evidence solidified that estrogen is indeed an effective treatment for osteoporosis, this became the impetus for the recommendation of long-term estrogen treatment for the prevention of osteoporosis, rather than short-term symptomatic therapy.[54]

Evidence is strong from several lines of investigation that lack of estrogen plays a major role in the pathogenesis of osteoporosis. Bone is constantly being remodeled under the influence of the bone-forming osteoblasts and the bone-resorbing osteoclasts. Aging after about age 30 imparts a small imbalance that results in a gradual loss of bone (type II osteoporosis). However, with removal of estrogen, as, for example, at oophorectomy in premenopausal women, or by any other mechanism, the bone mineral density quickly falls.[55–57] This is known as estrogen-dependent, postmenopausal, or type I osteoporosis. It affects mainly the trabecular bone of the wrist and spine. At an average of 12 years after menopause, the bone mineral content drops below a critical fracture threshold.[58] After years of cumulative bone loss, the lifetime risk for developing a hip fracture is 15%. The risk for both spinal and wrist fractures is also 15% each by the age of 70 in white women.[59]

The protective effect on bone loss provided by estrogen is not hampered by the addition of progestins to the regimen.[60] Because of the antiestrogenic effects of progestins in general, this was an unexpected but welcome finding. Progestins may in fact confer some independent protection against osteoporosis.[61–63]

The exact mechanism of action for the protection of bone by the sex steroids in not yet known. It is possible that the increased calcium absorption induced by estrogen and the still-to-be-determined role of osteoblast estrogen receptors are involved in the process. Roles for the calcium regulating hormones are being clarified. A role for calcitonin in the pathogenesis of osteoporosis has not been identified.[64,65] However, it is noted that the sensitivity of osteoclasts to parathyroid hormone is reduced by estrogen,[66] and recent data suggest that estrogen replacement can decrease the set point of parathyroid hormone stimulation by calcium, resulting in less potential bone resorption.[67]

The ultimate proof of the effectiveness of estrogen in preventing osteoporosis is the reduction of fractures. There is a 50 to 60% decrease in arm and hip fractures with estrogen replacement.[53,68,69] Vertebral compression fractures can also be reduced up to 80% when the estrogen is supplemented with calcium.[70]

Taken alone, even high doses of calcium are not able to prevent osteoporosis.[71–73] Yet calcium is important as a prerequisite for bone maintenance. Supplementation of calcium is needed because postmenopausal women have impaired calcium absorption.[74] The daily calcium requirement averages approximately 1,500 mg.[75] Most women take in 500 mg of calcium in their diets each day, leading to the recommendation for 1,000 mg of calcium supplementation. However, such a dose is often not well tolerated because of flatulence and constipation. Fortunately, estrogen replacement aids calcium absorption significantly such that a woman on estrogen only needs to supplement 500 mg of elemental calcium each day. This dose is tolerated much better. Conversely, high doses of calcium can enhance estrogen's effectiveness such that lower doses of estrogen suffice if at least 1,500 mg of calcium are taken daily.

Other therapies to reduce fracture rates, including the addition of vitamin D, have not proven efficacious. Treatment with the bone formation stimulator sodium fluoride effectively increases bone mass, but produces bone that is fragile.[76] Other effective therapies that are not currently approved by the Food and Drug Administration (FDA) include low-dose parathyroid hormone to stimulate renal reabsorption and gut calcium absorption, and the active metabolite of vitamin D (1,25-dihydroxycholecalciferol) which also increases calcium absorption. Etidronate, a compound that inhibits osteoclast-mediated bone resorption, also has shown value in reducing new vetebral fractures in women with established osteoporosis.[77] "Coherence therapy" has shown initial promise.[78,79] This is an attempt to synchronize all bone-forming units using a variety of compounds, and then inhibit resorption such that a net bone gain prevails. Another promising approach is the use of calcitonin, which in one recent study proved as effective as estrogen in preventing postmenopausal bone decreases by inhibiting osteoclast activity.[80] More experience is needed with all of these therapies, and estrogen plus calcium remains the main osteoporosis prevention regimen.

Aggressive therapy is essential because once bone is lost, only a minimal amount can be regained with estrogen and calcium alone. Prevention of bone loss is the best strategy. It should be started early; otherwise the phase of accelerated bone loss that occurs soon after menopause will have an irrevocable effect. A serum estradiol level of about 50 pg/mL must be maintained for 6 to 24 months for bone formation to reach a steady state with bone resorption.

Other important features of a complete osteoporosis prevention schedule include consideration of risk factors (Table 22-2). Treatable factors should be searched for and corrected. Exercise is encouraged; as little as 30 minutes per day three times a week will increase bone mineral content in mature women.[81] An increased risk for osteoporosis is yet another reason to dissuade cigarette smoking. It induces conversion of estradiol to less effective metabolites[82] and increases hepatic metabolism of estrogen, resulting in lowered serum levels.[83] Inhibition by smoking of granulosa cell aromatase activity may help explain the earlier onset of menopause noted in smokers.[84]

TABLE 22-2. Risk factors for osteoporosis

Constitutional factors
Genetic (caucasians and orientals)
Low constitutional peak bone mass
Advancing age
Low body fat
Exogenous factors
Inadequate exercise
Smoking
Estrogen deficiency
Alcoholism
Altered calcium balance
Endocrinopathies—hyperparathyroidism, hyperthyroidism, hyperprolactinemia, acromegaly, Cushing's disease, diabetes, and hypogonadism

In the big picture of osteoporosis, the overall risk depends on the peak bone mass achieved by the age of about 35, and on the subsequent rate of bone elimination. Although estrogen slows the rate of bone loss, it must be continued indefinitely or a new accelerated phase of bone loss begins immediately.[85,86]

If a late postmenopausal woman has not yet been on hormonal replacement therapy in the past, there is nevertheless evidence of some benefit from starting late.[87] Estrogen used between the ages of 65 and 74 has been shown to reduce hip fractures.[70] When significant osteoporosis is already established, it is justified to vigorously treat with agents in addition to HRT and calcium in order to maximally prevent further loss. This may be an indication for the use of the active metabolite of vitamin D, calcitonin, or both together.

Cardiovascular Disease

One of the most exciting prospects for the use of HRT is its vast potential to significantly reduce morbidity and mortality from cardiovascular disease. Although cardiovascular disease has commonly been considered to predominate in males, the rise in rate of arteriosclerotic disease with age in women parallels that in males. The difference is that the initiation of the rise in rate in females starts about 10 to 12 years after that in the male.[88] Significant increases in the risk of coronary heart disease also do not become apparent until some 10 to 12 years after natural menopause.[89] The long-term toll of the elevated risk is readily apparent. Cardiovascular disease is now the number one cause of death in women in the United States. Ten times the number of women will die of ischemic heart disease alone than will succumb to osteoporotic fractures. Two out of three women will die from some facet of cardiovascular disease. Therefore, if the apparent protective effect of estrogen therapy holds true, this will mandate the continued aggressive use of replacement therapy.

The evidence to date convincingly supports the decreased risk of cardiovascular disease by estrogen users measured by a variety of end points. Of the nine case-control studies in the literature, only one small study did not support a beneficial effect of estrogen on cardiovascular disease. The statistical power of that study is low since it included only 17 cases, all of whom developed disease under the age of 50 and most of whom were smokers.[90–98]

Equally persuasive is the picture from the nine prospective cohort studies that have been published.[99–107] Only three did not clearly show the protective effects of estrogen against cardiovascular sequelae. One of these, the Walnut Creek Study,[99] also had the problem of low numbers. Twenty-six women from that cohort experienced infarctions, but only nine of them were estrogen users. The influential ongoing Framingham Study contended that there was a 50% increased risk for cardiovascular morbidity (not mortality).[100] However, responding to criticism, the authors reanalyzed their data and concluded that the data do not show a serious adverse effect of estrogen use on cardiovascular disease. Criticisms centered around the low number of subjects in each subcategory such that conclusions could not be justified. Also, dosages and durations of estrogen treatment were not recorded.

Of the remaining cohort studies that did find benefit from estrogen replacement, three are most regarded because of their sizes. The Lipid Research Clinics Follow-Up study was a multicenter investigation started in 1972. A cohort of 2,270 women was enrolled, which is over twice the size of the Framingham cohort. After 8.5 years of follow-up there was a resounding 63% reduction in the risk of death from cardiovascular disease in current estrogen users.[102]

The Leisure World Study was started in 1981 with 8,841 registrants as a longitudinal, prospective cohort study using the residents of a retirement community in Southern California. This is an affluent and stable population yielding relatively controlled study conditions. The latest report from the study in 1988 continued to show a substantially reduced relative risk of death from acute myocardial infarction in estrogen users of 0.59, and an overall relative risk of death from all causes for estrogen users of 0.80 compared to nonusers.[103]

The other large study is the Nurses' Health study, in which 121,964 nurses were followed prospectively. After adjustment for age, current users showed 70% less coronary disease than nonusers.[104]

Analysis of all the studies together has yielded overwhelming support for a protective benefit of estrogen against cardiovascular diseases.[108] Women so treated appear to have only one third to one half the risk of stroke or heart attack compared to untreated women.[89,102]

The mechanism of action of the reduction of cardiovascular compromise, including stroke, is being studied intently. Alterations of the lipid profiles contribute approximately 50% to the protective effect but are not the whole story.[102] As would be expected, estrogens increase high-density lipoproteins (HDLs) and very-low-density lipoproteins (VLDLs). There is a smaller reduction in low-density lipoproteins (LDLs) and a small increase in triglycerides. The increase in HDLs correlates with a decreased coronary mortality as does the reduction in LDLs, which is the most atherogenic fraction. The increase in triglycerides has been found to be of little consequence except as a risk factor in diabetics and in individuals with genetic defects in triglyceride metabolism.

Low-density lipoprotein particles serve to transport cholesterol from the liver to the peripheral cells. The cholesterol then enters cells under the control of LDL receptors. In animal studies, estrogen acts to increase these receptors in the liver, which results in increased degradation of the atherogenic LDL particles.

High-density lipoprotein particles slow the atherosclerotic progression by "scavenging" cholesterol from the peripheral cells and transporting it back to the liver for catabolism by hepatic lipase.[109] Estrogen decreases the hepatic lipase activity but androgens and progestins increase the enzyme's activity, leading to lower levels of the beneficial HDLs. The positive effects on the lipid profile induced by estrogen continue as long as the therapy is continued. So far this has been demonstrated for at least 10 years of use.[110]

Part of the favorable effect of estrogen on arteriosclerotic sequelae is independent of the lipid profile. Monkeys given enough progestin to lower their HDL levels (and fed a high-cholesterol diet) nevertheless had a decreased amount of coronary atherosclerosis when given estrogen.[111] The mechanism of this effect is as yet unsettled. It could be that even women with completely normal cholesterol profiles might enjoy additional protection from cardiovascular disease if administered estrogen replacement.

With the benefits of estrogen on cardiovascular risk showing so much promise, attention must be turned to the legitimate question of whether the addition of progestins diminishes the gains. Although preliminary, progestins added to combination replacement therapy have not been shown to influence blood pressure.[112] Careful monitoring of blood pressures, however, would be prudent until more data accumulate.

Progestins in general could possibly have adverse effects on cardiovascular risk by decreasing HDL levels. However, there seems to be both a structural and a dose-response relationship to this effect.[113–120] Therapeutic progestins are derived from one of two structural types. The 19-nortestosterone group includes norethindrone, levonorgestrel, desogestrel, and ethynodiol diacetate. The other group, the C_{21} progestins, include chlormadinone, megestrol acetate, and the commonly used MPA. The C_{21} compounds can lower HDL levels, but when used in reduced, therapeutic doses they appear to lower HDL less than the 19-nortestosterone agents.[121] Although far from conclusive, the majority of studies on the subject have failed to demonstrate an adverse lipid effect by sequential combination therapy using usual doses of C_{21} progestins.[115–117] This was the case in one study after 3 years of sequential MPA administration using the often-prescribed dose of 10 mg/day for 10 days.[122]

The important work remaining to conclude this story will include the determination of the optimal doses of progestational agents that can confer endometrial and perhaps breast protection without undermining the beneficial effects of estrogen during long-term replacement therapy.

Potential Risks of Hormonal Replacement Therapy

Perhaps based on the highly publicized negative experiences with high-dose oral contraceptives in the 1970s, many physicians are reluctant to prescribe estrogens. Fear of cancer and an inordinate apprehension about potential side effects are the most common reasons for the unwillingness of a large number of physicians to advocate long-term replacement therapy.[123] Review of the potential risks of HRT is in general reassuring because most adverse effects are rare and their dose-response relationships allow some control over them (Table 22-3). Some of these factors need to be considered.

Endometrial Cancer

Unopposed estrogen administration increases the risk of cancer of the endometrium some four to eight times above the baseline risk.[124,125] The risk is increased by higher doses of estrogen and longer periods of exposure. The risk is known to linger even after the estrogen has been discontinued.[124] Ample clinical documentation, including a huge and convincing recent prospective study,[126–129] is available that progestins can keep this risk to a minimum. Part of the mechanism is by depletion of nuclear estrogen receptor replenishment. In women previously treated for stage I adenocarcinoma of the uterus of any grade, no risk of recurrence was noted with estrogen replacement despite the estrogen dependence of the tumor.[130] It seems logical that a progestational agent added to the estrogen would be even more likely to antagonize recurrence. Neither was increased recurrence found in a group of posttreatment stage I and II cervical cancer patients who were treated for 5 years with estrogen replacement.[131]

TABLE 22-3. Potential side effects and risks of hormonal replacement therapy

Estrogen therapy
Endometrial carcinoma
Thromboembolic disease
Hypertension
Gallstones
Migraine headaches, depression
Breast cancer (?), mastodynia
Gastrointestinal symptoms
Resumption of vaginal bleeding
Weight change
Skin eruptions, scalp hair loss
Potential exacerbation of leiomyomata or endometriosis
Progestin therapy
Adverse effect on cardiovascular lipid profiles
Acne
Skin reactions
Mastodynia
Weight change
Depression

Breast Cancer

Because of the scope and serious nature of the problem, whether or not estrogen replacement therapy contributes to breast carcinoma is of major importance. Difficulties investigating this concern arise as a result of the apparent long latency. Studies must also control for the fact that most breast cancers do not manifest until after the menopause, and that the rise in incidence is progressive throughout life. Concern about the use of exogenous estrogen is justified because of the epidemiologic, endocrine, and molecular evidence of an estrogen dependence of breast cancer cells.[132]

It is difficult to extrapolate the data regarding the use of oral contraceptives and the risk for breast cancer because of the different doses given to younger women at disparate levels of risk. However, the data regarding replacement estrogen use and breast cancer are persuasive. One prospective and three case-control studies have been reported so far with enough numbers to investigate high-risk subgroups. Not one of the studies has recorded an increased risk of breast cancer with even high doses of estrogen replacement.[133–135] The prospective study showed no increased risk for breast cancer after 5 years of estrogen replacement.[135] It is reassuring to see no acceleration in the subclinical breast cancers that might be present while estrogen replacement is being administered.

Neither have progestins been shown to increase the risk of breast cancer in humans. In fact, there is encouraging evidence that progestins may even have a protective effect against breast cancer.[136,137] It may be that progestational agents cause estrogen receptor depletion in the breast cells just as they do in the endometrium.

Recently, a Swedish study published in the *New England Journal of Medicine*[138] prompted much public anxiety because of its conclusion that estradiol (but not conjugated estrogens) used long-term for replacement therapy may increase the risk of breast cancer. The study also suggested that added progestin may further increase the risk. The later inference, however, was based on only ten patients and the credibility of the former conclusion was roundly criticized for a variety of reasons. In particular, those women in the study who had breast cancer and were prescribed, yet never took replacement estrogen had a relative risk of 1.6 when compared to women who were never prescribed estrogen therapy. This relative risk was not different than that of the remaining 151 women in the cohort who took estrogen and later developed breast cancer. This suggests a significant selection bias and renders the conclusions unreliable.

Overall, the available data on breast cancer and hormone replacement do not justify a change in current prescribing habits.

Hypertension

The question arises regarding estrogen's effect on blood pressure. No association has been found between hypertension and the low estrogen doses used for replacement therapy in normo- or hypertensive women. Only on rare occasions does a legitimate idiosyncratic reaction to estrogen elevate blood pressure. Studies done so far have shown either no change or a sometimes significant fall in blood pressure with estrogen treatment.[112,139–142] Clinically, the hypertensive patients may be just the ones to actively treat with hormone replacement rather than witholding it for fear of exacerbating hypertension.

Thromboembolism

In doses used for estrogen replacement therapy, there is no evidence for any significant increase in thromboembolism or thrombophlebitis.[143] However, caution is advised in a patient with a history of thromboembolic disease because of the known risks associated with the higher dose oral contraceptives.[144]

Metabolic Complications

As with thrombotic complications, altered carbohydrate metabolism is not seen with the low doses of estrogen used for replacement therapy.[145] Leiomyomata and endometriosis are not contraindications to HRT since regrowth has only rarely been reported for each.

Gallbladder Disease

Oral contraceptives elevate the risk of gallbladder disease to a 1.5 to 2.0 relative risk and may do the same for estrogen replacement users.[146] Other studies do not support the relationship.[147] Even if estrogen was a true contributing variable to gallbladder disease, the contribution would be small. This prompted one group of investigators to conclude that the potential gallbladder risks should not enter into the decision of whether or not to use noncontraceptive estrogen.[148]

Tailoring a Hormone Replacement Program—Recommendations

A successful long-term approach to maximizing the benefits of HRT hinges on compliance. This relies in large measure on the working relationship and confidence a woman has with her physician. Health management in the menopause is a relatively new event for our society. The life span for women in the United States has doubled since the turn of the century, leaving women to wonder whether or not menopause is a medical condition or a natural phase that should be handled without special intervention. It pays to take the time to assess each woman's knowledge, fears, and attitude concerning the menopause and replacement therapy. The more educated a woman is about the risk-benefit ratios of taking estrogen replacement, the more likely she is to accept the support that the hormone offers.[149]

Contraindications

Absolute contraindications to estrogen replacement therapy include acute and chronic liver disease, pregnancy, undiagnosed abnormal genital bleeding, acute vascular thrombosis, neuroophthalmologic vascular disease, and untreated estrogen-dependent neoplasms. Judgment must be employed in some women with familial hyperlipidemias, migraine headaches, and seizure disorders as well.

Chronology of Estrogen Deprivation

The age of the woman being considered for HRT is important because it may affect the method of treatment. Almost nobody disagrees that replacement therapy is required for the woman less than 40 with gonadal dysgenesis and surgical castration. Her risk of cardiovascular disease is accelerated and because there is potentially a longer time for estrogen deprivation, her lifetime risk for osteoporotic fractures increases.[150] Probably younger patients treated with prolactin-elevating psychotropic medications also fall into this category, since their resultant amenorrhea indicates long-term estrogen deficiency. Other examples of these young estrogen-deficient women are amenorrheic athletes and women with anorexia nervosa.[151] In each case their lack of estrogen is truly pathologic. A sequential approach usually makes the most sense in these young women because monthly bleeding is often desired for psychological reasons. Frequently these young women need a higher dose than the usual 0.625 mg of conjugated estrogens, so the dose is titrated up until withdrawal bleeding is achieved. If oophorectomies have been performed because of endometriosis, recurrence is only rarely a problem on estrogen replacement. However, because carcinoma has been reported to develop from endometriosis implants in patients on unopposed estrogen,[152] this is an argument for adding a progestin to the regimen of such a patient even if a hysterectomy has been performed.

The most typical situation is the woman who has experienced perimenopausal DUB and has been placed on periodic progestins after the appropriate evaluation and biopsy. The logical transition is to begin early, aggressive estrogen replacement in addition to progestin therapy once withdrawal bleeding ceases or when vasomotor symptoms develop even as cyclic bleeding continues.

Another scenario that is still common is the untreated late postmenopausal woman with osteoporosis due to estrogen deprivation and aging. Estrogen-progestin-calcium administration is still beneficial in this older group to slow further bone loss and consequent fracture risk.[87] This is an indication for quantitative assessment of bone mass using either computed tomographic scanning or the dual-photon absorptiometry techniques for baseline and progress assessments. Older women with osteoporosis may require higher doses (1.25 mg conjugated estrogens or equivalent) to slow the progression of bone loss.

Choosing Hormones and Doses for Replacement Therapy

The most critical considerations in choosing an estrogen are the dose, duration, method and route of administration, and presence or absence of a concomitant progestin. One specific estrogen type offers no known clinical advantage over another. Estrogens can be classified into two broad categories, natural or synthetic. The estrogens considered "natural" (17β-estradiol, conjugated equine estrogens, piperazine estrone sulfate) metabolize to the same plasma products as those produced by the ovary. The synthetics (e.g., ethinyl estradiol, mestranol, and diethylstilbestrol) have structural changes sufficient to prevent their ready metabolism, making them relatively more potent. Conjugated equine estrogens are a mixture of estrone sulfate and equine estrogens similar in structure to 17β-estradiol. They are the most characterized form of commercially available estrogen and in general (depending on which estrogen effect is being measured) are somewhat more potent than the other natural estrogens.

Concerning the important end point of bone density, 0.625 mg of daily conjugated estrogens is sufficient to maintain bone mass.[153,154] This is equivalent to 1.0 mg of micronized estradiol. The equivalent dose of ethinyl estradiol has not been established for bone maintenance but is probably about 5 to 10 μg.

Route of Administration

Estrogens are usually given orally in the United States for reasons of convenience and dependability of effect. A variety of other methods have been used in an attempt to increase convenience and compliance, such as subcutaneous estrogen pellets, percutaneous gels, and the now widely available 17β-estradiol transdermal patches. Oral estrogens reliably and quickly affect liver metabolism, causing favorable changes in the lipoprotein levels.[155] They may also, however, have potentially adverse effects on liver metabolism. Increased production of renin substrate yields a theoretic potential for hypertension, and decreased antithrombin III levels could possibly lead to clotting abnormalities.[156] Neither effect has been a clinical problem at replacement doses. Thus, beneficial lipid effects outweigh the risks of oral estrogen use.

Even though transcutaneously delivered estrogens can control vasomotor symptoms and mount significant serum estradiol and estrone levels, they were found to have much less of the beneficial effect on lipid levels than did oral estrogens.[157,158] It was thought that this could be due to a "first-pass" effect of oral estrogens in which high concentrations pass directly from gut to liver through the portal system. In this manner advantageous enzymatic changes could be facilitated, whereas parenteral administration was thought to lack this benefit. However, the early studies of percutaneous administration of estrogen were short-term studies of less than 6 months' duration. It is now known that the maximum serum levels of estradiol and estrone take 6 months to achieve despite the use of high doses of percutaneous estradiol. By 12 months the decrease in LDLs and total cholesterol using percutaneous delivery is of similar magnitude to that induced by oral preparations.[159] It took 25 months to raise the HDL-cholesterol levels using the percutaneous route. It appears that percutaneous estrogen can indeed adequately lower LDL and total cholesterol levels but that the time course is much slower than that produced by the oral route. Furthermore, the dose of percutaneous estradiol used in this study was 3 mg daily as compared to the approximate dose of 0.1 mg delivered daily by commercially available transdermal patches.

Percutaneous estrogen has also been found to protect against osteoporosis. However, for this effect, as for the lipid-lowering benefit, effective doses for long-term prophylaxis have not yet been determined as they have for the oral route. Until this information is known it is best to continue using the proven oral route.

Progestins in the Postmenopausal Woman

Estrogen therapy alone was used until the mid-1970s, when the risk of endometrial neoplasia became apparent.[160] The protective action of MPA 10 mg/day for 10 days was demonstrated early on,[161] and has become standard, but the optimal dose and duration remain controversial. To be conservative, some recommend 12 to 14 days of progestin use each month despite the lack of adequate prospective documentation for the added exposure. Perhaps the most persuasive data come from a London study of 398 women.[162] These women are receiving various estrogens while their progestin exposure varies between 7 and 21 days per month. Each woman has submitted to endometrial curettage under general anesthesia each year. By

1985, after 9 years of therapy, hyperplasia developed in about 2% of those exposed to 7 days of progestin but not one woman using 10 days or more developed hyperplasia! Therefore, 10 days of added progestin each month should be the minimum until more precise data are collected. Although the onset of its protective mechanisms may be more dependent on duration of therapy than on dose,[163] data are again lacking to clarify the minimum daily dose of progestin needed to prevent hyperplasia. If side effects such as fluid retention, mastodynia, or depression develop, a dose of 5 mg of MPA usually corrects the symptoms but at an unknown jeopardy to endometrial protection. Since the standard of 10 mg per day of MPA has proven adequate to date, this dose should be the initial dose when used in a sequential regimen.

At the present state of knowledge, in women without a uterus we continue to add a cyclic progestin because of the potential benefits of reducing breast cancer[136,137] and reducing bone loss.[63] It is possible to monitor the lipid profile but impossible to retract the occurrence of breast cancer.

Methods of Administration

The favored sequence in the United States has been to give estrogen on days 1 through 25 of each month plus progestin starting on days 14 or 16 through day 25, then nothing for the remainder of the month. Because a variety of menopausal symptoms often develop during the 5 to 6 days off hormones, a system most used in Europe is gaining popularity in the United States. The progestins are started on day 1 and continued for 12 to 14 days, while the estrogen is given daily without a break. This is a simpler system to remember. In addition, the time to bleeding is used as a bioassay. If bleeding begins prior to cycle day 10, inadequate secretory change is likely and the dose of progestin is increased.[164]

Up to 90% of women on these methods resume regular bleeding, which is a bothersome drawback for most. Certainly menses are not essential to cancer prevention as long as an atrophic endometrium is maintained. One way to produce atrophic endometrium reliably, and to decrease bleeding, is a natural extension of the European sequential method. When continuous daily low doses of progestins (2.5 to 5 mg MPA, or 0.35 to 0.7 mg norethindrone) are added to the daily estrogen (0.625 mg conjugated estrogens, or 1.0 mg micronized estradiol), bleeding is seen in only a minority of women.[165,166] Since 2.5 mg of MPA daily appears sufficient to protect the endometrium, the monthly MPA intake is 75 mg compared to 100 mg or more in the sequential methods.[163,167] The continuous progestin has not been reported to reverse the estrogen benefit to bone.[168,169] Currently, studies differ on the effect of the regimen on lipid status,[170,171] so annual measurements of the lipid profiles are in order until clinical trials assess the long-term lipoprotein effects. The future may afford a convenient single combined pill to be taken daily.

Follow-up

A women can deal with the initial problems of starting HRT, such as heavy bleeding and breast tenderness, much better if she is forewarned about what she may encounter and informed that her regimen can be "fine tuned." Vaginal cytology and serum hormone levels are less satisfactory than tailoring dosages by symptoms and bleeding. After frequent visits during the period of adjustment, regular follow-up is important for continued compliance and safety.

If calcium supplementation greater than 500 mg/day is used, serum calcium and phosphorous should be measured each of the first 2 years in case the high doses of calcium unmask asymptomatic hyperparathyroidism. If the values are normal, no further monitoring is necessary. Continued annual cervical cytology and mammography are indicated. Blood pressure measurements and a yearly assessment of the serum cholesterol and, if available, HDL and LDL measurements are a good precaution.

It is not cost effective to routinely obtain

bone density measurements in an attempt to identify those at high risk of osteoporosis.[172] Rather, because of the multiple benefits, we offer HRT to almost all postmenopausal women, including those who are asymptomatic. If a woman needs to know whether she is at increased risk for osteoporosis before she will accept therapy, however, quantitative computerized tomography or dual-photon absorptiometry are sensitive methods. They are most worthwhile when used serially if the long-term adequacy of therapy in a known osteoporotic woman is in question.

Because of the rarity of detecting an abnormality on routine endometrial biopsy in asymptomatic women, a reasonable alternative to sampling all women starting HRT is to reserve the procedure for those at high risk. This would include those who have risk factors or manifestations of chronic estrogen exposure, such as DUB, along with those women who develop atypical, irregular bleeding during therapy.

Occasionally a woman will complain of decreased libido despite combined HRT. A psychosocial etiology is usually found, but when none is apparent help may be provided by the addition of an androgen (methyltestosterone, up to 5 mg daily).[173] Now available is a combination tablet containing 0.625 mg esterified estrogens with 1.25 mg methyltestosterone (Estratest HS). This combination does not prevent endometrial hyperplasia, so the addition of a progestin is still necessary. Both androgens and progestins can have an unfavorable effect on the lipoprotein profile, so this needs to be monitored. Hirsutism is an additional side effect of the product.

Other dilemmas that sometimes occur include situations in which estrogen is contraindicated. Medroxyprogesterone acetate can be given (10 to 20 mg daily orally, or 150 mg intramuscularly each month in depot form) to help relieve vasomotor symptoms, but other manifestations of estrogen deficiency are not improved. Supplemental calcium must be increased to 1,000 mg/day when estrogen cannot be used. If relief of vasomotor symptoms is inadequate, Veralipride (100 mg/day) is effective. This compound blocks hypothalamic dopamine receptors, so the common side effect is galactorrhea. Clonidine is also helpful in relieving hot flashes but is less effective than the other therapies and may have significant side effects, such as drowsiness.

Sometimes a woman simply cannot tolerate progestational therapy. In this instance it is legimate to give unopposed estrogen replacement therapy under close surveillance, with periodic endometrial biopsies.

Conclusion

The results of the last decade of research on the replacement of hormones have steadily confirmed a positive benefit-to-risk ratio. The relief of atrophic conditions and vasomotor distress are clear indications, and a significant beneficial impact on long-term bone preservation has become undeniable. It is highly likely that appropriate doses of estrogen can produce long-term cardiovascular risk reduction by producing favorable lipoprotein profiles. The recent recommendation of a consensus conference is well taken—that all postmenopausal women (without contraindications) should be made aware of the consequences of estrogen deficiency and should be offered the opportunity to receive estrogen therapy.[174] Initiation of estrogen should be timely and the addition of a progestin is imperative for those women with uteri. The lowest dose of estrogen consistent with reversal of estrogen deficiency is advisable. Ultimately, the decision to use replacement therapy rests with the well-informed patient. Under the best of circumstances, the replacement of hormones should be supplemental to sound personal health care habits that can minimize the amount of hormones required. These include a proper diet, exercise, and the cessation of smoking. The full potential of hormone replacement therapy can only be realized by its long-term administration.

References

1. Rybo G: Variations in mentrual loss. Res Clin Forums 1982;4:81.
2. Noyes RW, Hertig AW, Rock J: Dating the endometrial biopsy. Fertil Steril 1950;1:3.
3. Bartlemez GW: The phases of the menstrual cycle and their interpretation in terms of the pregnancy cycle. Am J Obstet Gynecol 1957;74:931.
4. Hsueh AJW, Peck EJ, Clark JH: Progesterone antagonism of the estrogen receptor and estrogen-induced uterine growth. Nature 1975;254:337.
5. Gurpide E, Gusberg S, Tseng L: Estradiol binding and metabolism in human endometrial hyperplasia and adenocarcinoma. J Steroid Biochem 1976;7:891.
6. Chimbira TH, Cope E, Anderson AB, et al: The effect of danazol on menorrhagia, coagulation mechanisms, haematological indices and body weight. Br J Obstet Gynaecol 1979;86:46.
7. Strickler RC: Dysfunctional uterine bleeding in ovulatory women. Postgrad Med 1985;77(1):235.
8. Smith SK, Abel MH, Kelly RW, et al: Prostaglandin synthesis in the endometrium of women with ovular dysfunctional uterine bleeding. Br J Obstet Gynaecol 1981;88:434.
9. Sheppard BL: The pathology of dysfunctional uterine bleeding. Clin Obstet Gynecol 1984;11(1):227.
10. Boyd ME: Dysfunctional uterine bleeding. Can J Surg 1986;29(5):305.
11. DeVore GR, Owens O, Kase N: Use of intravenous premarin in the treatment of dysfunctional uterine bleeding—a double-blind randomized control study. Obstet Gynecol 1982;59:285.
12. Anderson ABM, Haynes PJ, Guillebaud J, et al: Reduction of menstrual blood loss by prostaglandin synthetase inhibitors. Lancet 1976;1:774.
13. Hall P, Maclachlan N, Thorn N, et al: Control of menorrhagia by the cyclo-oxygenase inhibitors naproxen sodium and mefenamic acid. Br J Obstet Gynaecol 1987;94:554.
14. Fraser IS, Pearse C, Shearman RP, et al: Efficacy of mefanamic acid in patients with a complaint of menorrhagia. Obstet Gynecol 1981;58:543.
15. Halbert DR: Menstrual delay and dysfunctional uterine bleeding associated with antiprostaglandin therapy for dysmenorrhea. J Reprod Med 1983;28:592.
16. Nilsson L, Rybo G: Treatment of menorrhagia. Am J Obstet Gynecol 1971;110(5):713.
17. Nilsson L, Rybo G: Treatment of menorrhagia with epsilon aminocaproic acid: a double blind investigation. Acta Obstet Gynecol Scand 1966;45(4):429.
18. Fraser IS: Treatment of dysfunctional uterine bleeding with danazol. Aust NZ J Obstet Gynaecol 1985;25:224.
19. Loffer FD: Hysteroscopic management of menorrhagia. Acta Eur Fertil 1986;17:463.
20. Gimpelson RJ: Panoramic hysteroscopy with directed biopsies versus dilatation and curettage for accurate diagnosis. J Reprod Med 1984;29:575.
21. Zimmermann R: Dysfunctional uterine bleeding. Obstet Gynecol Clin North Am 1988;15(1):107.
22. De Jong P, Doel F, Falconer A: Outpatient diagnostic hysteroscopy. Br J Ob Gynecol 1990;97:299.
23. Bent AE, Ostergard DR: Endometrial ablation with the neodynium:Yag laser. Obstet Gynecol 1990;75:923.
24. DeCherney AH, Diamond MP, Lavy G, et al: Endometrial ablation for intractable uterine bleeding: hysteroscopic resection. Am J Obstet Gynecol 1986;155:574.
25. Goldrath MG, Fuller TA, Segal S: Laser photovaporization of endometrium for the treatment of menorrhagia. Am J Obstet Gynecol 1981;140(1):14.
26. Lomano JM: Photocoagulation of the endometrium with the Nd:YAG laser for the treatment of menorrhagia: a report of ten cases. J Reprod Med 1986;31:148.
27. Ngu A, Quinn MA: Dysfunctional uterine bleeding in women over 40 years of age. Aust NZ J Obstet Gynaecol 1984;24:30.
28. Spellacy WN: Abnormal bleeding. Clin Obstet Gynecol 1983;26(3):702.
29. Sherman MB, West JH, Korenman SG: The menopausal transition: analysis of LH, FSH, estradiol, and progesterone concentrations during menstrual cycles of older women. J Clin Endocrinol Metab 1976;42:629.
30. Einerth Y: Vacuum curettage by the Vabra[R] method. Acta Obstet Gynecol Scand 1982;61:373.
31. Kaunitz AM, Masciello A, Ostrowski M, Rovira EZ: Comparison of endometrial biopsy

with the endometrial Pipelle and Vabra aspirator. J Reprod Med 1988;33:426.
32. Grimes DA: Diagnostic dilation and curettage: a reappraisal. Am J Obstet Gynecol 1982;142:1.
33. Judd HL, Shamonki IM, Frumar AM, Lagasse LD: Origin of serum estradiol in postmenopausal women. Obstet Gynecol 1982;59:680.
34. Meldrum DR, Davidson BJ, Tataryn IV, Judd HL: Changes in circulating steroids with aging in postmenopausal women. Obstet Gynecol 1981;57:624.
35. Eriksen EF, Colvard DS, Berg NJ, et al: Evidence of estrogen receptors in normal human osteroblast-like cells. Science 1988;241:84.
36. Coope J, Thomson JM, Poller L: Effects of "natural oestrogen" replacement therapy on menopausal symptoms and blood clotting. Br Med J 1975;4:139.
37. Meldrum DR: The pathophysiology of postmenopausal symptoms. Semin Reprod Endocrinol 1983;1:11.
38. Schiff I, Regestein Q, Tulchinsky D, Ryan KJ: Effects of estrogens on sleep and psychological state of hypogonadal women. JAMA 1979;242:2405.
39. Campbell S, Whitehead M: Estrogen therapy and the menopausal syndrome. Clin Obstet Gynecol 1977;4:31.
40. Jensen J, Christiansen C: Dose-response and withdrawal effects on climacteric symptoms after hormonal replacement therapy. A placebo-controlled therapeutic trial. Maturitas 1983;5:125.
41. Iosif CS, Batra S, Ek A, Astedt B: Estrogen receptors in the human female lower urinary tract. Am J Obstet Gynecol 1981;141:816.
42. Hammond CB, Maxson WS: Current status of estrogen therapy for the menopause. Fertil Steril 1982;37:5.
43. Semmens JP, Wagner G: Estrogen deprivation and vaginal function in postmenopausal women. JAMA 1982;248:445.
44. Riggs LA, Hermann H, Yen SSC: Absorption of estrogens from vaginal creams. N Engl J Med 1978;298:195.
45. Wilson PD, Faragher B, Butler B, et al: Treatment with oral piperazine oestrone sulphate for genuine stress incontinence in postmenopausal women. Br J Obstet Gynaecol 1987;94:568.
46. Rud T: The effect of estrogens and gestagens on the urethral pressure profile in urinary continent and stress incontinent women. Acta Obstet Gynecol Scand 1980;59:265.
47. Sarrel PM: Estrogen replacement therapy. Obstet Gynecol 1988;72(5)(suppl):3S.
48. Sarrel PM: Sexuality in the middle years. Obstet Gynecol Clin North Am 1987;14:49.
49. Semmens JP, Wagner G: Effects of estrogen therapy on vaginal physiology during menopause. Obstet Gynecol 1985;66:15.
50. Dennerstein L, Burrows GD: Hormone replacement therapy and sexuality in women. J Clin Endocrinol Metab 1982;11:661.
51. Vandenbroucke JP, Witteman JCM, Valkenburg HA, et al: Noncontraceptive hormones and rheumatoid arthritis in perimenopausal and postmenopausal women. JAMA 1986;255:1299.
52. Cauley JA, Petrini AM, LaPorte RE, et al: The decline of grip strength in the menopause: relationship to physical activity, estrogen use and anthropometric factors. J Chronic Dis 1987;40:115.
53. Ettinger B, Genant HK, Cann CE: Long-term estrogen replacement therapy prevents bone loss and fractures. Ann Intern Med 1985;102:319.
54. Lindsay R, Hart DM, Forrset C, Baird C: Prevention of spinal osteoporosis in oophorectomised women. Lancet 1980;2:1151.
55. Richelson LS, Wahner HW, Melton LJ III, Riggs BL: Relative contributions of aging and estrogen deficiency to postmenopausal bone loss. N Engl J Med 1984;311:1273.
56. Genant HK, Cann CE, Ettinger B, Gordan GS: Quantitative computed tomography of vertebral spongiosa: a sensitive method for detecting early bone loss after oophorectomy. Ann Intern Med 1982;97:699.
57. Nilas L, Christiansen C: Bone mass and its relationship to age and the menopause. J Clin Endocrinol Metab 1987;65:697.
58. Ettinger B: Preventing postmenopausal osteoporosis with estrogen replacement therapy. Int J Infertil 1986;31(suppl):15.
59. Ettinger B: Prevention of osteoporosis: treatment of estradiol deficiency. Obstet Gynecol 1988;72(5)(suppl):13S.
60. Christiansen C, Nilas L, Riis BJ, et al: Uncoupling of bone formation and resorption by combined oestrogen and progestagen therapy in postmenopausal osteoporosis. Lancet 1985;2:800.
61. Christiansen C, Riis BJ: 17-β Estradiol and continuous norethisterone: A Unique treat-

ment for established osteoporosis in elderly women. J Clin Endocrinol Metab 1990; 71:836.

62. Lindsay R, Hart DM, Purdie D, et al: Comparative effects of oestrogen and a progestogen on bone loss in post menopausal women. Clin Sci Mol Med 1978;54:193.
63. Gambrell RD: Osteoporosis: Benefits and risks of estrogen-progestogen replacement therapy. J Miss State Med Assoc 1986; 22(10):267.
64. Tiegs RD, Body JJ, Wahner HW, et al: Calcitonin secretion in postmenopausal osteoporosis. N Engl J Med 1985;313:1097.
65. Prince RL, Dick IM, Price RI: Plasma calcitonin levels are not lower than normal in osteoporotic women. J Clin Endocrinol Metab 1989;68(3):684.
66. Riggs BL: Osteoporosis—a disease of impaired homeostatic regulation? (editorial). Miner Electrolyte Metab 1981;5:265.
67. Boucher A, D'Amour P, Hamel L, et al: Estrogen replacement decreases the set point of parathyroid hormone stimulation by calcium in normal postmenopausal women. J Clin Endocrinol Metab 1989;68(4):831–836.
68. Weiss NC, Ure CL, Ballard JH, et al: Estimated incidence of fractures of the lower forearm and hip in postmenopausal women. N Engl J Med 1980;303:1195.
69. Kiel DP, Felson DT, Anderson JJ, et al: Hip fracture and the use of estogens in postmenopausal women: the Framingham Study. N Engl J Med 1987;317:1169.
70. Riggs BL, Seeman E, Hodgson SF, et al: Effect of the fluoride/calcium regimen on vertebral fracture occurrence in postmenopausal osteoporosis. N Engl J Med 1982;306: 446.
71. Riis B, Thomsen K, Christiansen C: Does calcium supplementation prevent postmenopausal bone loss? N Engl J Med 1987; 316:173.
72. Ettinger B, Genant HK, Cann CE: Postmenopausal bone loss is prevented by treatment with low-dosage estrogen with calcium. Ann Intern Med 1987;106:40.
73. Stevenson JC, Whitehead MI, Padwick M, et al: Dietary intake of calcium and postmenopausal bone loss. Br Med J 1988;297:15.
74. Heaney RP, Recker RR: Distribution of calcium absorption in middle-aged women. Am J Clin Nutr 1986;43:299.
75. Heaney RP, Recker RR, Saville PD: Menopausal changes in calcium balance performance. Lab Clin Med 1978;92:953.
76. Riggs BL, Hodgson SF, O'Fallon WM, et al: Effect of fluoride treatment on the fracture rate in postmenopausal women with osteoporosis. N Engl J Med 1990;322:802.
77. Watts NB, Harris ST, Genant HK, et al: Intermittent cyclical etidronate treatment of postmenopausal osteoporosis. N Engl J Med 1990;323:73.
78. Lindsay R: Managing osteoporosis: current trends, future possibilities. Geriatrics 1987;42:35.
79. Eastell R, Riggs BL: New approaches to the treatment of osteoporosis. Clin Obstet Gynecol 1987;30:860.
80. MacIntyre I, Stevenson JC, Whitehead MI, et al: Calcitonin for prevention of postmenopausal bone loss. Lancet 1988;1:900.
81. Chow RK, Harrison JE, Brown CF, Hajek V: Physical fitness effect on bone mass in postmenopausal women. Arch Phys Med Rehabil 1986;67:231.
82. Michnovicz JJ, Hershcopf RJ, Naganuma H, et al: Increased 2-hydroxylation of estradiol as a possible mechanism for the anti-estrogenic effect of cigarette smoking. N Engl J Med 1986;315:1305.
83. Jensen J, Christiansen C, Rodbro P: Cigarette smoking, serum estrogens, and bone loss during hormone replacement therapy early after menopause. N Engl J Med 1985;313:973.
84. Barbieri RL, McShane PM, Ryan KJ: Constituents of cigarette smoke inhibit human granulosa cell aromatase. Fertil Steril 1987;46:232.
85. Lindsay R, MacLean A, Kraszewski A, et al: Bone response to termination of estrogen treatment. Lancet 1978;1:1325.
86. Horsman A, Nordin BEC, Crilly RG: Effect on bone of withdrawal of estrogen therapy. Lancet 1979;2:33.
87. Quigley MET, Martin PL, Burnier AM, Brooks P: Estrogen therapy arrests bone loss in elderly women. Am J Obstet Gynecol 1987;156:1516.
88. Castelli W: Epidemiology of coronary heart disease: the Framingham Study. Am J Med 1984;76:4.
89. Colditz GA, Willett WC, Stampfer MJ, et al: Menopause and the risk of coronary heart disease in women. N Engl J Med 1986; 316:1105.

90. Rosenberg L, Armstrong B, Jick H: Myocardial infarction and estrogen therapy in postmenopausal women. N Engl J Med 1976; 294:1256.
91. Pfeffer RI, Whipple GH, Kurosake TT, Chapman JM: Coronary risk and estrogen use in postmenopausal women. Am J Epidemiol 1978;107:479.
92. Jick H, Dinan B, Rothman KJ: Noncontraceptive estrogens and non-fatal myocardial infarction. JAMA 1978;239:1407.
93. Rosenberg L, Sloane D, Shapiro S, et al: Noncontraceptive estrogens and myocardial infarction in young women. JAMA 1980; 224:339.
94. Ross RK, Paganini-Hill A, Mack TM, et al: Menopausal oestrogen therapy and protection from death from ischaemic heart disease. Lancet 1981;1:585.
95. Bain C, Willett W, Hennekens CH, et al: Use of postmenopausal hormones and risk of myocardial infarction. Circulation 1981; 64:42.
96. Adam S, Williams V, Vessey MP: Cardiovascular disease and hormone replacement treatment: a pilot case-control study. Br Med J 1981;282:1277.
97. Szklo M, Tonascia J, Gordis L, Bloom I: Estrogen use and myocardial infarction risk: a case-control study. Prevent Med 1984; 13:510.
98. Sullivan JM, Vander Zwaag R, Lemp GF, et al: Postmenopausal estrogen use and coronary atheroslerosis. Ann Intern Med 1988;108:358.
99. Petitti DB, Wingerd J, Pellegrin F, Ramcharan S: Risk of vascular disease in women: smoking, oral contraceptives, non-contraceptive estrogens, and other factors. JAMA 1979;242:1105.
100. Gordon T, Kannel WB, Hjortland MC, McNamara PM: Menopause and coronary heart disease: the Framingham Study. Ann Intern Med 1978;89:157.
101. Lafferty FW, Helmuth DO: Postmenopausal estrogen replacement: the prevention of osteoporosis and systemic effects. Maturitas 1985;7:147.
102. Bush TL, Barrett-Connor E, Cowan LD, et al: Cardiovascular mortality and noncontraceptive use of estrogen in women: results from the Lipid Research Clinics Program Follow-Up Study. Circulation 1987;75: 1102.
103. Henderson BE, Paganini-Hill A, Ross RK: Estrogen replacement therapy and protection from acute myocardial infarction. Am J Obstet Gynecol 1988;159:312.
104. Stampfer MJ, Willett WC, Colditz GA, et al: A prospective study of postmenopausal estrogen therapy and coronary heart disease. N Engl J Med 1985;313:1044.
105. Burch JC, Byrd BF, Vaughn WK: The effects of long-term estrogen on hysterectomized women. Am J Obstet Gynecol 1974;118:778.
106. Hammond CB, Jelovsek FR, Lee KL, et al: Effects of long-term estrogen replacement therapy: I. Metabolic effects. Am J Obstet Gynecol 1979;133:525.
107. Wilson PWF, Garrison RJ, Castelli WP: Postmenopausal estrogen use, cigarette smoking, and cardiovascular morbidity in women over 50. The Framingham Study. N Engl J Med 1985;313:1038.
108. Speroff T, Dawson N, Speroff L: Is postmenopausal estrogen use risky? Results from a methodologic review and information sythesis. Clin Res 1987;35:362A.
109. Miller GJ, Miller NE: Plasma-high-density-lipoprotein concentration and development of ischaemic heart-disease. Lancet 1975; 1:16.
110. Hart DM, Farish E, Fletcher DC, et al: Ten years postmenopausal hormone replacement therapy—effect on lipoproteins. Maturitas 1984;5:271.
111. Adams MR, Clarkson TB, Koritnik DR, Nash HA: Contraceptive steroids and coronary artery atherosclerosis in cynomolgus macaques. Fertil Steril 1987;47:1010.
112. Hassager C, Christiansen C: Blood pressure during oestrogen/progestogen substitution therapy in healthy post-menopausal women. Maturitas 1988;9:315.
113. Hirvonen E, Malkonen M, Manninen V: Effects of different progestogens on lipoproteins during postmenopausal replacement therapy. N Engl J Med 1981;304:560.
114. Mattsson L, Cullberg LG, Samsioe G: Influence of esterified estrogens and medoxyprogesterone on lipid metabolism and sex steroids. A study in oophorectomized women. Horm Metab Res 1981;14:602.
115. Silferstolpe G, Gustafsson A, Samsioe G, Syanborg A: Lipid metabolic studies in oophorectomized women: effects on serum lipids and lipoproteins of three synthetic progestogens. Maturitas 1983;4:103.

116. Ylostalo P, Kauppila A, Kivinen S, et al: Endocrine and metabolic effects of low-dose estrogen-progestin treatment in climacteric women. Obstet Gynecol 1983;62:682.
117. Mattson L, Cullberg G, Samsioe G: A continuous estrogen-progestogen regimen for climacteric complaints. Acta Obstet Gynecol Scand 1984;63:673.
118. Ottosson UB, Carlstrom K, Damber JE, von Schoultz B: Serum levels of progesterone and some of its metabolites including deoxycorticosterone after oral and parenteral administration. Br J Obstet Gynaecol 1984;91: 1111.
119. Wren B, Garrett D: The effect of low-dose piperazine oestrogen sulphate and low-dose levonorgestrel on blood lipid levels in postmenopausal women. Maturitas 1985;7:141.
120. Ottosson UB, Johansson BG, von Schoultz B: Subfractions of high-density lipoprotein cholesterol during estrogen replacement therapy: a comparison between progestogens and natural progesterone. Am J Obstet Gynecol 1985;151:746.
121. Hirvonen E, Lipasti A, Mälkönen M, et al: Clinical and lipid metabolic effects of unopposed oestrogen and two oestrogen-progestogen regimens in postmenopausal women. Maturitas 1987;9:69.
122. Ravnikar V, Murin V, Nutkik J, et al: Blood lipid levels in postmenopausal women on hormone replacement therapy. Paper presented at 35th Annual Meeting of the Society of Gynecological Investigation, 1988.
123. Jaszmann LJB: Epidemiology of the climacteric syndrome in Campbell S (ed): The Management of the Menopause and Postmenopausal Years. International symposium Proceedings, The University of London, November 1984. Baltimore, University Park Press, 1976, pp 11–23.
124. Mack TM, Pike MC, Henderson BE, et al: Estrogens and endometrial cancer in a retirement community. N Engl J Med 1976; 294:1262.
125. Ziel HK, Finkle WD: Increased risk of endometrial carcinoma among users of conjugated estrogens. N Engl J Med 1975;293: 1167.
126. Thom MH, White PJ, Williams RM, et al: Prevention and treatment of endometrial disease in climacteric women receiving estrogen. Lancet 1979;2:455.
127. Whitehead MI, Townsend PT, Pryse-Davies J, et al: Effects of estrogen and progestins on the biochemistry and morphology of the postmenopausal endometrium. N Engl J Med 1981;305:1599.
128. Gambrell RD, Babgnell CA, Greenblatt RB: Role of estrogens and progesterone in the etiology and prevention of endometrial cancer: a review. Am J Obstet Gynecol 1983; 146:696.
129. Persson, I, Adami HO, Leif B, et al: Risk of endometrial cancer after treatment with oestrogens alone or in conjunction with progestogens: results of a prospective study. Br Med J 1989;298:147.
130. Creasman WT, Henderson D, Hinshaw W, Clarke-Pearson DL: Estrogen replacement therapy in the patient treated for endometrial cancer. Obstet Gynecol 1986;67: 326.
131. Ploch E: Hormonal replacement therapy in patients after cervical cancer treatment. Gynecol Oncol 1987;26:169.
132. Adams JB: Molecular endocrinology of breast cancer (editorial). Pathology 1987; 19:213–215.
133. Kaufman DW, Miller DR, Rosenberg L, et al: Noncontraceptive estrogen use and the risk of breast cancer. JAMA 1984;252:63.
134. Wingo PA, Layde PM, Lee NC, et al: The risk of breast cancer in postmenopausal women who have used estrogen replacement therapy. JAMA 1987;257:209.
135. Buring JE, Hennekens CH, Lipnick RJ, et al: A prospective cohort study of postmenopausal hormone use and risk of breast cancer in U.S. women. Am J Epidemiol 1987; 125:939.
136. W.H.O. Collaborative Study of Neoplasia and Steroid Contraceptives: Breast cancer, cervical cancer, and medroxyprogesterone acetate. Lancet 1984;2:1207.
137. Gambrell RD, Maier R, Sancers BI: Decreased incidence of breast cancer in postmenopausal estrogen-progestogen users. Obstet Gynecol 1983;62:435.
138. Bergkvist L, Adami H-O, Persson I, et al: The risk of breast cancer after estrogen and estrogen-progestin replacement. N Engl J Med 1989;321:293.
139. Lind T, Cameron EC, Hunter WM, et al: A prospective, controlled trial of six forms of hormone replacement therapy given to postmenopausal women. Br J Obstet Gynaecol 1979;86:(suppl 3):1.

140. Pfeffer RI, Kurosaki TT, Charlton SK: Estrogen use and blood pressure in later life. Am J Epidemiol 1979;110:469.
141. Lutola H: Blood pressure and hemodynamics in postmenopausal women during estradiol-17β substitution. Ann Clin Res 1983; 15:(suppl 38):9.
142. Wren BG, Routledge AD: The effect of type and dose of oestrogen on the blood pressure of postmenopausal women. Maturitas 1983;5:135.
143. Gambrell RD: The menopause: Benefits and risks of estrogen-progestogen replacement therapy. Fertil Steril 1982;37:457.
144. Stolley PD, Tonascia JA, Tockman MS, et al: Thrombosis with low-estrogen oral contraceptives. Am J Epidemiol 1975;102:197.
145. Thom M, Chahravarti S, Onam DH, Studd JWW: Effect of hormone replacement therapy on glucose tolerance in postmenopausal women. Br J Obstet Gynaecol 1977;84:776.
146. Boston Collaborative Drug Surveillance Program: Surgically confirmed gallbladder disease, venous thromboembolism, and breast tumors in relation to postmenopausal estrogen therapy. N Engl J Med 1974; 290:15.
147. Scagg RKR, McMichael AJ, Seamark RF: Oral contraceptive, pregnancy and endogenous estrogen in gallstone disease—a case-control study. Br Med J 1984;288:1795.
148. Kakar F, Weiss NS, Strite SA: Non-contraceptive estrogen use and the risk of gallstone disease in women. Am J Public Health 1988;78:564.
149. Ferguson KJ, Hoegh C, Johnson S: Estrogen replacement therapy—a survey of women's knowledge and attitudes. Arch Intern Med 1989;149:133.
150. Svanberg L: Effects of estrogen deficiency in women castrated when young. Acta Obstet Gynecol Scand [Suppl] 1982;106:11–15.
151. Rigotti NA, Nussbaum SR, Herzog DB, et al: Osteoporosis in women with anorexia nervosa. N Engl J Med 1984;311:1601.
152. Frohlich EP, Koller AM, Van Blerk PJ, Margolius KA: Adenocarcinoma from endometriosis causing urinary tract obstruction in a patient on oestrogen replacement therapy after hysterectomy. A case report. S Afr Med J 1988;74(12):638.
153. Genant HK, Cann CE, Ettinger B, Gordan GS: Quantitative computed tomography of vertebral spongiosa: a sensitive method for detecting early bone loss after oophorectomy. Ann Intern Med 1982;97:699.
154. Lindsay R, Hart M, Clark DM: The minimum effective dose of estrogen for prevention of postmenopausal bone loss. Obstet Gynecol 1984;63:759.
155. Jensen J, Nilas L, Christiansen C: Cyclic changes in serum cholesterol and lipoproteins following different doses of combined postmenopausal hormone replacement therapy. Br J Obstet Gynecol 1986;93:613.
156. DeLignieres B, Basdevant A, Thomas G, et al: Biological effects of estradiol-17 beta in postmenopausal women: oral versus percutaneous administration. J Clin Endocrinol Metab 1986;62:536.
157. Fahraeus L, Larsson-Cohn U, Wallentin L: Lipoproteins during oral and cutaneous administration of oestradiol-17β to menopausal women. Acta Endocrinol 1982;101:597.
158. Basdevant A, DeLignieres B, Guy-Grand B: Differential lipemic and hormonal responses to oral and parenteral 17β-estradiol in postmenopausal women. Am J Obstet Gynecol 1983;147:77.
159. Jensen J, Riis BJ, Strom V, et al: Long-term effects of percutaneous estrogens and oral progesterone on serum lipoproteins in postmenopausal women. Am J Obstet Gynecol 1987;156:66.
160. Weiss NS, Szekely R, Austin DF: Increasing incidence of endometrial cancer in the United States. N Engl J Med 1976;294:1259.
161. Gambrell RD: Role of hormones in the etiology and prevention of endometrial and breast cancer. Acta Obstet Gynecol Scand [Suppl] 1982;106:37.
162. Varma TR: Effect of long-term therapy with estrogen and progesterone on the endometrium of postmenopausal women. Acta Obstet Gynecol Scand 1985;64:41.
163. Gibbons WE, Moyer DL, Lobo RA, et al: Biochemical and histologic effects of sequential estrogen/progestin therapy on the endometrium of postmenopausal women. Am J Obstet Gynecol 1986;154:456.
164. Padwick ML, Pryse-Davies J, Whitehead MI: A simple method for determining the optimal dosage of progestin in postmenopausal women receiving estrogens. N Engl J Med 1986;315:930.
165. Mattsson L, Cullberg G, Samsioe G: Evalu-

ation of a continuous oestrogen-progestogen regimen for climacteric complaints. Maturitas 1982;4:95.
166. Prough SG, Aksel S, Wiebe RH, Shepherd J: Continuous estrogen/progestin therapy in menopause. Am J Obstet Gynecol 1987; 157:1449.
167. Weinstein L: Efficacy of continuous estrogen-progestin regimen in the menopausal patient. Obstet Gynecol 1987;157:1449.
168. Munk-Jensen N, Nielsen SP, Obel EB, Eriksen PB: Reversal of postmenopausal vertebral bone loss by oestrogen and progestogen: a double blind placebo controlled study. Br Med J 1988;296:1150.
169. Williams SR, Frenchek B, Speroff T, Speroff L: A study of combined continous ethinyl estradiol and norethindrone acetate for postmenopausal hormone replacement. Am J Obstet Gynecol 1990;162:438.
170. Mattsson L, Cullberg G, Samsioe G: A continuous estrogen-progestin treatment in climacteric women. Obstet Gynecol 1983; 62:682.
171. Jensen J, Riis BJ, Stom V, Christiansen C: Continous oestrogen-progestogen treatment and serum lipoproteins in postmenopausal women. Br J Obstet Gynaecol 1987;94:130.
172. Hall FM, Davis MA, Baran DT: Bone mineral screening for osteoporosis. N Engl J Med 1987;316:212.
173. Sherwin BB, Gelfand MM, Brender W: Androgen enhances sexual motivation in females: a prospective, crossover study of sex steroid administration in the surgical menopause. Psychosom Med 1985;47:339.
174. Progestagen use in postmenopausal women. First International Consensus Conference on Progestagen Use in Postmenopausal Women, Naples Florida, September 1988. Lancet 1988;2:1243.

23

Management of Uterine Leiomyomata

ALBERT ALTCHEK

The Scope of the Problem

Uterine leiomyomata (also called leiomyomas, myomas, or fibroids) are the most common solid female pelvic tumor. They occur in about 20 to 40% of all women and develop during the menstrual years. They are more common and tend to occur at an earlier age in black women and in certain families. Most fibroids are small, multiple, and asymptomatic and are discovered on routine physical examination. Many series report that 50% of women with fibroids are symptomatic; however, this is probably an inflated incidence related to patient selection and surveys of hospitalized cases. Recent reviews indicate that in the United States, there are about 175,000 hysterectomies (27% of a total of 680,000 hysterectomies) and 18,000 myomectomies done annually for fibroids. Among women with hysterectomies 8 to 15% require transfusions and 24 to 42% have febrile morbidity; there is 0.1 to 0.2% fatality. Myomectomy cases have an 18% chance of receiving blood or blood products and an increased chance of mortality and morbidity, and 10 to 30% end up with an unplanned hysterecctomy. The need for transfusion is related in part to the extent of preoperative anemia. About 8% of transfusion cases develop a biochemical hepatitis.[1] About 15 to 30% of myomectomy cases have a recurrence of symptomatic fibroids and about 50 to 75% of these (or 10 to 15% of those with the first myomectomy) require a second major operation.[1,2]

New Considerations in Management and Consumerism

Management of fibroids has been modified recently because of:

1. Consumerism—listening to the wishes of the patient
2. Having the patient participate in management decisions
3. Women having careers in business and professions wishing to preserve fertility or menses in the "elderly" age group
4. The availability of gonadotropin-releasing hormone (GnRh) analogs
5. A new appreciation of and microsurgical philosophy of myomectomy
6. The ability of sonography and magnetic resonance imaging (MRI) to determine that pelvic masses are leiomyomatous and not ovarian
7. Increased use of the hysterogram and hysteroscope to identify submucous leiomyoma
8. The legal right ("patients' bill of rights") of the patient to accept or reject recommended treatment
9. The developing technique of hysteroscopic myomectomy
10. Fear on the part of physicians of accusation of unnecessary surgery

"Consumerism," or the extent to which the patients' wishes are followed, is a complex matter that depends on: (a) the patient understanding the medical problems and risks, (b) the patient participating in the management decision, (c) the responsibility of the patient in follow-up, and (d) the extent to which the physician willl be flexible and allow latitude. The physician is not obliged to care for a patient if he/she feels that the patients' wishes cannot be followed without endangering her health and if the patient can go to another health care facility.

Symptoms

Most myomas are asymptomatic and slow growing. The usual symptoms of myomas are menorrhagia, pelvic pain and pressure, and the palpation of a pelvic mass. The myomas are usually located in the uterine fundus, where they are usually intramural and subserous and rarely intraligamentous. About 5% are submucous. They may be sessile or polypoid when submucous or subserous. About 30% of symptoms are due to menorrhagia, which may occur with intramural fibroids. If the fibroids are submucous, the menorrhagia is often more severe, with "gushing" episodes. Usually, there is no metrorrhagia. The latter may occur with cramps when there is an aborting, pedunculated, submucous necrotic myoma. Myomas are infrequently located in the cervix or rarely in the round ligament of the uterus.

A low anterior wall myoma may compress the bladder to reduce its volume capacity and result in urinary frequency, or rarely compress the urethra to make voiding difficult. Posterior wall myomas may press the rectosigmoid to cause constipation and pressure. Intraligamentous laterally growing myomas may compress the ureter (and also a large uterus may compress the ureters at the pelvic brim), causing hydroureter and hydronephrosis. Acute or subacute pain and fever may result from myoma degeneration (often a vascular accident) or from torsion of a pedunculated myoma. There may be dysmenorrhea and dyspareunia. Infertility, miscarriage, and midtrimester abortion may be due to fibroids, especially a large submucous fibroid; however, other factors have to be considered. Infrequently, a large cervical or lower segment fibroid may obstruct labor. Myomas near the cornual angles may obstruct the fallopian tubes. Rarely, there may be diffuse uterine stromal leiomyomatosis, intravenous leiomyomatosis, or benign metastasizing leiomyomatosis. The incidence of sarcoma arising from a leiomyoma is probably less than 0.5%. Primary sarcoma of the uterus is rare but may be misdiagnosed as a fibroid.

Diagnosis

The diagnosis of uterine leiomyomata is suggested by a positive family history; symptoms of menorrhagia, pressure, and pain; and laboratory findings of iron-deficiency (blood loss) anemia and elevation of the erythrocyte sedimentation rate (ESR) if there is myoma degeneration. On physical examination, there may be a hard abdominal pelvic mass palpable. On bimanual pelvic and rectovaginal examination, the uterus may form a confluent, multinodular, hard, smooth mass measured clinically by comparison to gestational size (measured from the first day of the last menstrual period). Moving the abdominal mass causes movement of the cervix, indicating a rigid connection, whereas a palpable ovarian mass will not transmit movement to the cervix. The usual enlarged myomatous uterus is mobile except if it is wedged in the true pelvis, or if it is fixed by associated pelvic inflammatory disease or endometriosis. Pedunculated subserous fibroids may feel like hard ovarian masses. The fibroid uterus usually has an enlarged uterine cavity that may be measured by sounding the uterus and that also confirms the mass to be uterine myomata rather than ovarian. Nevertheless, an adherent ovarian mass may simulate a fibroid uterus.

Sonography is ideal to confirm the pelvic mass as leiomyomata of the uterus and to make precise measurements. Ideally, in addition to a pelvic scan, there should be a simultaneous abdominal scan to rule out other pathology (gallstones, liver tumor, obstructed kidney, etc.). Sonography is readily available, noninvasive, without x-ray radiation, and relatively inexpensive. Computed tomography (CT) and MRI scans are rarely done. The former is expensive and involves x-rays. The latter is extremely expensive, difficult to obtain, and sometimes emotionally disturbing to the patient, but does give excellent visualization. The hysterogram is the "gold standard" for identification of a submucous fibroid, although an early intrauterine pregnancy or a large endometrial polyp may give a similar picture. Usually, a hysterosalpingogram is done to visualize the fallopian tubes at the same time. Ideally, some dye should be made to enter the peritoneal cavity to be certain of tubal patency. A determination of the ESR is often made prior to hysterography because salpingitis may be provoked. Hysteroscopy may be used to identify a submucous fibroid and other intrauterine pathology.

A normal-sized uterus may harbor a significant submucous fibroid. The aborting submucous polypoid myoma presents a foul, necrotic, bloody mass through a dilated cervix. Rarely, fibroids are associated with polycythemia, ascites, and hyperprolactinemia.

Ideally, uterine leiomyomata should be followed by gynecologists. Many otherwise excellent physicians never learned how to do a pelvic examination and may misinterpret an ovarian neoplasm for a uterine fibroid, or miss an ovarian neoplasm developing with myomas present. Also, some do not ask about excessive menstrual flow, which could explain anemia and be a clue to myomas.

The differential diagnosis includes:

1. Diffuse or localized adenomyosis of the uterus, which does not "shell out" surgically like leiomyomata, and which may cause progressive acquired dysmenorrhea and menorrhagia. Even the MRI may not be able to distinguish localized adenomyosis from leiomyomata.
2. Ovarian neoplasms, which may hug the uterus to form one conglomerate mass, or be smooth and hard.
3. Uterine sarcoma, which is very rare but may be rapid growing, contain histologic large vessels, and have necrotic areas.

Gonadotropin-Releasing Hormone (GnRH) Agonist Analog Therapy

Background

The gonadotropin-releasing hormone is also referred to as gonadotropin-releasing substance, luteinizing hormone–releasing hormone (LH-RH), and luteinizing hormone–releasing substance (LH-RS). It is a decapeptide secreted by the hypothalamus in pulsatile fashion and transported by a portal capillary blood circulation to the anterior pituitary, where, also in pulsatile fashion, the gonadotropic cells secrete follicle-stimulating hormone (FSH) and luteinizing hormone (LH). The structure and function of GnRH have been reviewed.[2] Amino acids 1, 6, and 10 are associated with binding to the gonadotropic cells. The second (histidine) and third (tryptophan) amino acids initiate calcium influx, which together with calmodulin calcium receptor effects gonadotropin release in secretory granules. Gonadotropin synthesis is also stimulated. The half-life of GnRH is only 2 to 8 minutes. It is taken into the gonadotrope cell and degraded with peptidase cleavage between amino acids 6-7 and 9-10. The normal physiology is that the single GnRH is able to cause cyclic patterns of two gonadotropins (FSH and LH) in part by variations of the frequency and intensity of the GnRH pulsations. Continous exogenous high-dose GnRH administration, when first used experimentally to increase gonadotropin secretion, was found to cause an initial increase but shortly thereafter a profound shutdown of gonadotropic secretion. This paradoxical biphasic gonadotropin response has been explained by two mechanisms—desensitization and down-regulation. The first refers to an uncoupling or disconnection of GnRH receptor binding from subsequent gonadotropin release. The latter refers to a decreased number of receptor sites.

Gonadotropin-releasing hormone agonist analogs are made in the laboratory by changes at the sixth and tenth amino acids, thereby enabling the molecule to resist degradation by pituitary peptidases. Thus, the analogs have half-lives of 80 to 480 minutes. The agonists have a high affinity for the binding sites and keep them bound. This down-regulation does not leave many unoccupied receptor sites. Thus, the agonist analog acts like a continous exogenous high-dose GnRH infusion. There is an initial stimulation of gonadotropin followed by a prolonged (but reversible) depression and secondary hypogonadism. Some reports of hypogonadism despite apparently normal levels of gonadotropin after GnRH treatment are thought to be due to the secretion of immunoactive but biologically inactive gonadotropins. The result is a reversible medical menopause.[1,2]

In the pathogenesis of fibroids, a single cell undergoes a change and proliferates.[3] Estrogen must be present, and it also increases the rate of growth of fibroids. Fibroids of themselves have a local increased estrogen milieu and have an increased concentration of estrogen receptors compared to the surrounding myometrium.[4] Correspondingly, the myomas would be susceptible to estrogen deprivation and become smaller. In addition, leiomyoma have binding sites for GnRH and therefore the analog therapy might also act directly on the leiomyoma.[5] Gonadotropin-releasing hormone therapy causes a reduced binding of epidermal growth factor to leiomyomas, which might play a partial role in myoma reduction.[6] Although it has not been specifically studied, it is assumed that the number of cells of the myoma remain the same during the time of GnRH shrinkage treatment and the subsequent enlargement.

Obviously, individual myomas may vary greatly in their response to estrogen deprivation. Whereas cellular myomas would react, myomas with previous degeneration and with residual connective tissue, scarring, and calcification would not be expected to get smaller.

Although a synthetic GnRH antagonist would have the theoretical advantage of immediate cessation of gonadotropin secretion rather that the GnRH agonist action of initial stimulation followed by suppression, GnRH antagonists so far have been too toxic for clinical use.

Leuprolide

Until recently, there was only one GnRH analog being marketed commercially, leuprolide acetate (Lupron; TAP Pharmaceuticals). The Food and Drug Administration (FDA)-approved use is for the palliative treatment of advanced prostatic cancer. Therefore, the packaged dosage is for that purpose. It is packaged for use as a single daily subcutaneous injection of 1 mg (0.2 mL) in a 2.8-mL multiple-dose vial (Lupron). It is also packaged for use as a single monthly (every 28 days) intramuscular injection of 7.5 mg (1 mL diluent to mix with 7.5 mg of lyophilized microspheres to make a suspension) (Lupron Depot). The usual dose for gynecologic research to treat myomas is half of the prostate cancer dose. Most researchers used a daily subcutaneous dose of 0.5 mg (0.1 mL) or a monthly dose of 3.75 mg (but sometimes up to 7.5 mg).[1]

The clinical gynecologist in practice who might decide to use leuprolide in managing myomas would find it easier to use the depot preparation rather than the daily injection. The package insert indicates that since the product does not contain a preservative, the suspension should be discarded if it is not used immediately, although it is stable for 24 hours following reconstitution. Thus, the gynecologist will either inject half of the dose and discard the other half (unless there are two patients with similar schedules) or inject the entire amount in one patient. Most gynecologists do not realize the dosage differences and use the entire depot injection for one patient, thereby giving the usual patient twice the dose of the usual reported research cases. Fortunately, GnRH is relatively harmless in overdosage. Present clinical use might reflect the higher dose in increased incidence of hot flashes and other side effects, better reduction of myoma size, and reduced incidence of bleeding during treatment. Since the FDA approval is only for prostatic cancer, the manufacturer is not permitted to indicate the usual dose for other (gynecologic) purposes. It is anticipated that leuprolide will be approved for use with myomas, and with this approval the depot vial dose will be reduced by half and the present approximate price of $315 per vial will also be halved.

The FDA Drug Bulletin of April 1982 noted that, under the Federal Food, Drug and Cosmetic (FDC) Act, a drug approved for marketing may be labeled and advertised only for those uses that the FDA has approved. These are referred to as "approved uses." "The FDC Act does not, however limit the manner in which a physician may use an approved drug. Once a product has been approved for marketing, a physician may prescribe it for uses or in treatment regimes or patient populations that are not included in approved labeling. Such 'unapproved' or, more precisely 'unlabeled' uses may be appropriate and rational in certain circumstances, and may, in fact, reflect approaches to drug therapy that have been extensively reported in medical literature. The term 'unapproved uses' is, to some extent, misleading . . . accepted medical practice often includes drug use that is not reflected in approved drug labeling."[7]

"If your application for a drug is medically sound and the patient gives informed consent, lack of FDA approval should not prevent you from using your individual judgment in treating her."[8] An informed consent procedure for leuprolide therapy for uterine fibroid tumors is described.[1] The general consensus is that GnRH analog therapy for myomas, although not approved for this purpose by the FDA, is appropriate for selected cases with informed consent.[9] In addition, despite lack of FDA approval, the extensive literature makes valid a request to have insurance pay for the medication as standard practice.

In clinical trials, there has been a remarkable agreement among investigators regarding the effect of GnRH analog on uterine myomas.[9,10] There is a reversible menopause and amenorrhea induced by anterior pituitary blocking of gonadotropins and hypogonadism. The cessation of menorrhagia permits a rise in hemoglobin. Friedman et al. treated 18 women with 3.75 mg intramuscular leuprolide depot every 4 weeks for 24 weeks.[11] There was a mean reduction in pretreatment volume of 40% from 505 ± 93 cm^3 to 305 ± 57 cm^3 achieved after 12 weeks, and to 307 ± 57 cm^3 after 24 weeks. All 18 women had hot flashes, and 2 discontinued therapy because of this. Four to six had irregular vaginal bleeding, peripheral edema, insomnia, depression, joint pain, or headache. There were small increases in total cholesterol and decreases in high-density lipoprotein cholesterol (similar to postmenopausal women) as well as in calcium, albumin, phosphate, serum glutamic oxaloacetic transaminase, and alkaline phosphatase of unknown mechanism. There were no changes in bone density as measured by single-photon absorptiometry of the ultradistal radius. After treatment the mean uterine volume increased to 88% of pretreatment size in 3 months. All cases except one had a return of menses within 85 days of the last injection (average 71 days).[11]

There has been general agreemeent that after 3 months of GnRH analog therapy, the maximal reduction of uterine volume is achieved, usually 40 to 60%.[10] The reduction in uterine volume parallels the degree of hypoestrogenism.[12] There is also general agreement that by 3 to 4 months after cessation of GnRH analog therapy the myomas have already regrown to about 80% of their original size. Nevertheless, recurrence of symptoms may be slower to return.

Doppler measurement of uterine arterial blood flow velocity waveforms of patients with large myomas being treated with GnRH agonist (nasal Buseralin) showed significant increases of resistance index, indicating a reduction in blood flow.[13] Friedman et al. noted a strong correlation between total intraoperative blood loss and preoperative uterine volume with myomectomy surgery.[14] They reported a significant reduction in intraoperative blood loss in patients treated with intramuscular depot leuprolide 3.75 mg every 4 weeks for 12 weeks if the uterine volume was large (600 cm^3 or larger) before leuprolide treatment. They attributed this to both the smaller size of and decreased blood flow to the uterus and the leiomyoma compared to an equivalent untreated group. They found significant degeneration of leiomyoma in three of nine myomectomy cases treated with leuprolide. The leiomyoma did not "shell out" neatly and required piecemeal removal; however, this did not significantly lengthen the time for surgery. Friedman et al. believe that depot leuprolide therapy should be considered to correct menorrhagia-induced anemia and to reduce the size of large uteri and thereby reduce intraoperative blood loss. These effects would reduce the need for blood transfusion. Their leuprolide-treated patients had a return of menses within 60 days (range from 26 to 74 days). Their myomectomy technique utilized a broad ligament tourniquet and ovarian vascular clamps but did not use a vasoconstrictor. The authors did not explain whether this was their usual technique or whether it was modified for their study. Their series was not large enough to determine whether there was an advantage of leuprolide for the only modestly enlarged uterus.[14] Others have also reported reduced blood loss at myomectomy[15,16] and hysterectomy for myomas after presurgical treatment with GnRH analog.[17]

There has been concern about osteoporosis with GnRH analog treatment longer than 6 months. Most reports indicate no or only insignificant and reversible bone loss.[18,19]

A fascinating new pilot study involved the use of daily leuprolide injection alone for 3 months to reduce fibroid size and then continued with the addition of oral conjugated estrogen (days 1 through 25) and medroxyprogesterone acetate (days 16 through 25) for 2 years. The fibroids remained small, there was no bone loss, and hot flashes were relieved.[20] Medroxyprogesterone acetate given simultaneously with GnRH analog prevents hot flashes and permits increase of hemoglobin but prevents reduction in fibroid size.[21]

There have been a few cases of restoration of fertility after GnRH analog therapy, perhaps by reducing myoma pressure on a fallopian tube or reduction of a submucous fibroid.[22,23] It is not generally recommended as the sole treatment because the fibroids tend to grow back after treatment and during pregnancy.

Although uterine staining is not unusual during treatment, Friedman reported three cases of heavy vaginal bleeding due to hyaline degeneration of submucous fibroids requiring blood transfusions and emergency myomectomies.[24]

A disadvantage in treating uterine leiomyoma of the uterus is the possibility of unrecognized leiomyosarcoma. In a case reported by Meyer et al., a 46-year-old woman with a history of uterine leiomyoma, a slowly progressive uterine growth from a 12- to 14-weeks' size over 2 years, menorrhagia, and anemia desired an alternative to abdominal hysterectomy.[25] Transvaginal sonography showed several uterine solid homogeneous masses, with the largest 8 cm. None were tender. Endometrial sampling was benign. She was treated with daily subcutaneous injections of leuprolide and oral iron for 3 months. The menorrhagia ceased and the anemia was cured but the uterus remained 14 weeks in size. At myomectomy, an 8-cm soft, fleshy, fundal myoma was removed and found to have individual areas of hemorrhage and necrosis on bisection. The microscopic appearance was leiomyosarcoma. The clinical picture was characteristic of leiomyoma until the myomectomy, although reduction in uterine and myoma size would have been expected. Nevertheless, the authors pointed out that not all myomas get smaller. There was no repeat sonography prior to the myomectomy, so one cannot be certain whether the areas of necrosis could have been identified or whether the GnRH analog therapy caused the necrosis. Progestin therapy of myomas is known to cause degeneration, increased cellularity, and mitotic activity in myomas. This also raises the question of whether benign myomas may develop a pseudomalignant appearance after GnRH analog therapy.[25] There is a different prognosis in sarcoma developing in a leiomyoma versus a primary leiomyosarcoma.

Tamoxifen is a nonsteroidal antiestrogen used to treat metastatic and postmenopausal breast cancer. Chronic therapy has been associated with an estrogenic vaginal effect, endometrial stimulation causing postmenopausal bleeding, endometrial hyperplasia, endometrial and endocervical polyps,[26] endometrial carcinoma,[27,28] and postmenopausal reactivation of endometriosis.[29] It is possible, therefore, that tamoxifen might stimulate myomatous growth in the postmenopausal woman.

It has been suggest that GnRH agonist administration may help identify those patients who might benefit from surgical castration in benign metastasizing leiomyoma. A case report showed improvement of lung leiomyoma metastasis.[30]

Consensus Regarding Use

The current consensus is that GnRH analog therapy for fibroids, despite the lack of specific FDA approval, may be considered as a temporary medical alternative for 3 to 6 months prior to surgery. The therapy will stop heavy menstrual bleeding, thereby allowing correction of anemia from menorrhagia and permitting autologous transfusion of the patient donating her own blood prior to her surgery. The therapy will reduce total uterine size by about 50%, reduce uterine artery blood flow, and reduce blood loss at surgery. Thus, it may increase the chance of myomectomy versus hysterectomy, permit a Pfannenstiel abdominal incision rather than a vertical incision, permit consideration of vaginal versus abdominal hysterectomy, and permit consideration of a hysteroscopic resection of a submucous fibroid.[10,31] To a lesser extent GnRH analog therapy may be used to defer surgery for a more convenient time.

Consideration of therapy to delay surgery in the perimenopausal woman in anticipation of natural menopause is an experimental use. The problem is that therapy is usually limited to 6 months and it may be uncertain when natural menopause will occur.

Most clinicians believe that infertile women should have a myomectomy after therapy before attempting pregnancy because of rapid regrowth of fibroids and because of possible pregnancy stimulation of myomas.

The usual side effect of GnRH agonist therapy is hot flashes, which may occur in most patients. Their incidence can be reduced by oral medroxyprogesterone 20 mg daily, but this has been found to prevent the shrinkage of the fibroids.[21] Abnormal vaginal bleeding occurs in up to 67% of cases, usually in the first month. Less common side effects are vaginal dryness, dyspareunia, dry mouth, and rarely headache, nasal congestion, insomnia, reduction in breast size, decreased libido, joint swelling, mood swings, and depression.[2]

The contraindications to GnRH agonist therapy include a rapidly enlarging fibroid uterus (growth of 3 or more gestational weeks during a 6-month period); suspicion of uterine, ovarian, or cervical malignancy; osteoporosis; untreated atypical adenomatous hyperplasia; pregnancy; and undiagnosed vaginal bleeding.[2] Barrier contraception is advised during therapy.

Nafarelin

Synarel (nafarelin acetate) nasal spray solution has just been approved (1990) by the FDA for use in the management of endometriosis. It is a synthetic agonist analog of naturally occurring GnRH D-Napthylalanine has been substituted for the amino acid glycine in position 6. This makes the molecule very hydrophobic, which favors its interaction with biologic membranes and hydrophobic carrier proteins and retards cleavage. The serum half-life is 3 hours. Each 0.5-

ounce bottle contains 10 mL Synarel nasal solution as 2 mg/mL (as nafarelin base). There is a metered spray that delivers 200 μg of Nafarelin per spray. With the usual daily dose of 400 μg, one bottle is intended to provide a 30-day (60 sprays) supply. It is administered as one spray into one nostril in the morning and one spray in the other nostril in the evening.[32] Treatment should be started between day 2 and day 4 of the menstrual cycle. If the 400-μg daily dose does not produce amenorrhea after 2 months of treatment, the dose may be increased to 800 μg daily given as one spray into each nostril in the morning and again in the evening, making a total of four sprays daily. The recommended duration of treatment is 6 months. Safety data on retreatment are not available. The manufacturer recommends that if further treatment is contemplated bone density should be assessed.[32]

At the end of 6 months of initial treatment for endometriosis, 60% of those who took Synarel 400 μg daily were symptom free, 32% had mild symptoms, 7% had moderate symptoms, and 1% had severe symptoms.[32] Of the 60% who had complete relief, 50% were symptom free 6 months after treatment was discontinued, 33% had mild symptoms, and 17% had moderate symptoms. Pregnancy rates were unaffected. There are apparently no drug interactions or overdose reactions. There was a 0.2% immediate hypersensitivity reaction. There was a small loss of bone density and a slight increase in mean total cholesterol.

At the onset of treatment nafarelin stimulates the release of pituitary gonadotropins, resulting in a temporary increase of ovarian steroidogenesis; however, by 4 weeks there is a decreased secretion of gonadal steroids. Nonhormonal contraception is advised even though theoretically the treatment should be contraceptive because missing successive doses may result in breakthrough bleeding or ovulation. The contraindications include hypersensitivity to GnRH, GnRH analogs, or excipients (benzalkonium chloride, acetic acid, and sorbital); undiagnosed abnormal vaginal bleeding; pregnancy (anomalies due to hormonal changes were found in rats); and breastfeeding. Aside from amenorrhea, there may be a 90% incidence of hot flashes, a 20% incidence of headaches, emotional lability, decreased libido, vaginal dryness, and about a 10% incidence of acne, myalgia, and reduction in breast size. Ten percent have nasal irritation. Normal pituitary-gonadal hormonal function is usually restored in 4 to 8 weeks after treatment is discontinued.[32]

Like Lupron depot, nafarelin may be used by the physician at his/her discretion and with the understanding of the patient for use in other conditions. Although one previous study[2] of nasal spray GnRH analog therapy found it less reliable than administration by injection for fibroids, with consciencious use it should give similar results. The approximate cost for Lupron depot monthly injection is about $315, whereas the Synarel (nafarelin nasal spray) for 1 month is about $280.

Nafarelin nasal spray 400 μg twice daily for a total daily dose of 800 μg for 6 months, when used in women with myomas, resulted in a decrease in serum gonadotropins and estradiol, a mean decrease of large myomas of 46 ± 9%, and a decrease of uterine volume of 57% ± 7% (as determined by MRI).[16] Clinically, there was amenorrhea, restoration of hemoglobin, and relief of pelvic pressure symptoms. The response of individual myomas was variable, ranging from 7% to over 90%. Nevertheless, even when myomas responded poorly to nafarelin, the myometrium almost always decreased in thickness. The decrease in total uterine volume was sometimes much greater than the decrease in myoma size. On pelvic examination, the uterus decreased in size in 8 of 11 cases. Posttreatment myomectomy resulted in easier separation of myomas from surrounding myometrium and with only little blood loss. One patient discontinued treatment after 5 months because of depression, which resolved after the treatment was stopped. All cases experienced hot flashes. Myomas often reenlarge after treatment is stopped.[16]

Indications for Surgery

The Nonpregnant Patient

Buttram and Reiter formulated a general guideline for leiomyomata[33] but it must be individualized for each case. For the symptomatic patient (abnormal bleeding, pain, pressure) who desires fertility and has no other cause of infertility, then multiple myomectomy is offered. There is special concern for intramural and submucous fibroids 2 cm or larger and for subserous tumors which may compress a fallopian tube.

The symptomatic patient not desirous of fertility with "uncontrolled or debilitating symptoms" should have a dilation and curettage of the uterus and hysterectomy. Exceptions might be relief of symptoms after vaginal removal of a pedunculated submucous leiomyoma or the patient about to go into menopause.

For asymptomatic patients with the usual slow-growing leiomyomata with a uterus less than 10 to 12 weeks' gestational size, only 6-month observation intervals are necessary regardless of fertility status. For those with a uterus 10 to 12 weeks' size or larger, or if the uterus is growing rapidly (6 weeks' size enlargement within a year), management depends on the desire for fertility. If the patient wishes to be pregnant then a 6- to 12-month observation trial is given. If the patient wishes to delay pregnancy for a year or longer and has a uterus 10 to 12 weeks or larger in size or has a rapidly growing fibroid uterus, then myomectomy is advised. The results of myomectomy in smaller uteri are better than in very large uteri regarding subsequent fertility, blood loss, adhesions, and the need to do a hysterectomy. Patients who have been followed for a long period, have a stable-sized uterus, and wish to delay pregnancy may be simply observed even if the uterus is 10 to 12 weeks' size or larger.

For the asymptomatic patient who does not desire fertility, and has a 10- to 12-week or larger uterus or a rapidly growing uterus, hysterectomy is considered. The reasons are the risks of impingement on adjacent structures and of concealing ovarian pathology. This is especially true of the postmenopausal woman. For the asymptomatic patient under age 30 who does not desire pregnancy, with a 10- to 12-weeks' size nongrowing uterus, observation is appropriate.[33]

In former years, myomectomy rather than hysterectomy was reserved for the woman under age 35 who was married with a fertile husband and who had a strong desire for pregnancy. At present, marriage is not essential and there does not appear to be an absolute age limit. Women may wish to retain the uterus for fertility or for emotional reasons to continue menstruation.

At menopause, there will not be further growth and often some regression is noted. Nevertheless, infrequently, postmenopausal bleeding may occur with myometrial shrinkage, causing the myoma to become submucous.[34] Occasionally, one encounters an elderly woman with an over 12-weeks' size asymptomatic uterus (often with calcification) that has been documented as being the same size for many years. Depending on the informed patient's wishes, it may be reasonable to continue to carefully observe the patient and do periodic sonograms to detect any ovarian neoplasm.

The usual indications for surgery for myomas include:

1. Abnormal bleeding
2. Pain
3. Pressure symptoms
4. Size greater than 12 weeks' gestation
5. Rapid growth
6. Reproductive disturbance
7. Confusion with ovarian neoplasm
8. Coincidental

These indications are relative and variable and require judgment. The bleeding is usually menorrhagia associated with submucous fibroids, often with sudden "gushing" during the menses. The amount of blood loss may be estimated by the number of pads and extent of the soaking and by the hemoglobin. Severe menorrhagia can result in marked anemia despite oral iron, and patients who refuse care sometimes collapse with menses. Menorrhagia can be a cause of anemia of unknown origin if there is an inadequate history.

Pain may vary from slight discomfort to intense pain similar to that of an acute surgical abdomen.

A midline lower anterior myoma may compress the bladder and reduce its capacity. Intraligamentous myoma may compress a ureter and silently cause hydronephrosis.

The traditional size limit of 12 weeks is because a larger uterus fills the pelvis, making it difficult to palpate any ovarian enlargement, and because of an increased chance of a vascular accident to a myoma. Depending on the informed patient's desires and the philosophy of the physician, some patients are simply carefully observed for even the larger sized uterus. Rapid growth or growth after menopause requires surgery.

Pedunculated subserous and intraligamentous myomas may simulate ovarian neoplasms on physical examination and sonography, and surgery may be appropriate.

In select circumstances surgery is done because of another primary indication. Recently, I performed a subtotal hysterectomy for a large myomatous uterus at the time of a colon carcinoma resection since the bulky uterus would have made subsequent follow-up examination difficult. Hysterectomy or myomectomy may be considered when surgery is done for ovarian neoplasm, endometriosis, or tubal reconstruction.

Myomectomy for Infertility

Most infertility is not due to myomas. Among black women, fibroids were the sole cause of infertility in 12%, whereas among whites they were the sole cause in 3%. Postponing pregnancy will increase the incidence of leiomyomata-associated infertility. It is generally agreed that submucous myomas may predispose to infertility, spontaneous abortion, premature labor, abnormal fetal presentation, and infection. A large cervical or lower segment fibroid may obstruct labor. Fibroids may obstruct fallopian tubes and cause infertility.[33]

The indication for myomectomy in the treatment of infertility remains unclear except for cases with submucous fibroids or cases in which the fibroid obstructs a fallopian tube and when other causes of infertility have been excluded.[35] For large submucosal leiomyomas with no other cause of infertility, myomectomy gave a corrected pregnancy rate of 61.5%, usually within 11 months after surgery. Only one case conceived as late as 3 years. Submucosal leiomyomas were considered a cause of infertility and were amendable to surgical treatment with restoration of fertility.[36]

A report from Johns Hopkins described 46 cases (34 primary infertility, 12 secondary infertility) with no other detectable cause for infertility except myomas; all patients had myomectomy. After surgery, 38% of primary and 50% of secondary infertility patients had full-term pregnancies. Preoperative distortion of the endometrial cavity was not impressively correlated with the postoperative prognosis.[37] A report from Yale of 50 cases of myomectomy showed that 25 of the patients subsequently conceived. Of those who had surgery for a pelvic mass, 68% conceived, whereas only 16% conceived if there had been a normal infertility evaluation. Women over age 30 who became pregnant had fewer and smaller fibroids. It was thought that myomectomy might even decrease fertility by causing adhesions.[38] Rosenfield reported 23 cases with otherwise unexplained infertility with myomectomies for subserous or intramural myomas (none were submucosal). Of these patients 15 (65.2%) conceived, all but one within the first year. There were 18 term births in 13 patients. It was not possible to predict which cases would be successful.[39] Other series emphasize the prognostic value of duration of infertility prior to myomectomy.[40]

Thus, it seems that there is a 50% chance of pregnancy after myomectomy, but there are many confounding factors, including other known and unknown causes of infertility. There is an increased chance of cesarean section after myomectomy, especially for large submucous or intramural myomas, when the endometrial cavity is entered, and when there is a postoperative uterine infection.

Leiomyomata in Pregnancy

The incidence of uterine leiomyomas in pregnancy is underestimated because most are asymptomatic and most women do not have routine sonography. To make the situation more complex, it is difficult to distinguish localized uterine contractions from myomas in early pregnancy by sonography. The incidence of leiomyomas is increased in black and older women.

Most fibroids in pregnancy are asymptomatic and do not grow. A sonographic study of 29 women with leiomyoma during pregnancy showed that in 78% there was no size increase. In 22%, the fibroids increased in size but not more than 25%.[41] Another sonographic study of 113 women found that in the second trimester, small fibroids increased and large fibroids decreased in size. Most cases of fibroids in pregnancy do not cause any problem. In the third trimester, fibroids generally decreased in size. Significant degeneration of fibroids in pregnancy is associated with severe pain and a characteristic sonographic pattern (heterogeneous, or anechoic/cystic spaces). Torsion of a pedunculated subserous fibroid may also cause pain. When problems do occur, there may be a tendency for lower segment fibroids to be associated with retained placenta and increased cesarean section rate. Fibroids of the uterine corpus, especially submucous, may be associated with early abortions. Multiple fibroids may increase the incidence of malpresentation and premature contractions.[42] Occasionally, large submucous fibroids may cause fetal head deformation.

Carneous degeneration is the most common complication of fibroids in pregnancy. The usual treatment is conservative—bed rest, analgesics, and observation. Myomectomy for an intramural necrotic fibroid usually causes severe bleeding and abortion.

A retrospective study of myomas in pregnancy reported an incidence of 8.6 cases per 10,000 deliveries (106 in 123,492 deliveries).[43] Of the 106, 62% were asymptomatic, 13% had premature labor, and 25% were hospitalized because of pain. Laparotomy was done on 14 (13%) because of symptoms or uncertain diagnosis, and of these 5 had only an exploratory laparotomy. Six had myomectomies because of an abdominal mass and pain for pedunculated (subserous) myomas with stalk widths from 2 to 5 cm, without fetal loss. "Only myomata that are pedunculated should be considered for removal."[43] The authors agreed that most authorities caution against myomectomy at cesarean section; however, they reported 13 cases of incidental myomectomy with only one case complicated by uterine artery ligation and transfusion. Their conclusion was that "elective myomectomy at Cesarean delivery may be safe in carefully chosen patients." These fibroids were submucous, intramural, and subserosal and 9 of the 13 had been asymptomatic and were an incidental finding.[43]

Traditionally obstetricians only perform myomectomies at cesarean section if the myoma is subserous and on a narrow pedicle or if it is in the line of the incision and prevents closure; "the removal of intramural leiomyomata from the pregnant uterus (at Cesarean section) is inadvisable because of the recognized difficulty in controlling blood loss."[34] It is true that occasionally a submucous fibroid may become necrotic and infected in the puerperium.

Myomectomy

Abdominal Myomectomy

Philosophy and Preparation

Stangel's philosophy is to resect all fibroids of significant size or location that would compromise the uterine cavity or tubal function.[44] If necessary, some fibroids are allowed to remain. Even if the fallopian tubes cannot be salvaged, the uterus may be preserved for possible future in vitro fertilization. If necessary to control bleeding to avoid a hysterectomy, then bilateral ascending uterine artery ligation is done.[44] Stangel's philosophy is important to understand because in the past nonfunctioning fallopian tubes were considered a reason not to do a myomectomy, and hysterectomy was done if surgery was necessary for symptoms. The patient should be aware of this newer concept, and the patient should participate in the decision by discussion in advance of the surgery.

The patient should be aware of the possibility of the need for transfusion and, if desired and feasible, of donating blood for herself in advance (see "Gonadotropin-Releasing Hormone Agonist Analog Therapy").

The patient should be aware that myomectomy may not be possible because:

1. Unexpected conditions may be discovered at surgery, such as:
   a. an ovarian neoplasm may be adherent to the uterus and simulate uterine myomas
   b. the uterine myoma may be diffuse adenomyosis or even localized adenomyosis (which does not "shell out")
   c. the uterine myoma may be a malignant sarcoma
2. The fallopian tubes may not be able to be repaired (depending on the wishes of the patient)
3. Dangerous hemorrhage may develop during the surgery

The long range planning should consider the patient's age, symptoms, wishes, future pregnancy, and alternative approaches. The immediate planning should consider work and home responsibilities. The surgery is best done shortly after a menstrual period is completed to reduce bleeding tendency.

The diagnosis should be confirmed by sonography for the total size of the uterus and ideally by hysterosalpingography to detect a filling defect of a submucous fibroid and to test the fallopian tubes. Hysteroscopy has been done less often. Although MRI can detect the submucous location of myomas, it is not readily available, is expensive, and may be emotionally disturbing to patients with claustrophobia. Vaginal sonography is helpful for deep pelvic masses.

Technique

Preoperative Considerations

Most myomectomies are done through the abdominal route for access to and evaluation of the entire pelvis. Although in the past some institutions routinely advised a dilation and curettage just prior to any laparotomy for diagnostic purposes, such a procedure may traumatize the endometrium and introduce bacteria and is best avoided. Correspondingly, some also avoid hysteroscopy or hysterosalpingography immediately preceding the myomectomy. These procedures are done in advance to determine the presence of submucous myomas and whether they are on the anterior or posterior walls.

Just prior to the myomectomy, Stangel injects diluted methylene blue through the cervix to stain the endometrium to simplify reconstruction after removal of a large submucous myoma.[44] Others use a diluted indigo-carmine dye through the cervix or transfundally if tubal surgery is required in order not to stain the tissue. There are differences of opinion as to whether the endometrium should be stained.

Many use prophylactic antibiotics, usually broad spectrum. If there is the possibility of entering the endometrial cavity, some also add an antianaerobic preparation such as metronidazole. Some use doxycycline routinely.

Usually general endotracheal anesthesia rather than conduction anesthesia is used because of the possibility of hemorrhage and for time flexibility. "Hypotensive anesthesia" is used by some.[34] The patient is examined under anesthesia.

Stangel advises packing the vagina to elevate the uterus for exposure and to stretch the uterine arteries to reduce bleeding.[44] A Foley catheter is made indwelling in the bladder. Some advise cervical dilation to facilitate postoperative drainage.[34]

Usually a low transverse abdominal incision is used, such as a Pfannenstiel or, if more room is needed, a Maylard incision. For large myomas (over 12 weeks' size) or if the diagnosis is uncertain, a vertical incision is considered.

Operative Planning and Philosophy

On opening the abdomen a careful evaluation is made of the pelvis, including the size, number, and location of myomas; the condition of the fallopian tubes and ovaries; whether the uterus is retroverted; and whether there is endometriosis. Other structures are checked, such as the appendix, and the upper abdomen is palpated. Usually the appendix is not removed to minimize the chance of infection. In leisurely fashion, a plan is made of the surgical approach, although once the myomectomy itself is started expeditions surgery will keep blood loss to a minimum. Some place a moist Kerlex gauze in the posterior cul-de-sac to lift the uterus and to prevent accumulation of blood.[33]

A microsurgical philosophy is used—minimum and only delicate handling of tissues, minimum trauma, having all powder washed of the gloves, avoiding abrasive wiping and pads, keeping tissues moist with irrigating solution, meticulous hemostasis, and using fine nonreactive sutures. The purpose is to avoid postoperative adhesions, bleeding, and infection. In general, the two major problems are prevention of bleeding and avoiding adhesions.

In former years, the tourniquet technique was used routinely. This involved making small symmetrical openings through the anterior and posterior leaves of the broad ligaments on each side of the cervix, through which was passed a rubber urethral catheter or a 0.25-inch Penrose drain, which was tied or clamped posteriorly sufficiently tightly to occlude the uterine arteries. Sometimes, to avoid bladder trauma, especially when there were anterior wall myomas, and to later cover the anterior wall of the uterus, the vesicouterine fold of peritoneum was incised transversely and the bladder mobilized downward. Depending on the location of the myomas, the infundibulopelvic ligaments with their vessels were sometimes occluded by a Penrose drain tourniquet using the broad ligament hole, or by a vascular clamp.

Although the tourniquet technique is still recommended by many authors, in recent years there has been a tendency to avoid it if possible as part of a microsurgical philosophy to keep trauma to a minimum and thereby reduce the possibility of postoperative adhesions. Some recommend the tourniquet technique and hypotensive anesthesia only for extensive myomectomies.[34]

Older techniques that have been avoided in recent years as part of a microsurgical philosophy include: routine mobilization of the bladder flatp peritoneum to cover incisions of the anterior uterus, suturing peritoneal omental grafts to cover uterine incisions, and suturing sigmoid colon over posterior uterine incisions. These techniques had been used to reduce adhesions but are self-defeating.

Hemostasis is significantly improved by injection of diluted vasopressin (Pitressin; Parke-Davis) into the myometrium surrounding the myoma with a fine needle prior to incision. Some inject the branches of the uterine arteries bilaterally instead of or in addition to injection around the myomas. Pitressin is supplied as an ampule containing 0.5 mL (10 pressor units) or 1.0 mL (20 pressor units). It is a synthetic 8-arginine vasopressin of the posterior pituitary. It causes contraction of smooth muscle of all parts of the vascular bed, especially capillaries, small arterioles, and small venules. It will not stop bleeding from arteries. It also has an antidiuretic action. When used it is diluted with 20 to 30 mL of Ringer's lactate or saline, and only the smallest effective volume is given to blanch the tissues. The anesthetist should be alerted since there may be side effects and it may be contraindicated with certain heart disease. Claims of reduced blood loss with laser or an electric knife may be partly related to the use of vasopressin. Vasopressin must be differentiated from oxytocin (Pitocin). Whereas sometimes oxytocin (Pitocin) is given in a 1-mL ampule intravenous bolus, vasopressin (Pitressin) is never given that way.

Myomectomy

The major decision is where to incise the uterus. In general, anterior wall myomas are removed from an anterior midline incision and posterior myomas from a posterior midline incision. Sometimes, preoperative hysterograms may be confusing or the myomas may be lateral. A simple but reliable method that I use is to find the proximal attachment of the tubes (and adjacent round ligament of the uterus and round ligament of the ovary) to the lateral uterine fundus and draw an imaginary line between these two sites. If most of the bulge of the uterine mass is anterior to these lateral sites, then the myoma is probably in the anterior wall. This approach is especially helpful for the solitary, huge, round myoma that appears to be in the center of the fundus. Ideally, the incision is midline to avoid the lateral tubes, and to reduce blood loss, and directly over a myoma or over several myomas that could be removed through the initial primary incision into the uterus and then through secondary deeper lateral incisions.

The incision may be made by cold steel scalpel, needle-tip electrode with minimal current, or laser, although the latter two cause less bleeding and possibly less trauma. The myoma is reached and then grasped with a tenaculum and upward traction is applied. The myoma is gently shelled out of the pseudocapsule of the surrounding compressed myometrium by blunt and sharp dissection or needle eleectrode or laser beam. If vessels in a pedicle base are identified, they are clamped and ligated with 3-0, fine, absorbable, nonreactive sutures. The surgeon's finger gently explores the defect to determine whether the endometrial cavity has been entered and to palpate for adjacent myomas. The latter are removed through deep secondary incisions.

Any opening into the uterine cavity is carefully closed by anatomic reconstruction using submucosal sutures, avoiding the endometrium if feasible. Meticulous hemostasis is accomplished by the use of diluted vasopressin injection into the surrounding myometrium, ligation of vessels, and suturing to close the myoma defect. The latter has been traditionally closed in three layers with 2-0 or 3-0 sutures. A large, heavy needle may be necessary to avoid breakage in a secondary defect. Buttram and Reiter indicated that figure-of-8 2-0 Vicryl (V) or Dexon (D) sutures may be necessary in some cases for hemostasis and to close the dead space. Their message is that "Interrupted V or D sutures spaced approximately 1 cm. apart are utilized in closing the myometrial incision in an effort to use as little suture material as possible. The sutures are placed just beneath the serosa, traverse the myometrium (and the endometrium if the uterine cavity has been opened) and are tied beneath the serosa. Layer closure as advocated by Bonney and others usually is necessary and increases the risk of myometrial necrosis and inflammation."[33] This is an important new concept of suturing for gentle approximation of tissue rather than for hemostasis, with multiple, tight layers causing ischemia. Many now use 3-0 sutures.

If necessary for hemostasis, each ascending uterine artery may be ligated with a single 0 mass suture through the broad ligament and incorporating the side of the cervix.[44] The excessive redundant serosa and surface myometrium may be excised, but keep in mind that it tends to shrink. The serosa may be closed with a 4-0, 5-0, or 6-0 fine, nonreactive absorbable synthetic suture in a subserosal imbricating, or "baseball" partial imbricating, fashion to reduce adhesion formation.

In general most recommend that posterior wall myomas be removed by a separate posterior wall incision (similar to that for anterior wall myomas) that is midline, vertical, and over the myoma(s) and through which all the posterior wall tumors can be removed by the primary or deeper secondary incisions. Where the endometrial cavity has already been entered for removal of anterior submucous fibroids, Buttram and Reiter advocated "transcavity enucleation" of posterior wall fibroids.[33] The purpose is to avoid a separate posterior serosal wall incision, which has a high risk of postoperative adhesions involving tubes and ovaries. If the endometrial cavity is not opened then they advise removal of posterior lower uterine segment myomas through a posterior midline incision. Upper posterior fundal myomas are removed if possible by an extension of the anterior fundal incision.[33]

Buttram and Reiter often perform a uterine suspension empirically to reduce adhesions after myomectomy by triplication of the round ligaments or a Gilliam uterine suspension and/or uterosacralplication. They also use already stretched round ligaments to cover the uterine incision.[33]

Lateral intraligamentous myomas require special care because as they enlarge they come to lie just over the ureter and lateral to the uterine vessels. The broad ligament is opened and the myoma carefully sharply dissected and often bisected in removal.

Cervical myomas comprise only about 5% of myomas but pose a technical challenge. Sometimes the bladder is mobilized downward and a midline anterior incision is made over the cervix and anterior fornix. Careful posterior cul-de-sac dissection may be needed for posterior cervical fibroids. Cervical fibroids are sometimes removed through the vagina.

A frequently used irrigating solution is Ringer's lactate alone or with 5,000 units of heparin/L. It is used to wash away clots, to precisely identify bleeding vessels, and to keep tissues moist. Some use corticosteroids in irrigating solutions. Some routinely use a Jackson-Pratt drain for the cul-de-sac, with a separate stab incision. The drain is removed after 1 to 2 days.

Techniques for Reducing Adhesions

With closure of the peritoneum some favor the instillation of 100 to 200 mL of 32% dextran 70 in glucose (Hyskon) or saline to reduce adhesions by hydroflotation and siliconizing effects and by reducing platelet adhesion. The large macromolecules absorb water into the peritoneal cavity and may cause hypovolemia. There is a transient prolongation of the bleeding time and a small risk of anaphylaxis hypersensitivity. Because of the latter, some have discontinued the use of dextran. Others have used intraperitoneal glucocorticoids, promethazine, nonsteroidal antiinflammatory drugs, and Ringer's lactate with and without heparin. Other methods that have been used to prevent postoperative adhesions include administration of dexamethasone 20 mg and promethazine 25 mg in various combinations of one or more doses each intramuscularly preoperatively, intraperitoneally, and/or intramuscularly postoperatively.

Interceed (Johnson & Johnson) absorbable adhesion barrier (oxidized regenerated cellulose fabric) has recently been marketed as an adjunct in gynecologic pelvic surgery to reduce postoperative pelvic adhesions after hemostasis is achieved. It reduced adhesions after pelvic sidewall surgery. It is not a hemostatic agent and its effectiveness is reduced when saturated with blood. Its safety and effectiveness in combination with other adhesion prevention treatment has not been established. Investigation is now going on regarding its use in covering myometrial incisions. Other research involves calcium channel blockers and plasminogen activator.

Laser Myomectomy

A recent invovation in abdominal myomectomy has been the use of the laser to replace the cold steel scalpel. Although it has not received general acclaim, proponents claim decreased blood loss, less tissue destruction, shorter operating time, and good reproductive results.[45,46] Reported series are small and uncontrolled. A microsurgical philosophy and dilute vasopressin injection of themselves would be expected to improved results.

The most frequently used laser for gynecology is the carbon dioxide laser, which is used for vaporization for ablation or excision. The beam is absorbed by water or water-containing tissue in the first 100 μm of tissue depth. Deeper coagulation is related to the heat of vaporization. High power densities and rapid incisions result in decreased coagulation, hemostasis, and necrotic tissue. With low power density and defocusing there is an increase in coagulation effect. The hand-held laser gives a precise hemostatic incision with coagulation of small vessels up to 2 mm in diameter, can evaporate small myomas, can be used to dissect out myomas that are held under traction, and in defocused beam can lightly treat the pseudocapsule and bed to reduce bleeding. The special training required reminds the surgeon to use careful technique. Lasers are expensive and special precautions are used, including eye protection against reflected laser beams. In contrast, proponents of electric needle surgery appear happy with their results, and there is no need for expensive equipment and special training.

Surgical Outcome

After myomectomy for infertility about 40% of women become pregnant, usually within the first 2 years. There may be other causes of infertility to confound the data. The preoperative size of the uterus is important, with a better chance of pregnancy if the uterus is 10 weeks in size or smaller versus the 12- to 16-weeks size. The total spontaneous abortion rate was reduced from 41% preoperatively to 19% postmyomectomy. There was an 81% chance of reduction or resolution of menorrhagia.[33] After myomectomy, there is about a 15% chance of recurrence of fibroids and 11% overall of patients will require subsequent surgery for myomas.[33]

Trends in Abdominal Myomectomy

The current trends in abdominal myomectomy emphasizes:

1. A microsurgical philosophy of minimal trauma, meticulous hemostasis, and avoidance of postoperative adhesions.
2. If possible, avoiding the trauma of an elastic tourniquet around the cervix to occlude the uterine vessels, and vascular clamps on the infundibulopelvic ligament.
3. Use of a single midline vertical incision on the anterior or posterior walls over the myomas to avoid the fallopian tubes and minimize bleeding, and if necessary and feasible using secondary deeper incisions.
4. Consideration of injection of diluted vasopressin to reduce small vessel operative bleeding.
5. Consideration of the use of unipolar electric needle or laser for incising the uterus, dissecting out the myoma, and hemostasis of small vessels.
6. If possible, avoiding a tight, layered closure of the uterine defect, which may cause ischemic necrosis and infection; instead use a minimum of interrupted, fine (2-0 or 3-0), nonreactive, synthetic, absorbable sutures.
7. Closure of the serosa with an imbricating fine (4-0 to 6-0), absorbable suture, and not suturing an omental-peritoneal graft or colon over the uterine incision.
8. Consideration of a prophylactic broad-spectrum antibiotic.
9. Participation of the patient in the decision for surgery and whether to do a myomectomy if the fallopian tubes are not functional.

Vaginal Myomectomy

All experienced clinicians have had the unexpected and unnerving experience during a uterine curettage of a ring clamp (polyp forceps, ovum forceps) catching on a small (2- to 3-cm) submucous sessile myoma so that the clamp could not be removed without tearing out the myoma. Usually, there is no excessive bleeding.

Traditionally, the aborting pedunculated submucous fibroid is removed by vaginal myomectomy. There is usually a history of cramps and irregular bleeding. On vaginal examination, the cervix is dilated by an often necrotic myoma. There are often other fibroids in the uterus. When the myoma is removed, great care is required to avoid pulling excessively on the pedicle lest the uterine fundus invert and be cut when the pedicle is amputated. Very often, even if hysterectomy is planned, it may be wise to delay further surgery until the expected endometritis is healed.

Among a group of 46 women with symptomatic prolapsed pedunculated submucous myomas, vaginal myomectomy was successful in 43 and 3 required an abdominal operation. The usual case was simple and quick, with an uneventful postoperative course. After a median follow-up of 5.5 years, only 8.8% of 34 cases required a repeat vaginal myomectomy, and only 5.9% needed a hysterectomy.[47]

A new approach to management of the submucous fibroid is the use of laminaria to achieve a gradual large dilatation of the cervix without laceration, followed by transcervical grasping and removal of the fibroid. The procedure was successful in 83 of 92 cases, thereby avoiding traditional abdominal myomectomy or hysterectomy with minimal morbidity.[48] Nevertheless, this aggressive approach requires experience and judgment since removal of a large sessile submucous myoma could result in heavy bleeding. It would be interesting to have further studies and follow-up since the trauma might cause intrauterine synechiae.

Hysteroscopic Myomectomy

Diagnostic hysteroscopy may be used to verify a submucous fibroid and also to visualize endometrial polyps or pathology, congenital anomalies, uterine tubal ostia, and intrauterine devices. Operative hysteroscopy in selected situations may be used to remove a submucous myoma or for endometrial ablation for severe menorrhagia.

Hallez et al. in Paris used a fine intrauterine resectoscope with a mobile electrical loop and continuous flow to resect 61 submucous leiomyomas. They also used the technique for uterine septa and synechiae.[49] Others have used the hysteroscopic resectoscope for women who refused or were poor risks for hysterectomy or myomectomy. For those with menorrhagia due to submucous fibroids or polyps, 91% had resumption of normal menses. For infertility patients, the term pregnancy rate was 33%. The technique is also used for endometrial ablation for menorrhagia of unknown cause, with a 90% improvement rate. In a group of 90 women, there was one uterine perforation and one case of endometritis.[50]

Operative hysteroscopy is considered by Neuwirth if there is a submucous myoma causing menorrhagia, if the uterus sounds less than 10 cm, and if the uterus is not palpable abdominally.[51] A preoperative hysterogram or observation hysteroscopy is done to identify the myoma. The patient understands that an emergency laparotomy for myomectomy or hysterectomy may be necessary. Prophylactic antibiotics are used. Simultaneous observation laparoscopy is done to avoid uterine perforation. The cervix is held with a tenaculum, the uterine cavity is sounded, and the cervix is dilated to 6 cm. Neuwirth, in an experimental technique, uses a resectoscope hysteroscope to shave the submucous fibroid in a manner similar to that of the urologist who shaves the prostate in a transurethral resection.[51] This requires skill and experience. There is a wire diathermic loop with 30- to 40-W cutting or coagulating current that can be moved back and forth. A large sessile submucous myoma may be shaved down to the level of the endometrial cavity, with the remainder of the myoma left in place in the wall of the uterus. Small or pedunculated submucous myomas may be shaved off and/or extracted directly. Neuwirth uses Hyskon as a distending medium. Shavings are evacuated. Slight bleeding is controlled by direct electrocoagulation. Severe bleeding is controlled by intrauterine 60- to 80-mL balloon tamponade, which may be left in place for several days. Although saline or Ringer's lactate may be used as a distending medium for observation hysteroscopy, the electric wire of the resectoscope requires a nonconducting fluid. Excessive absorption of Hyskon (dextran) may cause an increase in intravascular volume, hyponatremia, and congestive heart failure, or allergic reaction. (Other distending fluids, such as water, glycine, or sorbitol, may also cause hyponatremia.) The patient is kept on bed rest for 12 to 18 hours, the balloon is deflated, the patient ambulates, and then the balloon is removed. She is discharged on the first or second postoperative day. Oral estrogen is prescribed for 10 days to encourage endometrial growth in patients who wish to retain fertility. Pregnancy is feasible even though the myoma has been resected only to the endometrial plane. "The results to date in this series (55 cases) have been excellent, but this approach is not for the novice."[51]

De Charney and his colleagues at Yale allow conception 2 months after surgery and permit vaginal deliveries (see Chap. 16). Others have used hysteroscopic lasers and scissors to treat submucous fibroids. Operative hysteroscopy may be done with the carbon dioxide, neodymium:yttrium-aluminum-garnet argon, and potassium–titanyl phosphate crystal lasers. However, lasers are expensive and complex and require special training.

Hysteroscopic myomectomy is an evolving concept in technology, indications, complications, and long-term results. There is a need for more study, standardization of technique and training, and evaluation of preoperative GnRH analog therapy.

Hysterectomy

Incidence

Hysterectomy is the most frequently performed operation in the United States. Incidence peaked in 1975 with 724,000 hysterectomies, representing a rate of 8.6 per 1,000 women. This has subsequently declined; in 1987 there were 653,000 hysterectomies, representing a rate of 6.7 per 1,000 women. The rates for ages 15 through 44 were 7.0, for 45 through 64 were 8.0, and for 65 and over were 3.4. The rate is lowest in the Northeast. Approximately 25 to 30% are done vaginally and approximately 45% are done with oophorectomies. Women over age 40 are more likely to have oophorectomies. Three conditions account for over half of all hysterectomies: leiomyomas, dysfunctional uterine bleeding, and pelvic relaxation. Sterilization is no longer considered an indication for hysterectomy except when other circumstances make the operation necessary. By age 65, over half of all women will have had a hysterectomy.[52]

"Large differences in hysterectomy rates exist between communities within a single state as well as between regions within the United States."[53] This may be related to differences in training, style of practice, and availability of gynecologists and hospital beds. "Moreover, substantial disagreement persists over the appropriate indications for hysterectomy as compared with those for conservative surgery or alternative therapy."[53]

Indications

The American College of Obstetricians and Gynecologists (ACOG) gynecologic criteria for hysterectomy for leiomyomata[54] are:

1. Asymptomatic myomata associated with a uterine size equal to or larger than that after 12 weeks of gestation (transverse measurement of at least 8 cm. or weight of 280 g. or more), determined by physical examination or ultrasound examination
2. Excessive uterine bleeding evidenced by either a or b:
   a. bleeding for more than 8 days during more than a single cycle and profuse bleeding (for example, large clots, gushes, limitations on activity)
   b. anemia due to acute or chronic blood loss
3. Pelvic discomfort caused by myomata associated with a uterine size equal to or larger than after 12 weeks of gestation
   a. acute and severe
   b. chronic lower abdominal or low back pressure
   c. bladder pressure with urinary frequency not due to urinary tract infection
4. Rapid growth in size of uterus/myomata, to a point equal to or larger than uterine size after 12 weeks of gestation

"Hysterectomy is the procedure of choice" for the patient with myomas who has completed her family or who does not plan to bear children and who has symptoms or a uterine size of a 12-week gestation, or with a rapid growth in a myoma. The symptoms would include pain (degeneration, torsion, dysmenorrhea), bleeding (hypermenorrhea, anemia), and pressure. Sometimes hysterectomies are done at the time of pelvic surgery for other conditions, such as ovarian neoplasm or prolapse.[55]

Individualization is required. The asymptomatic, 12-weeks' size, perimenopausal uterus may get smaller after menopause. There are asymptomatic postmenopausal women with a uterus 12 weeks in size or larger (often calcified) who have not changed under observation for many years. "Leiomyomas that cause troublesome bleeding, pain, or compression of the ureters or that grow rapidly may be readily accepted as an indication for hysterectomy, but criteria for performing hysterectomy to remove asymptomatic leiomyomas are less clear."[53] The risk-benefit ratio for asymptomatic leiomyomas of different sizes is unknown."[53] With vague symptoms one should consider a coincidental psychiatric somatization disorder.[56]

The Normal Ovary

Individualization is also required regarding removal of normal ovaries at the time of hysterectomy. There is a general agreement that ovaries should be removed when the patient is within a few years of or already in menopause, usually at the age range of 45 to 49 or older. The patient's wishes and family history of ovarian cancer are important. Usually, ovaries are preserved below age 40. "There are still no uniformly acceptable and established criteria for removal or retention of normal ovaries at the time of abdominal hysterectomy"[34] (for leiomyomata).

Concerning the possibility of future malignancy in retained ovaries, it does no good to remove one ovary and leave the other. A large collected series showed only a 0.1% chance of ovarian cancer developing in retained ovaries. A statistical analysis suggested 1.5% of retained ovaries will develop an ovarian neoplasm, of which 60% will be malignant. However, there are patients who are predisposed to ovarian neoplasms, and prophylactic bilateral oophorectomy may be considered in the premenopausal patient with Peutz-Jeghers syndrome; a past history of breast, colon, or rectal cancer; or with a strong family history of ovarian cancer. Nevertheless, despite prophylactic oophorectomy, there have been a few patients who later developed intraabdominal malignancy with the histology of ovarian carcinoma, presumably from peritoneal epithelium.[34]

There is a residual ovary syndrome causing an adnexal mass, pain, dyspareunia, or pelvic discomfort thought to be due to impaired blood supply. It may be that 1 to 2% of women with one or both ovaries conserved will subsequently require another operation for a problem with an ovary or tube, such as a nonneoplastic cyst, small hydrosalpinx, or adhesions.[34] There is a subgroup of women with premature loss of ovarian function by unknown mechanism (mean age 45.4 ± 4.0 years versus 49.5 ± 4.04 years for normal loss of ovarian function) who had a hysterectomy at age 44 or less with ovarian preservation.[57]

Prophylactic Antibiotics

The concept of prophylactic antibiotics for elective hysterectomy for benign disease without infection has undergone change. About 25 years ago, most gynecologists did not use antibiotics for prophylaxis and were fearful of allergic reactions and/or of creating bacterial resistance. The present climate has changed with the recognition that postoperative infection is common even in "clean" cases and in part may be related to the presence of potentially pathogenic organisms in the normal endocervical canal. Furthermore, preoperative vaginal antiseptics have not helped much more than physical washing. In recent years, many gynecologists have started to use prophylactic antibiotics prior to surgery and/or for the first 24 hours afterward. Although of limited bacterial coverage, first-generation cephalosporins are found to be effective.

The ACOG Committee on Gynecologic Practice made the following statement on prophylactic use of antibiotics with abdominal hysterectomy[58]: "Based on a review of the literature, the following are important considerations in the prophylactic (24 hours or less) use of antibiotics with gynecologic surgery:

—The incidence of febrile morbidity and serious infection are significant for patients undergoing abdominal hysterectomy.

—Available data suggest that rates of febrile morbidity and infection are reduced by the prophylactic use of antibiotics with abdominal hysterectomy, but not as markedly as with vaginal hysterectomy.

—Although the prophylactic use of antibiotics with abdominal hysterectomy decreases overall morbidity, this usage may not be necessary for certain groups of patients with low risks for wound or pelvic infection.

—A variety of low-cost antimicrobial agents appear to be as effective as higher-cost products."

The Committee on Gynecologic Practice therefore supports the optional perioperative administration of short-course antibiotics as prophylaxis for patients undergoing abdominal hysterectomy.[58]

The Parkland Memorial Hospital of Dallas, Texas, found major operative site infection in 57% of patients after vaginal hysterectomy and 32% after abdominal hysterectomy.[59] The infections requiring parenteral antimicrobial therapy were pelvic cellulitis, infected pelvic hematoma, pelvic abscess, and infected abdominal incision. Single-dose prophylaxis of 1 g of cefazolin was as efficacious as multiple perioperative doses in preventing major postoperative infection. It made no difference whether the cefazolin was given by intramuscular or intravenous route. It reduced infection to 6.3% (vaginal) and 7.6% (abdominal) when elective hysterectomy was done for benign disease. Cefazolin is inexpensive and is not used as a therapeutic agent for acute pelvic infections. Nevertheless, single-dose second- or third-generation cephalosporin or piperacillin in prospective clinical trials may give even better protection (in the range of 3.4%).[59]

Technique

In the classic *Te Linde's Operative Gynecology* (6th ed., 1985), Mattingly and Thompson recommended two changes in the standard Richardson technique of abdominal hysterectomy. The vaginal vault cuff is now left open for drainage with or without a small Penrose drain. This permits drainage and reduces the incidence of cuff cellulitis, abscess, and hematoma. A continuous running locked suture is placed around the cut vaginal margin for hemostasis. It is interesting that the original Richardson technique utilized routine vaginal drainage (as did the Falk technique). Such techniques had been developed in preantibiotic days and were especially helpful where there was associated pelvic inflammatory disease. The second change is the use of a posterior culdeplasty to prevent later vaginal vault prolapse or enterocele. A suture is placed through the uterosacral ligament on one side, then through the cul-de-sac peritoneum into the posterior vaginal fornix lumen and then out to incorporate the opposite uterosacral ligament. When tied, it approximates the sacrouterine ligaments behind the vagina and pulls the posterior vaginal fornix in a posterior direction. Care is taken to avoid the ureters.[34]

Problems and Complications

Even with MRI it is not always possible to differentiate focal adenomyosis from leiomyomata. In one study only 10 of 12 leiomyomas were diagnosed by MRI, and in 2 differentiation was not possible.[60] Microscopic adenomyosis cannot be demonstrated with MRI.[60]

There are rare forms of "benign" leiomyomas—intravenous leiomyomatosis, metastasizing pulmonary leiomyomatosis (with dyspnea),[61] diffuse uterine leiomyomatosis, and disseminated intraperitoneal leiomyomatosis. Infrequently, polycythemia may be associated with fibroids. There is a rare familial leiomyomatosis cutis et uteri (Reed's syndrome) in which 45% of patients require hysterectomy before age 35.[62]

Although it would be expected that the preoperative diagnosis of leiomyoma of the uterus would be confirmed by pathology examination of the uterus, only 84% were confirmed.[63] This may be related to difficulties in examination of patients, diffuse enlargement of the uterus (especially in multiparas), other pathology (ovarian enlargement), and using leiomyoma as the preoperative diagnosis instead of menorrhagia. Increased use of sonography, second opinion, and departmental review may improve the accuracy of preoperative diagnosis.

Injury to the ureter during hysterectomy may occur, especially at the pelvic brim or adjacent to the cervix. Predisposing factors include endometriosis, ovarian neoplasm, adhesions, distorted anatomy, and previous bladder repair.[64] The rate is from 0.2 to 0.5 per 100 hysterectomies.[53]

When vesicovaginal fistula occurs, it is usually due to an unrecognized bladder injury with extravasation of urine into the vaginal cuff. It may result in severe abdominal pain, paralytic ileus, hematuria, and bladder irritabilty. Early recognition and treatment might prevent fistula development.[65] Bladder laceration ranges from 0.3 to 0.8 per 100 operations.[53]

"More than one patient in ten undergoing hysterectomy requires transfusion," although preexisting anemia may have been responsible.[53] Immediately after abdominal hysterectomy, there may be vaginal bleeding from the angles of the vaginal cuff, which may have everted during surgery and may have not been suture ligated. The usual treatment is vaginal suturing of the cuff. Within the first 24 hours, there may be slow or violent intraperitoneal bleeding that requires a repeat laparotomy. About 10 days postoperatively there may be vaginal bleeding due to sloughing of vascular pedicles.

The chance of death from hysterectomy overall is 1 or 2 per 1,000 operations, and over 600 hysterectomy-related deaths can be expected per year.[53] The mortality rates for hysterectomy were higher for cases associated with pregnancy or cancer (29.2 and 37.8, respectively, versus 6.0 per 10,000 hysterectomies), with increasing age, and among blacks (perhaps as a result of socioeconomic factors).[66] The major causes of death are postoperative complications, including cardiac arrest, pulmonary embolism, infection, and myocardial infarction, with occasional case of postoperative hemorrhage, intestinal obstruction, and subarachnoid hemorrhage.[34,53]

Although mortality is rare, morbidity is frequent, with 30 to 50% having fever. Operative infections include infected hematoma or abscess above the vaginal cuff, pelvic cellulitis, abdominal wound infection, and adnexal infection. The infections contain polymicrobial, mixed, aerobic and anaerobic bacteria of endogenous origin similar to organisms normally present in the cervix and colon. Infection tends to be more common in abdominal rather than vaginal hysterectomy, in women under age 40, in the indigent, if there was a conization of the cervix (more than 2 days and less than 6 weeks previously), if the vaginal vault cuff was closed, or if there was inadequate hemostasis.[34]

Note added in proof: Depot Lupron (leuprolide acetate) has just been approved by the FDA for the treatment of endometriosis and is provided now also as a 3.75 mg monthly injection. The cost has not been halved.

References

1. Friedman AJ, Barbieri RL: Leuprolide acetate: Applications in gynecology. Curr Prob Obstet Gynecol Fertil 1988;11(6):207–236.
2. Rein MS, Friedman AJ: Medical therapy for uterine leiomyomata, in Barbieri RL, Schiff I

(eds): Reproductive Endocrine Therapeutics. New York, Alan R. Liss, Inc., 1988, pp 198–222.

3. Townsend DE, Sparks RS, Baluda MC, McClelland G: Unicellular histogenesis of uterine leiomyomas as determined by electrophoresis of glucose-6-phosphate dehydrogenase. Am J Obstet Gynecol 1970;107:1168–1173.
4. Lumsden MA, West CP, Hawkins RA, et al: The binding of steroids to myometrium and leiomyomata (fibroids) in women treated with the gonadotropin-releasing hormone agonist Zoladex (ICI 118630). J Endocrinol 1989; 121:389–396.
5. Wiznitzer A, Marbach M, Hazum E, et al: Gonadotropin-releasing hormone specific binding sites in uterine leiomyomata. Biochem Biophys Res Commun 1988;152:1326–1331.
6. Lumsden MA, West CP, Bramley T, et al: The binding of epidermal growth factor to the human uterus and leiomyomata in women rendered hypo-oestrogenic by continuous administration of an LHRH agonist. Br J Obstet Gynaecol 1988;95:1299–1304.
7. Use of approved drugs for unlabeled indications, FDA Drug Bulletin, vol 12, no 1, Washington, DC, Department of Health and Human Services, 1982.
8. Kaunitz AM: Prescribing FDA-approved drugs in nonapproved ways. Contemp OB/GYN 1983;22:27–28.
9. Marut EL: Etiology and pathology of fibroid tumor disease: Diagnosis and current medical and surgical treatment alternatives. Obstet Gynecol Surv 1989;44:308–310.
10. Friedman AJ: Clinical experience in the treatment of fibroids with Leuprolide and other GnRH agonists. Obstet Gynecol Surv 1989;44:311–313.
11. Friedman AJ, Harrison-Atlas D, Barbieri RL, et al: A randomized, placebo-controlled, double-blind study evaluating the efficacy of leuprolide acetate depot in the treatment of uterine leiomyomata. Fertil Steril 1989;51:251–256.
12. Maheux R, Lemay A, Turcot-Lemay L: Dose-related inhibition of acute luteinizing hormone response during luteinizing hormone-releasing hormone agonist treatment for uterine leiomyoma. Am J Obstet Gynecol 1988; 158:361–364.
13. Matta WH, Stabile I, Shaw RW, Campbell S: Doppler assessment of uterine blood flow changes in patients with fibroids receiving the gonadotropin-releasing hormone agonist Buserelin. Fertil Steril 1988;49:1083–1085.
14. Friedman AJ, Garfield J, Rein MS, Doubilet PM, Harrison-Atlas D: A randomized, placebo-controlled, double-blind study evaluating leuprolide acetate depot treatment before myomectomy. Fertil Steril 1989;52:728–733.
15. Healy DL: Clinical applications of GnRH analogs, in Soules MR (ed): Controversies in Reproductive Endocrinology and Fertility. New York, Elsevier, 1989, pp 135–164.
16. Andreyko JL, Blumenfeld Z, Marshall LA, et al: Use of an agonistic analog of gonadotropin-releasing hormone (Nafarelin) to treat leiomyomas: assessment by magnetic resonance imaging. Am J Obstet Gynecol 1988;158:903–910.
17. Lumsden MA, West CP, Baird DT: Goserelin therapy before surgery for uterine fibroids. Lancet 1987;1:36–37.
18. Comite F: GnRH analogs and safety. Obstet Gynecol Surv 1989;44:319–325.
19. Waibel-Treber S, Minne HW, Scharla SH, et al: Reversible bone loss in women treated with GnRH-agonists for endometriosis and uterine leiomyoma. Hum Reprod 1989;4:384–388.
20. Friedman AJ: Treatment of leiomyomata uteri with short-term leuprolide followed by leuprolide plus estrogen-progestin hormone replacement therapy for 2 years: a pilot study. Fertil Steril 1989;51:526–528.
21. Friedman AJ, Barbieri RL, Doubilet PM, et al: A randomized, double-blind trial of a gonadotropin releasing-hormone agonist (leuprolide) with or without medroxyprogesterone acetate in the treatment of leiomyomata uteri. Fertil Steril 1988;49:404–409.
22. Kessel B, Liu J, Mortola J, et al: Treatment of uterine fibroids with agonist analogs of gonadotropin-releasing hormone. Fertil Steril 1988;49:538–541.
23. Gardner RL, Shaw RW: Cornual fibroids: a conservative approach to restoring tubal patency using a gonadotropin-releasing hormone agonist (goserelin) with successful pregnancy. Fertil Steril 1989;52:332–334.
24. Friedman AJ: Vaginal hemorrhage associated with degenerating submucous leiomyomata during leuprolide acetate treatment. Fertil Steril 1989;52:152–154.

25. Meyer WR, Mayer AR, Diamond MP, et al: Unsuspected leiomyosarcoma: Treatment with a gonadotropin-releasing hormone analogue. Obstet Gynecol 1990;75:529–532.
26. Neven P, DeMuylder X, Belle YV, et al: Tamoxifen and the uterus and endometrium (Letters to the Editor). Lancet 1989;1:375.
27. Fornander T, Rutqvist LE, Cedermark R, et al: Adjuvant tamoxifen in early breast cancer: occurrence of new primary cancer. Lancet 1989;1:117–119.
28. Matthew A, Chabon AB, Kabakow B, et al: Endometrial carcinoma in five patients with breast cancer on tamoxifen therapy. NY State J Med 1990;90:207–208.
29. Cano A, Matallin P, Legua V, et al: Letters to the editor. Lancet 1989;1:376.
30. Hague WM, Abdulwahid NA, Jacobs HS, Craft I: Use of LH-RH analogue to obtain reversible castration in a patient with benign metastasizing leiomyoma. Br J Obstet Gynaecol 1986;93:455–460.
31. Donnez J, Schrurs B, Gillerot S, et al: Treatment of uterine fibroids with implants of gonadotropin-releasing hormone agonist: Assessment by hysterography. Fertil Steril 1989;51:947–950.
32. Synarel (nafarelin acetate). Nasal solution 2 mg/ml. Product Monograph. Palo Alto, CA, Syntex Laboratories, 1990.
33. Buttram VC Jr, Reiter RC: Uterine leiomyomata, in Buttram VC Jr, Reiter RC (eds): Surgical Treatment of the Infertile Female. Baltimore, Williams & Wilkins, 1985, pp 201–228.
34. Mattingly RF, Thompson JD: Leiomyomata uteri and abdominal hysterectomy for benign disease, in Mattingly RF, Thompson JD (ed): Te Linde's Operative Gynecology, ed 6. New York, JB Lippincott Company, 1985, pp 203–255.
35. Dlugi AM: Uterine surgery, in De Cherney H, Polan ML (eds): Reproductive Surgery. Chicago, Year Book Medical Publishers, 1987, pp 155–166.
36. Garcia CR, Tureck RW: Submucosal leiomyomas and infertility. Fertil Steril 1984;42:16–19.
37. Babaknia A, Rock JA, Jones HW Jr: Pregnancy success following abdominal myomectomy for infertility. Fertil Steril 1978;30:644–647.
38. Berkeley AS, DeCherney AH, Polan ML: Abdominal myomectomy and subsequent fertility. Surg Gynecol Obstet 1983;156:319–322.
39. Rosenfeld DL: Abdominal myomectomy for otherwise unexplained infertility. Fertil Steril 1986;46:328–330.
40. Gatti D, Falsetti L, Viani A, Gastaldi A: Uterine fibromyoma and sterility: role of myomectomy. Acta Eur Fertil 1989;20:11–13.
41. Aharoni A, Reiter A, Golan D, et al: Patterns of growth of uterine leiomyomas during pregnancy. A prospective longitudinal study. Br J Obstet Gynaecol 1988;95:510–513.
42. Lev-Toaff AS, Coleman BG, Arger PH, et al: Leiomyomas in pregnancy: sonographic study. Radiology 1987;164:375–380.
43. Burton CA, Grimes DA, March CM: Surgical management of leiomyomata during pregnancy. Obstet Gynecol 1989;74:707–709.
44. Stangel JJ: Uterine construction by myoma resection: Laparotomy, in Stangel JJ (ed): Infertility Surgery. Norwalk, CT, Appleton & Lange, 1990, pp 224–235.
45. Reyniak JV, Corenthal L: Microsurgical laser technique for abdominal myomectomy. Microsurgery 1987;8:92–98.
46. Starks GC: CO_2 laser myomectomy in an infertile population. J Reprod Med 1988; 33:184–186.
47. Ben-Baruch G, Schiff E, Menashe Y, Menczer J: Immediate and late outcome of vaginal myomectomy for prolapsed pedunculated submucous myoma. Obstet Gynecol 1988;72:858–861.
48. Goldrath MH: Vaginal removal of the pedunculated submucous myoma: the use of laminaria. Obstet Gynecol 1987;70:670–672.
49. Hallez JP, Netter A, Cartier R: Methodical intrauterine resection. Am J Obstet Gynecol 1987;156:1080–1084.
50. Brooks PG, Loffer FD, Serden SP: Resectoscopic removal of symptomatic intrauterine lesions. J Reprod Med 1989;34:435–437.
51. Neuwirth RS: Uterine reconstruction by myoma resection: the hysteroscopic rectoscope, in Stangel JL (ed): Infertility Surgery. Norwalk, CT, Appleton & Lange, 1990, pp 236–242.
52. American College of Obstetricians and Gynecologists, Public Information Department: Media kit, from data obtained from National Hospital Discharge Survey, National Center for Health Statistics. Washington, DC, American College of Obstetricians and Gynecologists, 1969.
53. Esterday CL, Grimes DA, Riggs JA: Hysterectomy in the United States. Obstet Gynecol 1983;62:203–212.

54. American College of Obstetricians and Gynecologists: Criteria sets. Hysterectomy, abdominal or vaginal (leiomyomata). Quality Assurance in Obstetrics and Gynecology. Washington, DC, American College of Obstetricians and Gynecologists, 1989, p. 29.
55. Weingold AB: Pelvic mass, in Kase NG, Weingold AB, Gershenson DM (eds): Principles and Practice of Clinical Gynecology, ed 2. New York, Churchill Livingstone, 1990, pp. 545–581.
56. Waldemar G, Werdelin L, Boysen G: Neurologic symptoms and hysterectomy: A retrospective survey of the prevalence of hysterectomy in neurologic patients. Obstet Gynecol 1987;70:559–563.
57. Siddle N, Sarrel P, Whitehead M: The effect of hysterectomy on the age at ovarian failure: identification of a subgroup of women with premature loss of ovarian function and literature review. Fertil Steril 1987;47:94–100.
58. Committee on Gynecologic Practice, American College of Obstetricians and Gynecologists: Prophylactic use of antibiotics with abdominal hysterectomy. ACOG Committee Opinion no 82. Washington, DC, American College of Obstetricians and Gynecologists, 1990.
59. Hemsell DL, Hemsell PG: Antimicrobial prophylaxis for hysterectomy, in Abstracts of Current Clinical and Basic Investigation, 38th Annual Clinical Meeting. Washington, DC, American College of Obstetricians and Gynecologists, 1990, p 8.
60. Mark AS, Hricak H, Heinrichs LW, et al: Adenomyosis and leiomyoma: differential diagnosis with MR imaging. Radiology 1987; 163:527–529.
61. Lipton JH, Fong TC, Burgess KR: Miliary pattern as presentation of leiomyomatosis of the lung. Chest 1987;91:781–782.
62. Garcia-Muret MP, Pujol RM, Alomar A, et al: Familial leiomyomatosis cutis et uteri (Reed's syndrome). Arch Dermatol Res 1988; 280(suppl):S29–S32.
63. Lee NC, Dicker RC, Rubin GL, Ory HW: Confirmation of the preoperative diagnoses for hysterectomy. Am J Obstet Gynecol 1984; 150:283–287.
64. Daly JW, Higgins KA: Injury to the ureter during gynecologic surgical procedures. Surg Gynecol Obstet 1988;167:19–22.
65. Kursh ED, Morse RM, Resnick MI, Persky L: Prevention of the development of a vesicovaginal fistula. Surg Gynecol Obstet 1988; 166:409–412.
66. Wingo PA, Huezo CM, Rubin GL, et al: The mortality risk associated with hysterectomy. Am J Obstet Gynecol 1985;152:803–808.

24

Genital Prolapse

GIGLIA A. PARKER and DAVID H. NICHOLS

The uterus in genital prolapse is a passenger, not the perpetrator. Nevertheless, descent of the uterus or other portions of the female pelvic anatomy is and will continue to be a cause of major distress to more and more women as the population ages. With women living a third of their lives after menopause, quality of life issues are becoming increasingly more important. Older women are pressing their gynecologists for relief of conditions their mothers and grandmothers often considered the irremediable lot of their sex. Urinary and fecal incontinence, pelvic pressure, low backache, and dyspareunia are a few of the complaints that may be found in association with genital prolapse.

Contributing Factors

It is not known with certainty why some women develop genital prolapse and others do not. Factors that contribute to the problem can be identified in most women, but the degree to which each factor is responsible for the end result is sometimes less obvious, and the fact remains that many women whose pelvic supports have sustained apparently similar insults do not develop prolapse. Factors that contribute to prolapse fall into the following broad categories:

1. Congenital
   a. Genetic
   b. Developmental
2. Acquired
   a. Physiologic insult
   b. Anatomic insult

Genetic Factors

Probably a major underlying factor predisposing women to uterine prolapse is a basic connective tissue weakness. Racial differences in incidence, the tendency for pelvic relaxation problems to run in families, and the frequent concomitant finding of hiatal hernia and wide abdominal striae in individual patients with prolapse point to the possibility of inherited differences in connective tissue strength as an etiologic factor. The genetic determination of the bony configuration of the pelvis will determine also whether the forces of labor will place undue stress on certain pelvic supports and attachments. For example, a narrow pelvic arch causing a posterior displacement of the delivering fetal head may concentrate damage to the posterior vaginal wall attachments.

Developmental Factors

Among developmental factors that have been associated with various types of prolapse are the presence of congenital enterocele from failure of the fusion of the fascia of Denonvilliers and the development of uterine prolapse in patients with sacral agenesis or bladder extrophy with absence of the pubic symphysis. Some investigators have thought that there is an increased incidence of prolapse in women with spina bifida occulta. The anomaly may lead to defective pelvic innervation with resulting incompetence of the levator ani.[1]

Physiologic Insults

Primary among the physiologic insults that affect the integrity of the pelvic supports is estrogen deficiency. Not only does the incidence of prolapse increase markedly with the estrogen deprivation of menopause, but associated conditions such as urinary incontinence may sometimes be relieved by estrogen replacement. On the other hand, the high hormonal levels of pregnancy may predispose women to damage of the pelvic supports, subsequently actualized by the forces of delivery. Malpas[2] stated that prolapse in pregnancy often develops shortly after the first missed menstrual period. Additionally, an unexpectedly high incidence of urinary stress incontinence and pelvic relaxation has been found in women who underwent cesarean delivery without labor. Recent work suggests that androgens[3] as well as estrogens contribute to the maintenance of vaginal integrity.

Increased incidence of prolapse following periods of war has led to the postulation of a role for nutrition in the etiology of prolapse. Another explanation involves the increase in heavy manual labor for women during such periods. A final physiologic insult may be the significant exhaustion of reparative tissue replication that occurs around the age of 100 years.

Anatomic Insults

Certainly increased intraabdominal pressure from trauma or ascites, chronic cough with smoking or pulmonary disease, heavy lifting, or straining with chronic constipation contributes to initial prolapse as well as failure of repair. Childbirth is undoubtedly the most common insult to the anatomic integrity of pelvic supports, when tissues may be stretched beyond their limits, torn, or avulsed. Other trauma from accident or injury is less frequent, but unfortunately an iatrogenic contribution is not uncommon and may arise from such causes as less-than-ideal obstetric management, breakdown of episiotomy, failure to support the vaginal vault at hysterectomy, and change in the normal vaginal axis following retropubic urethropexy.

Pelvic Supports

Since the uterus is a passenger in genital prolapse, failure of it to be maintained in a normal position results from loss of integrity by the structures that support it. For comprehensive details of pelvic anatomy an appropriate text should be consulted, but a brief overview is relevant here.

The tissues that compose the supporting structures of the pelvis include bone and cartilage, smooth and striated muscle, and connective tissue with collagen and elastic fibers. Bone and cartilage are firm, strong, and inflexible. They resist sudden stresses and strains, but can remodel with a gradual change in architecture in response to prolonged strain and stress. Smooth muscle helps to maintain tone via rhythmic contractions, but can elongate under stress with no increase in tone until the point of maximal distention. Striated muscle, on the other hand, lacks inherent rhythmic contractions, tends to maintain a constant length on contraction in response to stress, and thus helps to restore and maintain equilibrium. Elastic fibers respond to stress by stretching, but seek to return to their original state much like a rubber band does. Collagen fibers do not stretch, but because they are flexible and permit movement without stretching, similar to a piece of string.

The number of muscle cells tends to remain constant with aging. However, the quantity of elastic tissue decreases. Collagen fibers swell, fuse, and become hyalinized. Changes in bony architecture appear to be related to both age and hormones.

These types of tissues make up the supporting structure of the pelvis, which, for the sake of simplicity, can be conceptualized into seven anatomic systems[4]:

1. Bony pelvis
2. Broad ligaments and subperitoneal retinaculum (upper)
3. Uterosacral ligaments (upper and middle)
4. Cardinal ligaments (middle)
5. Pelvic diaphragm (lower)
6. Urogenital diaphragm (lower)
7. Perineum (lower)

Bony Pelvis (Ultimate Support)

The bony pelvis is the ultimate structure to which the soft tissues of the pelvis are anchored. It can be defective from either congenital or acquired disorders, which must be taken into account when planning a repair.

Fasciae and "Ligaments" (Upper and Middle Pelvic Supports)

The existence of various fasciae in the pelvis has been the source of much debate throughout the history of gynecology. The main connective tissue supports of the uterus are usually spoken of as "ligaments." In fact, these so-called ligaments are essentially condensations of connective tissue following the principle routes of blood vessels, nerves, and lymphatic channels to and from the pelvic viscera.

The *broad and round ligament complex* provides for the entrance and egress of blood and lymphatic vessels supplying the uterus, tubes, and ovaries. There is a subperitoneal retinaculum of connective tissue that is a source of some strength, but generally contributes little to the support of the uterus, unless markedly accentuated by severe fibrosis or scarring, as from infection or endometriosis. The *round ligaments* assist in maintaining anteversion of the uterus.

The *uterosacral ligaments* are also important in uterine anteversion. Additionally, they help to maintain the position of the uterus and upper vagina over the levator plate so that these organs do not slip through the genital hiatus of the levator muscle complex. The connective tissue elements of the anterior or cervical portion of the uterosacral ligaments are enmeshed with those of the medial portion of the cardinal ligaments. Thus, they are at that point inseparable and can constitute a surgically useful complex in many repairs.

The *cardinal ligaments* provide the most important lateral supports of the uterus and upper vagina. As the vessels and lymphatics from the hypogastric plexus enter and leave the uterus and vagina along their lateral margins, they are surrounded by perivascular fibroareolar sheaths of considerable strength. This network has been called the horizontal connective tissue ground bundle by von Peham and Amreich.[5] Where the cervix meets the vagina, this lateral paracervical condensation of tissues turns anteriorly to follow the axis of the vagina. The vaginal fornices are thus attached through connective tissue bridges to the arcus tendineus arising from the fascia overlying the obturator internus muscle on each side of the pelvis (Fig. 24-1). The significance of these tissues to the support of the uterus and vagina was demonstrated by the classic experiments of Mengert using weights applied to a cadaver cervix. "Marked descent of the uterus amounting to actual prolapse never occurred so long as any part of the upper two-thirds of the paravaginal and/or lower two-thirds of the parametrial tissues were intact."[6]

Subuterine (Lower Pelvic) Supports

Support to the pelvis below the uterus is provided by the pelvic diaphragm, the urogenital diaphragm, and the perineum.

Pelvic Diaphragm

The pelvic diaphragm consists of the coccygeus and levator ani muscles. The *coccygeus* arises from the ischial spine and inserts along the lateral margin of the coccyx and lower sacrum. It follows the course of its aponeurosis, the sacrospinous ligament. It is so

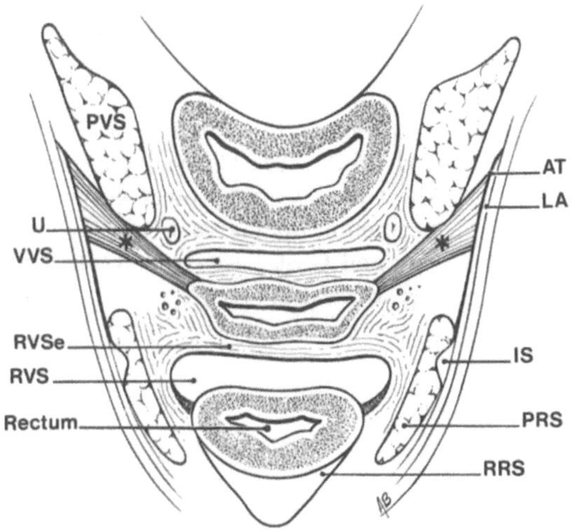

FIGURE 24-1. Connective tissue bridges (*), shown here exaggerated for illustrative emphasis, attach the vaginal fornices to the arcus tendineus (AT) arising from the fascia overlying the obturator internus. Note the relationships of the connective tissue spaces and septa of the pelvis to these bridges, to the levator ani (LA), to the ureter (U), and to the ischial spine (IS). The pelvic spaces include the paravesical space (PVS) and the pararectal space (PRS) on each side and the vesicovaginal space (VVS), rectovaginal space (RVS), and retrorectal space (RRS) in the midline. The rectovaginal septum (RVSe) forms the anterior limits of the rectovaginal space (RVS). (Modified from ref. 4, p. 33.)

intimately related to the superior aspect of this strong ligament that the two essentially constitute a single sacrospinous coccygeus complex.

The divisions of the *levator ani* take their names according to their origins and insertions. The medial and anterior division most important to the gynecologist is the *pubococcygeus*. Arising from the pubis, the paired pubococcygei sweep posteriorly along the sides of the urethra, vagina, upper portion of the perineal body, and rectum. Behind the rectum, they fuse to form the *levator plate*, which is attached to the coccyx.

The uterus and upper vagina normally rest on an empty rectum which, in turn, rests on the horizontally inclined levator plate (Fig. 24-2). If the integrity of the levator ani and its fascia is compromised so that they sag with the pull of gravity or increased abdominal pressure, the levator plate will no longer lie in a horizontal plane, and the uterus and cervix will tend to slide down it toward the genital hiatus of the pelvic diaphragm.

There is considerable variation in the attachments of the pubococcygeus to the organs it passes by. The most constantly distinct of these seems to be the development of the most medial and caudal portion of the muscle into a strong muscular sling that passes behind the rectum as the *puborectalis*, which is thought occasionally to be continuous with the deep external anal sphincter. Other investigators have sometimes described slips of muscle contributing to a posterolateral investment of the urethra (*pubourethralis*) and decussating prerectally to join the perineal body (*pubovaginalis*).

The portion of the levator ani that forms the intermediate portion of the pelvic diaphragm between the pubococcygeus and coccygeus is the *iliococcygeus*, which arises from the white line or arcus tendineus on the surface of the obturator internus on each side of the pelvis. Avulsion of fascial attachments to the arcus tendineus may contribute to paravaginal defects in the supports of the lateral vaginal fornices.

Urogenital Diaphragm

The urogenital diaphragm running between the inner surfaces of the ischiopubic rami closes the anterior portion of the pelvic outlet and adds support to the levator portal, which is naturally larger in women as it is pierced by both the urethra and vagina. The urogenital diaphragm is sandwich-like, with its superior and inferior fascial layers separated by the deep transverse perineal muscle.

Reflections of the superior and inferior fascial layers form the posterior and anterior *pubourethral ligaments*, which, along with the intermediate pubourethral ligament, suspend the urethra from the pubic bone (Fig. 24-3). This suspensory mechanism

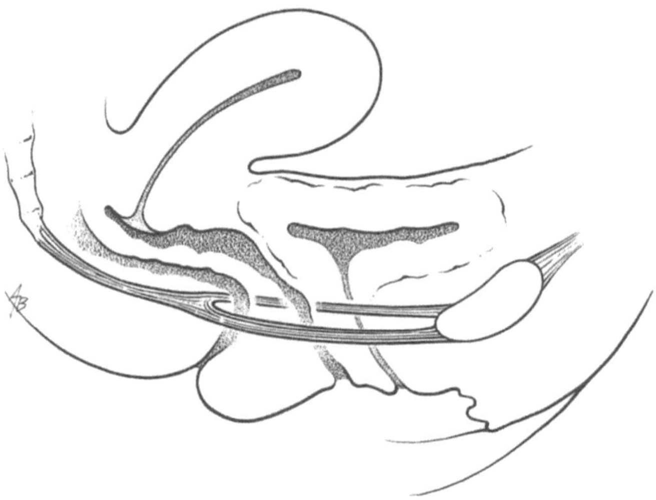

FIGURE 24-2. In the standing position, the upper vagina lies in an almost horizontal plane and rests on the rectum, which in turn rests on the levator plate (Modified from Nichols DH, Milley PS: Clinical anatomy of the vulva, vagina, lower pelvis, and perineum, in Sciarra J (ed): Gynecology and Obstetrics. New York, Harper & Row, 1977.)

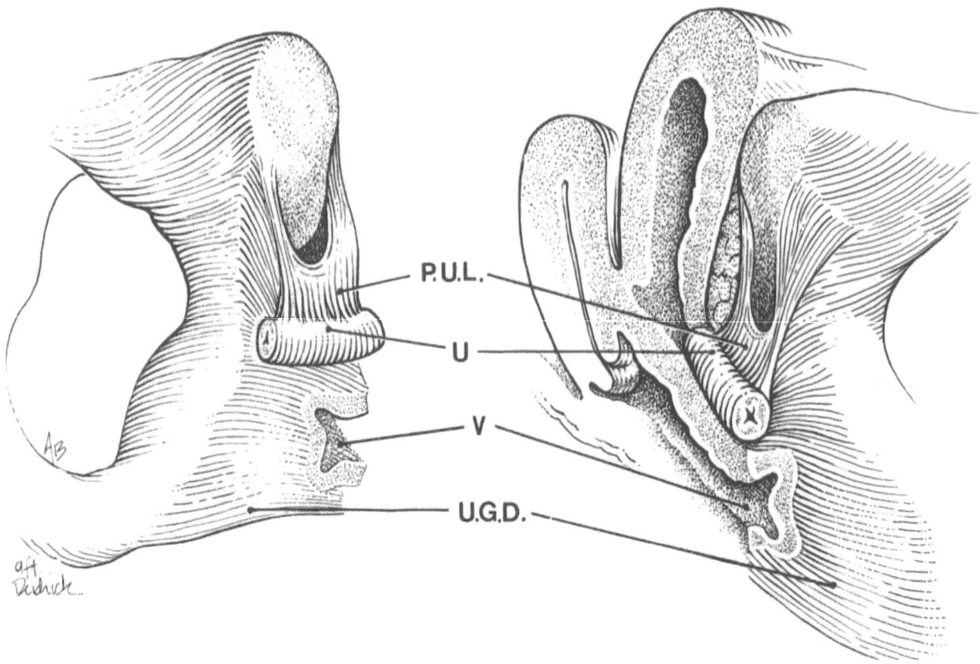

FIGURE 24-3. The pubourethral ligament (P.U.L.) suspends the urethra (U) from the pubic bone. The ligament is formed by reflections of the fascial layers of the urogenital diaphragm (U.G.D.), which closes the anterior portion of the pelvic outlet and at the level of the urethra is intimately related to the vagina (V). (Modified from Milley PS, Nichols DH: The relationship between pubourethral ligaments and urogenital diaphragm in the human female. Anat Rec Wistar Institute Press, 1971;170:281–284.)

maintains the proximal portion of the urethra within the abdominal portion of the pelvic cavity so that sudden increases in intraabdominal pressure will be transmitted to both the proximal urethra and bladder. Intraurethral pressure may therefore continue to exceed intravesical pressure so that urinary continence is maintained. Additionally, attachment of the posterior fibers of the urogenital diaphragm to the perineal body serves as another fixation point, lessening the tendency of the urethra and urethrovesical junction to rotate around the attachments of the pubourethral ligaments to the pubis.

Perineum

The most inferior support of the pelvis is the perineum. Its strength lies in the *perineal body*, which is a pyramidal fibromuscular elastic structure between the rectum and vagina on a line between the ischial tuberosities. Like the hub of a wheel, it is the focus of attachment for various muscles, including the deep and superficial transverse perineal muscles, the bulbocavernosi, the external anal sphincter, and some fibers from the pubococcygei, which insert into it like spokes into a wheel. Above, its apex joins the rectovaginal septum (the anterior layer of Denonvilliers' fascia, which is a peritoneal fusion fascia well formed in most subjects by the 14th week of fetal life). This significant support to the posterior vaginal wall (Fig. 24-4) in turn offers support to the anterior vaginal wall beneath the urethra.

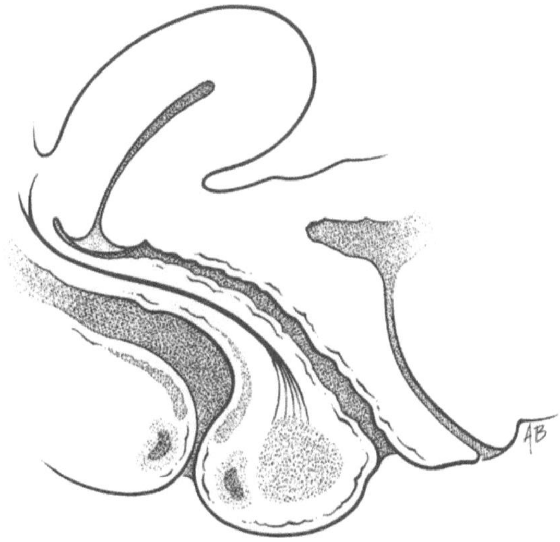

FIGURE 24-4. The rectovaginal septum, formed in part by the anterior layer of the fascia of Denonvilliers, fuses with the apex of the perineal body and offers significant support to the posterior vaginal wall. (Modified from ref. 4, p. 42.)

Types of Disorders

For years, gynecologists argued whether the position of the uterus and vagina in the pelvis is maintained by suspension from above, as advocated by Fothergill,[7] or support from below, as argued by Paramore[8] and Halban and Tandler.[9] Victor Bonney[10] urged the important contributions of both points of view: The type of prolapse depends on which systems are damaged. Injuries may consist of overstretching, tearing, and/or avulsion of support structures.

Genital prolapse generally falls into two categories: eversion of the upper vagina and eversion of the lower vagina. While uterine prolapse usually accompanies the former, it may be associated in later stages with the latter. The worst case would be total procidentia, in which the cervix and uterus lie completely outside the bony pelvis with the vagina turned inside out (Fig. 24-5).

Damage to Upper Support Systems

If upper support systems are damaged, with elongation of the round and/or uterosacral ligaments, the uterus may become retroverted. In this position, its axis is more aligned with that of the vagina instead of being at right angles to it. Thus, there is a greater tendency for the uterus to descend through the vaginal canal.

Damage to Middle Support Systems

Eversion of the upper vagina with descent of the cervix will occur if the middle pelvic sup-

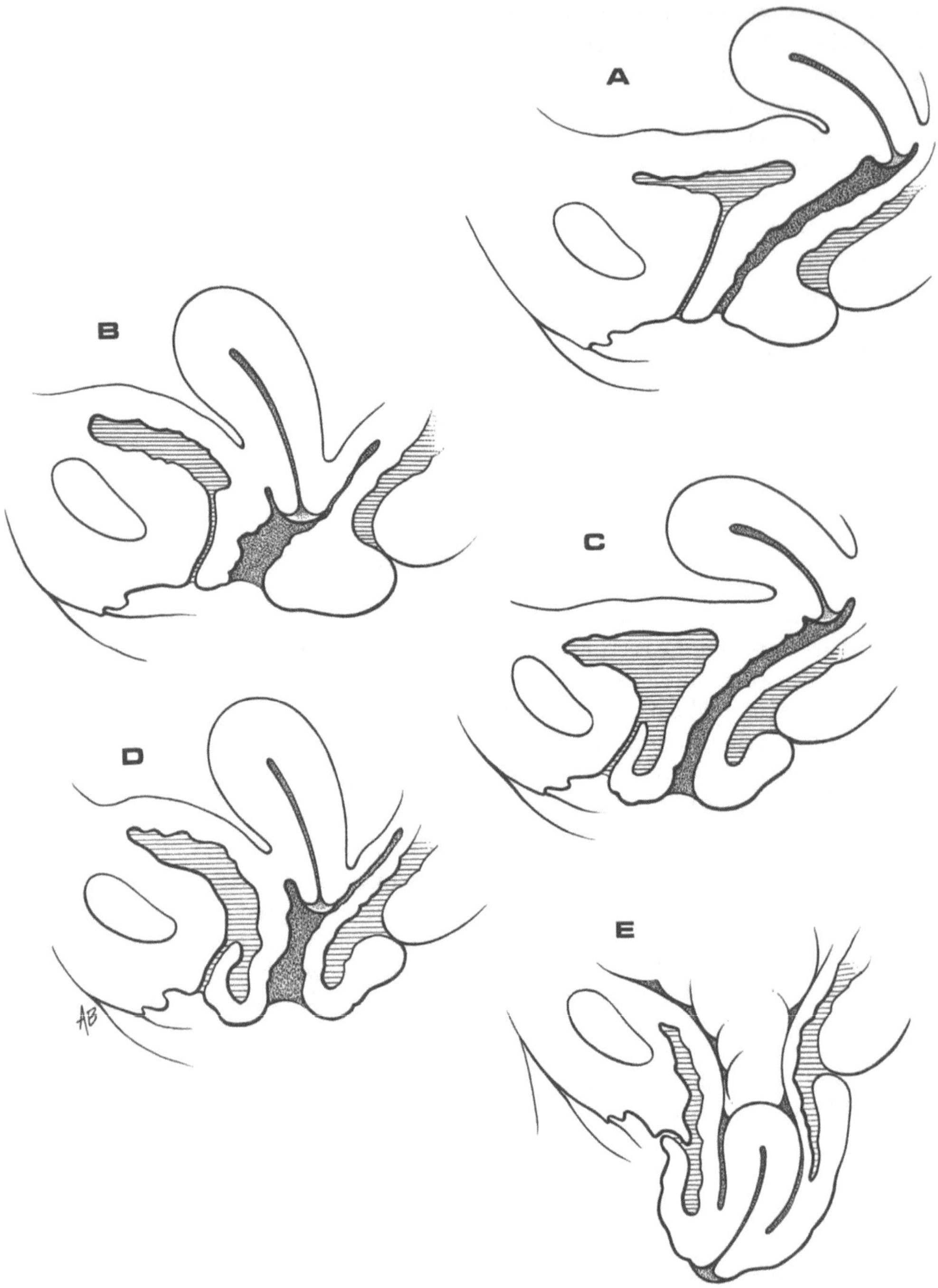

FIGURE 24-5. Damage to the various supports of the pelvic viscera results in different types of genital prolapse. (A) Normal relationships. (B) When uterine supports are damaged, the cervix begins to descend through the vaginal canal. (C) Damage to lower pelvic supports tends to result in eversion of the lower vagina with formation of a cystocele and rectocele. (D) When all supports are damaged descent of the cervix will accompany eversion of the lower vagina. (E) In the worst case, total procidentia, the uterus lies completely outside the pelvis. (Modified from Nichols DH: Types of genital prolapse. Postgrad Med 1969;46:183–187.)

ports are damaged. If the upper supports are strong, the corpus of the uterus will be held in its usual location, but the cervix will begin to elongate and descend through the vagina. An enterocele will tend to form with the elongating cervix. If the uterosacral ligaments are strong and the lower portion of the cardinal ligaments is weak, an enterocele will not tend to accompany the elongating cervix, and the cul-de-sac may be higher than expected at the time of vaginal hysterectomy for this uterovaginal-type prolapse.

If all upper and middle supports are weak, the uterus will tend to lie along the same axis as the vagina and to descend as a whole without elongation of the cervix. As long as the lower supports remain strong, the uterus and cervix will, because of the freedom of movement provided by the rectovaginal space, slide down the anterior wall of the intact rectum. The descent of the cervix brings the anterior wall of the cul-de-sac with it to produce a traction-type enterocele. Because the base of the bladder is attached to the upper vagina and cervix by the supravaginal septum, this type of prolapse may ultimately be accompanied by the bladder, causing a traction cystocele as well.

Damage to Lower Support Systems

Damage to the lower support systems results in eversion of the lower vagina. In isolated damage to the posterior segment—the perineum and anterior rectum—as may occur with a narrow pubic arch in a difficult second stage of labor, the vaginal and perineal attachments to the levator ani may be avulsed. A mid- or lower vaginal rectocele can be expected.

If the anterior segment of the lower support systems is damaged, as may occur with the shearing trauma of a descending fetal head beneath a wide-angle pubic arch, problems with support of the urethra and bladder are more likely. Avulsion or stretching of the pubourethral ligaments will permit a rotational descent of the proximal urethra and bladder neck. Stress urinary incontinence may result. A cystocele may result either from stretched or torn midline anterior vaginal supports, or from avulsed paravaginal attachments to the arcus tendineus, or both.

Cystocele and rectocele are invariably present in the less common generalized prolapse, which often develops as menopausal changes are superimposed on a congenital connective tissue weakness further complicated by obstetric damage to the pelvic diaphragm. As the cul-de-sac descends with the uterus in this situation, both the anterior and posterior peritoneal walls are involved and, therefore, enterocele may be absent.

Diagnosis

History

Evaluation of genital prolapse begins as does any evaluation in medicine, with a careful history and physical examination. What are the patient's symptoms? Does she complain of pelvic pressure, a bulging mass, backache, urinary incontinence, or constipation? When did the symptoms begin? How have they progressed over time? What makes them worse or better? To what degree do they interfere with her quality of life? Particular attention must be given to the genital, urinary, and gastrointestinal systems, because all three are often related to defects in pelvic support.

Urinary System

Does the patient have difficulty emptying her bladder? Must she elevate it with her finger in order to do so? Are frequency, urgency, nocturia, or burning present? Is there any urinary incontinence? If so, under what circumstances does it occur?

Genital System

Is the patient sexually active? Has there been any interference with coital comfort? Is vaginal dryness or obstruction of the vagina a problem? What is the status of an aging partner's erectile and penetrating ability?

Gastrointestinal System

If the patient complains of constipation, is she referring to the infrequency, hardness, or necessity of straining at bowel movements? How often does she use stool softeners or laxatives? What is her fluid intake? Are stools ribbon-like? Is there a sensation of aching or of something left inside after a bowel movement? Is digital manipulation necessary for complete evacuation? Is there any evidence of soiling or incontinence?

Other Aspects of the History

Important elements from the *gynecologic history* include previous surgeries, menopausal status, and the use of hormonal replacement. The *obstetric history* should include the number of pregnancies, the types of deliveries, the weights of the infants, and any difficulties with labor or delivery.

Questions regarding the *medical history* of an elderly patient who may come to surgery should pay particular attention to cardiovascular and pulmonary status. Are there any metabolic problems, such as diabetes mellitus, that may compromise soft tissue strength or healing? Are there orthopedic or arthritic problems that would make a lithotomy position difficult?

The *social history* should include special inquiry into the patient's living circumstances. Has her occupation required heavy lifting? What type of help would she have at home after surgery, should that be the chosen therapy? Would a visiting nurse or short-term stay in an extended care facility be helpful?

Family history should elicit whether any relatives have had similar disorders. Are there close relatives with hernias or connective tissue diseases? If not already covered, the *review of systems* should make inquiry about weight gain, which may have increased intraabdominal pressure, or weight loss, which may have decreased ischiorectal fat under the pelvic diaphragm with resultant of loss support. Is there a history of smoking or allergies leading to increased coughing or sneezing? How willing is the patient to follow through with recommended alterations in life-style?

Physical Examination

After a general physical examination has been performed with special reference to the patient's blood pressure, nutritional status, muscular development, skin turgor and striae, and presence of any cardiac or pulmonary disease, edema, or arthritis, attention may be focused on the pelvic examination. The patient should be evaluated in the standing as well as the lithotomy position. Details of findings should be accurately recorded, because there may be a substantial difference between the observations in an alert, cooperative patient in the examining room and an anesthetized one in the lithotomy position in the operating room. A perception of such differences is critical to the effectiveness of a surgical repair should one be undertaken.

With the patient in the lithotomy position, the vulva is examined for signs of atrophic changes. The integrity of the perineum is assessed, and note is made of the location of the anus. Should the anus have descended from its normal location in the sulcus between the gluteal muscles to become the most dependent portion of the perineum, substantial risk for the development of the *perineal descent syndrome* is present. When the syndrome results from a pudendal neuropathy, levator atrophy may be associated with a gradual progression of symptoms from obstipation and ribbon-like stools ultimately to a distressing fecal incontinence. Immediate steps to increase roughage in the diet and change bowel habits to eliminate the tendency to strain at stool are necessary.

Grading Prolapse

The patient is then examined for the degree of genital prolapse. Descriptive terms are best employed, but on occasion a system of classification is helpful. One may, by examination of the patient standing and straining as with a Valsalva maneuver, grade six sites of potential support loss on a scale of 0 to 4. The vaginal sites to be considered are the urethra, bladder, cervix or vaginal vault, cul-de-sac, rectum, and perineum. Urethrocele (or descent of the urethrovesical junction), cystocele, cervical (or vault) prolapse, and rectocele are graded as follows:

Grade 0—normal location
Grade 1—up to one half the distance to the hymen
Grade 2—between one half and the whole distance to the hymen
Grade 3—from the hymen to half the maximum distance beyond it
Grade 4—maximum possible descent

Old perineal lacerations are scored as:

Grade 0—intact
Grade 1—up to half the distance to the anal sphincter
Grade 2—between half and the whole distance to the anal sphincter
Grade 3—into the anal sphincter
Grade 4—involving the rectal mucosa

Grading of enterocele is according to dissection of the rectovaginal septum, not according to the descent of a cul-de-sac normally positioned relative to a prolapsing vault:

Grade 0—normal position
Grade 1—up to one fourth the distance to the hymen
Grade 2—from one fourth to one half the distance to the hymen
Grade 3—from one half to three fourths the distance to the hymen
Grade 4—from three fourths to the whole distance to the hymen

The patient is asked to bear down so that the observer may determine which of the displaced organs prolapses first. If complete procidentia is present, the gynecologist must replace the uterus within the pelvis before asking the patient to strain. If the cervix is first to appear at the introitus, the primary site of damage is likely to be the middle and upper suspensory systems, particularly the cardinal-uterosacral ligament complex. If cystocele and rectocele appear first, the primary site of damage is probably the lower support system, with secondary damage to the higher supports. Digitally correcting the primary defect with the fingers during the exaination can permit the gynecologist to assess the degree of secondary damage.

In the absence of complete prolapse, the vagina is examined for depth and axis. A Papanicolaou smear and bimanual examination are performed. The size, shape, position, and mobility of the internal genitalia are noted. Rectovaginal examination permits the best assessment of the uterosacral ligaments and enterocele. The prominence of the ischial spines and the strength of the coccygeus-sacrospinous ligament complexes should be ascertained. Performing clinical pelvimetry will help the surgeon predict the feasibility of certain vaginal procedures, such as the morcellation of a myomatous uterus.

The patient is then asked to contract her voluntary muscles so that the examiner, with one finger first in one lateral fornix and then the other, may determine the strength and symmetry of the pelvic diaphragm and especially of the pubococcygeus. Integrity of support of the anterior vaginal fornix is assessed by gentle palpation, and accentuation of the forniceal sulcus coincident with voluntary pubococcygeal muscle contraction is noted. The patient is asked to cough, and any loss of urine is noted. If urinary incontinence is demonstrated, the Marshall or Bonney test, elevating but not occluding the urethrovesical junction, should be carried out. Consideration should be given to performing a Q-tip test, office cystometry, and, especially if urinary incontinence is recurrent, a full urodynamic evaluation, including cystometrogram, urethral pressure profile, uroflometry, urethroscopy, and electromyography.

Paravaginal Defects

If a cystocele is present, a hint of its etiology may be obtained by examining the rugae of the anterior vagina. If the rugae are normal, the defect leading to the cystocele may be paravaginal, especially if the anterior lateral fornices are blunted and do not elevate with contraction of the pubococcygei. Paravaginal defects may also be suspected if elevation of the anterior lateral fornices individually or together with a curved ovum forceps or paravaginal elevator eliminates the cystocele or rotational descent of the urethrovesical angle. Absence of rugae in a bulging anterior vaginal wall may indicate a midline defect. Evaluation of either the anterior or posterior vaginal wall may be enhanced by using the bottom half of a disarticulated Graves speculum as a retractor.

Standing Examination

After examination is completed in the lithotomy position, the patient is asked to stand with one foot on a stool or the low shelf at the foot of the examining table (Fig. 24-6). The manual portion of the examination is then repeated first with the patient at rest; then she is asked to squeeze her pelvic diaphragm, to bear down with a Valsalva maneuver, and finally to cough. The descent of each site is again evaluated. Incontinence not present with the patient in lithotomy position may be demonstrable when she stands.

The examiner then places a thumb into the patient's vagina and an index finger gently into the rectum. The presence of a high, mid, or low rectocele, or of a defect in the perineal body, is noted. When the patient strains, palpation of bowel in a sac between the rectum and vagina confirms the presence of an enterocele.

Treatment

When the nature of the prolapse, including the degree of damage to each of the pelvic supports, has been fully assessed, various treatment options may then be discussed with the patient. These may be surgical, nonsurgical, or a combination of the two.

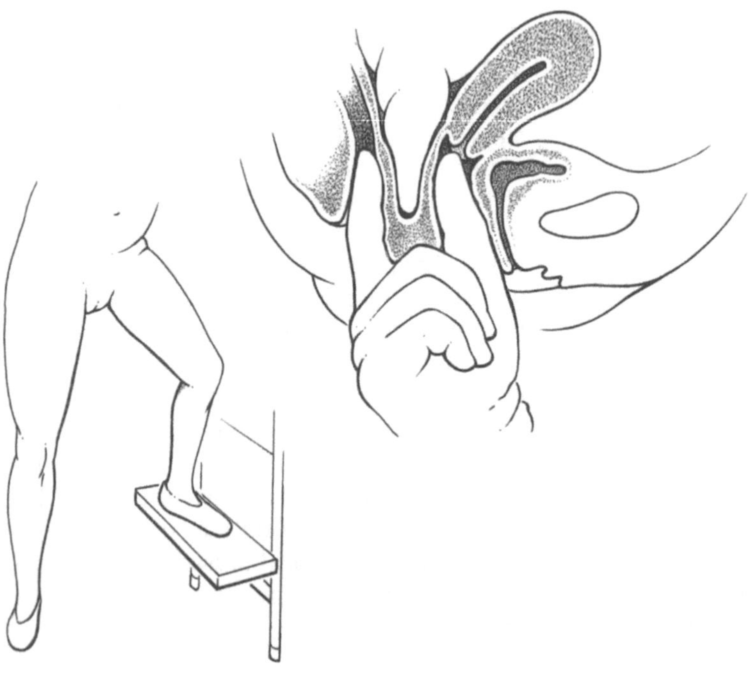

FIGURE 24-6. An enterocele is best detected on rectovaginal examination with the patient standing. The patient places one foot on a small stool to allow the examiner to insert a thumb into the vagina and an index finger into the rectum. Palpation of bowel in a sac between the rectum and vagina when the patient strains confirms the presence of an enterocele. (Modified from Nichols DH: Repair of enterocele and prolapse of the vaginal vault, in Barber H (ed): Goldsmith's Practice of Surgery. Philadelphia, JB Lippincott, 1981.)

Nonsurgical Treatment

Life-style Changes

The initial step is to help the patient pinpoint and change elements in her life-style that unnecessarily increase intraabdominal pressure. Losing excess weight, removing girdles, avoiding heavy lifting, stopping smoking, and treating allergies or pulmonary disease leading to sneezing or coughing may ultimately be as important as any surgical repair in preventing progression or recurrence of prolapse.

Hormone Replacement

When not contraindicated in a postmenopausal patient, hormonal replacement therapy, especially intravaginal estrogen, will not only reverse atrophic changes, but lessen the speed of prolapse progression, and in many patieents will relieve symptoms of urinary urgency and precipitancy and occasionally of incontinence. Restoration of vaginal thickness, improved blood supply, and enhanced tissue elasticity will also optimize conditions for a successful and lasting surgical outcome. Vaginal installation of 2 g of estrogen cream at bedtime for 2 weeks followed by gradual tapering to one application per week can be started 1 to 2 months preoperatively and continued indefinitely.

Exercises

Benefit may also be obtained from a sustained program of Kegel or perineal resistive exercises. The exercises improve the tone and function of the pelvic diaphragm. Some patients with mild stress urinary incontinence learn to contract the pubococcygei before a cough or sneeze to prevent involuntary urine loss. Patients having difficulty contracting the appropriate muscle groups can be helped by the use of a perineometer, which gives visual biofeedback on the strength of a pubococcygeal contraction.

Pessaries

In addition to their usefulness as a nonsurgical treatment of uterine retroversion or prolapse, pessaries can assist in the diagnostic differentiation of a gynecologic versus an orthopedic etiology for backache. A low backache arising from genital relaxation is usually absent on arising, but comes on and gets worse as the day progresses. A backache of orthopedic origin will often be worse when relaxed muscles allow for greater joint instability in the morning, but improve as the day progresses. If a patient's backache has both gynecologic and orthopedic components, one may wish to predict the degree of relief she could expect from surgery to correct her genital relaxation. Insertion of a properly fitting pessary that restores the pelvic viscera close to their normal relationships can aid in this determination (Fig. 24-7). If the patient experiences relief from wearing the pessary for 2 to 4 weeks, it is then removed to see if symptoms return. The difference in the patient's degree of comfort is what she reasonably could expect to gain by reconstructive surgery.

While pessaries are not a definitive solution for genital relaxation, they are very useful for temporizing until future surgical reconstruction or for relieving a patient who is not a surgical candidate. Preoperative use of a pessary in advanced degrees of prolapse may aid in decongesting the mucosa, improving circulation, and reestablishing vaginal tonicity, especially if combined with vaginal estrogen therapy in the postmenopausal patient. Symptomatic women who wish to complete childbearing before having a surgical repair may also be candidates for a pessary.

Pessaries useful in symptomatic uterine retroversion include the Smith, Hodge, and Risser pessaries. The Gehrung pessary is useful in cystocele, rectocele, and some procidentias. Pessaries helpful in uterine prolapse or vaginal vault eversion include the ring, Gellhorn, cube, and inflatable doughnut pessaries. The ring and Gehrung pessaries as well as the Smith, Hodge, and Risser pessaries, will permit coitus. Those pessaries that fill the vaginal canal will not.

A pessary should be checked by the physician during the first week after its inser-

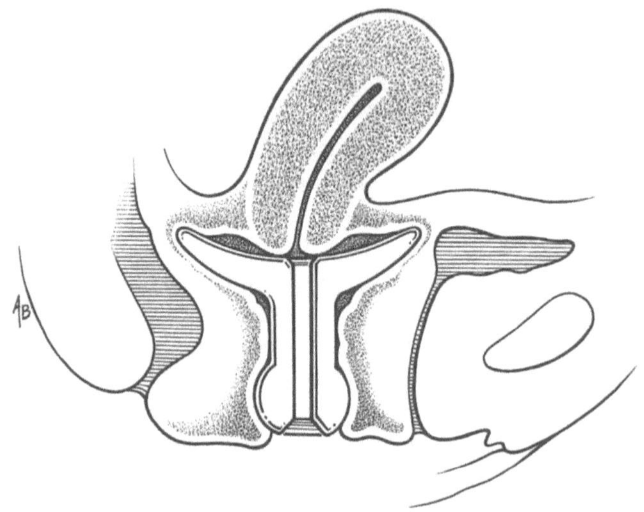

FIGURE 24-7. A vaginal pessary may aid in both the diagnosis and management of symptoms related to genital prolapse. (Modified from Nichols DH, Julian PJ: The vaginal pessary, in Nichols DH, Evrard JR (eds): Ambulatory Gynecology. Philadelphia, Harper & Row 1985, p. 197.)

tion and quarterly for as long as it is in use. Preferably, a patient or her caretaker should be able to remove and reinsert the pessary. If not, she should be seen monthly to remove and cleanse the pessary and to check the vagina for irritation or ulceration. Because the vagina will tend to enlarge with time to accommodate the pessary, the patient may need to be fitted with a larger size.

Surgical Treatments

Principles of Reconstructive Surgery

In planning reconstructive surgery for genital prolapse, the surgeon should keep in mind a number of important principles. Many prominent gynecologic surgeons concur with Burch's statement that "The golden rule of the pelvic surgery is the conservation of useful function." Te Linde believed that there are three reasons for gynecologic surgery: to save life, to relieve suffering, and to correct symptomatic anatomic deformities.[11] Most reconstructive surgery falls into the latter two categories, although in situations such as incarcerated procidentia, the first principle is called into play as well.

Since the uterus is a passenger, not the culprit, in genital prolapse, merely removing it will not cure the patient's problem. Hysterectomy for uterine disease is primarily a destructive procedure. The emphasis is on what is taken out. How one removes the uterus is of secondary importance as long as the diseased tissue is ablated. On the other hand, in reconstructive surgery, the critical concern is what is done with that which remains with the patient. Removal of an otherwise normal prolapsed uterus may be appropriate to permit adequate mobilization of the supporting tissues to be used in the repair, but hysterectomy is only one step in a very carefully planned operative reconstruction. There is no such thing as a standard patient, a standard prolapse, or a standard repair.

All sites of possible damage should be carefully evaluated. While special attention is directed toward the primary weakness, all defects should be repaired. Although symptoms of prolapse are generally associated with demonstrable weakness in pelvic supports, some anatomic defects may be asymptomatic. The relevance of aphorisms, such as "If it ain't broke, don't fix it," or Grandison Royston's dictum," A "bladder is as good as it works, not as bad as it looks," is real, for one is unlikely to make a totally asymptomatic patient better by operating on her. Equally true, the gynecologist should be cautious about recommending surgery to a symptomatic patient in the absence of substantiating pathology.

However, failure to repair all defects at the initial procedure is likely to favor recurrence and need for subsequent repair. For example, the anterior vagina rests and derives considerable support from a structurally intact posterior vaginal wall. Failure to perform a proper posterior colporrhaphy or perineorrhaphy will compromise the strength of the anterior repair. Neither the patient nor the health care system can afford the burden and tragic expense of necessary reoperation.

The goal of reconstructive surgery is, as far as possible, to restore normal anatomy. Sometimes the surgeon must overcorrect a primary weakness or create a compensatory defect, but, if so, special care must be taken to ensure that other structures are not thereby placed in jeopardy. For instance, overcorrecting a cystocele or performing a sacral colpopexy without adequate attention to support of the urethrovesical junction can lead to stress urinary incontinence. *Ventral suspension* of the vagina (Fig. 24-8) or *retropubic urethropexy* can change the vaginal axis and promote the development of an enterocele.

There is still a limited place among very old and infirm women with significantly increased medical risks for performing a *LaFort colpocleisis* or *vaginectomy*, but because these procedures distort normal anatomy they should be the exceptions rather than the rule. Another alternative is perineorrhaphy to permit such a patient to retain a pessary. Preserving the ability to function coitally, whether or not she chooses to do so, is very important to most women. Every attempt should be made to retain vaginal width and depth and to restore the upper vagina to its normal horizontal position resting on an intact levator plate.

Reconstructive Procedures

The surgical procedures that have been described for the cure of genital prolapse are legion. Some, such as the *Spaulding-Richardson composite operation*, the *Watkins transposition operation*, and the *Manchester (Donald or Fothergill) operation*, are primarily of historical interest. Other procedures are done much less frequently today than they were in the past. Among these are the procedures for uterine suspension, such as the *Gilliam suspension* or *Olshausen's op-*

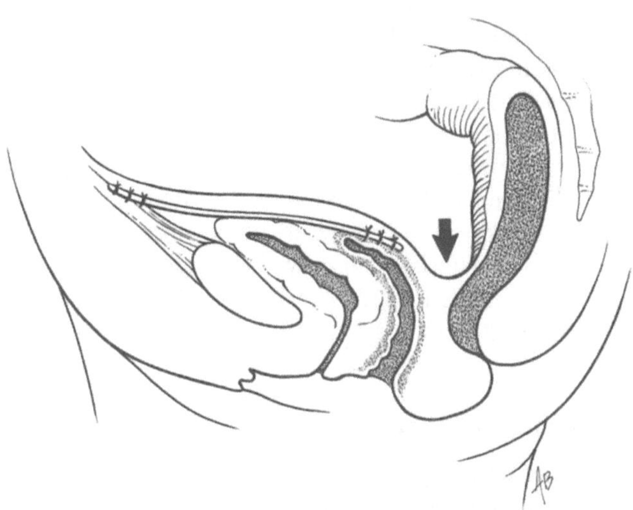

FIGURE 24-8. Ventral suspension of the vagina may lead to the development of an iatrogenic enterocele (arrow). (Modified from Nichols DH: Repair of enterocele and prolapse of the vaginal vault, in Barber H (ed): Goldsmith's Practice of Surgery. Philadelphia, JB Lippincott, 1981.)

eration, which fixed the uterus to the anterior abdominal wall, and the *Baldy-Webster procedure*, using the round ligaments to create a supporting posterior hammock for the uterus.

Uterine Suspension

Occasionally, especially in the presence of congenital malformations of the pelvis, the gynecologist may encounter a young nulliparous patient with marked uterine prolapse. Such a patient may be a candidate for a *sacral cervicopexy* such as that described by Stoesser,[12] in which fascia lata or synthetic mesh is used to create a cervicosacral ligament to reduce the prolapse. Patients who have undergone this procedure have carried a pregnancy to term and delivered vaginally without difficulty. However, the relative safety of cesarean delivery today warrants strong consideration for choosing abdominal delivery to preserve the integrity of the repair in these rare patients. Another reported alternative in young women who wish to preserve fertility is the transvaginal fixation of the proximal uterosacral ligaments to the sacrospinous ligament.[13]

Upper Support Damage

When the upper supports of the uterus are weakened, and the uterus becomes retroverted, there is seldom justification for surgical repair if the retroversion is an isolated finding. However, if a woman who desires future childbearing has clear-cut problems with uterine retroversion, and comes to laparotomy for some other indication, procedures such as a *modified Baldy-Webster procedure* or triplication of the round ligaments with plication of the uterosacral ligaments may be appropriate. Indeed, if the patient is having conservative surgery for endometriosis, or tubal surgery for relief of infertility, her treatment may well be incomplete without them.

Middle Support Damage

When the middle uterine supports are damaged while the upper supports remain intact, the cervix begins to elongate and descend through the vaginal canal. If this uterovaginal-type prolapse[2] progresses slowly over a long period of time, the cardinal and uterosacral ligaments, although elongated, may actually be hypertrophied. In this case, when *vaginal hysterectomy* is performed following the completion of childbearing, the cardinal-uterosacral complex can be shortened by medial excision and the remaining portion can be reattached to the posterior vaginal vault. The *New Orleans or McCall culdeplasty*[14] aids in assuring posthysterectomy support of the vaginal vault (Fig. 24-9). If an enterocele is not present, the peritoneal reflections may be much higher than anticipated, and there may be some difficulty entering the peritoneal cavity initially. If an enterocele is present, the redundant peritoneum should be resected and the sac ligated as high as possible. Widening of the posterior vaginal fornix may be reduced by excision of an appropriate wedge.

When both the upper and middle uterine supports are weakened, eversion of the upper vagina occurs with the descent of a usually retroverted, normal-sized uterus without cervical elongation. At vaginal hysterectomy, the uterosacral ligaments should be shortened, and the enterocele, which is often of the traction type, should be resected. If the uterosacral ligaments are of sufficient strength, several culdeplasty stitches may be placed to further obliterate the cul-de-sac and support the vaginal vault.

Alternatively, Watson has described a *vaginal hysterectomy-colporrhaphy*,[15] which deperitonizes completely the pouch of Douglas and approximates the posterior portions of the uterosacral ligaments together in front of the rectum. The redundant uterosacral ligaments, cardinal ligaments, and broad and round ligaments are reconstructed "to form a flat musculofascial roof, to which the anterior vaginal wall is sutured."

Lower Support Damage

When damage is to the lower pelvic supports, the result is usually eversion of the vagina

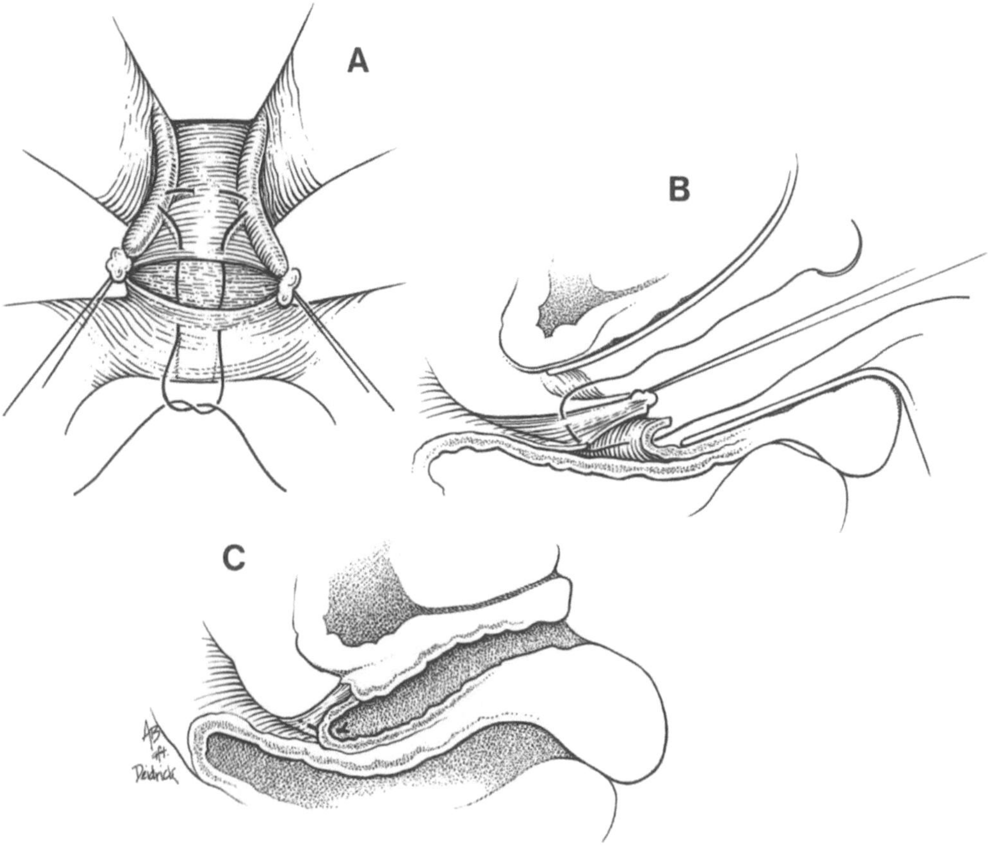

FIGURE 24-9. Support of the vaginal vault during a vaginal hysterectomy for uterovaginal prolapse can often be accomplished by means of McCall or New Orleans culdeplasty. (A) The absorbable suture is passed through the portion of the posterior vaginal wall that will become the new apex of the vagina, through the peritoneum, and through a point high on each uterosacral ligament. (B) The first half of the stitch shown in sagittal section. (C) After closure of the peritoneal cavity and tying of the McCall stitch, the vaginal apex is suspended cranial and posterior to the new cul-de-sac. (Modified from ref. 17, p. 223.)

with cystocele and/or rectocele. If the damage is midline, the repair is midline via *anterior* and/or *posterior colporrhaphy*. If the damage is lateral with avulsion of vaginal attachments to the arcus tendineus, a *paravaginal repair*, ideally done vaginally[16] rather than abdominally, will be necessary to correct the defect.

Rotational descent of the urethrovesical angle should be corrected by *plication of the posterior pubourethral ligament* portion of the urogenital diaphragm (Fig. 24-10); funneling of the bladder neck should be corrected by *Kelly plication sutures*. During *posterior colpoperineorrhaphy*, because the presence and strength of the pubovaginalis is inconstant, the need for "levator stitches," bringing together in the midline between the rectum and vagina a portion of the pubococcygei, is variable. More often than not, such a step will contribute to subsequent dyspareunia and is unnecessary to support adequately the posterior vaginal wall. Care should be taken, however, to resect any redundancy in the fascia of Denonvilliers and to reattach the latter to the perineal body.

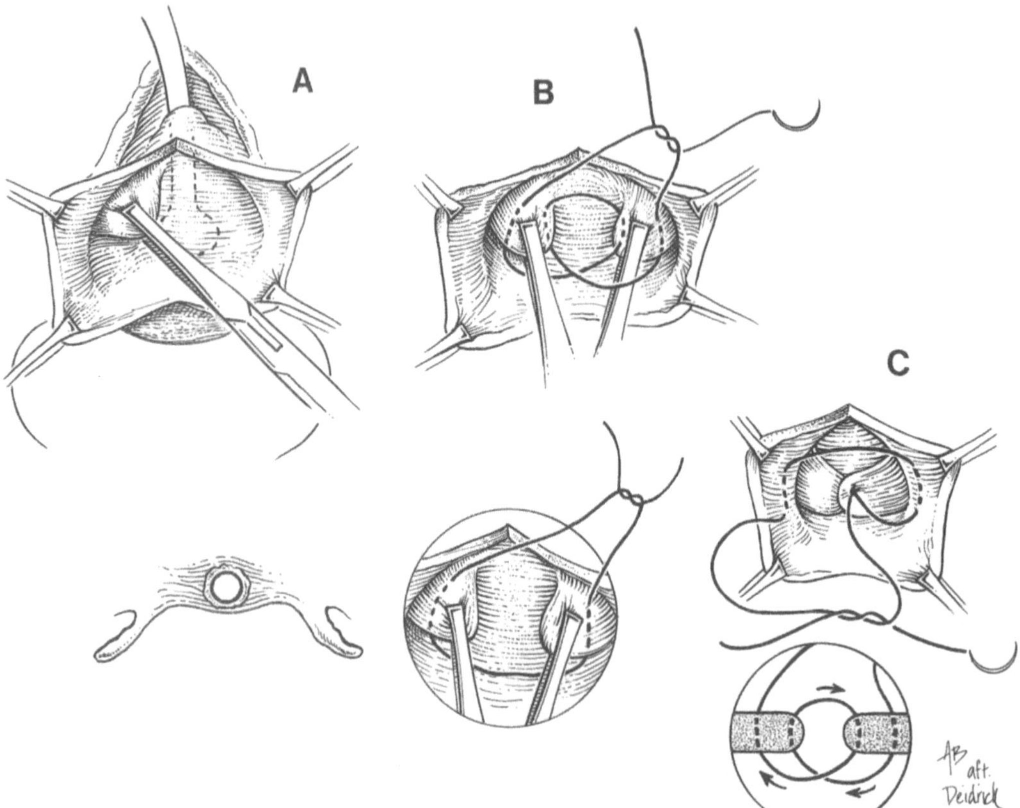

FIGURE 24-10. Plication of the posterior pubourethral ligament. In the absence of a urethrovaginal space corresponding to the vesicovaginal space, a proper depth of dissection is necessary to avoid devascularization of the urethra (lower left). The pubourethral ligament is gently grasped at the level of the urethrovesical junction (A). A long-acting absorbable suture plicates the ligaments in a far-near–near-far fashion (B and lower right). The suture is then passed through the under side of the vaginal wall to reestablish the fusion of the urogenital diaphragm with the vaginal wall (C). An alternate method is applicable when nonabsorbable suture is used (center circle). (Modified from ref. 4, pp. 256, 258.)

Procidentia

When there has been substantial damage to all three levels of support, the end result is usually procidentia, with the vagina turned completely inside out. In this case, there is rarely sufficient strength in the cardinal-uterosacral complex to support the vaginal vault when the uterus is removed. If one resorts to *colpocleisis* or *vaginectomy*, not only will coital function of the vagina be severely limited or destroyed, but enterocele will not be corrected, if one does not deliberately open the peritoneum and resect it. One may also create or exacerbate stress urinary incontinence.

Preferably, vaginal hysterectomy, extensive anterior colporrhaphy, and posterior colpoperineorrhaphy will be accompanied by a procedure that restores the vaginal depth and axis. This goal can be accomplished by either *sacrospinous colpopexy* (Fig. 24-11) or *sacral colpopexy* (Fig. 24-12). Since many of these patients are elderly and have a number of medical problems, the morbidity of the abdominally performed sacral colpopexy is less desirable unless there is some other reason to open the abdominal cavity. In expe-

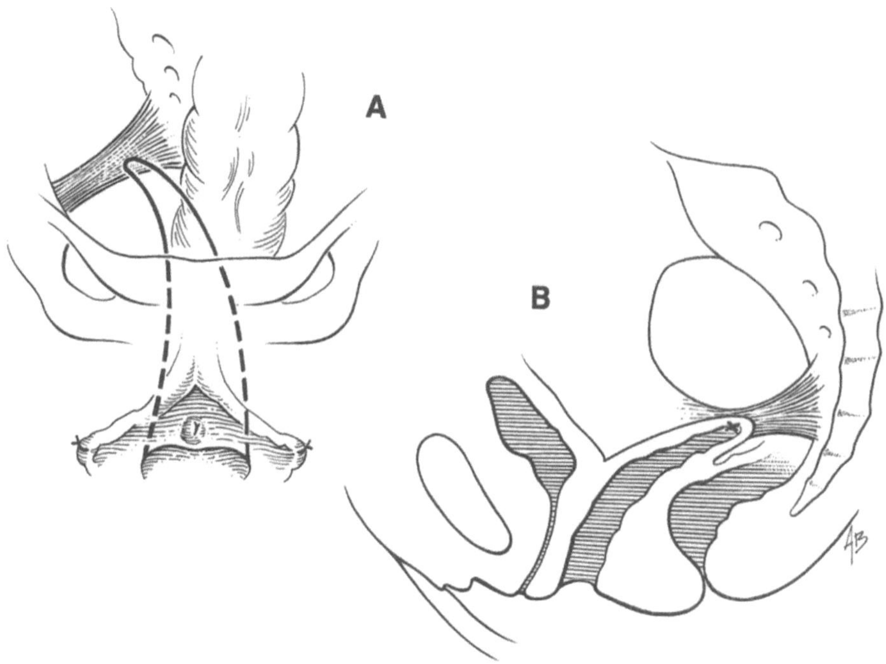

FIGURE 24-11. Transvaginal sacrospinous colpopexy, shown coronally (A) and sagittally (B), suspends a prolapsed vagina by attaching it to the sacrospinous ligament. (Modified from Randall CL, Nichols DH: Surgical treatment of vaginal inversion. Obstet Gynecol 1971;38:330.)

rienced hands, vaginal hysterectomy, colporrhaphy, and vaginally performed sacrospinous colpopexy can be accomplished within 2 hours. The patient may have the procedure under spinal anesthesia and be ambulatory the day after surgery.

Prophylaxis

Should the vaginal vault not be adequately supported at the time of hysterectomy, the prolapse will definitely recur, sometimes within days of the initial procedure. Even when hysterectomy is performed for a reason other than prolapse, principles of good surgical technique demand that careful thought be given to vault support to reduce the possibility of a future prolapse.

At vaginal hysterectomy, the uterosacral and cardinal ligaments, shortened if necessary, should be reattached to the vaginal cuff. The round ligaments, although not as strong, may provide additional support. The points of attachment to the cuff should be at the 4 or 5 and 7 or 8 o'clock positions on the vaginal vault. Some investigators recommend the routine use of a McCall culdeplasty stitch at the time of vaginal hysterectomy. Another approach, which may be especially helpful in elongating the anterior vaginal wall and in preventing a posthysterectomy anterior enterocele, is to suture the previously tied peritoneal closure stitch to the anterior vaginal cuff in the midline.[17]

At abdominal hysterectomy as well, the uterosacral and cardinal ligaments should be reattached to the vaginal cuff.[18] If the posterior vaginal vault seems particularly vulnerable, a stitch not unlike that used vaginally in a McCall culdeplasty may be placed to plicate the uterosacral ligaments in the midline and attach them to the vaginal cuff more caudally. If a deep cul-de-sac threatens to serve as a dilating wedge in the development of a posthysterectomy enterocele, it may be obliterated by a *Moschcowitz*[19] or a *Halban*[20] procedure (Fig. 24-13).

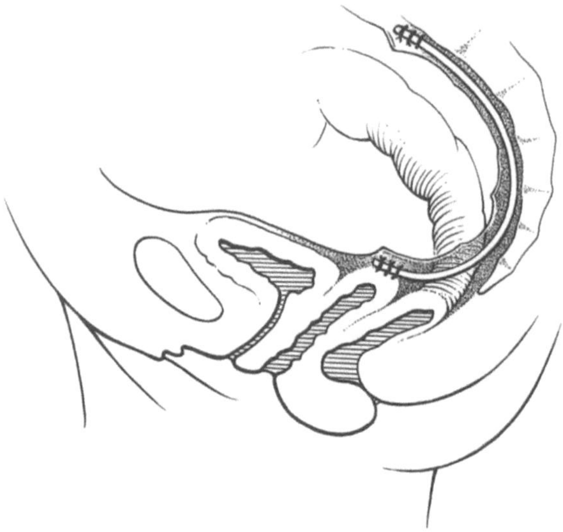

FIGURE 24-12. Transabdominal sacral colpopexy suspends a prolapsed vagina by attaching it to the sacrum via synthetic mesh or a fascial strip. (Modified from Nichols DH: Repair of enterocele and prolapse of the vaginal vault, in Barber H (ed): Goldsmith's Practice of Surgery. Philadelphia, JB Lippincott, 1981.)

Conclusion

Since quality of life is such an important ingredient in longevity, more and more women are seeking relief for distressing symptoms of genital prolapse that compromise their ability to remain active and involved. Advances in intraoperative and postoperative care have largely eliminated age as a contraindication when surgical management is the treatment of choice. With the rising cost of hospitalization, however, neither the patient herself nor society in general will look kindly upon the need for repeated procedures. The right operation must be chosen and properly performed the first time the patient goes to surgery.

References

1. Swash M: Histopathology of the pelvic floor muscles, in Henry MM, Swash M (eds): Coloproctology and the Pelvic Floor, Pathophysiology and Management. Boston, Butterworths, 1985, pp 129–150.
2. Malpas P: Genital Prolapse and Allied Conditions. New York, Grune & Stratton, 1955.

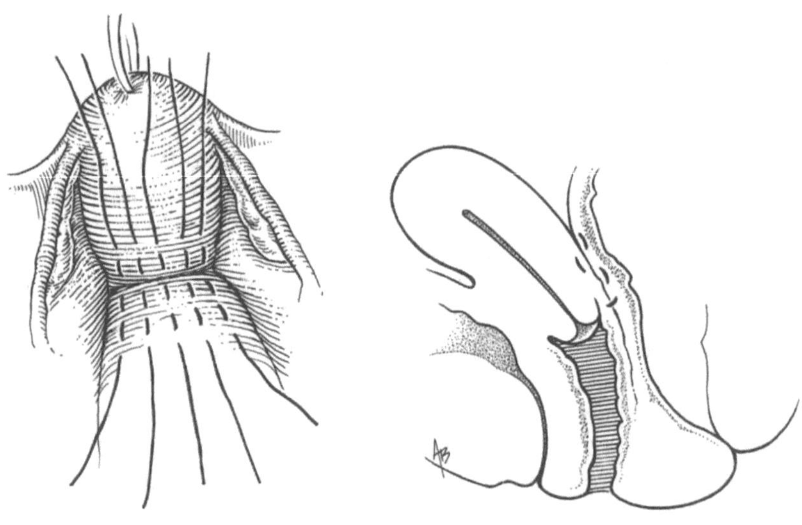

FIGURE 24-13. Halban culdeplasty may be used at the time of laparotomy to obliterate a deep cul-de-sac (as illustrated) or at the time of abdominal hysterectomy to prevent the possibility of posthysterectomy enterocele formation. In the latter case, similar sagittal rows of sutures attach the visceral peritoneum of the rectum to the peritoneal surface of the posterior vaginal wall. (Modified from Nichols DH: Repair of enterocele and prolapse of the vaginal vault, in Barber H (ed): Goldsmith's Practice of Surgery. Philadelphia, JB Lippincott, 1981.)

3. Riddick DH: The Perimenopausal Woman in Transition. Audio Digest: Obstetrics and Gynecology, 1988, vol 35, no 24.
4. Nichols DH, Randall CL: Pelvic anatomy of the living, in Vaginal Surgery. Baltimore, Williams & Wilkins, 1989, pp 1–45.
5. von Peham H, Amreich J: Operative Gynecology. Philadelphia, JB Lippincott, 1934.
6. Mengert WF: Mechanics of uterine support and position. Am J Obstet Gynecol 1936; 31:775–782.
7. Fothergill WE: On the pathology and the operative treatment of displacements of the pelvic viscera. J Obstet Gynaecol Br Emp 1907; 13:410–419.
8. Paramore RH: The supports-in-chief of the female pelvic viscera. J Obstet Gynaecol Br Emp 1908;13:391–409.
9. Halban J, Tandler J: Anatomie und Atiologie der Genitalprolapse beim Weibe. Vienna, Wilhelm Braumuller, 1907.
10. Bonney V: The sustentacular apparatus of the female genital canal, the displacements that result from the yielding of its several components and their appropriate treatment. J Obstet Gynaecol Br Emp 1914;45:328–344.
11. Mattingly RF, Thompson JD (eds): Te Linde's Operative Gynecology. Philadelphia, JB Lippincott, 1985, p 63.
12. Stoesser FG: Construction of a sacrocervical ligament for uterine suspension. Surg Gynecol Obstet 1955;101:638–641.
13. Miyazaki FS: Miya hook ligature carrier for sacrospinous ligament suspension. Obstet Gynecol 1987;70:286–288.
14. McCall ML: Posterior culdeplasty: Surgical correction of enterocele during vaginal hysterectomy. A preliminary report. Obstet Gynecol 1957;10:595–602.
15. Watson AL: The technique of vaginal hysterectomy for uterine prolapse. Am J Obstet Gynecol 1951;61-A:206–218.
16. White GR: Cystocele: A radical cure by suturing lateral sulci of vagina to white line of pelvic fascia. JAMA 1909;53:1707–1708.
17. Nichols DH, Randall CL: Vaginal hysterectomy, in Vaginal Surgery. Baltimore, Williams & Wilkins, 1989, pp 182–238.
18. Mattingly RF, Thompson JD: Malpositions of the uterus, in Mattingly RF, Thompson JD (eds): Te Linde's Operative Gynecology. Philadelphia, JB Lippincott, 1985, pp 541–567.
19. Moschcowitz AV: The pathogenesis, anatomy, and cure of prolapse of the rectum. Surg Gynecol Obstet 1912;15:7–21.
20. Halban J: Gynäkologische Operationslehre. Berlin, Urban and Schwarzenberg, 1932.

25

Contemporary Management of Cancer of the Cervix

GUNTER DEPPE and VINAY K. MALVIYA

Thirteen thousand new cases of invasive cervical cancer were reported in the United States in 1991 (incidence estimates based on rates from NCI SEER program 1985–1987) and approximately 4,500 deaths were attributed to this disease in the same period.[1] The frequency of invasive cervical cancer in the general population is not known, but the best incidence data indicates a rate of approximately 10 to 12 per 100,000 women per year. The incidence appears to change from one locality to another and is less frequent in rural areas than in metropolitan areas. The mean age for invasive cervical cancer is 52.2 years and the distribution of cases is bimodal, with peaks between 35 through 39 years and 60 through 64 years.[1]

Diagnosis

The most common symptom in patients with invasive cancer of the cervix is vaginal bleeding. This may occur either as postcoital bleeding or as irregular acyclic bleeding, which may be mistaken by younger women as menstrual irregularity. Among older patients with this disease postmenopausal bleeding is the primary symptom, and patients with advanced disease present with malodorous or bloody vaginal discharge, weight loss, or symptoms related to obstructive uropathy.

As the cancer enlarges, the bleeding episodes become heavier, more frequent, and of longer duration. Late symptoms suggestive of advanced disease include the development of pain referred to the flank or leg, usually secondary to the involvement of ureters, pelvic sidewall, or sciatic nerve roots. Many patients may complain of dysuria, hematuria, rectal bleeding, or obstipation resulting from bladder or rectal invasion. Persistent edema of one or both lower extremities, produced as a result of lymphatic and venous blockage, are late manifestations of primary disease and frequent manifestations of recurrent disease. Massive hemorrhage and development of uremia with profound inanition may also occur as preterminal events.

Clinical appearance of cervical cancer varies considerably and depends upon the growth pattern and the extent of regional involvement. Three categories of gross lesions have been traditionally described: exophytic, endophytic or ulcerative, and polypoid lesions. The exophytic configuration occurs more commonly in squamous cell cancers than the ulcerative lesion (38 and 33%, respectively). Twenty-four percent of the lesions may be endophytic, and 5% are polypoid lesions.[2]

Staging

Staging of primary cervical carcinoma is based on careful clinical examination prior to any definitive therapy. The clinical stage must not be changed on the basis of subsequent findings. The Cancer Committee of the International Federation of Gynecology and Obstetrics (FIGO) allows inspection, palpation, colposcopy, endocervical curettage, hysteroscopy, cystoscopy, proctoscopy, intravenous urography, and chest x-ray for clinical staging. Suspected bladder or rectal involvement requires confirmation by biopsy and histologic evidence.

Findings by examinations such as lymphangiography, computed tomography (CT) scan, magnetic resonance imaging (MRI), arteriography, venography, radionuclide scans, and laparoscopy frequently are valuable in the planning of therapy, but should not be used for changing the clinical stage (Table 25-1). Evaluation of paraaortic nodes by lymphangiography is associated with a false-positive rate between 20 and 40% and a false-negative rate between 10 and 20%.[3–6] The value of CT scanning is to identify enlarged retroperitoneal nodes, especially in the paraaortic area, and to demonstrate abdominal pelvic masses. In the patient with recurrent disease, it is useful to identify disrupted lateral pelvic tissue planes because of invasion of the pelvic musculature. The accuracy of CT scan is between 80 and 85%, with a false-negative rate of 10 to 15% and a false-positive rate between 20 and 25%.[7–9] Early MRI data are comparable.[10] Demonstration of fine-needle aspirate showing metastatic disease, under CT guidance will allow the radiation treatment field to be extended and obviates the need for exploratory laparotomy to determine the status of lymph nodes. When doubt exists regarding the correct stage, the patient should be allotted to a lower stage. The current FIGO staging system (1985) of cervical cancer is outlined in Table 25-2.[11]

Over the last several decades, there has been continued confusion about the stages of preclinical invasive carcinoma of the cervix. There also has been increased pressure to put measurements into the definition, resulting in the new FIGO definition. Based on a recent survey at the 1988 meeting of the Society of Gynecologic Oncologists (SGO), it appears that the majority of gynecologic oncologists in the United States continue to use the 1974 SGO definition in the management of their patients: "A microinvasive carcinoma of the cervix is a lesion that invades the cervical stroma to a maximum depth of 3 mm beneath the basement membrane without evidence of lymphatic or vascular involvement."

Stage Ia carcinoma should include minimal microscopically evidence stromal invasion as well as small cancerous tumors of measurable size. Stage Ia should be subdivided into those lesions with minute foci of invasion only microscopically visible (stage

TABLE 25-1. Pretreatment evaluation of a patient with invasive cancer of the uterine cervix

Routine	Optional
History, physical examination including nodal survey	Coagulation studies
Pelvic assessment	Tissue markers
CBC w/differential w/platelets, electrolytes, BUN, creatinine	
Liver enzymes, EKG	
Radiographic Studies	
Chest x-ray, PA & lateral	Barium enema (adenocarcinoma advanced stage or where otherwise indicated)
IVP or CT scan of the abdomen and pelvis w/contrast	
MRI studies in patients with contrast allergy	Radioisotope scans (liver, bone)
Endoscopic Studies (if indicated)	
Cystoscopy	Cystoscopy, retrograde pyelography
Proctosigmoidoscopy	Rectosigmoidoscopy
	Pulmonary function tests
	Exam under anesthesia

CBC = complete blood count, BUN = blood urea nitrogen, EKG = electrocardiogram, PA = posteroanterior, IVP = intravenous pyelogram.

TABLE 25-2. FIGO staging system for carcinoma of the cervix

Stage	
0	Carcinoma in situ, intraepithelial carcinoma.
I	Carcinoma strictly confined to the cervix; extension to the corpus should be disregarded.
Ia	Preclinical carcinoma of the cervix, i.e., those diagnosed only by microscopy.
Ia1	Minimal microscopically evident stromal invasion.
Ia2	Lesions detected microscopically that can be measured; the upper limit of the measurement should not show a depth of invasion of >5 mm taken from the base of the epithelium, either surface or glandular, from which it originates; a second dimension, the horizontal spread, must not exceed 7 mm; larger lesions should be staged as Ib.
Ib	Lesions of greater dimension than Stage Ia2, whether seen clinically or not. Preformed space involvement should not alter the staging, but should be specifically recorded so as to determine whether it should affect future treatment decisions.
II	The carcinoma extends beyond the cervix, but has not extended onto the pelvic wall; the carcinoma involves the vagina, but not as far as the lower third.
IIa	No obvious parametrial involvement.
IIb	Obvious parametrial involvement.
III	The carcinoma has extended onto the pelvic wall; on rectal examination, there is no cancer-free space between the tumor and the pelvic wall; the tumor involves the lower third of the vagina; all cases with hydronephrosis or non-functioning kidney should be included, unless they are known to be due to another cause.
IIIa	No extension onto the pelvic wall, but involvement of the lower third of the vagina.
IIIb	Extension onto the pelvic wall or hydronephrosis or non-functioning kidney.
IV	Carcinoma has extended beyond the true pelvis or clinically has involved the mucosa of the bladder or rectum. A bullous edema, as such, does not permit a case to be allotted to Stage IV.
IVa	Spread of the growth to adjacent organs.
IVb	Spread to distant organs.

Reprinted from ref. 11, with permission.

Ia1) and the macroscopically measurable microcarcinomas (stage Ia2) in order to gain further knowledge of the clinical behavior of these lesions. The term, "stage Ib occult" should be omitted. The diagnosis of both stages Ia1 and Ia2 should be based on microscopic examination of removed tissue, which must include the entire lesion.

As noted, the lower limit of stage Ia2 should be that it can be measured macroscopically, and the upper limit of stage Ia2 is given by measurement of the two largest dimensions in any given section. The depth of invasion should not be more than 5 mm taken from the base of the epithelium, either surface or glandular, from which it originates. The second dimension, the horizontal spread, must not exceed 7 mm. Vascular space involvement, either venous or lymphatic, does not alter the staging, but should be specifically recorded since it may affect treatment decisions in the future. Lesions of greater size should be staged as Ib. Since it is impossible to estimate clinically whether cancer of the cervix has extended to the corpus, extension to the corpus should be disregarded.

A patient with malignancy extending beyond the cervix, but not extending to the pelvic wall, should be allotted to stage II. At clinical examination, it may be impossible to decide whether a smooth, indurated parametrium is truly cancerous or only inflammatory; therefore the case should be placed in stage III only if the parametrium is nodular to the pelvic wall or the growth itself extends to the pelvic wall. The presence of hydronephrosis or nonfunctioning kidney due to stenosis of a ureter by cancer permits a case to be allotted to stage III even if, according to other findings, the case should be allotted to stage I or II.

The presence of bullous edema, per se, should not permit a case to be allotted to stage IV. Ridges and furrows into the bladder wall should be interpreted as signs of submucosal involvement of the bladder if they remain fixed to the growth at palpation (i.e., a vaginal or rectal examination during cystoscopy). A cytologic finding of malignant cells in washings from the urinary bladder requires further examination and a biopsy from the wall of the bladder.

Spread

Cervical carcinoma spreads predominantly by direct extension to the lower uterine segment, vagina, and parametrium, and by lymphatic embolization through parametrial channels that terminate in lymph nodes along the pelvic sidewalls (e.g., obturator, hypogastric, and external iliac lymph nodes). From these areas, malignant tumor emboli may involve the common iliac, paraaortic, and, ultimately, the supraclavicular lymph nodes.

Surgical Staging

Clinical staging for cervical cancer has been found to be highly inaccurate, with overall errors in from 25% of patients with clinical stage I to 50% or more with stages III and IV cancer.[12–14] The benefit of surgical staging is diagnosis of extrapelvic disease. Transperitoneal surgical staging procedures are associated with appreciable major complications (13.2%) and occasional deaths (0.4%).[2] Alternative methods for extraperitoneal or paraaortic node dissection or pelvic node dissection have been described by Schellhas[15] and Berman et al.,[16] respectively who noted significantly fewer complications with their techniques. The risk of metastases of pelvic nodes ranges from 12.4% in stage Ib[17] to 56% in stage IVa.[16] The incidence of involved paraaortic nodes ranges from 4.5% in stage Ib[18] to 43.5% in stage IV.[19] It appears that some patients with microscopic metastatic disease in the aortic nodes may be salvaged by extended field radiation; hence, it seems important to surgically stage young patients at high risk for aortic node metastases.

Approximately one third of patients with metastases to the paraaortic nodes will have involved scalene lymph nodes[20,21]; therefore, scalene node biopsy may be performed in the patient with positive paraaortic nodes to rule out evidence of systemic disease. At present, surgical staging is a research tool that allows prospective studies aimed at improving control of systemic disease, thereby resulting in improved survival.

Selection of Therapy for Cervical Cancer

Age of the patient, tumor volume, depth of invasion of the tumor, presence of vascular invasion on biopsy or conization, palpable extension into the parametrial tissue, demonstrable extrapelvic disease, and preservation of sexual function are several important factors in the selection of treatment of a patient with cervical cancer. Tumor volume is the single most important predictor of survival in patients with cancer of the cervix. It is apparent that clinical FIGO staging is not an accurate measure of tumor volume. In many instances, surgical staging procedures may improve the physician's ability to treat the disease.

Treatment of Stage Ia

The incidence of lymph node metastases appears to be mainly dependent on the depth of invasion (Table 25-3); therefore, this variable is used to determine the extent of the surgery and whether regional lymph nodes should be treated. Treatment is outlined in Table 25-4.

Treatment of Stages Ib and IIa

Five types of radical hysterectomy have been described to tailor the extent of the procedure to the size of the cervical tumor[22] (Table 25-5).

Radical hysterectomy with lymphadenectomy or radiotherapy can be used for pa-

TABLE 25-3. Stage I cancer of the cervix: nodal metastases and depth of invasion

Depth of invasion (mm)	No. of patients	Number of patients with positive nodes	Positive nodes (%)
0.0–3.0	43	0	0.0
3.1–5.0	26	1	3.9
5.1–10.0	17	1	5.9
10.1–15.0	19	4	21.0
15.0	41	18	43.9
Total	146	24	16.4

Simon NL, Gore H, Shingleton HM, et al: Study of superficially invasive carcinoma of the cervix. Obstet Gynecol 1986;68:19, with permission of the American College of Obstetricians and Gynecologists.

tients with stage Ib and Stage IIa carcinoma of the cervix. The usual practice is to perform a radical hysterectomy if the patient is less than 70 years old, not obese, and in good physical condition, with a cancerous lesion smaller than 4 cm in diameter. Radical hysterectomy should not be attempted when the size of the lesion precludes tumor-free margins. Five-year survival rates range between 80 and 85% and are identical to those achieved with curative radiotherapy.[23]

Any enlarged or suspicious paraaortic nodes are selectively excised and, if they are positive, radical hysterectomy is terminated. Aortic lymphadenectomy is not performed in the absence of proven positive pelvic lymph nodes on frozen section. The advantages and disadvantages associated with radical hysterectomy and radiation therapy are outlined in Table 25–6.

Treatment of Stages IIb through IVa

Radiation therapy is an important method of primary treatment of patients with cervical cancer. It is the therapy of choice for patients with stages IIb, III, and IVa cancer.

Radiation therapy is a local treatment and is curative for those patients whose tumor has spread beyond the radiation field. The tolerance of normal bowel and bladder tissue limits the amount of radiation. Complications of radiation therapy are listed in Table 25-7. External teletherapy and intracavitary brachytherapy are used in various combinations (Table 25-8).

Patients with larger tumors are usually treated by external radiation to shrink the tumors, allowing a better intracavitary application. Microscopic squamous cell cancer can be locally controlled with 40 to 50 Gy, whereas larger tumors require 60 Gy and more, depending on the tumor volume.

To decrease the exposure of personnel to radiation, most institutions use afterloading applicators. Radioactive material is placed in the uterus and vagina after successful placement of the applicators surgically and confirming their position by x-ray.

High-dose brachytherapy has been recently introduced in our institution for patients with cervical cancer. This technique

TABLE 25-4. Surgical management of stage I cancer of the cervix

Histologic findings	Terminology	Recommended procedure
Stromal invasion to 1 mm as isolated projection	Focal microinvasion (stage Ia1)	TAH or TVH or conization
Stromal invasion to 3 mm	Microinvasion (Stage Ia_2)	TAH, node assessment
Stromal invasion to 3–5 mm	Microcarcinoma (Stage Ia_2)	Modified radical hysterectomy TAH, pelvic node dissection
Stromal invasion to 5 mm	Invasive carcinoma (Stage Ib)	Radical hysterectomy, aortic node assessment, pelvic node dissection

TAH = total abdominal hysterectomy, TVH = total vaginal hysterectomy.

TABLE 25-5. Types of hysterectomy

Type	Description
I	Extrafascial hysterectomy
II	Removal of medial half of cardinal and uterosacral ligaments
III	Removal of most of cardinal and uterosacral ligaments and upper third of vagina
IV	Removal of periureteral tissue, superior vesicle artery and three-fourths of vagina
V	Removal of portions of distal ureters and bladder

Modified from ref. 22, with permission of the American College of Obstetricians and Gynecologists.

appears to accomplish similar cure rates but allows outpatient therapy with reproducibility of the position of the sources and employ a shorter, although frequent intracavitary treatment sessions.[24,25]

Additionally, other methods of brachytherapy have been proposed to encompass areas of disease by giving a more homogeneous dose, but maintaining flexibility in dose distribution. The afterloading perineal Syed & Nesbitt template allows placement of interstitial ^{192}Ir to allow individualization of cervical and parmetrial dosage, particularly in tumors with altered anatomy or in tumors that have not regressed during external radiation. Excellent pelvic control was reported by Surwit et al.[26] and Martinez et al.[27] Others[28] reported significant morbidity with its use, and presently the final status of this technique remains to be determined. Interstitial therapy may be of particular value in the treatment of carcinoma of the cervical stump. Complications of radiotherapy are related to the total dose and the treated volume, and usually occur several months to years after completion of therapy. These include bleeding, strictures, fistulae, and perforation of the bladder and occur in less than 15% of patients.

The cure rate with radiation therapy has been reported to be 60% for stage II, 30% for stage III, and 10% for stage IV patients. Larger tumors are associated with poorer control of the pelvic disease within the radiation field and have a greater incidence of metastases to paraaortic nodes (15% in stage II, 30% in stage III).

To increase the tumor cell kill without affecting the normal tissues, radiosensitizing agents such as hydroxyurea may be combined with radiation.[29] In an effort to improve the cure rate of patients with advanced cervical cancer, chemotherapy using cisplatin, 5-fluorouracil, mitomycin C, and other active agents is being administered prior to, or concomitantly with, radiation.[30,31]

Identification of paraaortic lymph node metastases by CT scan and fine-needle aspiration or surgical staging prior to treatment warrents the inclusion of the aortic area within the radiation field. Total dose to

TABLE 25-6. Stage IB cancer of the cervix: advantages and disadvantages of radical surgery versus radiation therapy

Advantages	Disadvantages
Radiation	
All patients eligible	Serious bladder or bowel damage (2–6%)
Survival rates equal those with surgery	Vaginal stenosis
	Sexual dysfunction
	Delayed complications
Radical hysterectomy and lymphadenectomy	
Ovaries preserved	Urologic complications: fistulae/stricture (1–2%)
Estimated extent of tumor	Other operative complications
Functional vagina	Patient selection
Complications early	Postoperative radiation therapy required in few
Psychological advantage	

TABLE 25-7. Complications of radiation therapy

Acute radiation reaction
- GI tract—nausea, vomiting, diarrhea
- GU tract—acute hemorrhagic cystitis, UTI, loss of ovarian function, vaginal distortion
- Brachytherapy—DVT, UTI, fever
- Hematologic/immunologic systems—infrequent

Delayed complications
- GI tract—proctosigmoiditis, rectovaginal fistula, small bowel obstruction, perforation and fistula
- GU tract—ureteral stricture, vulvovaginal and uterovaginal fistula, vaginal distortion
- Bond injury—uncommon
- Carcinogenesis—rare uterine and bone sarcoma

GI = gastrointestinal, UTI = urinary tract infection, GU = genitourinary, DVT = deep venous thrombosis.

the paraaortic area should not exceed 45 Gy because of increased risk of gastrointestinal complications. Long-term survival has occurred mostly in patients with microscopic metastases to the aortic nodes. Few patients with microscopically involved aortic nodes have benefited from extended pelvic radiation.

Results of Surgical Therapy

The survival of patients following radical hysterectomy is largely dependent upon adequate surgical margins, site of positive lymph nodes, and presence of extrapelvic disease. Parametrial extension appears to be an important prognostic factor in cervical cancer. Noguchi et al.[32] reported a 94.4% 5-year survival with negative parametrial margins on radical hysterectomy specimens, but a 40.8% 5-year survival rate if these margins were involved by tumor. They also confirmed the ominous nature of uterine body extension of the tumor, reporting an 82% 5-year survival rate of patients without extension to the endometrial cavity and a 45% 5-year survival rate with extension to the lower uterine segment. Most patients with margins of less than 1 cm will recur at a rate which is minimally reduced by administration of postoperative radiation. Patients with negative lymph nodes have approximately a 90% 5-year survival.[33] Various survival rates for patients with positive nodes range from 20 to 60%, depending upon the number of nodes involved, and the location and size of metastases.[34–36]

To improve survival, postoperative radiation therapy has been used for patients with deep cervical stromal or parametrial invasion, positive surgical margins, and lymph node metastases, or some combination of these factors. However, radiation therapy may benefit only patients with more than three positive lymph nodes, those with tumor extension, or those with close margins of resection without node metastases.

TABLE 25-8. Summary of combination of external and intracavitary irradiation iin carcinoma of the cervix with intact uterus

Stage	Whole pelvis (rads)[a]	Brachytherapy (mg-hr)	Parametrium[b]
Ib	4,000	6,000 (2 applications)	None
IIa	4,000	6,000 (2 applications)	
IIb	4,000–5,000	6,000 (2 applications)	
IIIa	4,000–5,000	6,000 (2 applications)	
IIIb	4,000–5,000	6,000 (2 applications)	May add 1,000 rads to side of major involvement
IVa	5,000	4,000 (1 application)	
IVb (Palliative)	1,000, 3 times 1 week apart	6,000 (2 applications)	

[a] Patients with larger lesions or poor vaginal geometry merit higher doses of external irradiation.
[b] Reduce parametrial irradiation by 500 rads with previous pelvic surgery or pelvic inflammatory disease.

Posttreatment Follow-up

The cytotoxic effects of radiation therapy continue after completion of therapy. Primary healing of the cervix includes regrowth of normal epithelial tissue or obliteration of the vaginal vault without evidence of ulceration or discharge. On rectovaginal examination, the residual induration is smooth. Lack of tumor shrinkage or actual tumor growth and persistance of nodularity of the cardinal or uterosacral ligaments are signs of disease. Active tumor discovered within the first 3 to 6 months following therapy is termed "persistent cancer," whereas that found thereafter is termed "recurrence." CT-directed needle biopsies of suspicious areas may be used to establish a diagnosis of persistent or recurrent cervical cancer.

Since more than three fourths of all recurrences occur within the first 2 years following initial therapy, posttreatment evaluation, including pelvic examination and examination of inguinal and supraclavicular lymph nodes, should be done every 3 to 4 months for the first 2 years, then every 4 to 6 months for the next 3 years. Chest x-ray, CT scan of the pelvis and abdomen, and intravenous pyelogram, when needed, may help in localizing areas of recurrent cervical cancer. Patients who have been treated initially with surgery may benefit from early diagnosis of recurrent cancer because they still might be cured by radiation therapy.

Serum Markers

Various serum markers (carcinoembryonic antigen, β-human chorionic gonadotropin, α-fetoprotein) to monitor the course of the disease have been measured, with inconclusive results, in patients with invasive cervical cancer. The most promising appears to be the squamous cell carcinoma antigen, a purified subfraction of tumor antigen TA-4. This glycoprotein, when elevated, was considered specific for presence of squamous cell carcinoma. It has since proved to be an unreliable screening test or an indicator of tumor regression. It may, however, be useful in the follow-up of patients after treatment for early detection of recurrent cancer.

Incidentally Found Cervical Carcinoma

Patients with invasive cervical carcinoma found incidentally in a uterus removed following "simple" hysterectomy for benign disease can be treated by radiotherapy or reoperation.

Adjuvant radiotherapy accomplished similar survival rates for patients with stage I and stage II disease as primary therapy.[37,38] Survival after radiation therapy depends upon the volume of disease, status of surgical margins, and length of delay from surgery to radiation therapy. Patients with microscopic disease have a 95 to 100% 5-year survival rate. The survival rate of patients with macroscopic disease-free margins is 82 to 84%, those with macroscopically positive margins 38 to 87%, and those with obvious residual cancer 20 to 47%.[39–41] Delay in treatment for a period greater than 6 months is associated with 20% survival.[38]

Reoperation may be indicated in younger patients with a small tumor to preserve ovarian function. Survival following radical reoperation with removal of pelvic nodal tissues, parametrium, and upper vagina is equal to that of radical hysterectomy for stage I tumor.[41] We have used both methods and favor reoperation in younger patients. This situation emphasizes that cervical cytology is mandatory before a hysterectomy is performed for any indication. Similarly, biopsy of any suspicious lesion is essential, regardless of the cytology report.

Sarcoma of the cervix may be treated surgically but, because sarcomas metastasize predominantly via the bloodstream, the enthusiasm for radical surgical procedure continues to be limited. Radiation therapy may be used to treat the primary tumor. Chemotherapy has not been effective.

Patients with verrucous carcinoma of the

cervix should be treated with total or modified radical hysterectomy as determined by the size of the lesion. Lymphadenectomy is not recommended because this tumor rarely metastasizes and radiation therapy is detrimental.[42]

Coexistent Pelvic Mass

The etiology of a pelvic mass must be clarified by appropriate investigation prior to treatment of cervical cancer. Some of the commonly encountered masses include leiomyoma, adnexal masses, pelvic kidney, diverticular abscess, and carcinoma of the colon. An enlarged, fluid-filled uterine cavity might represent a hematometra or pyometra. These should be drained by cervical dilatation and the patient appropriately treated with antibiotics prior to beginning definitive therapy. Repeated cervical dilatations may be necessary to ensure continuous drainage.

Definitive surgery for early-stage disease or radiation therapy for advanced cancers may begin when pyometra has cleared, the white blood cell count has returned to normal, and repeat physical examination or ultrasound evaluation has confirmed a shrinkage of the uterus. Uterine leiomyoma shrinks during radiation therapy and requires no further treatment.

Undiagnosed adnexal masses must be explored to exclude coexistent cancer of the ovary. In early-stage cancer, radical hysterectomy, bilateral salpingo-oophorectomy, and cytoreductive surgery of the ovarian cancer is indicated. If the cervical cancer cannot be treated surgically, however, the ovarian disease should be debulked without removal of the uterus, allowing subsequent radiation therapy. Cisplatinum-based chemotherapy may be administered concurrently with pelvic radiation therapy.

The coexistent second primary may required treatment modification. In many cases, aggressively treating cervical cancer may be justified except when the second cancer has an extremely poor prognosis, at which time treatment for cervical cancer should be palliative. The delivery of high-dose fractions (10 Gy), given three times at weekly intervals, offers an excellent palliative alternative.[43] Selected "spot" radiation treatment to sites of distant metastases may assist in controlling pain.

Cancer of the Cervical Stump

Cancer of the cervical stump is less commonly encountered today than in earlier years, when supracervical hysterectomy was an acceptable surgical procedure. Miller et al.,[44] however, reported that cervical stump cancer represented 4 to 5% of all patients treated during a 12-year interval ending in 1975. Early-stage disease is treated surgically, although, technically, the surgical procedure may be complicated by bowel adhesions and by the extension of the bladder over the cervical stump as part of previous reperitonealization. If the lesion is only focally microinvasive, vaginal trachelectomy may be an acceptable option.

An intracavitary tandem may be used if the endocervical canal is at least 2 cm in length, and vaginal ovoids may be used unless severe anatomic distortion exists. If the cervical tandem cannot be placed, treatment may be completed by external radiation therapy.

Radiation-related deaths of 3 to 7% in Miller et al.'s series[44] mandate pretherapy radiologic evaluation of the upper gastrointestinal tract, and consideration of a four-field box technique, thereby reducing the incidence of bladder and bowel complications. Rousseau et al.[45] and Miller et al.[44] reported similar survival, stage for stage, for patients with cervical stump cancer.

Barrel-Shaped and Bulky Cancer of the Cervix

The large, endophytic lesion that distorts and expands the endocervix and lower uterine segment to a barrel shape and tumors greater than 6 cm in diameter are associated with a poorer prognosis. Distortion of the lower uterine segment implies large tumor volume and, with expansion, clinical staging becomes difficult. Furthermore, the superior lateral aspect of the expanded lower uterine segment is often beyond the curative isodose curve. Fletcher and Montague[46] achieved decreased rates of central pelvic recurrence by administering 40 to 50-Gy whole pelvic teletherapy, followed by a single intracavitary application, prior to a conservative hysterectomy, although significant improvement in patient survival was not demonstrated. Galleon et al.,[47] in a nonrandomized study involving 79 patients with bulky or barrel-shaped lesions, reported a decrease in the incidence of pelvic recurrence from 19 to 2% and in extrapelvic recurrences from 15% to 7%. Increase in long-term survival was not reported.

Complications of combined radiation and extrafascial hysterectomy predominantly include bowel and urologic complications and vary widely from none reported by Galleon et al.[47] to 8% by Edinger et al.,[48] 12.3% by O'Quinne et al.,[49] 19.64% by Perkins et al.,[50] and 50% by Blake et al.[51] The risk of urinary fistulae was increased by avoiding the removal of the upper vagina and blunt dissection. At our institution, radiation responses are gauged carefully to make a decision concerning adjunctive hysterectomy after the patient has received 50-Gy whole pelvic teletherapy. If the response is excellent, the radiotherapy is continued. If, at the end of radiation therapy, the patient still has persistent disease, then exenterative surgery is used as the surgical procedure of choice rather than conservative hysterectomy.

Stage IVa Cancer of the Cervix

Consideration may be given to a primary exenterative procedure with radiation therapy for patients with advanced disease (stage IVa) in which the tumor has not reached the pelvic sidewall. The survival rate of patients with extension to the bladder was found to be 30% using primary radiation therapy, with a urinary fistula rate of 3 to 8%.[52] Perches et al.[53] reported a significant improvement in survival rates by combining radiation therapy with exenterative surgery in patients with stages III and IVa disease, if the tumor mass fails to respond rapidly to the radiation therapy, or if visible tumor remains in the central pelvis at the conclusion of radiation therapy. Rectal extension is less commonly observed but may require a diverting colostomy prior to therapy.

Small-Cell Cancer of the Cervix

Albores-Saavedra et al.[54] described small-cell cancer of the cervix as originating from the argyrophilic cells of the endocervical epithelium, which are part of the tumors of the APUD (*a*mine *p*recursor *u*ptake and *d*ecarboxylation) cell system. Pathologically, the diagnosis is aided by the finding of neuroendocrine granules on electron microscopy, as well as immunoperoxidase slides that are positive for a variety of neuroendocrine problems, such as calcitonin, insulin, glucagon, somatostatin, gastrin, and adrenocorticotropin.

Treatment of these cancers is controversial, because widespread vascular involvement leads to early metastases. Van Nagell et al.,[55] Silva et al.,[56] and Yoshida et al.[57] reported on the unpredictable behavior of these neoplasms and proposed that bone, liver, and brain scans be performed in these patients in addition to routine studies.

Local therapy alone gives no chance of cure. Forty-four percent of 41 patients in van Nagell et al.'s series died of widespread metastases and of recurrent cancer within 2 years.[55] Combination chemotherapy with vincristine, Adriamycin, or cisplatin or etoposide and cisplatin[58] in combination with radiation therapy or surgery may be employed to increase the cure rates in these patients.

Adenocarcinoma of the Cervix

In recent years, the incidence of adenocarcinoma of the cervix seems to be increasing, with some authors reporting an incidence of up to 18%.[59] In the past, approximately 5% of cervical cancers constituted adenocarcinomas.[60] Several subtypes of cervical adenocarcinoma have been described (Table 25-9).

Papanicolaou smear screening and colposcopy are often inadequate because of the absence of an exocervical lesion or the similar appearance of the malignant and normal glandular epithelium. An endocervical curettage should be done in patients who have abnormal bleeding or a palpably abnormal cervix. Endophytic growth pattern may be responsible for extensive clinically undetectable tumor volume; therefore, surgical staging laparotomy may more accurately determine the stage of the adenocarcinoma.

Conclusions about poorer prognoses of certain histologic types of adenocarcinoma[61] and variant radiosensitivity from squamous cell carcinoma are based on retrospective studies with heterogeneous groups and small numbers of patients. No difference in the prognosis for adenocarcinoma or squamous cell carcinoma was demonstrated in recent reports.[62,63]

At present, patients with adenocarcinoma should be treated with radical hysterectomy or radiation therapy for early-stage disease and radiation therapy for advanced-stage disease, with the same selection criteria used for squamous cell carcinoma.

TABLE 25-9. Histopathology of cervical adenocarcinoma

Well differentiated
Adenosquamous
Clear cell
Glassy cell
Minimally invasive (adenoma malignum)
Adenoid basal
Adenoid cystic

Cervical Cancer in Pregnancy

Cervical carcinoma in situ occurs in approximately 1:770 pregnancies.[64] To identify cervical intraepithelial neoplasia, all pregnant patients should have a Papanicolaou smear at their first antenatal visit, and any gross cervical lesion requires biopsy. If the smear is abnormal, recommended management is outlined in Figure 25-1.

Diagnostic conization or wedge resection of the cervix should be reserved for patients with inadequate colposcopy and/or a Papaniclaou smear suspicious for invasive cancer, and for patients with microinvasive cancer or possible cervical adenocarcinoma on colposcopically directed biopsy. If conization is required, it should only be done in the second trimester. Reported incidence of complications following a cone biopsy, including hemorrhage, premature labor, and/or abortion, ranged from 9 to 33%.[64]

Patients with cervical intraepithelial neoplasia and microinvasive carcinoma of less than 3 mm invasion beneath the basement membrane, with no evidence of lymphatic or blood vessel involvement, may be followed with Papanicolaou smears and colposcopy to term and delivered vaginally.

The reported incidence of invasive cancer during pregnancy is approximately 1:2,200 pregnancies.[64] Diagnosis can be accomplished by biopsy of a gross cervical lesion or, in the absence of a lesion, as outlined in Figure 25-1. Diagnosis often is delayed because bleeding may be assumed to be related to pregnancy complications. There is no conclusive evidence to suggest that pregnancy has any effect on the growth rate or inherent pathologic behavior of cervical cancer.

Invasive cancer of the cervix may prevent dilatation of the cervix and cause hemorrhage due to friability of the cervical lesion. The cancer has no detrimental effect on the fetus; however, fetal damage may be related to treatment involving radiation or chemotherapy. Pregnant patients with cancer of the cervix should undergo the same pretreat-

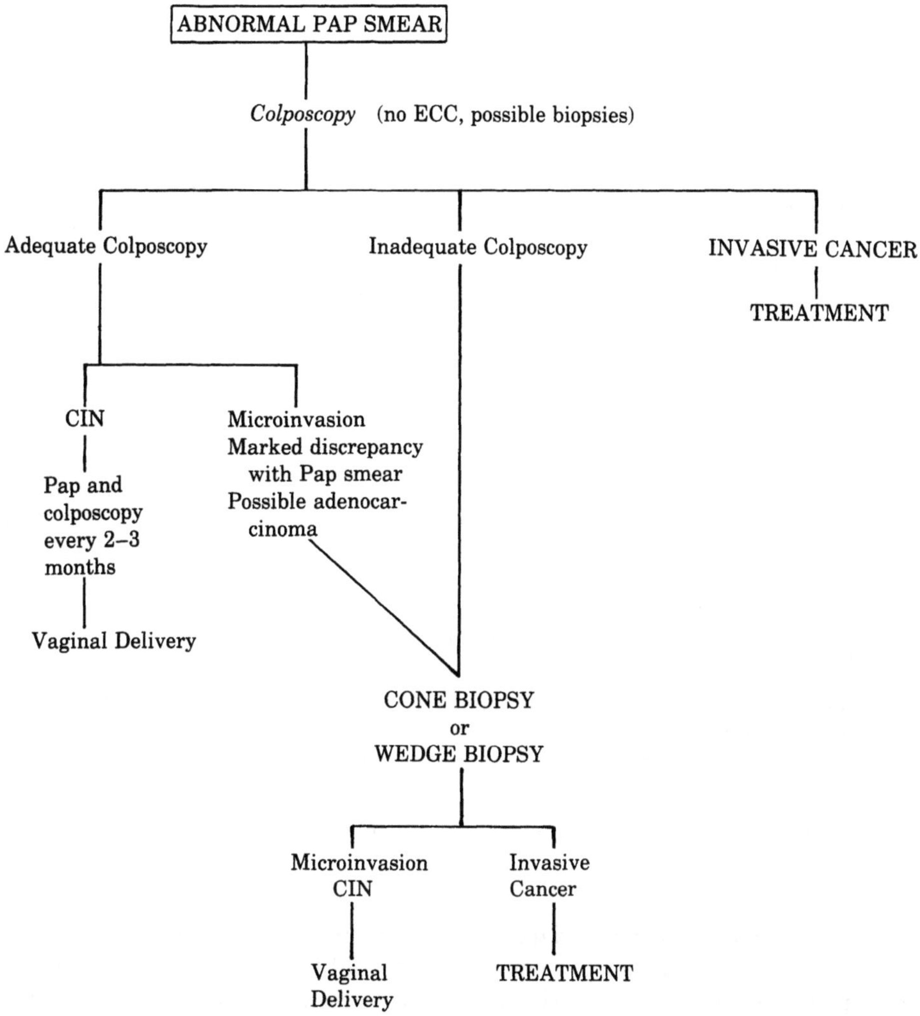

FIGURE 25-1. Management of abnormal Papanicolaou smear during pregnancy. ECC = endocervical curettage, CIN = cervical intraepithelial neoplasia.

ment evaluation as nonpregnant patients, omitting those tests that cause radiation damage to the fetus. Clinical staging may be less accurate because of edema and softness of broad ligaments, paracervical, and parametrial tissues; development of the lower uterine segment; and the presence of the fetus.[65]

Treatment

Prior to 24 weeks' gestation, there is no difference in the treatment, stage for stage, between pregnant and nonpregnant patients. Treatment usually will be associated with fetal demise (Fig. 25-2).

Surgery in pregnancy is facilitated by tissue edema, but may be complicated by increased vascularity of the pelvis, leading to potentially greater blood loss and longer operating time.[66] No significant difference in the incidence of surgical complications is reported in the literature.[64,65] The treatment of patients with cervical cancer 4 to 8 weeks prior to fetal viability requires individualization (Fig. 25-3). Factors to be considered are the availability of neonatal care and the patient's attitude toward fetal survival. Al-

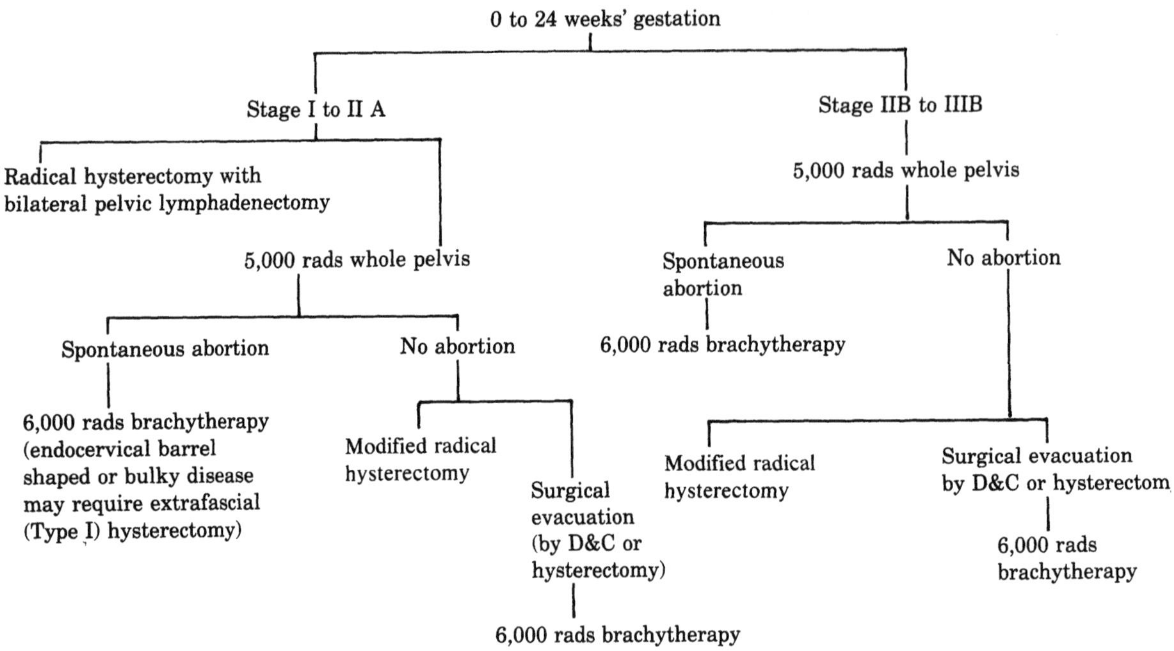

Stage IVA with involvement of bladder and rectum: anterior, posterior, or total pelvic exenteration with or without preoperative external radiation therapy.
Stage IVB: 1,000 rads external radiation therapy, 3 times a month for local control only, with or without chemotherapy.

FIGURE 25-2. Suggested therapy for cervical cancer during early pregnancy.

though there is no reported risk with delayed therapy,[67] treatment probably should not be postponed for more than 4 weeks. If the patient requests delay of treatment, fetal lung maturity should be assessed prior to delivery.

Inadvertent vaginal delivery through a cancerous cervix will likely cause profuse maternal bleeding and risk of infection. The risk of cancer dissemination and poor prognosis have not been proven.[68]

No significant difference was found in the overall prognosis for all stages of cervical cancer in pregnancy compared to that in

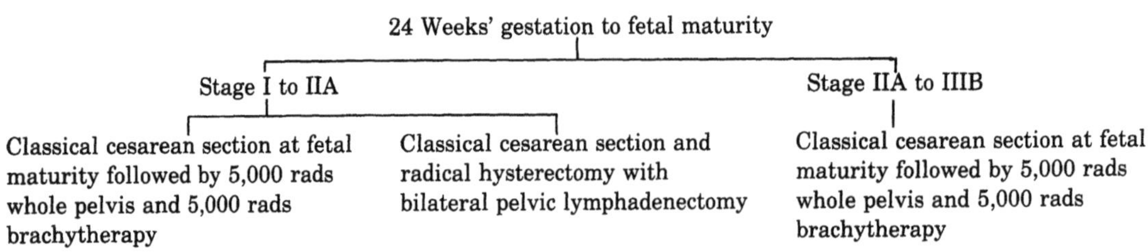

Stage IVA with involvement of bladder and rectum: may be treated with classical cesarean section at fetal maturity with or without 5,000 rads external radiation therapy postoperatively.

StageIVB: classical cesarean section at fetal maturity followed by 500–1,000 rads external radiation therapy, 3 times a month for local control only, with or without chemotherapy.

FIGURE 25-3. Suggested therapy for cervical cancer during late pregnancy.

nonpregnant women. For advanced stages, prognosis appears less favorable in pregnancy, probably because of difficulties with adequate radiation dosimetry.[64]

Recurrent Cervical Carcinoma

It is estimated that approximately 35% of patients treated for invasive cervical cancer will have recurrent or persistent disease following therapy, presenting with different symptoms (Table 25-10). Patients whose initial treatment consisted of surgery should be evaluated for radiation therapy. Possible surgical treatment may be considered for patients who failed radiation therapy. To date there exists no curative chemotherapeutic regimen for treatment of cervical cancer.

Surgical Treatment

A select group of patients with small-volume central pelvic recurrence after radiation therapy may be successfully threated with radical hysterectomy. A 20 to 50% rate of urinary tract complications (strictures, fistulae) has been reported to occur postoperatively.[64–71] Because of this high complication rate and the difficulty in obtaining adequate free surgical margins, exenterative procedures have been recommended for patients with central recurrences as the last chance for cure.

Pelvic exenterative procedures may be total (removal of bladder, rectum, vagina, uterus) or partial, with conservation of bladder (posterior exenteration) or conservation of rectum (anterior exenteration). Preoperative CT scan evaluations of the pelvis, abdomen, and lungs are often necessary to rule out metastatic disease. Signs of unresectable pelvic sidewall disease and other contraindications for exenterative surgery are listed in Table 25-11. In spite of a preoperative negative metastatic work-up, only one in three patients explored for recurrent disease will be able to undergo an exenterative procedure.

TABLE 25-10. Symptoms of recurrent cervical cancer

Weight loss (unexplained), anorexia
Leg edema (excessive and often unilateral)
Pelvic and/or thigh-buttock pain
Serosanguineous vaginal discharge
Progressive ureteral obstruction
Supraclavicular lymph node enlargement (usually left side)
Cough
Hemoptysis
Chest pain

TABLE 25-11. Preoperative contraindications to exenteration in patients with recurrent cancer of the cervix

Absolute
Extrapelvic disease
Triad of unilateral leg edema, sciatica, and ureteral obstruction
Tumor-related pelvic sidewall fixation
Bilateral ureteral obstruction
Severe, life-limiting medical illness
Inability of patient to care for herself
Jehovah's Witness
Inadequate hospital facilities to manage intraoperative and postoperative complications
Relative
Age over 70 years
Large tumor volume
Unilateral ureteral obstruction
Metastasis to the distal vagina

It is not justifiable to do a pelvic exenteration when adequate margins cannot be obtained. An anterior exenteration is justified in patients with cancer limited to the cervix and anterior upper vagina. The role of posterior exenteration for patients with isolated posterior vaginal recurrence is extrememly limited because of the inability to obtain a wide surgical margin of resection between cervix and bladder. Reanastomosis of the colon with the end-to-end anastomosis, thereby avoiding a permanent colostomy and creation of a neovagina, is possible in some patients who undergo an anterior or total pelvic exenteration.

Results

Since the original description of pelvic exenteration by Brunschwig,[72] the operative

TABLE 25-12. Exenteration for female genital tract cancer (predominantly carcinoma of the cervix), 1970–1989

Reference	Year	No. patients	Operative mortality (%)	Survival 5 yr or as indicated (%)
Ketcham et al.[73]	1970	162	17	38
Deckers & Ketcham[74]	1971	68	10	48
Spratt et al.[78]	1973	274	9	40
Symmonds et al.[75]	1975	198	8	33
Rutledge et al.[76]	1977	296	13	42
Barber[77]	1977	671	—	20
Curry et al.[79]	1981	37	5.4	43 (2 yr)
Orr et al.[80]	1983	125	9.6	45.3
Averette et al.[81]	1984	92	24	38.3
Bompiani et al.[83]	1985	44	4.5	34 (2 yr)
Rutledge[84]	1977–85	148	4.7	47
Talledo[85]	1985	42	15	39
Rutledge & McGuffee[88]	1955–86	448	10	53.7
Roberts et al.[89]	1987	38	5.3	43
Kraybill et al.[82]	1988	58	15.5	38
Anthopoulos et al.[86]	1989	20	5	58
Morley et al.[87]	1989	100	2	55
Soper et al.[90]	1989	69	7.2	48

mortality has decreased to approximately 5%, while the 5-year survival has increased to between 40 and 60% (Table 25-12). However, serious postoperative complications still make exenteration a formidable technical procedure (Table 25-13).

Chemotherapy

Cytotoxic chemotherapy for carcinoma of the cervix has been used most often for patients with recurrent disease after failure of their initial surgery and/or radiation therapy. Chemotherapy has not been very successful for these patients because of the following factors:

1. Limited bone marrow tolerance because of prior radiation therapy.
2. Difficulty of drug delivery because of poor vascularity of the tumor resulting from postsurgical and postradiation changes.
3. Difficulty of response evaluation because of scarring and fibrosis from previous treatments.
4. Decreased tolerance to nephrotoxic chemotherapeutic agents because of compromised renal function from ureteral obstruction.

Single-Agent Chemotherapy

Several drugs have been used as single agents in the treatment of cervical cancer. The literature has been reviewed recently and active chemotherapeutic agents were identified.[91,92] The most extensively tested drug has been cisplatin. The Gynecologic Oncology Group studied women with recurrent cervical cancer in a randomized trial, and three separate regimens utilizing cisplatin 50 mg/m^2, 100 mg/m^2, or 20 mg/m^2, daily for 5 consecutive days and repeated every 3 weeks, were compared. The 100 mg/m^2 regimen accomplished the highest response rate (31%) with no advantage in survival or time to progression. Nausea and vomiting were the most frequent side effects reported, and ototoxicity, nephrotoxicity, and peripheral neuropathy occurred infrequently in this study.[92] Similar activity was observed for two investigational drugs, dibromodulcitol (an α-ω–substituted hexitol)[93] and ifosfamide, an oxazaphosphorine deriv-

TABLE 25-13. Complications of exenterative procedures

Acute and immediate
Cardiovascular
Hemorrhage
Thrombophlebitis
Thromboembollism
Myocardial infarction
Infections
Pelvic cellulitis
Pelvic abscess
Wound infection
Pyelonephritis
Pneumonia
Gastrointestinal
Obstruction
Fistula/leak
Stomal necrosis
Miscellaneous
Psychosis
Pulmonary edema/failure
Urinary fistula/leak
Obstruction
Late complications
Intestinal
Obstruction
Stomal revision
Fistula
Hernia
Urinary
Infection
Stomal revision
Pelvic abscess/sinus
Fistula

Modified from: Shingleton HM, Orr JM: Cancer of the Cervix: Diagnosis and Treatment. Edinburgh, Churchill & Livingston, 1987, with permission.

ative and structural analogue of cyclophosphamide[94,95] (Table 25-14).

Combination Chemotherapy

Several combination chemotherapy regimens have been evaluated in the treatment of patients with advanced or recurrent carcinoma of the cervix, most involving small groups of nonrandomized patients. The most effective combination regimens have been those that combine cisplatin with other active agents.[91,92] Recently, a combination of cisplatin, bleomycin, and ifosfamide produced a 69% overall response rate; however, further studies are required to demonstrate

TABLE 25-14. Single-agent chemotherapy in cervical carcinoma

Drug	No. of evaluable patients	Response rate (%)
Ifosfamide	30	33
Cisplatin	166	31
Dibromodulcitol	55	29
Chlorambucil	44	25
Mitomycin	18	22
Porfiromycin	78	22
Melphalan	20	20
5-Fluorouracil (ICRF 159)	142	20
Hexamethylmelamine	64	19
Methotrexate	96	18
Vincristine	55	18
Carboplatin	176	17
Adriamycin	205	16
Iproplatin	176	11
Bleomycin	172	10

Modified from: refs. 91 and 92, with permission.

whether any combination regimen will accomplish a better response and longer survival than treatment with cisplatin alone.[95]

Several tentative conclusions may be assumed from the experience with chemotherapy for advanced carcinoma of the cervix:

1. Cisplatin is the best studied single agent. Its activity is at least equal to the other single drugs tested and is not schedule or dose dependent.
2. Numerous single agents appear to demonstrate moderate activity, the most active being ifosfamide and dibromodulcitol.
3. Combination chemotherapy has not shown a statistically significant increase of response or survival in controlled trials.
4. Combination chemotherapy should include cisplatin.

Adjuvant Chemotherapy

Several attempts have been made to improve the effects of radiation with concomitant radiosensitizing chemotherapeutic agents. Hydroxyurea increased response rate and survival when compared with radiation therapy administered alone in patients with advanced carcinoma of the cervix.[29,96] Currently, the Gynecologic Oncology Group is conducting a randomized adjuvant trial comparing hydroxyurea and radiation therapy with 5-fluorouracil and cisplatin plus radiation therapy in patients with advanced cervical cancer.

Summary

The incidence of invasive carcinoma of the cervix appears to be bimodal, affecting both younger women (35 through 39 years) and older women (60 through 64 years). It is often overlooked in younger women and mistaken for menstrual irregularities, whereas older patients, mistaking early symptoms as those of menopause, often display apathy, neglect, and denial until the signs are too frequent and frightening to ignore.

Although surgical staging is important in determining the effect of extrauterine disease, its use must be considered investigational. The most effective treatment for invasive cervical cancer remains early intervention with either radical surgery or radiation therapy. Research is ongoing regarding the use of chemotherapy for recurrent disease and in neoadjuvant settings, with either surgery or radiation therapy.

References

1. Boring CC, Squires TS, Tong T: Cancer statistics 1991. CA 1991;41:28–29.
2. Shingleton HM, Orr JW: Diagnosis, staging and selection of treatment for invasive tumors, in Shingleton HM, Orr JW (eds): Cancer of the Cervix: Diagnosis and Treatment. New York, Churchill Livingstone, 1987, pp 94–131.
3. Leman MG, Park RC, Barham ED, et al: Pretreatment lymphangiography in carcinoma of the uterine cervix. Gynecol Oncol 1975;3:354–360.
4. Pellier D, Chavanne G, Thu NT: Lymphographic evaluation of cervicouterine carcinoma. Bull Cancer 1979;66:515–518.
5. Piver MS, Barlow JJ: Paraaortic lymphadenectomy, aortic node biopsy, and aorta lymphangiography in staging patients with advanced cervical cancer. Cancer 1973;32:367–370.
6. DeMuylder X, Belanger R, Yauclair R, et al: Value of lymphangiography in stage IB cancer of the uterine cervix. Am J Obstet Gynecol 1984;148:160–163.
7. King LA, Talledo OE, Gallup DG, et al: Computed tomography in evaluation of gynecologic malignancies: a retrospective analysis. Am J Obstet Gynecol 1986;155:960.
8. Bandy LC, Clarke-Pearson DL, Silverman PM, Creasman WT: Computed tomography in evaluation of extrapelvic lymphadenopathy in carcinoma of the cervix. Obstet Gynecol 1985;65:73–76.
9. Hacker NF, Berek JS: Surgical staging, in Surwit E, Albert D (eds): Cervix Cancer. Boston, Martinus Nijhoff, 1987, pp 43–57.
10. Worthington HL, Balfe DM, Lee JKT, et al: Uterine neoplasms: MR imaging. Radiology 1986;159:725–730.
11. Beahrs OH, Henson DE, Hutter RV, Myers MH (eds): Manual for Staging of Cancer, ed 3. Philadelphia, JB Lippincott Co, 1988.
12. Chung CK, Nahhas WA, Zaino R, et al: Histologic grade and lymph node metastasis in squamous cell carcinoma of the cervix. Gynecol Oncol 1981;12:348–354.
13. Lagasse LD, Creasman WT, Shingleton HM, et al: Results and complications of operative staging in cervical cancer: experience of the Gynecologic Oncology Group. Gynecol Oncol 1980;9:90–98.
14. Sudersanam A, Charyulu K, Belinson J, et al: Influence of exploratory celiotomy on the management of carcinoma of the cervix. Cancer 1978;41:1049–1053.
15. Schellhas HF: Extraperitoneal para-aortic node dissection through an upper abdominal incision. Obstet Gynecol 1975;46:444–447.
16. Berman ML, Lagasse LD, Watring WG, et al: The operative evaluation of patients with cervical carcinoma by an extraperitoneal approach. Obstet Gynecol 1977;50:658–664.
17. Hsu C-T, Cheng Y-S, Su S-C: Prognosis of uterine cervical cancer with extensive lymph node metastases. Am J Obstet Gynecol 1972;114:954–962.
18. Berman ML, Keys H, Creasman W, et al: Survival and patterns of recurrence in cervical cancer metastatic to periaortic lymph nodes (a Gynecologic Oncology Group study). Gynecol Oncol 1984;19:8–16.
19. Hughes RR, Brewington KC, et al: Extended

field irradiation for cervical cancer based on surgical staging. Gynecol Oncol 1980;9:153–161.
20. Buchsbaum HJ: Extrapelvic lymph node metastases in cervical carcinoma. Am J Obstet Gynecol 1979;133:814–824.
21. Brandt B, Lifshitz S: Scalene node biopsy in advanced carcinoma of the cervix uteri. Cancer 1981;47:1920–1921.
22. Piver MS, Rutledge FN, Smith JP: Five classes of radical hysterectomy of women with cervical cancer. Obstet Gynecol 1974;44:265–272.
23. Delgado G: Stage IB squamous cancer of the cervix: the choice of treatment. Obstet Gynecol Surv 1978;33:174–183.
24. Bauer M, Scholz-Wendtland R: The radiobiology of gynecologic HDR and LDR afterloading techniques. Rontgen-Blatter 1989;42:54–56.
25. Utley JF, Von Essen CF, Horn RA, et al: High dose rate afterloading brachytherapy in carcinoma of the uterine cervix. Int J Radiat Oncol Biol Phys 1984;16;2259–2263.
26. Surwit EA, Manning MR, Aristizabal SA, et al: Interstitial thermoradiotherapy in recurrent gynecologic malignancies. Gynecol Oncol 1983;15:95–102.
27. Martinez A, Cox RS, Edmundson GK: A multiple site perineal applicator (MUPIT) for treatment of prostatic, anorectal, and gynecologic malignancies. Int J Radiat Oncol Biol Phys 1984;10:297–305.
28. Ampuero F, Doss LL, Khan M, et al: The Syed-Neblett interstitial template in locally advanced gynecological malignancies. Int J Radiat Oncol Biol Phys 1983;9:1897–1903.
29. Piver MS, Barlow JJ, Vongtama V, et al: Hydroxyurea as a radiation sensitizer in women with carcinoma of the uterine cervix. Am J Obstet Gynecol 1977;129:379–383.
30. Thomas G, Dembo A, Beale F, et al: Concurrent radiation, mitomycin C and 5-fluorouracil in poor prognosis carcinoma of the cervix: preliminary results of a phase I-II study. Int J Radiat Oncol Biol Phys 1984;10:1985–1990.
31. Malviya VK, Deppe G, Kim Y, et al: Concurrent radiation therapy, cisplatinum and mitomycin C in patients with poor prognosis cancer of the cervix: A pilot study. Am J Clin Oncol 1989;12:434–437.
32. Noguchi H, Shiozawa K, Tsukamoto T, et al: The postoperative clarification of uterine cancer and its clinical evaluation. Gynecol Oncol 1983;16:219–231.
33. Martembeau PW, Kjorstad KE, Iverson T: Stage IB carcinoma of the cervix in the Norwegian Radium Hospital: results of treatment and major complications II: Results when pelvic nodes are involved. Obstet Gynecol 1982;60:215–218.
34. Morrow CP: Is pelvic radiation beneficial in the postoperative management of stage Ib squamous cell carcinoma of the cervix with pelvic node metastases treated by radical hysterectomy and pelvic lymphadenectomy? Gynecol Oncol 1980;10:105–110.
35. Inoue T, Chihara T, Moita K: The prognostic significance of the size of the largest node on metastatic carcinoma from the uterine cervix. Gynecol Oncol 1984;19:187–193.
36. Pilleron JP, Durand JC, Hamelin JP: Prognostic value of node metastasis in cancer of the uterine cervix. Am J Obstet Gynecol 1974;119:458–462.
37. Davy M, Bentzen H, Jahren RR: Simple hysterectomy in the presence of invasive cervical cancer. Acta Obstet Gynecol Scand 1977;56:105–108.
38. Heller PB, Barnhill DR, Mayer AR, et al: Cervical carcinoma found incidentially in a uterus removed for benign indications. Obstet Gynecol 1986;67:187–190.
39. Durrance FY: Radiotherapy following simple hysterectomy in patients with stage I and II carcinoma of the cervix. Am J Roentgenol Radium Ther Nucl Med 1968;102:165–169.
40. Andras EJ, Fletcher GH, Rutledge F: Radiotherapy of carcinoma of the cervix following simple hysterectomy. Am J Obstet Gynecol 1973;115:647–655.
41. Orr JW Jr, Ball GC, Soong SJ, et al: Surgical treatment of women found to have invasive cervix cancer at the time of total hysterectomy. Obstet Gynecol 1986;68:353.
42. Kawagoe K, Yoshikaioa H, Kawana T, et al: Verrucous carcinoma of the uterine cervix. Nippon Sanka Fujinka Gakkai Zasshi 1984;36:617–622.
43. Chafe W, Fowler WG, Currie JL, et al: Single fraction palliative pelvic radiation therapy in gynecologic oncology: 1000 rads. Am J Obstet Gynecol 1984;148:701–705.
44. Miller BE, Copeland LJ, Hamberger AD, et al: Carcinoma of the cervical stump. Gynecol Oncol 1984;18:100–108.
45. Rousseau J, Fenton J, Debertrand P, et al:

Carcinoma of the cervix. Radiology 1972;103:413–418.
46. Fletcher GH, Montague ED: Does adequate irradiation of the internal mammary chain and supraclavicular nodes improve survival rates? Int J Radiat Oncol Biol Phys 1978;4:481–492.
47. Gallion HH, Van Nagell JR Jr, Donaldson ES: Combined radiation therapy and extrafascial hysterectomy in the treatment of stage IB barrel-shaped cervical cancer. Cancer 1985;56:262–265.
48. Edinger DD Jr, Watring WG, Anderson B, et al: Residual tumor following radiotherapy for locally advanced carcinoma of the uterine cervix: prognostic significance. Eur J Gynaecol Oncol 1984;5:90–94.
49. O'Quinn AG, Fletcher GH, Wharton JT: Guidelines for conservative hysterectomy after irradiation. Gynecol Oncol 1980;8:68–79.
50. Perkins PL, Chu AM, Jose B, et al: Posthysterectomy megavoltage irradiation in the treatment of cervical carcinoma. Gynecol Oncol 1984;17:340–348.
51. Blake PR, Lambert HE, MacGregor WG, et al: Surgery following chemotherapy and radiotherapy for advanced carcinoma of the cervix. Gynecol Oncol 1984;19:198–203.
52. Million RR, Rutledge F, Fletcher GH: Stage IV carcinoma of the cervix with bladder invasion. Am J Obstet Gynecol 1972;113:239–243.
53. Perches RD, Lebaton AT, Garces MC: Radiotherapy combined with surgery as treatment for advanced cervical cancer. Int J Radiat Oncol Biol Phys 1983;9:1785–1788.
54. Albores-Saavedra J, Rodriguez-Martinez HA, Larraza-Hernandez O: Carcinoid tumors of the cervix. Pathol Annu 1979;1;273–291.
55. Van Nagell JR Jr, Donaldson ES, Wood EC, et al: Small cell carcinoma of the cervix. Cancer 1977;40:2243–2249.
56. Silva EG, Kott MM, Ordonet NG: Endocrine carcinoma. Intermediate cell type of the uterine cervix. Cancer 1984;54:1705–1713.
57. Yoshida A, Yoshida H, Fukunishi R, et al: Carcinoid tumors of the uterine cervix: a light and electronmicroscopic study. Virchows Arch 1984;402:331–336.
58. Oldham PK, Greco FA: Small cell lung cancer, a curable disease cancer. Cancer Chemother Pharmacol 1980;4:173–177.
59. Berek JS, Hacker NF, Fu YS, et al: Adenocarcinoma of the uterine cervix: histologic variables associated with lymph node metastasis and survival. Obstet Gynecol 1985;65:46–52.
60. Ireland D, Hardeman P, Monaghan JM: Adenocarcinoma of the cervix: a study of 73 cases. Obstet Gynecol 1985;65:82–85.
61. Hopkins MP, Schmidt RW, Roberts JA, et al: The prognosis and treatment of stage I adenocarcinoma of the cervix. Obstet Gynecol 1989;72:915–921.
62. Kilgore LC, Soong SJ, Gore H, et al: Analysis of prognostic features in adenocarcinoma of the cervix. Gynecol Oncol 1988;31:137–148.
63. Kleine W, Rau K, Schivoeorer D, et al: Prognosis of adenocarcinoma of the cervix uteri: a comparative study. Gynecol Oncol 1989;35:145–149.
64. Hacker NF, Berek JS, Lagasse LD, et al: Carcinoma of the cervix associated with pregnancy. Obstet Gynecol 1982;59:735–746.
65. Sall S, Rini S, Pineda A: Surgical management of invasive carcinoma of the cervix in pregnancy. Am J Obstet Gynecol 1974;118:1–5.
66. Thompson JD, Caputo TA, Franklin EW III, et al: The surgical management of invasive cancer of the cervix in pregnancy. Am J Obstet Gynecol 1975;121:853.
67. Greer BE, Easterling TR, McLennan DA, et al: Fetal and maternal considerations in the management of stage Ib cervical cancer during pregnancy. Gynecol Oncol 1989;34:61–65.
68. Lee RB, Neglia W, Park RC: Cervical carcinoma in pregnancy. Obstet Gynecol 1981;58:584–589.
69. Mattingly RF, Boyd A, Frable WJ: The vaginal irrigation smear: a positive method of cervical cancer control. Obstet Gynecol 1967;29:463–470.
70. Bjornstahl H, Johnsson JE, Lindberg LG: Hysterectomy in central recurrence of carcinoma of the uterine cervix. Acta Obstet Gynecol Scand 1977;56:227–231.
71. Adcock LL: Radical hysterectomy preceded by pelvic irradiation. Gynecol Oncol 1979;8:152–163.
72. Brunschwig A: Surgical treatment of cancer of the cervix uteri: radical operation for cancer of the cervix. Bull NY Acad Med 1948;24:672–683.
73. Ketcham AS, Deckers PJ, Sugarbaker EV, et al: Pelvic exenteration for carcinoma of the uterine cervix: a 15-year experience. Cancer 1970;26:513–521.
74. Deckers PJ, Ketcham AS, Sugerbaker EV, et

al: Pelvic exenteration for primary carcinoma of the uterine cervix. Obstet Gynecol 1971;37(5):647–659.
75. Symmonds RE, Pratt JH, Webb MJ: Exenterative operations: experience with 198 patients. Am J Obstet Gynecol 1975;121:907–918.
76. Rutledge FN, Smith JP, et al: Pelvic exenteration: analysis of 296 patients. Am J Obstet Gynecol 1977;129:881–892.
77. Barber HR: Exenterative procedures in pelvic malignancy, in Gynecology and Obstetrics City, Publisher, 1977, vol 3, chap 63, pp 1–11.
78. Spratt JS Jr, Butcher HR Jr, Bricker EM: Exenterative surgery of the pelvis, in Dunphy JE (ed): Major Problems in Surgery. Philadelphia, WB Saunders, 1973, vol 12.
79. Curry SL, Nahhas WA, Jahshan AE, et al: Pelvic exenteration: a 7-year experience. Gynecol Oncol 1981;11(1):119–123.
80. Orr JW Jr, Shingleton HM, Hatch KD, et al: Gastrointestinal complications associated with pelvic exenteration. Am J Obstet Gynecol 1983;145:325.
81. Averette HE, Lichtinger M, Sevin BU, et al: Pelvic exenteration: a 15-year experience in a general metropolitan hospital. Am J Obstet Gynecol 1984;150:179–183.
82. Kraybill WG, Lopez MJ, Bricker EM: Total pelvic exenteration as a therapeutic option in advanced malignant disease of the pelvis. Surg Gynecol Obstet 1988;166:259–263.
83. Bompiani A, Bennetti-Panici P, et al: Pelvic exenteration in gynecologic oncology. Eur J Gynaecol Oncol 1985;6:165–169.
84. Rutledge FN: Gynecologic Cancer: Diagnosis and Treatment Strategies. Austin, University of Texas Press, 1987.
85. Talledo OE: Pelvic exenteration—Medical College of Georgia experience. Gynecol Oncol 1985;22:181–188.
86. Anthopoulos AP, Manetta A, Larson JE, et al: Pelvic exenteration: a morbidity and mortality analysis of a seven-year experience. Gynecol Oncol 1989;35:219–223.
87. Morley GW, Hopkins MP, Lindenauer SM, et al: pelvic exenteration, University of Michigan: 100 patients at 5 years. Obstet Gynecol 1989;74:934–942.
88. Rutledge FN, McGuffee VB: Pelvic exenteration prognostic significance of regional lymph node metastases. Gynecol Oncol 1987;26:374–390.
89. Roberts WS, Cavanagh D, Bryson SCP, et al: Major morbidity after pelvic exenteration. A seven-year experience. Obstet Gynecol 1987;69:617–621.
90. Soper JT, Berchuck A, Creasman WT, et al: Pelvic exenteration: factors associated with major surgical morbidity. Gynecol Oncol 1989;35:93–98.
91. Thigpen T, Vance R, Lambuth B, et al: Chemotherapy for advanced and recurrent gynecologic cancer. Cancer 1987;60:2104–2116.
92. Bonomi P, Yordan E: Chemotherapy of cervical cancer, in Deppe G (ed): Chemotherapy of Gynecologic Cancer. New York, Alan R. Liss, 1990.
93. Stehman FB, Blessing JA, McGehee R, et al: A phase II evaluation of mitolactol in patients with advanced squamous cell carcinoma of the cervix: a Gynecologic Oncology Group study. J Clin Oncol 1989;7:1892–1895.
94. Coleman RE, Harper PG, Gallagher C, et al: A phase II study of ifosfamide in advanced and relapsed carcinoma of the cervix. Cancer Chemother Pharmacol 1986;18:280–283.
95. Buxton EJ, Blackledge G, Mould JJ, et al: The role of ifosfamide in cervical cancer. Semin Oncol 1989;16(suppl 3):60–67.
96. Hreshchyshyn M, Aron BS, Boronow RC, et al: hydroxyurea or placebo combined with radiation to treat stages IIIb and IV cervical cancer confined to the pelvis. Int J Radiat Oncol Biol Phys 1979;5:317–322.

26

Endometrial Adenocarcinoma

HUGH R. K. BARBER and JACQUELINE C. JOHNSON

The upper genital tract consists of the endometrium, ovary, and breast. They share many features epidemiologically and whatever hormonal relationships that exist. The number of estimated new cases of carcinoma of the endometrium for 1991 is 33,000, with 4,500 deaths.[1] About 2.3% of newborn girls, 1 of every 32, will develop carcinoma of the endometrium sometime during their lives. Endometrial carcinoma is a disease of the suburban and the affluent patient and afflicts mostly mature women; diagnosis is usually made between the ages of 59 and 69. It occurs in women younger than 40 years of age in only 5% of cases. More than 70% of the cases occur after age 50, and 20% occur between ages 40 and 50. An increasing number of patients in the older age groups develop adenosquamous cancer of the endometrium. The emergence of this neoplasm is of considerable importance because of the short duration of symptoms and its poor response to radiation therapy as well as the dismal survival rates.[2]

Reports indicate that there is a higher incidence of carcinoma of the endometrium occurring at an early age among patients with continuous unopposed estrogen stimulation of the endometrium, as with the Stein-Leventhal syndrome, and a long history of anovulation, amenorrhea, or both.

Adenomatous hyperplasia should be considered a precursor lesion, particularly if there is any atypia present. There is a prevalence in women of perimenopausal age with dysfunctional bleeding, which is the first and most significant sign. In that age group, a serious problem should be suspected if there is any irregular bleeding, and a fractional curettage should be carried out.[3]

Adenocarcinoma of the endometrium contains squamous elements in about 30% of cases. These elements may be benign or malignant. Specific terminology has been used previously to designate such variants (i.e., adenoacanthoma, adenosquamous carcinoma). Other histologically definable patterns include serous papillary adenocarcinoma (10% of cases), clear cell adenocarcinoma (1 to 2%), mucinous carcinoma, and squamous carcinoma. Endometrial epithelial malignancy is found in 15 to 25% of women with ovarian endometrioid cancer, although this almost always represents a simultaneous neoplastic transformation rather than a metastasis.

The International Federation of Gynecology and Obstetrics (FIGO) staging was based on a survey comparing treatment from different centers since 1971. This clinical staging was subjective and often inaccurate. Stage II, which was determined by the endocervical curettage, was notoriously inaccurate. There was also variation in the histology of the curetted material, the hysterectomy specimens in cell types, and the tumor grade. Since several authors had reported the same discrepancies, a task force studied the problem and a new staging method has been advanced.

Since the prognostic factors—that is, myometrial penetration and cervical environment—are best identified through the hysterectomy specimen, the trend and primary treatment has been toward hysterectomy with or without pelvic lymphadenectomy and paraaortic node biopsies. These procedures represented a staging operation and were often therapeutic.

Although protocols had been examined in the absence of clear advantages for particular management policy, treatment should be individualized. The best regimen, involving surgery, radiation therapy, hormone therapy, and chemotherapy, has not yet been determined.

Epidemiology: Familial Tendency Versus Diet

Cancer epidemiology is the study of the incidence and distribution of human cancer in relation to a variety of environmental and intrinsic factors. It seeks correlation between the occurrence of a disease and particular conditions in the environment or the characteristics of the people affected.

Lynch has reported that there is some evidence of an autosomal-dominant inheritance of susceptibility to endometrial carcinoma. Warthin's famous cancer family "G" studies, following familial cancer tendencies since 1893, now include more than 650 blood relatives reviewed in detail. Eighteen members over age 40 have had carcinoma of the endometrium and 53 have had colonic carcinomas.[4] The question naturally raised is whether or not there is a cancer-susceptible genome interacting with an oncogenic virus. The counterargument to this is that families tend to have the same diet, and since many of the women are single or nulliparous, living at home, they are also exposed to the same diet.

There has been great interest in the possibility of endocrinopathy in the background of patients with carcinoma of the endometrium because of (a) the observations concerning obesity, nulliparity, and infertility in many of these patients; (b) the frequency of prior failure of ovulation and dysfunctional bleeding; and (c) the obvious proliferative effect of estrogen upon the endometrium in all age groups. In addition, there are many other epidemiologic factors that have to be addressed in a study devoted to cancer of the endometrium. Many of these studies have reported a frequent association of diabetes mellitus and carcinoma of the endometrium. Hypertension has often been reported with endometrial cancer. Cancer of the breast and of the ovary tend to occur more frequently in women with carcinoma of the endometrium than expected by chance.

Family tendency, race, endometrial hyperplasia, functioning ovarian tumors, endometrial polyps, and menstrual disturbances have all been associated with an increase in the incidence of carcinoma of the endometrium. Carcinoma of the endometrium is rare among Orientals, especially the Japanese. It is more common among Jews and those who have few children and eat a Western diet with saturated fats.

Currently nutrition is receiving a great deal of study in all forms of carcinoma. There is significant evidence that high fat consumption is linked to an increased incidence of certain common cancers, including carcinoma of the endometrium. Women who are obese and nulliparous and have an increase in the amount of fat in their diet are at risk. Aleem et al. reported that the total plasma estrogen level was significantly higher in patients with endometrial carcinoma and hyperplasia than in a postmenopausal control group and in a group of patients having uterine bleeding without any detected histologic endometrial lesions.[5] Furthermore, analysis of the estrogen levels in the endometrial carcinoma group reveal that when the patient was obese the estrogen level was significantly higher than those in the nonobese and controlled groups. By reducing dietary fat to about 25% of total caloric intake, total plasma estrogen level can be decreased.

Fiber consumption is high in most nations

where carcinoma of the endometrium and carcinoma of the breast is low. The fiber may act independently to protect against breast and carcinoma of the endometrium or may lower fat levels in the body by preventing absorption of dietary fat, or may protect by increasing estrogen excretion. Or, fiber's contribution may be strictly nutritional, in the sense that it decreases the appetite and promotes satiety, so that the patient does not need to eat a lot of fatty food to achieve a full feeling. It is recommended that high-risk patients should follow a diet that includes a 25% reduction in saturated and unsaturated fats, an increase in fiber-rich whole-grain foods, daily consumption of yellow and green-leafy vegetables (rich in β-carotene and vitamin C), reduction in salt-cured and smoked foods, and moderation in the consumption of alcohol.

Initiator-Promoter Concept of Cancer

The key basic discovery that has important implications for dietary control efforts is that cancer develops in two stages, over a period of time. The first stage is what cancer research scientists call the early, or initiation, stage. At that time, a cancer-producing chemical or perhaps a carcinogen or viral particle interacts with the genetic material of the body cells (the DNA) to produce an abnormal cell. This is called a mutation. It is the first step in developing cancer. It may be initiated by many factors, including the presence of dangerous chemicals in their environment or contact with them, viruses, and a multitude of other carcinogens yet unknown. Development of cancer is also related to the ability of the individual cells to repair the mutations that occur. The later period in the development of a cancer is known as the promotional stage. The cancer enlarges from abnormal cells to a palpable mass. A microscopic tumor may be present for years but will not develop into a frank cancer unless the body permits and promotes its profligate growth. The promotional stage may go on for as long as 15 to 20 years or perhaps longer before a frank cancer is present.

Estrogens and Endometrial Cancer

Until 1975 it was generally believed that there was no increased risk of endometrial carcinoma in patients on estrogen therapy, and there were even optimistic reports of a decreased incidence of the disease, but these studies were all inadequate when judged by modern standards of epidemiology. Since then there have been original studies by Smith et al.[6] and Ziel and Finkle.[7] Following this there have been several case-control (retrospective) studies that have shown a risk ratio of between 5 and 15. However, Horwitz and Feinstein,[8] using an alternative and analytic method for case-control studies, showed that the risk ratio showed very little difference between users and nonusers of estrogen.

Smith et al.[6] and Ziel and Finkle[7] used the risk factor in their studies. Horwitz and Feinstein[8] reported that the odds ratio used to express results of a case-control study as obtained from these patients is a ratio of two ratios: the exposure ratio in cases with carcinoma of the endometrium divided by the exposure ratio in the controls. The odds ratio is used as a substitute risk ratio; it is the rate at which the disease occurs in exposed women, divided by the rate of its occurrence in nonexposed women. These two rates of occurrence are very small, and if no distortions have occurred in the four groups that comprise the case-control study, the odds ratio would be approximately equal to the risk ratio. If the odds ratio exceeds 1 and is statistically significant either in a direct test of significance or by demonstrating that the value of 1 is not contained in the associated confidence interval, the investigator concludes that a causal relation may exist between the agent and the disease.

In their 1978 paper, Horwitz and Feinstein reported on a case-control study of estrogens and endometrial cancer in which alternative sampling methods were used to eliminate the detection bias that arises from the increased diagnostic attention received by women with uterine bleeding.[9] In a set of cases and controls chosen by conventional procedures, the odds ratio was 11.98. In an alternative set of cases and controls at the same institution, consisting of patients who had a dilatation and curettage (D&C) or hysterectomy because of uterine bleeding, the odds ratio was 1.7. In addition, they reported that methodologic analysis demonstrates detection biases stemming from the pattern of hospital referral and shows how biases were neglected or increased by the alternate procedure. The magnitude of the association between estrogens and endometrial cancer has been greatly overestimated because of detection bias. When appropriate compensation for a bias is introduced the odds ratio approaches a value much closer to 1.

The 1975 papers may have served to alert the medical profession to the possibility of endometrial cancer to encourage a more aggressive effort to identify this type of cancer in its early stages. The report of Horwitz and Feinstein[8] may serve to allay some of the fears that followed the reports by Smith et al.[6] and Ziel and Finkle.[7]

There are as yet no good long-term cohort (prospective) studies that will provide a conclusive answer to this controversy. Estrogen is a drug and should be used with all the indications and contraindications of any drug. It should be given when needed and indicated and should be withheld when there is a contraindication to its use.

Gambrell has shown that the addition of progesterone when estrogen therapy is used gives predictable protection against the development of carcinoma of the endometrium.[10] In addition, he has shown that endometrial hyperplasia is a precancerous lesion in some women and can be effectively reversed with 10 to 13 days of progestogen monthly in at least 98% of patients. The progestogen challenge test has been devised to identify postmenopausal women at greater risk for adenocarcinoma. It should be administered to all postmenopausal women with an intact uterus. This includes asymptomatic women, patients receiving estrogen replacement therapy, and women being evaluated for hormone therapy. If there is a positive response to the progestogen challenge as manifested by withdrawal bleeding, then the progestogen should be continued for 13 days each month for as long as withdrawal bleeding results. If there is no response then the progestogen test should be repeated at each annual examination. Universal use of the progestogen challenge test should prevent nearly all endometrial cancers.

Endometrial Hyperplasia and Endometrial Metaplasia[11]

Endometrial hyperplasia is seen so frequently that its relationship to cancer of the endometrium is debatable. However, endometrial hyperplasia in an active form in the postmenopausal patient may be a precursor of cancer. This is especially so if it is related to a feminizing tumor or perhaps to exogenous estrogen use.

The differential diagnosis of endometrial hyperplasia in early carcinoma is sometimes difficult. A variety of classifications give rise to conflicting terminology that is often based on inappropriate methodology and inadequate follow-up in large series of cases. Poorly differentiated carcinomas do not pose a diagnostic problem. Grade 1 tumors, however, which are often focal and composed of fairly regular glands lined by cells with slight to moderate atypia, can be harder to assess. There is general agreement that stromal invasion determines the presence of invasive cancer. It is important here to distinguish between the myometrium and the endometrial stroma. It is impossible to confirm myometrial invasion on curettage unless a specimen of myometrium is obtained, but it is possible to identify stromal invasion.

Stromal invasion occurs in several forms. In one, the stroma disappears between adjacent glands, producing a back-to-back pattern. In the second, the stroma becomes fibrotic, so that collagen and fibroblasts, rather than cellular endometrial stroma, are found between glands. In the third, the stroma becomes necrotic, and smaller foci of necrotic debris and polymorphonuclear leukocytes are found both between glands and extending into gland lumina. In addition, the stroma may undergo foam cell transformation. The latter is not necessary pathognomonic, since it is sometimes found in hyperplasia and even in normal endometria. However, the presence of stroma foam cells should alert the clinician to consult with the pathologist to analyze more sections so that invasive cancer can be ruled out.

A high percentage of patients with adenomatous hyperplasia can be cured with fractional curettage. This is particularly important for the perimenopausal patient. If uncomplicated adenomatous hyperplasia is confirmed on fractional curettage, our policy is simply to observe the patient. The popular choice is to administer progestational agents, but we believe that this may disguise the underlying disease, which may be controlled temporarily but may emerge later in a more advanced stage. Therefore, once uncomplicated adenomatous hyperplasia is diagnosed, we recommend performing another fractional curettage in about 6 months. If the result shows benign endometrium, serial endometrial biopsies—not endometrial aspirations—may then be performed as an office procedure.

However, for menopausal and postmenopausal patients in whom adenomatous hyperplasia persists or reappears at the time of the second curettage, we recommend a hysterectomy and bilateral salpingo-oophorectomy. It is important to weigh the danger of morbidity and mortality from a hysterectomy against the risk of developing endometrial carcinoma. A patient with atypical adenomatous hyperplasia can be managed according to the same regimen. If the patient is at unusually high risk, it is advisable to forego long-term follow-up and proceed with hysterectomy. An extremely obese patient with diabetes, hypertension, and arteriosclerosis with a compromised pulmonary system can be controlled by a progestational agent such as medroxyprogesterone acetate. The patient must be carefully monitored, however, with endometrial biopsies at 6-month intervals.

The appropriate management of endometrial hyperplasia depends on ruling out the possibility of metaplasia on the one hand and stromal invasion on the other. Carcinoma of the enodmetrium is diagnosed primarily from the glandular component. Thus, if the glands appear benign and do not invade their own stroma, it is fairly safe to assume that the overall lesion is benign.

The differential diagnosis of endometrial carcinoma includes not only the well-known hyperplasias but also a group of epithelial metaplasias, some appearing after known etiologic stimuli and other occurring without known antecedents. These have been classified into seven categories: morules and squamous metaplasia, papillary metaplasia, mucinous metaplasia, hobnail metaplasia, and clear cell metaplasia.[12] All of the metaplastic processes share the dubious distinction of being of absolutely no clinical significance unless they are mistaken by the pathologist for carcinoma. The first four are the most common and thus the most likely to be confused with cancer.

Squamous metaplasia and morules, in which the metaplastic foci are epidermoid in appearance but lack intracellular bridges, are most likely to be confused with adenoacanthoma, in which benign-appearing squamous cells or morules are seen within an adenocarcinoma. It is important to note that an adenocarcinoma, like any other carcinoma of the endometrium, is diagnosed primarily from the glandular component, and thus if the glands are benign in appearance and do not invade their own stroma, it is safe to assume that the overall invasion is benign.

Papillary metaplasia has also been called syncytial surface metaplasia because it tends to occur at or near the endometrial surface and is characterized by cells that form papillary projections and syncytia. The appearance resembles microglandular hyperplasia of the endocervix, but this endometrial lesion is usually seen in patients receiving exogenous estrogens rather than progestins. The surface location, the blandness of the nuclei, and the frequent penetration by neutrophils lead away from the diagnosis of papillary carcinoma. Papillary metaplasia may coexist with invasive cancer.

In tubal ciliated cells or eosinophilic metaplasia, occasionally glands are lined by cells with bland nuclei and brightly colored eosinophilic cytoplasm. Cilia are usually easily identified, but the term "eosinophilic metaplasia" may be used. The lesion resembles the metaplasia of a type of epithelium that is identical to that seen in the fallopian tube, and it is most likely to be confused with atypical hyperplasia or adenocarcinoma in situ. The distinction is made largely by the presence of ciliated cells in tubal metaplasia in the absence of nuclear stratification and atypia.

Endometrial Adenocarcinoma

Natural History

Adenocarcinoma of the endometrium is either localized or diffuse. Carcinoma arising in the endometrium tends to remain within the uterus for a long period, spreading by local extension within the endometrium. It may produce uterine enlargement. As it grows it invades the myometrium and advances toward the isthmus and endocervix. From there it spreads to the paravaginal and paracervical tissues, or it may invade directly through the myometrium to the serosa and the peritoneal cavity. Since the lymphatics increase in amount as the serosa of the uterus is approached, deep penetration of the myometrium is associated with an increased incidence of positive node involvement.[13] The lymphatic drainage of the upper part of the body of the uterus is along the course of the infundibular ligament to the aorta, cava, lumbar, and high common iliac nodes. This is usually seen late in the disease. Once the lymphatic system in the cervical area has been involved, the spread of cancer parallels that of cervical cancer.

Lymphatic anastomosis between the body of the uterus and the round ligament may be responsible for the rare metastases occurring in the inguinal area. Approximately 10 to 15% of patients with stage I cancer of the endometrium and 36 to 40% of those with stage II carcinoma will have lymph node metastases in the pelvis. The depth of myometrial invasion is also directly related to the percentage of lymph node metastases.

In summary, patients at high risk for extrauterine spread are those with involvement of the lower uterine segment or cervix, histologic grade 3 lesions, or myometrial invasion at least one quarter to one third of the depth of the myometrium.

Diagnosis[14]

Symptoms

The first and most important symptom in all age groups is abnormal bleeding. It is usually bright red. A patient who presents with postmenopausal bleeding without an obvious vaginal or cervical lesion must be considered to have endometrial carcinoma until proven otherwise. Finding a benign condition such as a cervical polyp does not rule out a malignant lesion; the patient may have both.

Screening

Cytologic screening programs aimed at identifying early changes in the cervix have yielded better results than those identifying early changes in the endometrium. The cytologic smear taken in the conventional manner at best may detect approximately 50% of cases of carcinoma of the endometrium. The cytologist can ascertain whether the endometrium is benign or malignant but has difficulty detecting "in-between groups" such as adenomatous hyperplasia, dysplasia, or carcinoma in situ. However, this can be done with a high degree of accuracy if tissue is obtained for paraffin block histologic examination.

Fractional Curettage

Early diagnosis may be delayed, unfortunately, by both patient and physician when carcinoma of the endometrium is present. There is an old wives' tale that abnormal bleeding is a normal sequence of menopause. Bleeding in the menopause is still looked upon by some patients as a rejuvenation and a reprieve from the ravages of aging. A fractional curettage should be carried out in every postmenopausal patient with a bleeding problem. No menopausal patient should receive treatment for menstrual abnormality without a diagnostic fractional curettage. Fractional curettage may be delayed because the Papanicolaou smear is negative. Although present, carcinoma may be unable to escape through a stenosed cervix.

Curettage of each sector of the uterus is done independently: first the endocervical canal and then each wall of the uterine cavity and the fundus. This technique enables the physician to identify and evaluate the extent of any existing malignancy and to discriminate between endocervical and endometrial carcinoma. Fractional curettage should be done with great caution. The endocervix is best curetted before the patient goes to sleep. With the patient awake it is easier to tell if the curette is going through the internal os, thus giving a false reading about the endocervix.

The percentages of detection by the different techniques for detecting endometrial carcinoma are: vaginal pool, 50%; ectocervical, 70%; endometrial brush technique, 75%; uterine aspiration smears, 80%; intrauterine lavage, 87%; and Gravelee jet washer, 93%. The usual technique used is fractional curettage, and it is well known that, even in totally benign cases, the specimens separately submitted as endometrial and endocervical have a likelihood only slightly greater than dictated by the laws of chance of actually consisting exclusively of the elements indicated on the respective specimen labels. Thus, the mere finding of carcinoma in the endocervical curettage specimen should not lead to the designation of the case as stage II.

Staging

Since corpus cancer is now surgically staged, procedures previously used for determination of stages are no longer applicable, such as the binding of fractional D&C to differentiate between stage I and stage II. However, it still has relevance in patients with endometrial cancer who are treated solely by radiation therapy and gives the resident in training an appreciation for doing a systematic curettage.

Clinical Staging[15]

The staging of these small number of cases that will be treated by radiation should be based on careful clinical examination and should be performed before any definitive therapy. It is desirable that the examination be performed by an experienced examiner with the patient under anesthesia. The clinical stage must under no circumstances be changed on the basis of subsequent findings. When it is doubtful to what stage a particular case should be allocated, the case must be referred to the earlier stage.

The following examinations are permitted for clinical staging of carcinoma of the endometrium: palpation, inspection, endocervical curettage, hysteroscopy, cystoscopy, proctoscopy, hysterography, and x-ray examination of the lungs and skeleton. Findings on examinations such as lymphangiography, arteriography, venography, and laparoscopy are of value for the planning of therapy, but because these are not yet generally available and also because the interpretation of results is variable, the findings of such studies should not be the basis for assigning or changing the clinical stage.

Proposed Classification, International Society of Gynecological Pathologists (ISGP)

Epithelial Tumors
A. Hyperplasia

B. Carcinoma
1. Endometrioid carcinoma
Variance
a. With squamous differentiation
I. Adenocarcinoma with squamous metaplasia adenoacanthoma
II. Adenosquamous carcinoma
b. Secretory
c. Ciliated cell
2. Mucinous adenocarcinoma
3. Serous adenocarcinoma
4. Clear cell carcinoma
5. Squamous cell carcinoma
6. Undifferentiated carcinoma
7. Mixed carcinoma
8. Unclassifiable

Under this classification (a correlation of the FIGO, International Union Against Cancer (UICC) and American Joint Committee on Cancer (AJCC) nomenclatures) staging is as shown in Table 26-1.

Cases of carcinoma of the corpus should be graded with regard to the degree of differentiation of the adenocarcinoma as follows:

Grade 1—Highly differentiated adenomatous carcinoma.
Grade 2—Moderately differentiated adenomatous carcinoma with partly solid areas.
Grade 3—Predominantly solid or entirely undifferentiated carcinoma.
Grade 4—Grade not assessed.

Surgical-Pathologic Classification[16]

Based on clinical staging, three out of four patients will be assigned to stage I. Also, errors in clinical staging (FIGO) occur in at least 15% percent of cases. Prognostic factors other than clinical stage are important if the treatment is to be individualized. The factors permitting a surgical-pathologic classification are: histologic differentiation; myometrial penetration; lymph node metastases (pelvic and paraaortic); occult involvement in the cervix; unexpected adnexal metastases; and positive peritoneal cytology.

Histologic differentiation is of primary significance in predicting survival. Using the FIGO designation, grades 1, 2, and 3 tumors have respective 5-year survival rates of 81, 74, and 50%. Such a strong correlation qualifies cytology as a key prognostic factor.

Myometrial penetration has predictive value particularly among the elderly patient. In the elderly patient the uterine myometrium is smaller in diameter so that the cancer cells can reach the lymphatics and the serosa at an earlier stage, giving rise to positive nodes and often peritoneal spread. The most difficult determination in this regard is the distinction between penetration into preexiting tongues of endometrium and the superficial myometrium and invasion into islands of adenomyosis, which is a more ominous manifestation of the same phenomena and represents true infiltration of the

TABLE 26-1. Clinical staging of endometrial adenocarcinoma

Stage 0	Atypical endometrial hyperplasia, carcinoma in situ. Histological findings are suspicious of malignancy. (Cases of Stage 0 should not be included in any therapeutic statistics.)
Stage I	The carcinoma is confined to the corpus.
Stage Ia	The length of the uterine cavity is 8 cm or less.
Stage Ib	The length of the uterine cavity is more than 8 cm.
Stage II	The carcinoma has involved the corpus and the cervix, but has not extended outside the uterus.
Stage III	The carcinoma has extended outside the uterus, but not outside the true pelvis.
Stage IV	The carcinoma has extended outside the true pelvis or has obviously involved the mucosa of the bladder or rectum. A bullous edema as such does not permit a case to be allotted to Stage IV.
Stage IVa	Spread of the growth to adjacent organs as urinary bladder, rectum, sigmoid, or small bowel.
Stage IVb	Spread to distant organs.

It is appreciated that there may be a small number of patients with endometrial carcinoma who will be treated primarily with radiation therapy. If that is the case, the clinical staging adopted by FIGO in 1971 would still apply, but designation of that staging system would be noted.

myometrial stroma. With penetration, the nests of tumor within the myometrium or around it are of roughly uniform size, sometimes with residual benign and endometrial stroma and/or glands at their periphery. A peripheral lymphoid response is generally absent or mild. In true invasion, the tumor nests are variable in size and shape, have angular margins, and are not surrounded by benign endometrium. Reactive changes can be observed in the surrounding myometrium, including stromal fibrosis, rarefaction, and marked inflammatory responses (usually lymphoid).

Lymph node involvement, occult cervical involvement, and unexpected adnexal metastases decrease the prognosis.

Positive peritoneal cytology may be the only evidence of tumor extension outside the pelvis. Reports from the literature indicate that approximately 10% of patients with stages I and II tumors have positive peritoneal cytology. We have reported that peritoneal cytology must be evaluated relative to the time at which curettage and exploratory laparotomy with pelvic washings are performed. Positive cytology results have a much greater significance if the laparotomy is carried out immediately after the curettage than if the procedures are separated by 72 hours or more, giving the cells time to migrate into the peritoneal cavity.

Identification of true stage II disease is often a problem. There are certain patterns that appear in stage II endometrial cancer, however: tumor invading the endocervical tissues and its overlying endocervical epithelium; tumor in the endocervical stroma without overlying glandular epithelium; tumor separate from endocervical tissue; and tumor only, in which the specimen consists entirely of carcinoma.

As anticipated, the 5-year survival is best when tumor is seen separate from the endocervical tissue and worst when tumor has invaded endocervical tissue, with or without overlying benign epithelium.

Surgical-Pathologic Classification FIGO[17]

Since there is general dissatisfaction with clinical staging and the backbone of treatment has been hysterectomy, a new surgical-pathologic classification was adopted in 1988 (Table 26-2).

All stages should be subgrouped with regard to the histologic type of adenocarcinoma as follows:

Grade 1—Highly differentiated adenomatous carcinoma.

TABLE 26-2. Surgical-pathologic staging of endometrial adenocarcinoma

Stage I	The carcinoma is confined to the corpus.
Stage Ia G123*	Tumor limited to the endometrium.
Stage Ib G123	Invasion up to and including ½ myometrial thickness.
Stage Ic G123	Invasion greater than ½ myometrial thickness.
Stage II	The carcinoma involves corpus and cervix.
Stage IIa G123	Endocervical gland involvement only.
Stage IIb G123	Cervical stromal invasion.
Stage III	
Stage IIIa G123	Tumor invades serosa and/or adnexa and/or positive peritoneal cytology.
Stage IIIb G123	Vaginal metastasis.
Stage IIIc G123	Metastasis to pelvic and/or paraaortic lymph nodes.
Stage IV	
Stage IVa G123	Tumor invasion of bladder and/or bowel mucosa.
Stage IVb	Distant metastasis including intra-abdominal and/or inguinal lymph nodes.

* Grades 1, 2, or 3 according to histological type (see text).

Reprinted from ref. 18, with permission.

Since carcinoma of the endometrium is now surgically staged, procedures used previously for determination of stages are no longer applicable, such as the finding of a fractional D&C to differentiate between Stage I and Stage II. Ideally, the width of the myometrium should be measured along the width of tumor invasion.

Grade 2—Differentiated adenomatous carcinoma with partly solid areas.
Grade 3—Predominantly solid or entirely undifferentiated carcinomas.

Also, the abbreviation My is used for no residual tumor at hysterectomy; My0 indicates no muscle invasion; My1 indicates inner one third invasion; My2 indicates middle one third invasion; and My3 indicates outer one third muscle invasion.

Rules Related to Staging

1. Since corpus cancer is now surgically staged, procedures previously used for determination of stages are no longer applicable, such as the finding of fractional D&C to differentiate between stage I and stage II.
2. It is appreciated that there may be a small number of patients with corpus cancer who will be treated primarily with radiation therapy. If that is the case, the clinical staging adopted by FIGO in 1971 would still apply but designation of that staging system would be used in the patient's notes. Therefore, it is important to ascertain the facility in which a fractional curettage records which staging method was used.
3. Ideally, width of the myometrium should be measured along with width of tumor invasion.

Aggressive Subtypes[18]

Adenosquamous Carcinoma

Adenosquamous carcinoma has both a malignant glandular and a squamous cell component. It is one of the more aggressive forms of endometrial carcinoma. This subtype at times present problems in histologic diagnosis, especially when the specimen consists only of curettings. It must be separated from adenoacanthoma with atypical squamous component on the one hand, and from the poorly differentiated adenocarcinomas on the other. In about 50% of cases the glandular component is dominant, and in the other 50% there is dominance of the squamous component. The squamous component varies in its differentiation. As in squamous cell cancer of the uterine cervix, the squamous component is usually large cell, less often keratinizing, and rarely small cell. The associated glandular component is similar to that observed in endometrial adenocarcinoma. Venous involvement is observed in 50% of cases and vascular and transtubal routes may be involved in the dissemination of the neoplasm. The mean age at detection is 65.5 years. The duration of symptoms tends to be short, and abdominal and distant spread is frequently observed. These tumors respond poorly to ionizing radiation. Five-year survival is 19 to 20%. The treatment is exploratory laparotomy, washings for cytology, node sampling of the paraaortic and pelvic areas, total hysterectomy, and bilateral salpingo-oophorectomy.

Glassy Cell Carcinoma

Although glassy cell carcinoma of the endocervix is well documented as being an exceptionally aggressive type of cancer, experience with similar tumors of the endometrium is much too scant for generalizations about aggressiveness. Nonetheless, awareness of the lesion will eventually bring to light additional cases of this most interesting carcinoma. Since the original publications, it has been shown that the glassy cell changes appear to be in the glandular component. At present all cases have also had large-cell, nonkeratinizing squamous components as well.

Clear Cell Carcinoma

Clear cell carcinomas occur throughout the female genital tract. Originally they were considered to be of mesonephric origin, but since they occur as a primary carcinoma in the endometrium they are now considered to be of müllerian and not mesonephric origin. Histologic grading of clear cell carcinoma of the endometrium has little meaning and a nuclear grading has not given a good correlation with prognosis. The recognition of the clear cell nature of the tumor and the tumor extent are the most important factors in predicting outcome.

Clear cell carcinoma is frequently accompanied by other histologic subtypes. The characteristic cells are polygonal with abundant clear or slightly eosinophilic cytoplasm. The cytoplasmic envelope is at times missing, especially in papillary and tubular types, giving a hobnail appearance to the cells. This feature is important, together with the presence of other clear cell patterns, in differentiating these tumors from non–clear cell papillary carcinomas, a distinction that is important to make because of the prognostic implications. The nuclei are usually poorly differentiated, that is, of high grade with large, prominent nucleoli. The high nuclear grade, as well as the pattern, readily separate these tumors from secretory carcinoma, with which they have been confused. These tumors have a very poor prognosis.

Papillary Serous Carcinoma

The papillary serous tumor has been known for a long time, but its pathologic features and clinical course have been appreciated only recently. Studies with the electron microscope show that these tumors have features intermediate between those of ovarian serous carcinoma and classical endometrioid carcinoma of the endometrium.

Histologically, these tumors are characterized by the presence of papillary structures with broad fibrovascular connective tissue cores. These are lined by regularly stratified cells that form secondary papillary structures and interconnecting arches. Exfoliation of small clusters of cells is quite common. Foci of necrosis are frequently seen. Since the tumors tend to be moderately to poorly differentiated, solid or large undifferentiated cells are almost invariably present. Psammoma bodies are detectable in approximately 30% of these tumors. A very characteristic feature of this lesion is its tendency to infiltrate the myometrium with lymphatic or vascular channels. This is often an unexpected microscopic finding, occurring in a small, atrophic uterus. Mostly normal-appearing uterine adnexae also frequently show extensive tumor emboli. As is expected from this pattern of dissemination, the prognosis is extremely poor, and this tumor should be treated aggressively with adjuvant therapy.

Management

Management According to Clinical Stage[19]

Stage 0 (carcinoma in situ): The management should be tailored to fit the needs of the patient, her age, and her desire for future childbearing. In selected young women with minimal disease, an alternative to hysterectomy may be a repeat curettage in 3 to 4 months or a course of progestational agents followed by repeat curettage in 6 months.

Stage Ia: Total hysterectomy and bilateral salpingo-oophorectomy are carried out in patients with stage IaG1 lesions. No external radiation is given, but some clinicians elect to insert a radiation source against the vaginal vault. In patients with histologic grade 2 or 3 lesions, external pelvic radiation and vaginal radiation should be given. If there is moderate to extensive myometrial invasion beyond one third or more of the myometrial cavity, postoperative external pelvic radiation and vaginal radiation should be given.[20]

Stage Ib: In stage IbG1, G2, or G3, total hysterectomy and bilateral salpingo-oophorectomy are followed by postoperative external pelvic radiation and vaginal radiation.

Stage II: Total hysterectomy and bilateral salpingo-oophorectomy are carried out, followed by postoperative external pelvic radiation and vaginal radiation.

Stage III: Total hysterectomy and bilateral salpingo-oophorectomy are carried out, followed by postoperative external pelvic radiation and vaginal radiation.

Stage IV: Primary therapy is structured to take out the uterus, tubes, and ovaries. External pelvic radiation and vaginal radiation are then given.

Distant metastases: Primary progestational therapy is indicated. Localized disease without response to progestational therapy is treated with x-ray therapy. In the presence of widespread disease and no response, chemotherapy should be given.

Radiation Therapy

A number of clinicians prefer to give preoperative external radiation or intrauterine radiation. The results in the literature do not show any differences between those whose patients receive preoperative versus postoperative x-ray therapy. However, if the uterus is large and knobby and seems thin walled, it is our opinion that preoperative external radiation is indicated.

There is no evidence providing the superiority of preoperative radiation in the management of carcinoma of the endometrium. Intracavitary radiation is effective in decreasing the incidence of vaginal metastases, and reports indicate that the difference is significant among those who are treated with vaginal radiation. The recurrence rate at the vaginal vault in untreated patients is as high as 18%, whereas among those treated with radiation it drops to 2%. There is less evidence that the 5-year survival rate is affected favorably. External megavolt radiation is also effective in decreasing pelvic (local and lateral) recurrence but it does not increase survival even in patients with grade 3 lesions. The morbidity rate from extending the radiation fields to the paraaortic area is uncertain. However, when treating the physically more favorable patient with cervical carcinoma, serious complications (i.e., small bowel injury) occur in only 6% when the dose does not exceed 4,400 rads and operative dissection of the paraaortic nodes is limited. Extensive dissection followed by doses exceeding 4,400 rads results in a higher complication rate.

The radiation therapy technique described here is a preoperative method, but if hysterectomy is the initial therapy, the same principles are observed in the use of radiation therapy postoperatively.[22]

Stage IaG1 (uterine cavity less than or equal to 8 cm, or a well-differentiated malignancy, or both):
- Burnet applicator loaded with 30-20-30 mg of cesium inserted for 48 hours 6 weeks postoperatively.

Stages IaG2 or G3 and IbG1, G2, or G3 (uterine cavity greater than 8 cm, or a poorly differentiated malignancy, or both):
- External radiation to pelvis, including the upper vagina
  - Dose—4,000 rads in 4 weeks
  - Field—opposing anterior and posterior fields as outlined under treatment of cervical cancer
- Burnet applicator loaded with 30-20-30 mg of cesium inserted for 48 hours 6 weeks postoperatively

Stages II, III, and IV: Treated as stage IaG2 or G3.

Recurrent Adenocarcinoma of the Endometrium

Localized symptomatic cancer is treated with radiation unless it is possible to excise it with a good margin using surgery. Selected patients may be candidates for pelvic exenteration. Although there are fewer indications for pelvic exenteration among patients with cancer of the endometrium than among those with cervical cancer, pelvic exenteration has a place among the methods of treatment used as a definite therapy for this cancer. Progestational agents have been used successfully to control metastases in the lung and in the vagina, but have been less successful in the pelvis.[21]

Distant Metastases

Primary progestational therapy is indicated. Medroxyprogesterone acetate (Depoprovera) 400 mg intramuscularly/day for 7 days and then every 7 to 10 days is administered. In addition, oral megestrol acetate (Megace) is given in doses up to 40 mg three or four times per day. Localized disease that does not respond to progestational therapy is treated with x-ray therapy. In the presence of widespread disease and no response to the hormone treatment suggested above, chemotherapy is indicated. However, chemotherapy has not been particularly effective. Combination therapy employing cyclophosphamide, Adriamycin, *cis*-platinum or carbo-platinum as well as mitomycin and 5-fluorouracil has shown some promise.

Progesterone therapy has been used for approximately three decades. The conclusions from studies over this period are that the response rate of patients with advanced and recurrent disease is 33%, and the response rate with well-differentiated tumors is 30 to 50%, as opposed to 0 to 15% for poorly differentiated tumors. The best responders are patients in whom treatment was continued for more than 6 months. Unfortunately, at this time the optimum dose is not yet defined, nor is the length of time that it should be given. Progesterone receptors are present in greatest numbers in well-differentiated tumors. This may be an explanation of the diminished response of the poorly differentiated cancers to progesterone therapy. The value of antiestrogen (Tamoxifen) is yet unproven. Side effects are minimal and tumor responses are seen in about 30% of unselected patients. Since Tamoxifen blocks the estrogen receptors on the tumor and progesterone may act in a similar manner, they should not be given together but rather in sequence.

Follow-up

Sixty percent of recurrences occur within 2 years and 90% occur within 5 years. Undifferentiated tumors tend to recur earlier. Recurrences may be in the vagina (uncommon if radiation therapy has been given) or pelvis or they may be distal. Plasma carcinoembryonic antigen (CEA) values may be of use in predicting recurrence. Pretreatment CEA values have been raised in 22% of well-differentiated tumors, 29% of moderately differentiated tumors, and 67% of undifferentiated tumors.[23]

After treatment of any stage of carcinoma of the endometrium, follow-up visits should be every 3 months for the first year, every 4 months for the next 2 years, every 6 months thereafter for 5 years and then annually. The following procedures should be performed at follow-up: Papanicolaou smear at each visit, determination of plasma CEA level at each visit, chest x-ray at 6 to 12 months, and ultrasound, CT scan, needle aspiration biopsy, and so forth as indicated.

Summary

The incidence of endometrial carcinoma is increasing in the United States, and it is now the most common gynecologic cancer. However, it will cause just over one half the number of deaths compared to cervical cancer and only one third the number from ovarian cancer.

High-risk factors include obesity, nulliparity and low parity, history of oligomenorrhea anovulation, infertility, late menopause, hypertension, diabetes, hormone-secreting tumors of the ovary, polycystic ovary syndrome, Turner's syndrome and other chromosomal mosaics, hormonal replacement therapy, exposure to ionizing radiation, prolonged estrogen use, sequential oral contraceptives, and a family history of endometrial and breast cancer.

The most common presenting symptom is abnormal bleeding. Cervical-vaginal smears are unreliable in detection and methods of sampling the endometrial cavity must be carried out to increase the diagnostic accuracy.

The number of carcinomas of the endometrium reported has increased. Whether this is due to better diagnostic measures or a change in pathology due to different carcinogens, viruses, or unknown factors is not known at this time. The ISGP proposed a new classification for epithelial tumors, which is outlined in the chapter.

The FIGO staging has been the sole

method for comparing treatment from different centers since 1961. However, there is a great deal of error due to the clinical staging, and studies have been undertaken to propose a new classification. Since a variety of prognostic factors, such as myometrial penetration, can only be resolved by a surgical approach, the trend in primary management has been toward hysterectomy with or without pelvic lymphadenectomy and paraaortic node biopsy. The general dissatisfaction with clinical staging lead to a new surgical-pathologic classification being adopted in 1988.

Since there is an absence of clear advantages for particular management policies, treatment for carcinoma of the endometrium should be individualized. The optimal regimens include surgery, radiation therapy, hormone therapy, and chemotherapy, but the ideal combination is yet to be determined.

Three out of four patients will be found to have stage I cancer. To permit individualization of management in stage I endometrial cancer, the following factors have prognostic value:

Histologic differentiation and cell type
Myometrial penetration
Lymph node metastases (pelvic and paraaortic)
Occult involvement of the cervix
Unexpected adnexal metastases
Positive peritoneal cytology
Ploidy
Estrogen-progesterone receptors
A proposed protocol of management of carcinoma of the endometrium by stage is presented.

Thirty-five percent of recurrences, either local or distant, occur within the first year, and 76% within 3 years. Ninety percent recur within 5 years and this is not related to the primary treatment. Recurrences to the vaginal vault are extremely rare now that the vault is irradiated either by external or intravaginal radiation.

It is now believed that if the carcinoma of the endometrium has been cured by conventional treatment, there is no contraindication to the use of estrogen replacement therapy for menopausal symptoms and protection for the cardiovascular and skeletal systems.

References

1. Cancer Facts and Figures in 1991. Atlanta, American Cancer Society, 1991.
2. Schottenfeld D, Fraumeni JF Jr: Cancer Epidemiology and Prevention. Philadelphia, WB Saunders Company, 1982.
3. Barber HRK: Manual of Gynecologic Oncology, ed 2. Philadelphia, JB Lippincott Company, 1989, pp 230–244.
4. Lynch HT, Krish AJ: Cancer family "G" revisited 1895–1970. Cancer 1971;27:1505.
5. Aleem FA, Moukhtar MA, Hung HC, et al: Plasma estrogen in patients with endometrial hyperplasia and carcinoma. Cancer 1976;38:2101.
6. Smith DC, et al: Association of exogenous estrogen and endometrial carcinoma. N Engl J Med 1975;293:1164.
7. Ziel HK, Finkle WD: Increased risk of endometrial carcinoma among users of conjugated estrogens. N Engl J Med 1975;293:1167.
8. Horwitz RI, Feinstein AR: Alternative analytic methods of case-control studies of estrogen and endometrial cancer. N Engl J Med 1978;299:1089.
9. Horwitz RI, Feinstein AR: Methodologic standards and contradictory results in case-control research. Am J Med 1979;66:556.
10. Gambrell RD Jr: The role of hormone in the etiology and prevention of endometrial cancer. Clin Obstet Gynecol 1986;13:695.
11. Silverberg SG: New aspects of endometrial carcinoma. Clin Obstet Gynecol 1984;11:189.
12. Norris HJ, Connor GP, Kurman RJ: Preinvasive lesions of the endometrium. Clin Obstet Gynecol 1986;13:725.
13. Sommers SC: Gross and microscopic pathology, in Barber HRK, Sommers SC (eds): Carcinoma of the Endometrium. Etiology, Diagnosis and Treatment. New York, Masson Publishing USA Inc, 1981, pp 67–80.
14. Barber HRK: Symptoms, signs and diagnosis, in Barber HRK, Sommers SC (eds): Carcinoma of the Endometrium. Etiology, Diagnosis and Treatment. New York, Masson Publishing USA Inc, 1981, pp 119–123.

15. Pettersson F (ed): Annual Report on the Results of Treatment in Gynecological Cancer. Stockholm, International Federation of Gynecology and Obstetrics, 1986.
16. Barber HRK: A surgical/pathologic classification for endometrial cancer: the time is now (editorial). The Female Patient 1988;13:9.
17. FIGO Stages. 1988 Revision Corpus Cancer Staging. Gynecol Oncol 1989;35:125.
18. Christopherson WM: The significance of the pathologic findings in endometrial cancer. Clin Obstet Gynecol 1986;13:673.
19. Barber HRK: Treatment of endometrial carcinoma, in Barber HRK, Sommers SC (eds): Carcinoma of the Endometrium. Etiology, Diagnosis and Treatment. New York, Masson Publishing USA Inc, 1981, pp 155–159.
20. Aalders J, Abeler V, Kolstad P, et al: Postoperative external irradiation and prognostic parameters in stage I endometrial carcinoma. Clinical and histopathologic study of 540 patients. Obstet Gynecol 1980;56:419.
21. Jones HW: Treatment of adenocarcinoma of the endometrium. Obstet Gynecol Surv 1975;30:147.
22. Barber HRK: Carcinoma of the endometrium, in van Nagell JR Jr, Barber HRK (eds): Modern Concepts of Gynecologic Oncology. Boston, John Wright PSG Inc, 1982, pp 213–238.
23. van Nagell JR, Donaldson ES, Wood EG, et al: The prognostic significance of carcinoembryonic antigen in the plasma and tumors of patients with endometrial adenocarcinoma. Am J Obstet Gynecol 1979;128:308.

27

Childhood Rhabdomyosarcoma of the Vagina and Uterus

DANIEL M. HAYS AND HAROLD M. MAURER

Vaginal rhabdomyosarcoma has been recognized as an entity for a century because of its arresting presentation as a protruding mass at the introitus in an infant or small girl, classically described as "botryoid" (i.e., "grapelike") in appearance. Early local excisions of these tumors were uniformly followed by local recurrence and eventual dissemination. With the advent of radical surgical procedures, total hysterectomy-vaginectomy or forms of pelvic exenteration were carried out with an occasional survivor.[1] When these procedures were combined with pelvic radiotherapy or single-agent chemotherapy regimens, success relative to survival was minimally improved.[2] In the 1960s, it was recognized that rhabdomyosarcoma was responsive to multiagent chemotherapy regimens, notably to vincristine, actinomycin-D, and cyclophosphamide (VAC),[3–5] and patients with tumors in all anatomic sites, including those with vaginal lesions, were treated with these agents.[5] It soon became apparent that the vagina is one of the anatomic sites in which primary tumors are highly chemotherapy responsive.[6,7] The combination of anterior or total pelvic exenteration, followed by local irradiation and 2 years of chemotherapy (usually VAC) was successful in preserving life in over 90% of children with vaginal rhabdomyosarcoma.[8,9] During the 1970s, a new goal became evident worldwide: to limit surgery and preserve both the bladder and rectum.[10–13] In respect to primary vaginal tumors this goal has been largely achieved, which makes them unique among primary pelvic rhabdomyosarcomas; in neither the case of primary bladder-prostate nor that of primary uterine tumors has the same degree of success been achieved. At present, there is little indication for pelvic exenteration in the management of patients with vaginal tumors, except in patients with extensive local recurrence and/or complete failure of standard therapies.

There are major differences between rhabdomyosarcoma arising in the vaginal walls versus the uterus.[9,14] These include the range of age at diagnosis, the forms of gross presentation, histology, and prognosis—albeit the number of primary uterine rhabdomyosarcomas is so small that these differences may be difficult to establish statistically. In the small child whose vagina is filled with a tumor mass, this anatomic distinction may be difficult to make. However, when such patients respond to therapy, the receding tumor almost inevitably retracts into the vaginal wall, leaving the cervix clear. Furthermore, when anterior exenterations were common in such patients, the site of the primary tumor identified in the specimen was almost always vaginal among younger patients. Currently, there is little reason to combine rhabdomyosarcomas in these two sites as an entity, as practiced in the past.[14,15] In patients below the age of 5 years, the primary site can be assumed to be vaginal unless there is evidence to the contrary. In this review, tumors primary in these two sites are described in separate sections.

Vaginal Rhabdomyosarcoma

Infants or small children with primary vaginal tumors are initially seen because of vaginal bleeding or discharge, tumor protrusion, or the passage of tumor fragments.[12,14] In only approximately 10% of the patients is the tumor visible at the vaginal orifice. Despite its characteristic gross appearance, the diagnosis is not always established by the first biopsy specimen. In the Intergroup Rhabdomyosarcoma Study (IRS), a number of patients with this tumor have a history of prior "negative" biopsies, and, in some cases, of fulguration of vaginal "polyps."[9] Approximately 25% of these patients with vaginal primaries have a palpable abdominal mass, which may extend superiorly as far as the umbilicus. Disseminated disease at diagnosis, including lesions in the lungs and/or bones, is seen in less than 10%.

Gross and Microscopic Pathology

Vaginal rhabdomyosarcomas do not ordinarily appear as a gross polyp with a "stalk" (in contrast to uterine-cervical rhabdomyosarcoma), but rather extend from a broad base in the vaginal wall, which is usually anterior or anterolateral. They invade the bladder more frequently than the rectum. Almost all vaginal rhabdomyosarcomas are of the embryonal subtype, and the majority have botryoid histologic features.[14] Biopsies usually reveal a narrow submucosal layer of malignant-appearing tissue covering a stroma with a loose "mixoid" appearance. If the biopsy includes only the latter tissue, misdiagnosis is common.[16] The appearance of this mixoid tissue, however, is characteristic and should lead to further biopsies. Superficial infection of the vaginal mucosa may also make histologic diagnosis difficult.

Differential Diagnosis and Staging

The most common clinical condition simulating the presence of a vaginal rhabdomyosarcoma is the presence of a vaginal foreign body, with secondary infection. However, once the lesion is established as a malignant tumor the diagnostic possibilities are limited, with all other histologic types far less common than rhabdomyosarcoma.[16,17]

Endodermal sinus tumors are seen in the vaginal walls in this age group, as well as in older females, and can be identified by appropriate marker studies as well as histologic features.[18,19] Adenocarcinomas of the vagina associated with interuterine exposure to hormone therapy are ordinarily found in an age group beyond 14 years,[20] whereas vaginal rhabdomyosarcomas are very uncommon in children older than 5 years.

Staging procedures in patients with vaginal rhabdomyosarcomas should include surveys to exclude pulmonary and osseous metastatic disease. The extent of pelvic, and possibly abdominal, involvement can be determined by computed tomography (CT) or ultrasound, and these will provide a "baseline" to assess the effects of subsequent chemotherapy. Cystoscopic examination to assess bladder involvement is ordinarily carried out at the time of vaginal biopsy. Some patients with large tumors have secondary ureteral obstruction revealed by intravenous urography or other studies. Some have significant compression of the sigmoid colon. In unusually large tumors, temporary urinary diversion or even temporary colostomy may be required, but an initial laparotomy is not necessary to determine the extent of disease.

Therapy

Primary therapy for all stages of vaginal rhabdomyosarcoma, including those that are small and apparently localized, is chemotherapy.[21–24] Rhabdomyosarcoma in this site is responsive, and even partial failure to respond is uncommon. Relatively massive tumors that fill the pelvis and lower abdomen are usually markedly reduced in size following several courses of therapy. Current standard therapy is repetitive "pulse" VAC, and early response to this regimen occurs in approximately 90% of children with these tumors.[14] Cisplatin and/or doxorubicin have been added for patients in IRS-III on an investigative basis. Local irradiation was formerly a standard element in therapy and is still employed in some patients. The age of these patients makes its use undesirable because of secondary effects on growth as well as ovarian function.

The management of vaginal tumors with brachytherapy, either before or following chemotherapy, is advocated in some European centers. In the largest series employing this approach, seven patients were treated with iridium-192 implants, and only one required subsequent hysterectomy-vaginectomy.[25] There was one therapy-related death. Long-range effects of irradiation in infancy on vaginal and uterine cervical tissues is unknown.

The response of vaginal tumors to successive courses of chemotherapy may be followed by serial CT, ultrasound, or, in some clinics, repeated examinations under anesthesia with vaginoscopy (cystoscopy if the bladder has been involved) and bimanual palpation. In the IRS protocols, reevaluation, usually including laparotomy, is recommended after 16 to 20 weeks of chemotherapy.[21] In some of these patients, despite no clinical evidence of a mass, positive or equivocal vaginal biopsies persist. Some of these biopsies reveal an apparent "maturation" of neoplastic cells, suggesting that their malignant potential has been modified. Despite a clinical complete response, however, a number of local recurrences have occurred in these patients when no form of local therapy was employed.[14]

After complete response or impressive partial response has been achieved, in most patients in the IRS a surgical approach has been employed consisting of simple hysterectomy–partial vaginectomy for proximal lesions, and reexcision of the site in the vaginal wall without hysterectomy for those situated more distally. Oophorectomy has not been performed. In some children with distal vaginal lesions, a secondary local resection has been carried out without laparotomy. Despite a clinical complete response status, the specimens from these procedures, with or without hysterectomy, have ordinarily contained apparently malignant tissue.[9,14] If this tissue has extended to the margins of resection, radiotherapy is given in the IRS. Otherwise, a continuation of the chemotherapy regimen for a total of 2 years has been standard.

Among patients in whom regression of the tumor is less impressive, the indications for surgery are more definite. The same surgical approaches are employed, and the same histologic findings seen. The mass excised at surgery in these patients usually consists of fibrous tissue and apparently necrotic tumor, with foci of malignant cells. The rare patient who fails to achieve even a partial response after five or six courses (months) of therapy can still ordinarily be managed by an approach short of pelvic exenteration—that is, with hysterectomy and partial vaginectomy followed by irradiation and continued chemotherapy. This may be attempted even when a gross excision of the tumor mass is all that appears to be feasible. However, if even a gross total excision would appear impossible, anterior exenteration will ordinarily remove the tumor mass and permit survival. This approach has not been necessary since the introduction of "pulse" VAC in the IRS. The addition of cisplatin and doxorubicin to the chemotherapy regimen or use of other "salvage" therapy such as ifosfamide and etoposide should be considered prior to radical surgery.

Management of Local Recurrence

Local recurrence was relatively common following the therapy regimens employed in IRS-I and -II,[9,14] and this had been noted in smaller series.[26] Therapy for local relapse is frequently successful, and recurrence should be regarded as an indication for a second attempt at multimodality therapy rather than the use of experimental chemotherapeutic regimens or pelvic exenteration. Local recurrence is usually determined at surgery, and here, again, an attempt to remove all gross tumor should be made, usually without anterior exenteration. If radiotherapy has not been employed initially, it should be at this juncture. Combinations of reexcision, radiotherapy, and alternative chemotherapies have resulted in long-term survival, particularly when local relapse had occurred following less intensive chemotherapy regimens. In patients who had been off therapy for some interval at the time of relapse, reinstitution of repetitive "pulse" VAC, with or without additional agents, has been effective.

Stage IV Disease

Patients with vaginal lesions and dissemination, particularly those in which this is confined to pulmonary metastases or a positive marrow, are potentially curable. Following the response of pulmonary lesions and the elimination of apparently malignant cells from the marrow by chemotherapy, the surgical approaches would be similar to those previously described. Isolated remaining pulmonary lesions may be excised at the time of the pelvic surgical procedures.

Uterine Sarcomas

Sarcomas of the uterus in patients less than 21 years of age are rare, and of heterogeneous histologic types.[9,14] Almost all patients with uterine lesions are beyond puberty, and many are young adults. Direct visualization of the cervix for determination of the primary site is routine in all of these patients.

In the IRS, a diagnosis of uterine rhabdomyosarcoma was made in a number of patients by institutional pathologists, and these patients were treated on the IRS chemotherapy/radiotherapy protocols. Subsequent examination by IRS review pathologists changed these diagnoses to other sarcomas (i.e., mixed müllerian sarcoma or mixed mesodermal sarcoma). These patients are mentioned in this review because it is probable that such errors may occur in the future.

Despite the overall rarity of these tumors, patients with uterine rhabdomyosarcomas should be divided clinically into two distinct groups: those that present with an isolated (frequently large) polyp extending from or through the cervix and those presenting with involvement of the body of the uterus.[9,14,25,26] Both groups of patients have presented with vaginal bleeding or discharge, sometimes associated with irregular or excessive menses. Tumor protrusion from the vagina was not seen in the patients in the IRS series, nor were abdominal masses recognized prior to bimanual examination by a physician. Diagnosis in all cases was by transvaginal biopsy (polyps) or dilation and curettage.

Treatment of the patients with polypoid lesions in the IRS has consisted of polypectomy, dilation and curettage, and occasionally simple hysterectomy.[9,14] These patients have been treated with relatively less intensive forms of chemotherapy, primarily standard VA ± C, without irradiation.[9] Although the numbers are small, to date 6/6 of the patients in this category in the IRS have survived, and the experience in several centers is similar.[26]

Patients with rhabdomyosarcomas of the body of the uterus have presented at a more advanced stage, with invasion of surrounding structures. The response to standard therapy regimens in the IRS-I and -II has been disappointing, although, again, the numbers involved are small.[9,14] It is probable that a recommended approach to these lesions should be an initial hysterectomy, because there appears to be little possibility of retaining the uterus. Because most patients with uterine lesions are at an age when growth is near completion, the routine use of relatively intensive radiotherapy is indicated. The ovaries, per se, have rarely, if ever, been found to be involved by this tumor. It should be possible to preserve an ovary by transposition out of the proposed irradiation field. Chemotherapy in this instance should be intensive, attempting to reverse what appears to be the poor prognosis seen in this small group of patients with nonpolypoid uterine rhabdomyosarcoma.

References

1. Hilgers RD, Ghavimi F, D'Angio G, et al: Memorial Hospital experience with pelvic exenteration and embryonal rhabdomyosarcoma of the vagina. Gynecol Oncol 1973;1:262–270.
2. Pinkel D, Pickren J: Rhabdomyosarcoma in children. JAMA 1961;175:293–298.
3. Sutow WW, Sullivan MP, Reid HL, et al: Prognosis in childhood rhabdomyosarcoma. Cancer 1970;25:1384.
4. Wilbur JR, Sutow WW, Sullivan MP, et al: Successful treatment of inoperable embryonal rhabdomyosarcoma. In Abstracts of the American Pediatric Society and Society for Pediatric Research, Atlantic City, NJ, April, 1971. Pediatr Res 1971;5:408.
5. Heyn RM, Holland R, Newton WA, et al: The role of combined chemotherapy in the treatment of rhabdomyosarcoma in children. Cancer 1974;34:2128–2142.
6. Hays DM, Raney Jr RB, Lawrence Jr W, et al (for the Intergroup Rhabdomyosarcoma Study Committee): Rhabdomyosarcoma of the female urogenital tract. J Pediatr Surg 1981;16:828–834.
7. Raney B, Hays D, Maurer H, et al (for the IRS Committee of the CCSG and POG): Primary chemotherapy ± radiation therapy (RT) and/or surgery for children with sarcoma of the prostate, bladder, or vagina: preliminary results of the Intergroup Rhabdomyosarcoma Study (IRS)-II, 1978–1982. In Proceedings of the American Society for Clinical Oncology, Abstract #C-293, 1983. J Clin Oncol 1983;2:75.
8. Grosfeld JL, Smith JP, Clatwothy Jr W: Pelvic rhabdomyosarcoma in infants and children. 1972; J Urol 107:673–675.
9. Hays DM, Shimada H, Raney RB Jr, et al: Sarcomas of the vagina and uterus: the Intergroup Rhabdomyosarcoma Study. J Pediatr Surg 1985;20:718–724.
10. Rivard G, Ortega J, Hittle R, et al: Intensive chemotherapy as primary treatment for rhabdomyosarcoma of the pelvis. Cancer 1975;36:1593–1597.
11. Voute PA, Vos A: Combination chemotherapy as primary treatment in children with rhabdomyosarcoma to avoid mutilating surgery or radiotherapy. Proceedings of the 13th meeting of the American Society for Clinical Oncology, March 1977, Abstract #C-244, p 327.
12. Jaffe N, Murray J, Traggis D, et al: Multidisciplinary treatment for childhood sarcoma. Am J Surg 1977;133:405–412.
13. Ortega JA: A therapeutic approach to childhood pelvic rhabdomyosarcoma without pelvic exenteration. J Pediatr 1979;94:205–210.
14. Hays DM, Shimada H, Raney RB Jr, et al: Clinical staging and treatment results in rhabdomyosarcoma of the female genital tract among children and adolescents. Cancer 1988;61:163–173.
15. Kumar APM, Wrenn Jr EL, Fleming ID, et al: Combined therapy to prevent complete pelvic exenteration for rhabdomyosarcoma of the vagina or uterus. Cancer 1976;37:118–122.
16. Duckett JW Jr, Cromie WB, Bongiovanni AM: Genitourinary tumors in children. Semin Oncol 1974;1:71–75.
17. Davos I, Abell MR: Sarcomas of the vagina. Obstet Gynecol 1976;47:342–350.
18. Anderson WA, Sabio H, Durso N, et al: Endodermal sinus tumor of the vagina. The role of primary chemotherapy. Cancer 1985; 56:1025–1027.
19. Kohorn EI, McIntosh S, Lytton B, et al: Endodermal sinus tumor of the infant vagina. Gynecol Oncol 1985;20:196–203.
20. Herbst AL, Cole P, Colton T, et al: Age-incidence and risk of diethylstilbestrol-related clear cell adenocarcinoma of the vagina and cervix. Am J Obstet Gynecol 1977;128:43–50.
21. Maurer HM, Gehan EA, Beltangady M, et al: The Intergroup Rhabdomyosarcoma Study–II. Cancer (in press).
22. Flamant F, Chassage D, Cosset JM, et al: Embryonal rhabdomyosarcoma of the vagina in children. Conservative treatment with curietherapy and chemotherapy. Eur J Cancer 1979;15:527–532.
23. Pratt CB, Hustu HO, Kumar APM, et al: Treatment of childhood rhabdomyosarcoma at St. Jude Children's Research Hospital, 1962–

1978. Natl Cancer Inst Monogr 1981;56:93–101.
24. Ghavimi F, Exelby PR, Lieberman PH, et al: Multidisciplinary treatment of embryonal rhabdomyosarcoma in children: a progress report. Natl Cancer Inst Monogr 1981;56:111–120.
25. Copeland LJ, Gershenson DM, Saul PB, et al: *Sarcoma botryoides* of the female genital tract. Obstet Gynecol 1985;66:262–266.
26. Brand E, Berek JS, Nieberg RK, Hacker NF: Rhabdomyosarcoma of the uterine cervix. Cancer 1987;60:1552–1560.

28

Legal Principles

MAX BORTEN

Little doubt exists in the minds of health care providers that medical malpractice litigation has had an impact on the way care is delivered to patients. In the not so distant past, comments on the medicolegal aspects of a case were considered amusing anecdotes of peripheral educational importance. Commentators and writers on the subject were careful to point out that their work was only the result of scholarly interest and in no way intended to affect the practice of medicine. Much has changed recently. Institutions have established quality assurance and risk management departments to review the care provided to patients. Protocols to follow before, during, and after certain procedures have been enacted. Continuing medical education requirements now include risk management training, which in some states is mandatory before applying for relicensure.

Even more evident—some may find it disturbing—are the frequent allusions to the medicolegal implications of any actions taken by physicians in all kinds of medical meetings and conventions. One can hardly attend any type of medical conference without this subject entering the conversation in one way or another. Medical journals and textbooks include medicolegal discussions regularly. Graduate and postgraduate training programs have incorporated this material into their curriculums.

It is my conviction that knowledge of the medicolegal principles that come into play is beneficial for all concerned parties. Continued discussion and comprehension of some basic legal principles that involve the daily practice of medicine should alleviate a physician's feelings toward patients and the practice of medicine in general. While it would be preposterous to attempt to describe all aspects of legal medicine in a book chapter, the selected subjects were chosen for their frequency and importance of their application.

Informed Consent

Some physicians still believe that the whole idea of informed consent is a fictional concept created by and for the members of the legal system. In support of their contentions, they charge that patients do not understand, cannot understand, and do not want to understand the intricacies of the medical procedures done to them. To further substantiate their contention, these physicians point out that the doctrine of informed consent is an attempt to redefine the doctor-patient relationship and acts as a deterrent to the fine interaction evidenced by the practice of medicine up to date. The fallaciousness of such arguments is obvious. Informed consent is not meant to interfere with or transform the physician-patient relationship; to the contrary, it rather fortifies interactions between them. Most practicing physicians agree that a well-informed patient is a better patient than one who is misinformed or not informed at all. The doctrine of informed consent rests upon the supposition that a properly informed patient will weigh the risks against the benefits of the treatment proposed by her physicians and compare them with the risk-benefit ratios of alternative treatments. Effective informed consent requires the patient to be informed about the diagnosis, contemplated treatment, risks inherent in such treatment, alternative methods of therapy, and effective prognosis with or without the recommended treatment.

The process of obtaining informed consent should be more of a dialogue than a monologue. It requires active participation of the physician and the patient. The patient should be told and be permitted to comprehend what is being considered. The patient should be encouraged to ask questions and ample time should be alloted for these to be thoroughly answered. Because informed consent is a process and not just a signature authorizing the physician to perform a procedure, if doubts exists about the patient's comprehension of the matter(s) discussed, a supplementary session should be scheduled at a later time. The physician must be convinced the patient understands the nature of what the planned procedure entails.

Clear documentation of the informed consent process is essential. The physician ought not to depend on a signed preprinted treatment consent form. The validity of such general authorization permitting the physician to exercise his or her judgment has been rejected for cases in which no emergency requiring immediate action to preserve the patient's life or health existed. Generalized forms tend to be too vague and ambiguous to be considered valid. The medical record must reflect what in reality has transpired. A specific note to that effect is advisable. The note ought to state that the patient is aware of her condition, is fully informed about the procedure, and above all understands the risk and possible complications of the planned treatment. The note must also indicate that the patient understands what she has been told and agrees to the treatment as outlined.

The extent of the risks that should be disclosed to the patient remains controversial. A physician's duty as it relates to disclosure is determined by the standards of the medical profession. This requires that disclosure made to patients conform to the general practice of the specialty. Because it is widely recognized and accepted that full disclosure of risks may be harmful to a particular patient, the physician may exercise the privilege of withholding disclosure. This privilege is subordinate to the duty to disclose and its exercise requires adequate substantiations. The privilege to withhold disclosure is not applicable just because a physician considers that such knowledge will prompt the patient to refuse permission for a procedure that the physician believes is indicated. Circumspection is in order when applying this principle.

A physician owes the patient a duty to disclose in a reasonable manner all significant medical information that he or she possesses, or reasonably should possess, to allow the patient to make an intelligent decision. It is essential to remember that the patient is entitled to know what a reasonable and prudent practitioner of the specialty ought to be aware of and not just what the physician claims as sufficient knowledge. It is the definition of what is significant medical information that has created enormous controversy. As a general guiding principle (variations exists among states), significant information is that which is material for a particular patient to know in order to make an informed decision. Materiality, as defined by some courts, is the significance a reasonable person would attach to the disclosed risk or risks in deciding whether or not to submit to a particular procedure or treatment. Some states go to the extent of defining as material the significance that a particular patient has placed on the preliminary discussions in deciding to undergo or forgo the procedure.

The doctrine of informed consent has been and will continue to be the subject of debate and controversy. The inherent difficulties of its definition and application are evident. Undoubtedly, forthcoming litigation will provide further refinement of this doctrine. Until then, physicians should continue to involve patients in the decision-making process to the full extent of their abilities and interest. It is essential to keep in mind that the right to self-determination is an inalienable right each patient possesses that is not subject to usurpation in the absence of a life-threatening emergency.

Consultations

The practice of gynecology involves the representation by a physician to the public at large that he or she possesses the required knowledge and necessary skills that can aid in the prevention, diagnosis, and treatment of diseases affecting the female reproductive organs. States, which as institutions are charged with the protections of their citizens, have instituted the process of licensing of professionals to protect the public and foster professionalism. To that effect, they require an applicant to have completed in a satisfactory manner a prescribed course of studies and to have passed a variety of state-approved examinations. The aforementioned process applied to physicians in general without specific delineation of specialty. The latter is usually delegated to the credentialing process within each individual institution. The hospital's credentialing process can be viewed as another method employed by the state in assuring quality medical care in an institutional setting.

At the present time, evolution of medical care has removed many surgical procedures from the hospital environment. Colposcopy, hysteroscopy, dilation and uterine curettage, and laser surgery to the uterus are commonly performed in the physician's office and/or ambulatory same-day surgical care centers. Thus, institutional guidelines are not applicable and not necessarily followed in all circumstances. Review boards and mandatory consultations do not intervene, and guidelines for quality assurance at times are circumvented. Responsibility for proper and adequate care is left with each individual physician. The decision to involve the opinions of additional consultants is then evaluated on a case-by-case approach.

When the services of a consultant are required, it is advisable to clearly document the problem for which the consultation is being called and the rationale for the request. Likewise, it is important to state if the consultation is limited to evaluation and recommendation for treatment, or if the consultant is also expected to care for the affected patient. In the latter case, the primary physician should further specify if treatment is to be established in conjunction with the referring clinician or is to become the full responsibility of the consultant. It is important to remember that abdication of responsibility for care to another physician requires the patient's specific acknowledgment and consent. Such transfer of responsibility should be clearly documented in the patient's medical record and specified in a referral note or consultation request to the consultant.

When it is anticipated that a consultant will be involved in the care of a patient, it is the responsibility of the primary physican to tell the patient beforehand about the issues that require resolution and the reasons that prompted the need to involve additional physicians. Common courtesy and professional respect require the referring clinician to introduce the consultant to the patient if she is hospitalized or to send an introductory letter and pertinent copy of the necessary records to the consultant, preferable in advance of the patient's visit. To avoid conflicting messages and confusing information, the record and/or note must clearly identify who is in charge of the patient's care.

Although it is not mandatory to follow the consultant's recommendations, the referring physician should have substantive arguments as to why he or she chooses not to do so. Reasons for choosing an alternative plan of management must be clearly documented. The rationale for a particular plan of action should be explained, elaborating the logic of the decisions. The patient must be made aware of the reasons supporting an alternative approach and given the option to follow the consultant's recommendations. An honest error in judgment usually proves supportable if adequate and appropriate methods were used to arrive to that decision and the patient has been an active participant in the decision-making process.

Postoperative Risk Management

Preventive risk management is not limited to the process of informed consent, performance of a flawless procedure, or the preparation of material for litigation if a legal action emerges. It is generally accepted that most surgical procedures performed on the uterus are accompanied at times by certain unavoidable risks. Uterine perforation and/or laceration and uncontrollable bleeding requiring a laparotomy or a hysterectomy are just some of the unexpected outcomes of a previously planned minor procedure. Actions by the physician following the occurrence of such a complication have an enormous impact on the eventual disposition of and action taken by the patient.

Notwithstanding the existence of a well-documented informed consent obtained prior to surgery, occurrence of any type of complication during the surgical procedure requires that certain steps be taken by the operating surgeon. Soon after recovery from anesthesia, the patient must be fully advised that an unexpected complication has occurred. A frank and forthright discussion about the complication(s) and their probable causes and mechanisms should take place with the patient and a family member (if any is included in her care). The discussion must include therapeutic options and the risks and benefits of each as well as the anticipated prognosis of each alternative. If actions were taken while the patient was unconscious, it is essential for the surgeon to share with the patient the thought process leading to the chosen actions. Documentation of the exchange of information is critical. Failure to make such notes handicaps the record, and such deficiency may prove injurious to a proper defense at a later stage. A detailed description of all postoperative events reflects good care and caring. It is essential to date, time, and identify the inscriber of all notations made into the medical record.

When dictating a discharge summary, the surgeon must remain objective and limited to a factual account of the events that transpired during the course of the hospitalization. Only information that can be readily found within the medical records should be included. The discharging physician must refrain from including in the report any opinions or conjectural statements that could be interpreted as self-serving; their value is doubtful and at times counterproductive by placing into question the physician's credibility.

Postoperative risk management entails more than just reporting to the institution or the insurance carrier that a complication has occurred. The unexpected mishap should be analyzed in depth by the physician and such honest evaluation shared with the appointed risk manager. This is best done whenever feasible soon after the complication has surfaced, thus ensuring that recollections of events are still fresh in the minds of the participants. Defense counsel may be involved if one has been appointed. Forthrightness and honesty with the risk manager and/or legal counsel concerning the most likely events leading to the occurrence of the complication is crucial. Caution must be exercised not to infringe upon the doctor-patient confidential relationship when disclosing relevant information.

Negligence

Whenever an unfavorable outcome arises as a result of medical care provided to a patient, the contemporary mindset questions: Did malpractice take place? Did the physician do something he or she was not supposed to do, or did they fail to do what was expected of them? Is there liability present and should somebody be held responsible? Was it malpractice or maloccurrence?

In medical malpractice litigation, cases stand or fall most commonly on the basis of the allegation that the physician was negligent. Several independent elements must be considered before a physician's true liability as a result of negligence is established. These elements are duty of care, breach of such duty, damages as a result of the breach of the established duty, and causation, which is the link between the breach of duty and subsequent damages.

A *duty of care* exists when it can be proven that a doctor-patient relationship had been established. The nature and extent of the physician's duty is defined by the legal obligation (contract) imposed on anyone who agrees to enter such a relationship. Once a physician consents to care for a patient, he or she has a duty to use reasonable care and diligence in carrying out the necessary steps to provide diagnosis and treatment programs. A physician will then be held responsible for any injury resulting from a deficiency of the requisite skill and knowledge or the failure to exercise reasonable care.

Standard of care is usually defined as the action(s) to be taken by a reasonable and prudent person under similar circumstances. Albeit a general or family practitioner is held to the minimal standard of care applicable to all reasonably careful and prudent physicians, a physician offering services as a specialist is required to practice at a higher level—namely, at the level of the standards of a specialist practicing in a similar field anywhere in the country. This uniform standard, now in force almost without exception throughout the United States, signifies that all gynecologists are expected to be knowledgeable about those advances within the profession that have been widely accepted and incorporated into practice.

At times, a special situation arises when the harm suffered by the patient (plaintiff) may be inferred to have been caused by negligence of the doctor (defendant) on the basis of circumstantial evidence. This doctrine is known as *res ipsa loquitur* and is seldom invoked in medical malpractice. To satisfy its requirements, the plaintiff must prove that: (a) the event does not ordinarily occur in the absence of negligence; (b) all contributory causes, including conduct of the plaintiff or third parties, are eliminated by the evidence; and (c) the alleged negligence is encompassed within the scope of the defendant's duty to the plaintiff. Contrary to most other cases, in which the plaintiff has to prove negligence, whenever this principle is applied the burden of proof shifts to the defendant to establish lack of culpability.

Before negligence by the defendant can be established, it is necessary for the plaintiff to show that the defendant's action(s) or omission(s) were the actual cause of the adverse outcome. If it can be shown that the resulting damage would have occurred regardless of the physician's conduct or even if the care had been optimal, then the defendant's conduct cannot be faulted as the cause. This is known as the "but for test" or the "*sine qua non* rule." The central question is not whether similar harm would have occurred if the defendant had not acted negligently, but whether the "exact same harm" would have materialized.

When an actual cause-effect relation has been established between the physician's conduct and the adverse outcome for the patient, the physician's act must be shown to be the proximate cause of the harm to the patient. This is known as *proximate (legal) cause*. This principle confines the defendant's liability to the resulting harm that was a foreseeable consequence of the negligent (omission or commission) act. By limiting liability only to foreseeable harm, this prong of the causation test attempts to eliminate cases in which, notwithstanding that a negligent act has been committed by the defendant, the injury sustained by the plaintiff is fortuitous in nature or the result of the misconduct of a third person or a malevolent turn of nature. While negligence by a third person may not exonerate the physician from liability, when overwhelming it can reduce his or her contribution. The defendant could also be held liable for any harm that follows in unbroken sequence from his or her negligent act if it can be shown that the defendant should have been able to anticipate that the negligence would be compounded by the ensuing negligence of a third party.

Under our system of civil law, a victim of medical malpractice is to be made whole economically by the aware of monetary damages to compensate for the resulting injury. Substantiation of damages is required before the victim can receive compensation. *Compensatory damages* are usually awarded when significant harm has resulted as a result of the defendant's negligence. They are known as special and general damages.

Special damages are intended to reimburse the plaintiff for actual economic loss and proven out-of-pocket expenses. These include all medical care (physicians, nurses, hospital, medications, and transportations) required to recover from the resulting injury. At times loss of wages are included within this category, but such compensation must be reasonable and not speculative in nature.

General damages encompass a variety of areas, including emotional injury (pain and suffering) and mental anguish (anxiety, tension, nervousness, and grief). Existence, severity, and duration of the aforementioned harm must be proven by expert testimony. Albeit the degree of damage is speculative, courts rely on the expert's opinion to set a monetary value on these injuries. Recently, loss of consortium, including loss of love, affection, compassion, sympathy, solace, and comfort in the conjugal fellowship between husband and wife, has been a growing area of recovery for plaintiffs.

Conclusion

Allowing room for disagreement, it is my belief that some positive aspects can be extracted from the prevailing situation. Physicians and hospitals have reviewed their procedures and required indications. Questions have been raised about long-standing practices that resulted in improved modifications. Self-evaluation and periodic reevaluation forced the abandonment of some established practices while promoting incorporation of up-to-date technology and management.

Speculation will continue as to who deserves the credit. Some suggest the result would have been the same without the involvement of the legal profession. In support of their contention, they point to the progress medicine has accomplished over centuries much before the above-mentioned legal doctrines came into effect. To the contrary, some will argue that the complexity of current medical treatment requires that provisions be made to protect the rights of patients.

Because the aforementioned controversy is likely to continue unresolved for the foreseeable future, it is in the best interest of physicians to familiarize themselves with the legal concepts more frequently invoked. A better understanding of the prevailing rights of patients and physicians can only help to reduce apprehension and defervese antagonism. After all, the goal is a common one. Patients want to remain healthy or regain their health, and physicians aim to facilitate them to succeed.

Index

Italicized page numbers indicate figures or tables on page.

MIX
Papier aus verantwortungsvollen Quellen
Paper from responsible sources
FSC® C105338

If you have any concerns about our products,
you can contact us on
ProductSafety@springernature.com

In case Publisher is established outside the EU,
the EU authorized representative is:
Springer Nature Customer Service Center GmbH
Europaplatz 3, 69115 Heidelberg, Germany

Printed by Libri Plureos GmbH
in Hamburg, Germany